# THE OXFORD FRANCIS BACON · XII

*Director:* Graham Rees

*Editorial Advisory Board*

THIS volume belongs to the new critical edition of the complete works of Francis Bacon (1561–1626). The edition presents the works in broadly chronological order and in accordance with the principles of modern textual scholarship. The works printed below, the *Historia ventorum* and *Historia vitæ & mortis*, belong to Part III—whose overall title was *Historia naturalis et experimentalis*—of the colossal six-part *Instauratio magna*. The texts are presented in the original Latin with new facing-page translations.

THE OXFORD FRANCIS BACON · XII

# The *Instauratio magna* Part III:
## *Historia naturalis et experimentalis: Historia ventorum* and *Historia vitæ & mortis*

EDITED WITH INTRODUCTION, NOTES, COMMENTARIES,
AND FACING-PAGE TRANSLATIONS BY

GRAHAM REES

WITH MARIA WAKELY

CLARENDON PRESS · OXFORD

# OXFORD

UNIVERSITY PRESS

Great Clarendon Street, Oxford OX2 6DP

Oxford University Press is a department of the University of Oxford.
It furthers the University's objective of excellence in research, scholarship,
and education by publishing worldwide in

Oxford New York

Auckland Cape Town Dar es Salaam Hong Kong Karachi
Kuala Lumpur Madrid Melbourne Mexico City Nairobi
New Delhi Shanghai Taipei Toronto

With offices in

Argentina Austria Brazil Chile Czech Republic France Greece
Guatemala Hungary Italy Japan Poland Portugal Singapore
South Korea Switzerland Thailand Turkey Ukraine Vietnam

Oxford is a registered trade mark of Oxford University Press
in the UK and in certain other countries

Published in the United States
by Oxford University Press Inc., New York

British Library Cataloguing in Publication Data

Data available

Library of Congress Cataloging in Publication Data

Data available

Typeset by RefineCatch Limited, Bungay, Suffolk
Printed in Great Britain
on acid-free paper by
Biddles Ltd., King's Lynn, Norfolk

ISBN 978-0-19-926500-8

1 3 5 7 9 10 8 6 4 2

*For*
*Ben and Ruth Wakely*
*and Emma Rees*

# PREFACE

LIKE other volumes of the edition, this one takes its place within a larger chronological plan. But it should be noted that here the chronological principle has been interpreted flexibly and in the light of Bacon's stated plans for his texts, i.e. this is one of a sequence of volumes (IX–XIII) which contains only the works of Bacon's great, unfinished meta-work, the *Instauratio magna*, and excludes ones of the same period which were not written in fulfilment of that ambitious six-part scheme. The works here excluded will of course appear in other volumes of the edition.

As I have said elsewhere in *The Oxford Francis Bacon*, the production of a printed book (especially one as complex as this is) is very much a cooperative enterprise and, in this case, one to which many individuals and institutions have contributed.

Many librarians helped me in the work, and in my efforts to put together photocopies and microfilm of exemplars of the printed copy-texts used in the preparation of this volume. Particular thanks are due (yet again) to the staff of the university and college libraries of Cambridge, London, and Oxford, and to the staff of the British Library. I am also deeply indebted to the Folger Shakespeare Library and to the Huntington Library for giving me access to their precious and substantial holdings of Baconiana.

I have received generous grants from the Modern Humanities Research Association, the Bibliographical Society, and the Isaac Newton Trust, as well as timely subventions from the British Academy, to whose portfolio of research projects *The Oxford Francis Bacon* belongs. The Arts and Humanities Research Council deserves full acknowledgement: it provided the medium-term funding which allowed me to employ first-rate researchers whose work has hastened completion of earlier volumes of this edition and of ones still to come. In connection with institutions, I must once again thank my own college—Queen Mary, University of London—which has furnished me with splendid working conditions, and many brilliant friends and colleagues to advise and support me.

As for individuals, I have already recorded the fact that members of our Editorial Advisory Board have made major contributions to the work, and to the rapid consolidation of our project. I would again have thanked our quondam chairman, the late Professor J. B. Trapp, had not death intervened and robbed the project of a scholar whose wisdom and good sense did so much to establish *The Oxford Francis Bacon* as a going concern. For all kinds of help, advice, and constant support, I am once again very much indebted to Dr Maria Wakely. She has done much to lay the analytical-bibliographical foundations of this volume, and of others in the edition—especially in the collation of copy-text exemplars. She checked repeatedly the computer transcriptions of the copy-texts which served as the basis for the edited texts, and advised on the content of the commentaries. Along with Dr Stephen Pigney, she gave time to the exacting task of checking the volume before it went to press, and to correcting the proofs afterwards. The exacting task of proof-reading was also undertaken by Carlo Carabba, Marta Fattori, and Sophie Weeks. Their thoroughness and scholarship has my warmest thanks. Thanks are also due to Janet Norton and Liza Verity of the National Maritime Museum, to Massimo Bianchi, James Binns, Constance Blackwell, David Colclough, Harold Cook, David Gants, Andrew Hadfield, David McKitterick, Randall McLeod, Philip Ogden, Richard Serjeantson, and Charles Webster. Without the support and encouragement of people associated with Oxford University Press this volume would have been much the poorer. In particular Andrew McNeillie has given unstintingly of his patience, knowledge, and wholehearted support.

G.R.

*Queen Mary,*
*University of London,*
*September 2007*

# CONTENTS

*Contents*

# APPENDICES

# LIST OF PLATES

*All plates are reproduced by permission of the British Library*

# REFERENCES AND
# ABBREVIATIONS

Volumes of *The Oxford Francis Bacon* are designated by the abbreviation *OFB* followed by a volume number in roman numerals.

The following abbreviations are used for the titles of Bacon's works, parts of works, and plans:

| | |
|---|---|
| *AC* | *Aphorismi et consilia* |
| *AL* | *Advancement of learning* |
| *ANN* | *Abecedarium nouum naturæ* |
| *CDNR* | *Cogitationes de natura rerum* |
| *CDSH* | *Cogitationes de scientia humana* |
| *CF* | *Calor et frigus* |
| *CHP* | *Catalogus historiarum particularium* |
| *CS* | *Commentarius solutus* |
| *CV* | *Cogitata et visa* |
| *DAS* | *De augmentis scientiarum* |
| *DFRM* | *De fluxu et refluxu maris* |
| *DGI* | *Descriptio globi intellectualis* |
| *DIN* | *De interpretatione naturæ sententiæ xii* |
| *DO* | *Distributio operis* |
| *DPAO* | *De principiis atque originibus* |
| *DSV* | *De sapientia veterum* |
| *DVM* | *De vijs mortis* |
| *HDR* | *Historia densi & rari* (Rawley's version) |
| *HDR(M)* | *Historia densi & rari* (manuscript version) |
| *HGL* | *Historia grauis & leuis* |
| *HIDA* | *Historia & inquisitio de animato & inanimato* |
| *HNE* | *Historia naturalis et experimentalis* |
| *HSA* | *Historia soni et auditus* |
| *HSMS* | *Historia sulphuris, mercurii et salis* |
| *HV* | *Historia ventorum* (published in *HNE*) |
| *HVM* | *Historia vitæ & mortis* |
| *IDM* | *Inquisitio de magnete* |
| *IL* | *Inquisitio legitima de motu* |
| *IM* | *Instauratio magna* |
| *IVDI* | *Indicia vera de interpretatione naturæ* |
| *NA* | *New Atlantis* |

| | |
|---|---|
| *NO* | *Novum organum* |
| *PA* | *Prodromi sive anticipationes philosophiæ secundæ* |
| *PAH* | *Parasceue ad historiam naturalem* |
| *PhU* | *Phænomena universi* |
| *PID* | *Partis instaurationis secundæ delineatio* & *argumentum* |
| *RPh* | *Redargutio philosophiarum* |
| *SI* | *Scala intellectus sive filum labyrinthi* |
| *SS* | *Sylva sylvarum* |
| *SSWN* | *Sylva sylvarum* (working notes) |
| *TC* | *Thema cœli* |
| *TDL* | *Topica inquisitionis de luce et lumine* |
| *TPM* | *Temporis partus masculus* |
| *VT* | *Valerius terminus* |

The following abbreviations are used for earlier editions and translations of Bacon's works:

| | |
|---|---|
| *Ess* | Sir Francis Bacon, *The essayes or counsels, civill and morall,* ed. Michael Kiernan, Oxford, 1985 (as from 2000, volume XV of *OFB*) |
| *LL* | James Spedding, *The letters and life of Francis Bacon,* 7 vols., London, 1861–74 |
| *Sc* | *Scripta in natvrali et vniversali philosophia. Amstelodami, Apud Ludovicum Elzevirium,* 1653 |
| *SEH* | *The works of Francis Bacon,* ed. James Spedding, Robert Leslie Ellis, and Douglas Denon Heath, 7 vols., London, 1859–64 |

In this and all other volumes of this edition, the edited texts will be presented with their copy-text signatures (printed books) or folio numbers (manuscripts) inserted at the appropriate points in the outer margins of edited texts. These signatures or folio numbers are the key to the edition's internal system of reference. They will be used to refer to all Bacon works cited in the apparatus of all volumes. In the case of works not presented in this volume or in volumes of the edition not so far published, this method of reference is supplemented by parenthetical references to the corresponding pages in *SEH*. This will make it easier (pending completion of the new edition) for readers to find references to other Bacon works. As further volumes of the edition are published the need for the rather cumbersome references to *SEH* will grow less. As nearly all the secondary literature on Bacon uses *SEH* as a primary source, Appendix II of this volume provides a list in which the signatures or folio numbers attached to the volume's edited texts are translated into *SEH* references.

Other references:

| | |
|---|---|
| *BJHS* | *British journal for the history of science* |
| BL | British Library |
| BNF | Bibliothèque nationale de France |
| Bodl. | Bodleian Library, Oxford |
| Briquet | C. M. Briquet, *Les filigranes: dictionnaire historique des marques du papier*, 4 vols., repr. New York, 1966 |
| *CCC* | *Commentarii collegii conimbricensis . . . in qvatvor libros de coelo, meteorologicos, & parua naturalia . . . Venetiis, M D CII. Apud Jacobum Vincentium, & Riccardum Amadinum* (the individual commentaries are abbreviated as *CCC de coelo, met.*, and *parua naturalia* respectively) |
| CUL | Cambridge University Library |
| *DM* | William Gilbert, *De mundo nostro sublunari philosophia nova*, Elzevier: Amsterdam, 1651 |
| *DNB* | *Dictionary of national biography* |
| *DRN* | Bernardino Telesio, *De rerum natura*, ed. Luigi De Franco, vol. I, Casa del Libro: Cosenza, 1965; vol. II, Casa del Libro: Cosenza, 1971; vol. III, La Nuova Italia: Florence, 1976 |
| *DSB* | *Dictionary of scientific biography* |
| Gibson | R. W. Gibson, *Francis Bacon: a bibliography of his works and of Baconiana to the year 1750*, Scrivener Press: Oxford, 1950 |
| Gibson(S) | R. W. Gibson, *Francis Bacon: a bibliography of his works and of Baconiana to the year 1750: Supplement*, privately issued: Oxford, 1959 |
| Heawood | E. Heawood, *Watermarks mainly of the 17th and 18th centuries*, Hilversum, 1950 |
| *HLQ* | *Huntington library quarterly* |
| *IELM* | P. Beal, *Index of English literary manuscripts 1450–1625*, I, parts 1 and 2, London, 1980 (the Bacon manuscripts are identified alphanumerically: the prefix BcF followed by a number) |
| *OCD* | *Oxford Classical Dictionary*, 3rd edn., ed. Simon Hornblower and Antony Spawforth, Oxford University Press: Oxford, 1996 |
| *OED* | *Oxford English Dictionary* |
| *OFB* | *The Oxford Francis Bacon* (i.e. the volumes of this edition) |
| *STC* | *A short-title catalogue of books printed in England, Scotland, & Ireland and of English books printed abroad 1475–1640*, ed. A. W. Pollard and G. R. Redgrave, 2 vols., 2nd edn. revised and enlarged by W. A. Jackson, F. S. Ferguson, and K. F. Pantzer, London, 1976–86; vol. III *Indexes*, compiled by Katharine F. Pantzer and Philip R. Rider, London, 1991 |

In addition:

| | |
|---|---|
| aph. | aphorism |
| *c-t(s)* | copy-text(s) |
| *cmt(s)* | commentary(ies) |
| lc | lower case |
| *nld* | normalized |
| *om* | omitted or omission |
| para(s) | paragraph(s) |
| *tns* | textual footnotes |

# INTRODUCTION

## I

### The Latin Natural Histories in Context

*(a)* *Historia naturalis* and *Historia vitæ*: Their Dating and Place in *Instauratio magna*

The *Historia naturalis* (the generic title of the volume containing the *Historia ventorum*), and *Historia vitæ & mortis* were the most formidable contributions that Bacon explicitly composed for Part III of his colossal six-part sequence of works, the *Instauratio magna*.[1] I have written about the plan of this sequence or meta-work elsewhere in the *Oxford Francis Bacon*,[2] so I confine myself to the briefest summary of it here. Part I was to be a work on the partitions of the sciences, a general survey of current knowledge. The survey aimed to identify the deficiencies of learning and, once they had been noted, Bacon intended to help with the task of making them good either by giving directions or by doing some of the remedial work himself. The requirements of this part were fulfilled to a degree, and retrospectively, by the *De augmentis scientiarum* (1623). Part II was to purge the mind of obstacles to the new philosophy, and give precepts for the 'Interpretation of Nature' which would yield an active philosophy productive of works. In short, Part II was to have been given over to all that would have been set out in the *Novum organum* had the latter not been left unfinished. Part III would have been devoted to 'Phænomena Universi & Historiam', i.e. to collections of natural-historical data the assembling of which Bacon regarded as a *sine qua non* for the reconstruction of the sciences.

Of the remaining three parts, IV was to have demonstrated the precepts of Part II in action. Part V was to be of a temporary nature, a repository of provisional theories and conclusions which Bacon had arrived at by the ordinary use of reason. These were nevertheless worth preserving because they were more likely to be true than the offerings of other philosophers. During at least one stage of his career Bacon believed that the anticipations of Part V would be eligible for

---

[1] The histories tell us on their title-pages that they were so destined; see p. lix below.
[2] *OFB*, VI, pp. xvii–xix; XI, pp. xix–xxii; and XIII, pp. xix–xxv.

promotion to Part VI were they able to pass muster when tested against whatever results the implementation of the legitimate method might bring.[3]

Part III, the part to which the edited texts of this volume belong, became the principal focus of Bacon's philosophical attentions following the publication of the unfinished *Novum organum*. Not only did he want '*rather to advance the whole work of the* Instauration *in many things than to perfect it in a few*' but he made the far stronger claim that the *Novum organum*, even if finished, would not push forward the renewal of the sciences very much without natural history, whereas natural history without the *Novum organum* would advance it not a little.[4] In fact natural history was the foundation of Bacon's entire programme for rebuilding knowledge. As *Distributio operis* has it, the new natural history was designed 'to illuminate the discovery of causes and nourish philosophy with its mother's milk'. The new history was also to be massive in scope, demanding cooperative work carried out over a period of many years, and consequently (as Bacon reminded James I in 1620) it was an enterprise that needed state funding.[5]

Now with *Novum organum* published, it was time for Bacon to produce exemplary natural histories in the hope of ensuring that people would not mistake his conception of this part of his enterprise. How much progress did he make towards fulfilling this aim? The fact is that he knew he could not get very far with Part III. Indeed he knew that it would be impossible for him to produce, single-handed, history of the bulk and quality required. In effect, all he could do was to produce simulations or approximations of the kind of functional natural history that his programme demanded. In *Historia naturalis* he proposed writing six specimen histories—selected for their nobility, usefulness, and differences from each other—to be published at the rate of one a month. The *Historia naturalis* listed their titles thus:

*Historia Ventorum.*
*Historia Densi & Rari, nec-non Coitionis, & Expansionis Materiæ per spatia.*
*Historia Grauis & Leuis.*
*Historia Sympathiæ, & Antipathiæ Rerum.*

---

[3] This was according to the early *PID* (*c.*1607), though this possibility was not mentioned in the 1620 plan of *IM*, the *DO* of 1620; see *OFB*, VI, p. xix; XI, p. xxi.

[4] *HNE*, p. 12 below.

[5] See *OFB*, XI, pp. 8, 36–43. For more on the *literary* relations between Bacon and James, see Maria Wakely and Graham Rees, 'Folios fit for a king: James I, John Bill, and the King's Printers, 1616–1620', *HLQ*, **68**, 2005, pp. 467–95.

*Historia Sulphuris, Mercurij, & Salis.*
*Historia Vitæ & Mortis.*

Of these only the first and last were finished and published in Bacon's lifetime. A third, the *Historia densi & rari*, came close to completion but waited for more than thirty years after its author's death to be published by William Rawley, Bacon's quondam chaplain and secretary.[6] Of the remaining three items nothing is extant save their prefaces. Published for the first time in the *Historia naturalis*, edited texts of these prefaces appear in this volume.

*Historia naturalis et experimentalis* (which is an introduction to Part III in general and the *Historia ventorum* in particular) and *Historia vitæ & mortis* are fairly easy to date. *Historia naturalis* was entered in the Stationers' Register on 20 September 1622, and was probably ready for sale before the end of November.[7] *Historia vitæ*, all copies of which are dated 1623, was entered on 18 December 1622, fifty-four days after which John Chamberlain noted (10 February 1623) that he 'lately bought' both that and *Historia naturalis*.[8] In other words the second history had probably become available to the public not much more than two months after the first.

## (*b*) Natural History and the End of Erudition

Against whom or what did Bacon shape his concept of natural history, and develop it in practice? The beginnings of an answer come *inter alia* from the authorities he mentioned. From his earliest to latest works he spoke of Agricola, Albert the Great, Cardan, Gesner, Gilbert, and Della Porta as moderns who had dealt in empirical and/or experimental findings; among the ancients he named Aristotle, Theophrastus, Dioscorides, and of course Pliny. He also threw in nameless writers 'ex Arabibus' for good measure.[9] Of the named authorities let us look at just four whose work served Bacon as imperfect models for, or (with qualifications) antitypes of, natural history—Aristotle, Pliny, Gesner, and Agricola.[10]

---

[6] For *HDR* and the circumstances of its publication see *OFB*, XIII, pp. lxxiii–lxxxiii.

[7] See *STC* 1155; also see *LL*, VII, p. 395 for Bacon's probable allusion to *HNE* in a letter to the Marquis of Buckingham, dated 24 Nov. 1622.

[8] See *STC* 1156; also see *LL*, VII, p. 399.

[9] For the Arabs, see *DAS*, E4ʳ (*SEH*, I, p. 456). For the other authorities, see below *passim*.

[10] These four are mentioned along with Dioscorides in the early *CDSH* (*c.*1603?) in BL Add. MS 4258, fos. 214–27, fo. 217ʳ (*SEH*, III, p. 189): 'Itaque prima Narratio est ea, cui Naturalis Historiæ communis appellatio tribuitur: cujusmodi est Aristotelis, Plinij, Dioscoridis, Gesneri, Agricolæ, reliquorum.'

Taking Aristotle first, in *The advancement of learning* of 1605, and its Latin translation, the *De augmentis scientiarum* of 1623, Aristotle is placed in a class by himself and elevated above all other natural historians:

So in naturall Historie, wee see there hath not beene that choise and iudgement vsed, as ought to haue beene, as may appeare in the writings of *Plinius, Cardanus, Albertus*, and diuers of the Arabians, being fraught with much fabulous matter, a great part, not onely vntryed, but notoriously vntrue, to the great derogation of the credite of naturall Philosophie, with the graue and sober kinde of wits; wherein the wisedome and integritie of *Aristotle* is worthy to be obserued, that hauing made so diligent and exquisite a Historie of liuing creatures, hath mingled it sparingly with any vaine or fayned matter, and yet on thother side, hath cast all prodigious Narrations, which he thought worthy the recording into one Booke: excellently discerning that matter of manifest truth, such wherevpon obseruation and rule was to bee built, was not to bee mingled or weakened with matter of doubtfull credite: and yet againe that rarities and reports, that seeme vncredible, are not to be suppressed or denyed to the memorie of men.[11]

Thus was Aristotle's *Historia animalium* exempted from the general censure that Bacon visited on the heads of other natural historians. Aristotle could at least tell the difference between fact and fancy—fact of the kind represented in that history, and in his *Problems*, and fancy of the kind that he confined to the pages of *De mirabilibus auscultationibus*.[12]

In fact Bacon rather envied Aristotle and the royal subventions allegedly given him by Alexander the Great to compile *Historia animalium*. Bacon's dearest, unfulfilled wish was to enjoy similar support from James I.[13] Yet in *Novum organum* he warned readers not to be deceived by the fact that in Aristotle's 'books on Animals, in his Problems, and in other tracts of his [e.g. the *Parva naturalia*] he often deals in

---

[11] *OFB*, IV, pp. 26–7, cf. *DAS*, E4ʳ (*SEH*, I, p. 456). The clause 'such wherevpon obseruation and rule was to bee built' is given a more Baconian turn in the Latin, viz. 'quæ tanquam Basis Experientiæ solida, Philosophiæ & Scientiis substerni possint'.

[12] The *Problems* and *De mirabilibus auscultationibus* are not now thought to be by Aristotle himself; *OFB*, XI, p. 517.

[13] See *DAS*, K4ʳ (*SEH*, I, p. 489): 'Si enim Alexander magnam vim pecuniæ suppeditauit Aristoteli, quâ conduceret Venatores, Aucupes, Piscatores, & alios, quo instructior accederet ad conscribendam Historiam Animalium; certè maius quiddam debetur ijs, qui non in Saltibus Naturæ pererrarunt, sed in Labyrinthis Artium viam sibi aperiunt.' For the fable of Alexander's support for Aristotle's research see Brian W. Ogilvie, *The science of describing: natural history in Renaissance Europe*, University of Chicago Press: Chicago and London, 2006, pp. 94–5.

experiments'. Aristotle generally did not take experience into proper account when he framed his 'decrees and axioms', but bent it to his opinions and dragged it 'about in chains', i.e. he rushed up to generalities on too narrow an empirical basis and so, in Bacon's terms, used natural history not as the foundation of his philosophy but as a *post hoc* justification of it. Aristotle's intellectual vices, dissected at length in *Novum organum*, usually undid most of the good which, in Bacon's view, he had undoubtedly accomplished.[14] This judicious judgement on Aristotelian natural history—good data implicated in a false relationship with a dogmatic natural philosophy—of course allowed Bacon to reject the philosophy while co-opting materials mined from the *Problems* and *Historia animalium* for his own ends in the *Historia ventorum* and *Historia vitæ*—as even a cursory inspection of the commentaries on our edited texts will show.[15]

As for Pliny, we have already seen that Bacon contrasted the Roman's natural-historical work unfavourably with Aristotle's; Pliny's 'being fraught with much fabulous matter, a great part, not onely vntryed, but notoriously vntrue'.[16] According to the *Descriptio globi intellectualis* (1612), Pliny had done the right thing when he had interwoven in his narratives materials concerning the three main branches of natural history—history of generations, pretergenerations, and arts[17]—for he had 'an ideal of natural history worthy of the name'. But his was an ideal 'which he quite failed to live up to in practice'.[18] He was the only person, as the *De augmentis* has it, who undertook a history in a manner appropriate to its dignity, though his performance was not equal to his conception.[19] According to the *Parasceue* (1620) Pliny had a true idea of the scope and scale of natural history but derived too much of his material from 'hearsay or reading'.[20] For all that, Bacon took even more material from Pliny than he did from Aristotle—as we shall see in due course.[21]

---

[14] *OFB*, XI, pp. 98–101; see also pp. 88, 106–8, 517, 519–20. For praise of the *Parva naturalia* see *CDSH*, fo. 216ʳ (*SEH*, III, p. 188).

[15] See, for instance, pp. 391, 393–4, 397, 421, 424 below.

[16] *OFB*, IV, pp. 26–7, cf. *DAS*, E4ʳ (*SEH*, I, p. 456).

[17] For this threefold distribution of natural history, see pp. xxviii–xxxii below.

[18] *OFB*, VI, pp. 102–5.

[19] *DAS*, Ll¹ᵛ (*SEH*, I, p. 497): 'qui Historiam Naturalem solus pro dignitate complexus est; sed complexam, minimè vt decuit, imò potiùs indignis modis tractauit'.

[20] *OFB*, XI, p. 467. For early Renaissance criticisms of Pliny, criticisms of considerable influence, see Ogilvie, *The science of describing*, pp. 121–33.

[21] See Introduction, pp. xliv–xlv below.

Turning now to the natural-historical practices of Bacon's contemporaries and near-contemporaries, the reader will excuse me if I say
rather more about them than I have about Bacon's assessments of
Aristotle and Pliny. I do so because the work of Conrad Gesner
(1516–65) and Georgius Agricola (1494–1555) embodied attitudes to natural history and its purposes which were not only still current but
represented ways of thinking and writing which Bacon was determined
should not contaminate or become confused with his readers' understandings of what he was about.

Comparing Pliny and Gesner in the early *Cogitationes de scientia
humana*, Bacon noted the flaws of both: 'Namqua multus Plinius in
fabulis, antiquitate et Censurâ Morum: Gesnerus autem hæreditatem
historiæ suæ ex multis partibus Philologiæ, ex paucis Philosophiæ . . .'.[22]
Gesner for Bacon was above all a man of scholarly words, words, and
more words still; and that was not an altogether unfair criticism. The
very titles of Gesner's works suggest as great a devotion to *verba* as to
*res*, and that he was rather too keen to rebut the charge of prolixity.
Gesner insisted (truthfully) that he weighed particulars and opinions
'ex innumeris authoribus', from personal contacts and observation,
and that he would have produced a much smaller volume had he not
paid proper attention to matters that would bring joy to grammarians,
matters like grammar itself, as well as proverbs, similitudes, the sayings
of the poets, and generally things which had to do with 'words rather
than the things themselves'.[23]

In practice, Gesner was as good as his word. Take for instance his
account of the elephant. This (like accounts of every other animal) is
divided into eight alphabetically labelled sections (A to H) with a first
section on the names of the animal and a couple of sections on the
animal's geographical distribution, sizes, colours, teeth, tusks, and
internal organs (here Gesner draws, as one would expect, on extensive
classical Latin sources on the subject).[24] Section D, on their faculties,
affections, and habits, presents, for example, a lot of material on the

---

[22] Fo. 226ʳ; as *SEH* (III, p. 191) points out, 'Namqua' should read 'Namque'.

[23] *Historiæ animalium lib. I. de quadrupedibus uiuiparis. Opvs philosophis, medicis,
grammaticis, philologis, poëtis, & omnibus rerum linguarúmque uariarum studiosis, utilissimum
simul iucundissimúmque futurum . . . Tigvri apvd Christ. Froschovervm, M. D. LI.*, β1ʳ⁻ᵛ: 'sed
hanc quoqu*e* diligentiam, si no*n* admodum utilem, iucundam tamen grammaticis & alijs
quibusdam fore spero . . . Philologiam autem appello, quicquid ad grammaticam, & linguas
diuersas, prouerbia, similia, apologos, poëtarum dicta, deniqu*e* ad uerba magis quàm res ipsas
pertinet.'

[24] *Ibid.*, pp. 409–42.

relations (mostly bad) between the elephant and the dragon.[25] Sections E to G concern the uses of elephants, and their edible and medicinal parts.[26] Finally comes the long section H on philology which includes etymology (the Greeks noticed that elephants were large and so named them after ὁ λόφος (the hill)), and proverbial lore (elephants are terrified of mice).[27]

The unicorn also gets the A to H treatment, and we are told the stories that had been told a hundred times before—for instance that lions are enemies to unicorns but unicorns are friendly to virgins.[28] Dragons too are not neglected. Indeed they are treated in much the same way that the elephant or, for that matter, the dog or horse, but this time not in Gesner's volume on viviparous quadrupeds but in his fifth and enormous volume which was on serpents. This volume, like all Gesner's writings, was derived from encyclopaedic reading, and the understanding that springs 'ex usu, experientia & exercitatione', as well as from the depths of natural philosophy, which is of course 'inter omnes philosophiæ humanæ partes pulcherrima, iucundissima, præstantissima atque vtilissima'.[29]

To accuse Gesner of prolixity, unselectivity, and lack of critical judgement is to miss the point—as Bacon well understood, for he, Bacon, wanted to sideline this kind of natural history, not to abolish it. To be comprehensive—to leave no stone unturned, no opinion unvoiced, and no word unspoken—was Gesner's very aim. It is also no use to say that his work was bookish and authority-ridden for it was meant to be. Says Gesner, 'diligenter & religiosè caui, ne usquam nomen authoris omitterem'.[30] As Laurent Pinon has said, Gesner's work, 'is actually an inventory of every element of natural knowledge since antiquity'. And, one might add, Gesner thinks of this as history because it is an assemblage of all *particulars* about this or that creature. His natural history was compilative and erudite in the highest degree, and the work of a man who may have been more afraid of the accusation of leaving a few things out rather than of including too

---

[25] *Ibid.*, pp. 420–9.     [26] *Ibid.*, pp. 429–42.

[27] *Ibid.*, pp. 427, 438, 441. On Gesner's natural-historical writings, see Laurent Pinon, 'Conrad Gessner and the historical depth of Renaissance natural history', in *Historia: empiricism and erudition in early modern Europe*, ed. Gianna Pomata and Nancy G. Siraisi, MIT Press: Cambridge, Mass., 2005, pp. 241–67.

[28] *Historiæ animalium lib. I.*, pp. 781–6.

[29] *Historiæ animalium lib. V. qui est de serpentium natura. . . Tiguri in officina Froschoviana.* M. D. LXXXVII., *2*ʳ, fo. 43 ff. Also see Pinon, 'Gessner', pp. 249–50.

[30] *Lib. I. de quadrupedibus uiuiparis*, β1ʳ.

many. How could you not talk of the dragon with the same seriousness and attention that you paid to the elephant when good authors had spoken of both and you could provide pictures of them, and when you wanted to make the reader (in one of Gesner's favourite words) *iucundissimus*, i.e. thrilled to bits?[31]

As for Agricola, to some modern eyes he may seem to belong to an intellectual world very different from Gesner's. With his practical advice on prospecting, his accounts of the different ways in which veins of ore occur in the ground; his understanding of mining finance, and his mass of precise, illustrated descriptions of mining technology— from simple hammers to pump and ventilation arrangements, all the way through to refining and assaying—Agricola accomplished in *De re metallica*[32] a tour de force in its way just as comprehensive as Gesner's massive volumes on animals. Yet it was a tour de force which seemed to look less to tradition than to a still-novel respect for art as much as for nature, and for practical utility of a kind which Bacon could not but have (with considerable reservations) admired.

Yet Gesner admired Agricola too,[33] and admired him not just because he had brought off a brilliant and thorough account of an industry and its machinery but framed it in a humanist conception of historical writing. For a man like Gesner, Agricola gave off the right signals. For instance, Book I contained compelling humanist *pro* and *contra* rhetoric in which everything and everyone from the Bible to Xenophon was pressed into service in defence of Agricola's project. Citation of authorities and quotation from the classical poets are not characteristic of Book I alone but are also to be found even in the deepest technological thickets of later books. Agricola could wield the citation and quotation, and could attend to the names of things in a variety of languages with all the confidence of the humanist-trained mind he possessed. Moreover, he did not confine himself to mining engineering;

---

[31] Pinon, 'Gessner', p. 250. In the generation after Gesner at least one natural historian, Ulisse Aldrovandi (1522–1605), made his reputation with his work on a dragon; see Paula Findlen, *Possessing nature: museums, collecting, and scientific culture in early modern Italy*, University of California Press: Berkeley, 1994, pp. 17–23. Findlen remarks (p. 22) that in 'his decision to interpret the dragon as a natural phenomenon, devoid of metaphysical implications but rich in anatomical meaning, Aldrovandi participated in a broad cultural trend to normalize the marvelous in the late sixteenth and seventeenth centuries'.

[32] *De re metallica libri xii. Quibus officia, instrumenta, machinæ, ac omnia denique ad metallicam spectantia, non modo luculentissimè describuntur . . .* Froben: Basle, 1556.

[33] *Lib. I. de quadrupedibus uiuiparis,* a2ʳ.

he took all things subterranean for his province; and that included underground animate beings, whether the mine demons (removable by prayer) spoken of in *De re metallica*, or the harmless dwarves inhabiting mines, wells, and rabbit holes which were discussed in Agricola's treatise *De animantibvs subterraneis*. Like Gesner he also had a serious interest in dragons, which, according to Agricola, fight with the eagle in some places and with the elephant in others.[34]

What did Bacon think of all this? He believed that his conception of natural history differed from his predecessors' in a number of ways: 'in its end or function, in its very mass, in its subtlety, and also in its selection and organisation for the procedures to come after it'.[35] So function, content, and scope made Bacon's conception of natural history novel—or so he thought. Let us look at these a little more closely.

Speaking of the function of natural history in the *Parasceue* Bacon asserted that 'neither *Aristotle*, nor *Theophrastus*, nor *Dioscorides*, nor *Pliny*, and still less the *Moderns*, have ever set themselves the goal of which we speak for natural history'.[36] In fact he had become so certain of this and so convinced that his conception would be misread in the light of received expectations that, when in 1612 he began the unfinished *Descriptio globi intellectualis*, a revision of the 1605 *Advancement of learning*, he inserted a new chapter into his text, a chapter which appeared repeatedly and in a number of guises in works produced thereafter.[37] This chapter, entitled '*The partition of Natural History according to its use and end*', was unequivocal:

the noblest end of natural history is this: to be the basic stuff and raw material of the true and legitimate induction, and to draw enough from the sense to furnish the intellect. For that other kind, which either delights by the charm [*jucunditate*] of its narratives or helps by the use of its experiments [*experimentorum usu*], and which is admitted for the sake of such pleasure or profit [*voluptatis, aut fructus gratiâ*], is undoubtedly of an inferior kind and in its very nature less valuable when set beside the kind which has the power and quality to be a proper preparative for the founding of philosophy. For, finally, the latter is the history which constitutes the solid and lasting basis of a true and active philosophy . . .'.[38]

---

[34] *De animantibvs subterraneis liber*, Froben: Basle, 1549, pp. 68–9, 77–8.

[35] *OFB*, XI, pp. 38–9.

[36] *Ibid.*, XI, pp. 455–7.

[37] This material appeared in *DO, NO, PAH, HNE*, and *DAS*; see *OFB*, XI, pp. 36–9, 154–61, 454–6; *DAS*, M3ʳ–M4ʳ (*SEH*, I, pp. 501–2); also see pp. xxvii–xxix below.

[38] *OFB*, VI, pp. 104–5.

Bacon could actually have been writing this critique of humanist natural history with Agricola's *De re metallica* or one of Gesner's stupendous volumes open in front of him. For Bacon the holy grail of *iucunditas*—of fulfilling the Gesnerian aim of delighting the reader—was simply irrelevant. As for Agricola, wonderfully diligent though his work was,[39] it would have suffered in Bacon's eyes from short-termism, and too great an emphasis on present usefulness, investment, and profit.[40]

The whole point of Bacon's natural history was to defer present gratification in order to approach a future in which natural history would be used to supply the material for a new natural philosophy which in turn would supply an unparalleled abundance of material benefits. Natural history was not to bypass the rebuilding of philosophy, and certainly not to supply the relatively poor returns which was all that extant natural history could afford. Bacon would have thought that Agricolan history did not aim to do more than 'mow the moss or green corn', or turn aside from the race to pick up Atalanta's golden apples.[41] The new natural history was to be completely subdued to a single, new, and severe function: it was to be the foundation and handmaiden of the new philosophy. The newness of the new natural history was that it existed only for this purpose, even if it did just happen to suggest new works by the way.[42]

Bacon's functional conception had profound consequences for the content of natural history. What he had in mind was quite different from a history compiled for pleasure or profit. As the *Descriptio* reminds us, the very bulk of existing natural history may be intimidating but the material was delightfully various, diligent to the point of curiosity, and fat with superfluities. But if you subtract 'the fables and antiquities, the citations and opinions of authorities, the empty squabbles and controversies, and finally the philology and embellishments (which are more appropriate to the table-talk and night-work of learned men than the building up of philosophy)', it will shrink to very little. The natural historians pride themselves on building a 'treasure-house of words

---

[39] *DAS*, Z3ʳ (*SEH*, I, p. 572): 'Georgius Agricola, Scriptor recens, diligentèr admodùm in Mineralibus'.

[40] On investment and the profit motive in Agricola, see *De re metallica*, pp. 19–29.

[41] *OFB*, XI, pp. 38–9.

[42] Bacon did not despise benefits that might spring from the process of compiling natural history; see for example *HV*, Q8ᵛ–R3ᵛ, pp. 128–30 below. He even recognized that natural history might have an operative part; see *DAS*, Z2ᵛ–Z3ʳ (*SEH*, I, pp. 572–3).

rather than a solid and reliable narrative of things', and they have squandered their energies describing the 'distinct varieties of flowers, of the iris or tulip, or again of shells or dogs or hawks', things which are the 'freaks and sports of nature, and come close to the nature of individuals', and provide 'slight and almost superfluous information for the sciences'.[43]

What Bacon wanted was an end to erudition—at least as far as the new model natural history was concerned. He was contesting an established discourse of creatures. Agricola and Gesner understood as well as Bacon did that *res* and *verba*, things and words, were different. Gesner also suspected that *verba* might run to superfluity, but for him natural history was not natural history without them.[44] Bacon, who might have been thinking of Gesner, Aldrovandi,[45] or any humanist natural historian in the quotations above, wanted to confine their erudition to the realm of post-prandial chat. Just as he had consigned the old logic to the sphere of opinion, and gave over the field of natural philosophy to a new logic, so he dumped the old natural history and devised the new natural history as a servant of a new philosophy. What the humanist natural historians relished Bacon junked. No more savouring of authorities; no more delight in the inherited jumble of *legenda*; no more cheerful devising of ingenious etymologies. Bacon served notice on cakes-and-ale natural history: something less comfortable, more strenuous, and entirely purpose-led was coming to take its place.

In fact Bacon nowhere launches a more devastating and unrelenting attack on the old erudition than in his accounts of his natural-historical programme. He inverts the usual model of the reader of the book of nature. The model reader is the illiterate child struggling with the ABC,[46] not the man of letters thoroughly at ease with the texts and whose fluent reading fails to get to grips with nature. People like Gesner, Cardan, and Scaliger were, as Anthony Grafton has said, commentators by vocation, who 'saw their duty not as discovering facts never before seen and drawing inferences from them but as assembling facts from reliable sources in a new and revealing order'. Research was a hunt for

---

[43]  *OFB*, VI, pp. 104–7.

[44]  *Lib. I. de quadrupedibus uiuiparis*, βr$^{r-v}$ (quoted in n. 23 above).

[45]  For Ulisse Aldrovandi's idea of natural history see, for instance, Findlen, *Possessing nature*, pp. 17–36.

[46]  For the alphabet metaphor in Bacon, see *OFB*, XIII, pp. 172–3, 305–6.

'the written records of scientific facts, ancient or modern', and most scientific writings 'resulted not in reports on controlled situations but in commentary or *bricolage*—the discussion of canonical texts, line by line', and the 'rearrangement of fragments from them into new treatises'.[47] For Bacon natural history was to be not a display of erudition but a display of the *lack* of it, a display which befitted, perhaps, a new non-referential conception of nature,[48] and was as austere and stripped down as could be. In Bacon's specimen histories even illustrations—which could have been helpful when he came to improving the performance of ships or windmills—are nowhere to be found. Indeed Bacon was sceptical of the value of visuals in natural history, though maybe the scepticism was fed more by pictures of tulips than by Agricolan technological illustration.[49]

Now Bacon's functional conception of natural history represented not just a discarding of superfluity, minute curiosity, and erudition but a transformation of what was left over, in particular of the scale and scope of natural history. The new natural history was to be a cooperative undertaking lasting for many years for it was to cover data concerning everything in nature.[50] Invoking a distinction which he had used repeatedly since at least *c.*1604, Bacon declared that natural history dealt with generations, pretergenerations, and arts.[51] The first of these records data gathered from nature in her ordinary course, i.e. working freely by secondary causes without impediments or constraints produced by accidental deviations or human intervention. According to the *Parasceue*—the text which was, with the *Catalogus historiarum*

[47] Grafton, 'Kepler as a Reader', *JHI*, 53, 1992, pp. 561–72 at 564–5. For Bacon on Cardan as natural historian, see *DAS*, E4ʳ (*SEH*, I, p. 456).

[48] See Peter Harrison, *The Bible, protestantism, and the rise of natural science*, Cambridge University Press: Cambridge, 1998, pp. 4–5.

[49] Bacon had no time for the 'abundant wealth we find in natural histories of descriptions and pictures of species, and the ingenuity lavished on their differences'; see *PAH* in *OFB*, XI, pp. 456–7. For discussion of ships and windmills, see *HV*, pp. 90–106 below and *cmts* thereon pp. 402–4 below. For non-Baconian instances of scepticism of illustration, see *OFB*, XI, p. 588. The passage in *PAH* is a revised version of the one in *DGI* (*OFB*, VI, pp. 104–7) quoted on p. xxv above, where Bacon does in fact refer to tulips. For early books with multiple pictures of tulips, see Anna Pavord, *The tulip*, Bloomsbury: London, 2000, pp. 64–5 *passim*.

[50] See, for instance, *OFB*, XI, p. 458.

[51] This threefold distribution was used in the early (*c.*1603) *CDSH* (fo. 217ʳ (*SEH*, III, p. 189)), in *AL* (in *OFB*, IV, pp. 63–4), and later appeared in *PhU* (*OFB*, VI, pp. 8–9). It was also used in *DGI* (*OFB*, VI, pp. 100–3, 384), *PAH* (*OFB*, XI, p. 454), and *DAS* (L4ʳ⁻ᵛ (*SEH*, I, p. 496)), the last of which was evidently based on the text of *DGI*.

*particularium,* meant to stand immediately before *Historia naturalis et experimentalis* in the sequence of *Instauratio magna*—'History of Generations' has five parts:

the first of the Ether and Heavenly Bodies; the second, of meteors and the Region of the air . . .; the third, of the Earth and sea; the fourth, of the elements (as they call them), flame or fire, air, water, and earth . . . not the primordia of things, but the greater masses of natural bodies . . . For these reasons I have grown used to calling bodies of . . . [this] kind *Greater Colleges,*. . . [which] make up part four of the history, going, as I have said, by the name of the elements . . . part five of this history embraces the *Lesser Colleges,* or *Species*— the matters with which natural history has mainly been concerned up to now.[52]

In his own view then, Bacon was expanding the scope of natural history to aspects of nature in her ordinary course far beyond the normal range of history of generations. The true extent of the expansion can be gauged by looking at his proposals for just the first of the five areas mentioned here, the proposals for a historical account of astronomical and astrological phenomena, proposals added at length to the text when *De augmentis scientiarum* was being prepared.[53]

Turning now to pretergenerations, the second member of the primary threefold distribution of natural history, the preternatural had, as Daston and Park have shown, long been established as a distinct ontological category suspended between the mundane and the super-natural.[54] For instance Aquinas had seen the preternatural as a product of secondary causes which happened to violate the ordinary course of nature by accident or unforeseen circumstances. They are the product of occult qualities which arise from the specific form of a substance impressed on them by the heavens.[55] While the contingent, accidental character of the preternatural led medieval philosophers to reject the study of the preternatural from natural philosophy, it seems that renewed interest in this realm in the Renaissance helped to expand the scope of natural philosophy to include marvels and to privilege the empirical and the magical, with the consequent introduction of new types of explanation into natural philosophy.[56]

---

[52] *OFB,* XI, pp. 458–61. In *CDSH* the distribution of natural history was four- not fivefold; *CDSH* does not mention the history of the elements; see fo. 217ʳ (*SEH,* III, p. 189).

[53] See *DAS,* V3ʳ–X4ʳ (*SEH,* I, pp. 551–60).

[54] Lorraine Daston and Katharine Park, *Wonders and the order of nature 1150–1750,* Zone Books: New York, 1998, pp. 14, 120–7.

[55] *Ibid.,* pp. 120–9.   [56] *Ibid.,* p. 137.

Bacon was less interested in these new types of explanation than in effecting a radical expansion of the scope of natural history. Rejecting explanations framed in terms of occult qualities, he wanted the same severity and rigour that he wished to be applied to any other part of natural history to be applied to pretergenerations. But then the question becomes *how* does one distinguish between the real and the fabulous? Severe inquiry cannot rule out what appears (to modern eyes) absurd: Bacon may have written little on wonders such as the dragon or unicorn but that does not mean he withdrew his credit from them entirely. Nor, for that matter, writing over a century later, did Linnaeus.[57] Bacon's position on marvels could not be anything other than an open-minded scepticism, coupled with the belief that verified data concerning pretergenerations could be invaluable to natural history.

Just how invaluable is made clear in the discussion of Instances with Special Powers in *Novum organum*.[58] Six types of such data were to be collected as a set of particular histories right from the start, because they all served to prevent the intellect being corrupted 'by the daily invasion of ordinary experience'. They purged the mind and let in 'the dry and pure light of true notions'.[59] Instances of Correspondence display the connections between bodies, and thus the fabric of the universe. Monadic Instances do not correspond with other things of the same kind: for example, loadstones among stones or quicksilver among metals. The point is to make such exceptions unexceptional by discovering explanations which comprehend them. Deviating Instances, nature's mistakes or monsters, also 'fortify the intellect against custom and reveal common forms', so we need a particular history of 'all the monsters and prodigious births of nature'. Frontier Instances exhibit bodies which seem to combine two species, and are good (*inter alia*) for pointing out the structure of things. Lastly, Instances of Power are the most perfect works of each art: 'rare and strange works of nature stir and raise the intellect to investigate and discover forms capable of encompassing them', but miracles of art do that even more for 'the manner of making and working such miracles of art is generally plain

---

[57] See *HVM*, N1$^v$, p. 234 below for the unicorn's horn as a tonic for the spirits. Bacon may not of course have believed that the unicorn was distinct from (say) the rhinoceros, which is what Albert the Great thought it was (see Harrison, *The Bible*, p. 87), and Bacon knew something of Albert's work (see *DAS*, E4$^r$ (*SEH*, I, p. 456)). Also see Carl Linnaeus, *Systema naturæ, sive regna tria naturæ proposita per classes, ordines, genera, et species*, Theodore Haak: Leiden, 1735.

[58] *OFB*, XI, pp. 272–446.     [59] *Ibid.*, XI, pp. 306–7.

to see, whereas in the miracles of nature it is often much less obvious'.[60] As Daston and Park put it, this group of Instances with Special Powers has a methodological function: to 'serve as an observational approximation of controlled experiments—or rather, as a record of the experiments nature performed on itself'.[61]

Let us turn now from the miracles of nature to the miracles of art, and to the third member of Bacon's primary distribution of natural history—mechanical and experimental history. Here we should remember that natural history had only begun to be constituted or invented as a discrete *genre* of scholarly activity just a couple of decades before Bacon's birth. Yet Bacon himself stretched this relatively new genre to comprise history of the mechanical arts and experimental history. It was not Agricola but Bacon who thought of *De re metallica* as a natural history, and he rightly believed that the inclusion of mechanical and experimental history in natural history was one of the most original features of his conception of this branch of learning. For him nature and the mechnical arts did not differ, as they had for Aristotle, in that the former had its own innate principle of change whereas the latter did not. The products of nature and art differed only in the efficient and not in their formal causes. Accordingly a mechanical and experimental history had to stand alongside and, indeed, above history of generations and pretergenerations. One recent writer has linked Bacon's view to a long history of debate on the art–nature polarity, debates which centred on alchemy and its critics.[62] Certainly, Bacon was well acquainted with Paracelsian ideas on art and nature and denied, as they (and other philosophical traditions) had, that artificial heats differed from natural ones.[63] He also saw the idea (Aristotelian in origin) that the art which perfected nature as a licence for focusing at length on experimental intervention in nature, and although he did not collapse the art–nature polarity, he thought that art and nature were the same in *essence*.[64]

---

[60] *Ibid.*, XI, pp. 288–304.

[61] *Wonders*, p. 239. They add (pp. 220–4) that, for Bacon, when nature wandered without the prodding of art, the marvels so produced mimicked the variability induced by art. Nature under compulsion resembled nature erring in that the variability of effects visible in both cases revealed possibilities hardly guessed at under ordinary conditions.

[62] William R. Newman, *Promethean ambitions: alchemy and the quest to perfect nature*, University of Chicago Press: Chicago, 2004, pp. 257–70.

[63] *OFB*, XI, pp. 310–15. For Bacon's Paracelsianism, see *OFB*, VI, pp. xlii–li.

[64] Newman, *Promethean ambitions*, pp. 258–60. Newman does not know whether he wants to see Bacon as an innovator or an inheritor in the matter of the art–nature question. His

Yet Bacon carried this principle well beyond anything envisaged by his predecessors. His insistence on the need for a mechanical and experimental history points in two complementary directions. In the first place, it points towards what Bacon's seventeenth- and eighteenth-century successors sometimes saw as a summons to construct a history of the trades, and the sort of activity which Bacon might have seen foreshadowed (if the short-termism and humanist learning were removed) in Agricola's greatest work. In the second place, this kind of history did something much more radical, namely the privileging of data acquired by 'vexing' nature, by exerting pressure on phenomena in controlled circumstances, and by so doing compensating for the defects of the senses. Take, for example, Bacon's fascination with the microscope, telescope, and thermoscope. Take too his understanding of the need to extend the senses by using instruments in combination: for instance, the water-filled-lead-ball and the phial-and-bladder experiments, with their deep and implicit denial that Aristotelian reliance on ordinary experience alone could suffice for scientific knowledge.[65] In short, the fully developed theory of experimentalism set forth in *Novum organum*[66] embodied the idea of using instruments separately or in combination to 'read' facts and then transmit them in a form accessible to the observer. A key point was that 'the subtlety of experiments is far greater than that of the sense . . . [and] I set little store by the immediate and peculiar perception of the sense, but carry the matter to the point where the sense judges only the experiment whereas the experiment judges the thing'.[67]

To sum up, if the Renaissance invented natural history as a literary

---

account suffers from a weakness of longitudinal studies of single idea-complexes—that they take insufficient account of the place the complex occupies in the heads of particular thinkers. For instance, Newman claims that it 'would not be too much too [*sic*] say that Bacon's entire program of reducing the distinction between nature and art to one of efficient causality is already to be found in the discussion of art and nature stretching from the High Middle Ages through the seventeenth century'. Newman's view neglects what Bacon made of this programme—especially, for instance, in his theorizing of experiment as a tool for acquiring knowledge. One does not find that kind of theorizing in the medieval *Book of Hermes*—for which see *ibid.*, pp. 259–60.

[65] *OFB*, XI, p. 374. For the highly artificial quantitative experimental work on the Peripatetic theory of the decuple proportionality of the elements, see *PhU* in *OFB*, VI, pp. 48–50; *NO* in *OFB*, XI, p. 82, ll. 18–21; *HDR* in *OFB*, XIII, pp. 38, 70, 277. Also see Bacon's proposed experiment on the heat of the Moon (*OFB*, XI, p. xli).

[66] See *OFB*, XI, pp. xliii, lxxxii–lxxxiii, xci, 560–1.

[67] See *ibid.*, pp. xl–xli, 34.

genre and a discipline,[68] Bacon reinvented it as the indispensable servant of a new philosophy. Standing behind the *Historia naturalis et experimentalis* is a post-humanist conception of natural history. Not only was its projected scope far greater than anything envisaged by Bacon's humanist predecessors or contemporaries, it embodied a radical, inclusive view of the content of natural history which made it at once functional and interventionist. It was interventionist in that it foresaw more than passive observation of the facts of nature; it foresaw the active, experimental inquisition of nature. It was functional in that it subserved a scientific, demonstrative end, not a philological one. Bacon did not want to identify animals named in classical sources or plants mentioned in biblical ones but to reconstruct the sciences for the material benefit of the human race. His natural history did not have erudite content because it had no erudite purposes. In fact he may well have been implicated in what one historian has called 'a new, non-symbolic conception of the nature of things', a conception allied to or rooted in Protestant attitudes to interpretation and, in particular, in a new view of nature and of texts, a view which drove a wedge between words and things and restricted the allocation of meanings to the former. He was a proponent of a literalism in which only words refer; the things of nature do not.[69] Bacon summed up and transcended emergent themes in early modern natural history: the new materiality, the passion for collecting as a means of organizing things, and the will to bring new kinds of data into the natural-historical enterprise.[70] But beyond all that was an epoch-making repositioning of natural history as an indispensable element of the new way, the way, at last, to reliable and productive knowledge.

---

[68] For the emergence of natural history as a coherent discipline with its own community of practitioners, see Ogilvie, *The science of describing*, pp. 25–138. Also see *idem*, 'Natural history, ethics, and physico-theology', in *Historia: empiricism and erudition*, ed. Pomata and Siraisi, pp. 75–103 at 80–1.

[69] Harrison, *The Bible*, pp. 114–15.

[70] For these themes, see *inter alia* Daston and Park, *Wonders*, pp. 13–14, 137–9, 159–61; David Freedburg, *The eye of the lynx: Galileo, his friends, and the beginnings of modern natural history*, University of Chicago Press: Chicago and London, 2002, *passim*. For the role of collecting and museums in the transformation of natural history see Findlen, *Possessing nature*, esp. pp. 50–63, 156–7, 165–71. For a useful attempt to identify the stages by which the discipline of natural history emerged and changed in the period 1490 to 1620, see Ogilvie, *The science of describing*, pp. 29–58.

(*c*)  *Historia naturalis et experimentalis:* Preliminaries to the
Natural Histories

Before we ask how Bacon put his natural-historical programme into
practice, we should recall that *Historia naturalis et experimentalis*, the
title of the 1622 volume that contains the *Historia ventorum*, was the
title of Part III of the *Instauratio magna* as a whole, or at least of Part III
as far as Bacon meant it to be represented (imperfectly) in the individual
histories he hoped to write over the coming months. In the sequence
of the *Instauratio magna* the *Historia naturalis* is preceded by items
published with the *Novum organum*, namely the two programmatic
natural-historical pieces, the *Parasceue ad historiam naturalem* and
*Catalogus historiarum particularium*, both of which Bacon actually
referred to in the preliminaries to the *Historia naturalis*.[71] Indeed in
these preliminaries—the dedicatory letter, a list of the six histories
which Bacon intended to publish in the coming months, the proemium
to *Historia naturalis*, and the 'Norma' or brief account of the kinds of
subject matter to be found in the natural histories—Bacon never lets
the reader forget the place of Part III in the *Instauratio magna* or, in
particular, the relationship between natural history and the doctrines
of *Novum organum*. A chief purpose of his from the title-page of the
*Historia naturalis*, and on through the preliminaries of that volume, was
to anchor and position it in the plan of the six-part meta-work. He
assumes that the reader is thoroughly acquainted with the plan, or
he wants to send readers not so acquainted back to the plan as it is
presented in the 1620 *Instauratio magna*.

The *Historia naturalis et experimentalis* is dedicated to Prince Charles.
The dedication, rich in biblical allusion, and as brief as it is skilful, is in
effect an attempt to control reader expectations. It plants the idea that
the natural history presented in the volume is a foreshadowing in little
of what will become a mighty project. We are offered only 'first fruits' or
'mustard seeds' which will grow into something which will provide
the 'keys both to knowledge and to works'. The fact that the volume
is but the beginning of an enterprise that will extend into the future is
immediately reinforced with a list of the titles which Bacon hoped
(fondly, as it happened) to publish at the rate of one a month over the
next six months.[72]

---

[71]  See *HNE*, C2$^{r-v}$, p. 12 below.    [72]  See *HNE*, A2$^{r-v}$, A4$^{r}$, pp. 4, 6 below.

The next preliminary, the proemium, represents, in tones reminiscent of the *Novum organum* on idols of the theatre, natural history as an antidote to ancient and modern philosophical sectarianism. It is a characteristic appeal for humility in the study of natural things. Recycling a favourite simile, one that reminds us that the *Abecedarium nouum naturæ* was originally destined for this volume, we are urged to become again as little children and apply ourselves to the business of learning nature's ABC, so that instead of stamping the image of our own mind on creation, we discover the image of things as they are. To this end Bacon explains that he has left the *Novum organum* unfinished in order to embark on Part III of the *Instauratio* so that there should be no doubt as to what he means natural history to be. Better, he says (in words echoed in the *Abecedarium*), to advance the whole work in many things than to perfect in a few. Then, after all the years spent developing it,[73] Bacon comes up with the arresting claim that the *Novum organum* without the natural history would advance the sciences much less than natural history without the *Novum organum*. There could be no clearer an indication of Bacon's post-*Novum organum* philosophical preoccupations than that.[74]

The final preliminary to the *Historia naturalis* is entitled 'Norma Historiæ præsentis', which purports to describe 'the rule and make-up of the history which I now attempt' more 'accurately and succinctly' than the *Parasceue*. However, key features of the Norma are absent from the *Parasceue*. Certainly Bacon at least begins by picking up on the *Parasceue* (aph. 10), by distinguishing the history of concrete bodies from the history of abstract natures (the study of the latter reserved for himself). He also promises 'a New Abecedarium of these which I have located at the end of this volume'.[75] Precisely what these abstract natures were, what the *Abecedarium* was, and why it did not appear at the end of the 1622 volume are matters dealt with at length in volume XIII of *The Oxford Francis Bacon*.

Bacon then explains the grounds (their especial dignity and difficulty) for selecting particular titles from the *Catalogus*. Yet of these six titles it is strange that two are not represented in the *Catalogus* at all,[76] two are not explicitly represented,[77] and only the *Historia ventorum* and

---

[73] See *OFB*, XI, pp. cxvii–cxviii. Also see *ANN* in *OFB*, XIII, pp. 172–3.

[74] *HNE*, B6ʳ–B8ᵛ, pp. 10–12 below.

[75] *HNE*, C2ʳ⁻ᵛ, p. 12 below; cf. *OFB*, XI, p. 472.     [76] *HSAR* and *HDR*.

[77] *HGL* is perhaps represented in items 22–5 of *CHP*, and *HSMS* by items 26–8, 32–3, and 128, although its title as such is absent from *CHP*; see *OFB*, XI, pp. 476, 484.

*Historia vitæ*, the two published in Bacon's lifetime, are actually listed therein.[78]

According to the Norma, each history is to begin with a series of particular topics. This item is not mentioned in the *Parasceue* unless one takes it to be a firming up of aphorism 9 of the *Parasceue* where Bacon proposes that questions should be put in natural history to prompt further investigation.[79] In fact the *Historia ventorum* and *Historia vitæ* are loosely organized in terms of fairly comprehensive lists of such questions which pave the way for the histories proper, and these, according to the Norma, are to start with the meat of natural history, i.e. tabular enumerations of historical data and experiments. The tables are to be accompanied by protocols for making good deficiences in existing data. Among these protocols are advice (*monita*) and directions (*mandata*) for 'experiments of light' and 'crucial instances' designed to get at the causes of things. This prescription is not even hinted at in the *Parasceue*, but seems once again to presuppose knowledge of the *Novum organum* on the part of the reader. How else would a reader in 1622 know what crucial instances were?[80]

The Norma then tells us that the histories will offer observations, speculations, and provisional rules. This is very important for the Norma aims to *differentiate* between the sorts of material which will occur in the histories—differences reinforced and marked out by typographical means[81]—and here it goes far beyond anything suggested in the *Parasceue* where, in aphorism 9, we are told that 'observations' would be inserted into natural history but not told what exactly Bacon meant by that.[82] However, it is clear that with these observations, speculations, and provisional rules Bacon is giving himself freedom to enter the realm of natural-philosophical explanation far more liberally than, for instance, his ideal of natural history as represented in the *Descriptio globi intellectualis* or *Novum organum* might allow. Such material does indeed appear copiously in the *Historia ventorum* and *Historia vitæ*, but more of that in a moment.

---

[78] *HVM* is represented in *CHP* by items 57–8; *HV* by item 6, and piecemeal by items 4–5 and 7–10; see *OFB*, XI, pp. 474, 480.

[79] *OFB*, XI, pp. 468–70.

[80] For particular topics see *DAS*, 2K1ʳ–2K3ᵛ (*SEH*, I, pp. 635–9). Also see *HNE*, C3ʳ⁻ᵛ, p. 14 below, cf. *OFB*, XI, pp. 468–9.

[81] See pp. lxiii–lxiv below.

[82] According to *PAH*, the insertion of 'observations'—by which Bacon seems to mean empirical generalizations—has a precedent in Pliny's practice; see *OFB*, XI, pp. 468–71.

The Norma concludes with a short paragraph which points out that the 'present history' not only contributes to Part III of the *Instauration*, but 'because of the titles from the *Abecedarium* and of the Topics', prepares the way for Part IV, and with 'the major observations, speculations, and rules' for Part VI. This is extraordinary in three ways. In the first place, even though the *Abecedarium* was very much on Bacon's mind when he was writing this, and even though that work and the preliminaries to the *Historia naturalis* have a great deal in common, the *Abecedarium* was not published with the *Historia naturalis*.[83] In the second place, Bacon yet again frames natural history in the meta-work; yet again he assumes a level of reader knowledge of the plan of that work which tempts one to think that he may have had an unfulfilled intention to publish a summary of the plan with the *Historia naturalis*. He also makes large claims about the contribution that his histories will make not just to the natural-historical parts of the meta-work but to other, later parts too. In the third and last place, the claim that speculative-theoretical materials in the specimen histories are effectively a contribution to Part VI of the *Instauratio* is a very strong claim indeed,[84] and one with an emphasis not found in Bacon's programmatic writings about the nature of the new natural history. He is saying that his *theorizing* about the data presented in the histories is actually powerful enough to count as a genuine strand of the new philosophy, a philosophy which elsewhere (Bacon insists) can arise only from the proper implementation of the procedures set out in *Novum organum*. So let's be ready in the specimen histories for kinds of discourse which are not strictly speaking natural-historical at all. Let's be ready for emphases in the *Historia ventorum* and *Historia vitæ* which the *Parasceue*, for instance, might not lead us to expect. The specimen histories have a pronounced theoretical thrust. When Bacon speaks of 'observations' in the Norma he does not mean in this context observational data; he means theoretical talk. And there are a lot of 'observations', 'speculations', and 'rules' in the two histories which are precisely of that kind.

---

[83] For instance, the Norma of *HNE* (C2$^r$–C6$^v$, pp. 12–16 below) and the Norma of *ANN* (*OFB*, XIII, pp. 220–25) have a great deal in common. Bacon seems to have withdrawn *ANN* from *HNE* at the last minute, mainly because he may have had second thoughts about its place and function in the meta-work which is the *IM*; see *OFB*, XIII, pp. xxi–xxiii.

[84] *HNE*, C6$^{r–v}$, p. 16 below.

## (*d*) *Historia ventorum*

After a brief preface,[85] Bacon proceeds (as promised in the Norma) to introduce the *Historia ventorum* with a list of particular topics or articles of inquiry which were meant 'both as light to the present and stimulus to future inquiry. For we command questions where we cannot command things'.[86] These topics fall into seven groups (some of them rather heterogeneous): the names and species of winds (topics 1–10), factors contributing to their production (11–15); the limits (16–18) and successions of winds (19–21); the motions of winds (22–7); powers of winds (28–31); the prognostics (32) and imitations (33) of winds. Let us take each of these sections not in turn but in the order they are dealt with in the rest of the history, for as Bacon himself said in this connection 'I do not follow the order of the questions slavishly, in case what was meant to be a help becomes a hindrance'.[87]

When addressing the question of the names of winds Bacon would have found a fair range of traditional names in authors well known to him—in Aristotle, Seneca, Pliny, Agricola, Cardan, the Coimbra commentators, William Gilbert, and Porta, to name but the most prominent.[88] But he sets 'little store by the interpretation of authors; for there is not much of substance in these same individuals'. Instead he seeks the certainty of fixed names.[89] This he achieves by subsuming the names current in the Graeco-Latin tradition beneath a thirty-two-point distribution of winds, the thirty-two points being the four cardinal points of the compass together with the seven lesser points between each of the cardinal.[90] That done, he gets down to the species of winds.

He identifies five species of winds: free, general, recurrent, prevailing, and marine.[91] Free winds get short shrift, for what Bacon has to say about them amounts to little more than the observation that however many points are assigned to the compass one is likely to find a wind blowing therefrom at one time or another. He has much more to say

---

[85] See *cmts*, p. 386 below.

[86] *HNE*, C3$^{r-v}$, p. 14 below. For particular topics, also see n. 80 above.

[87] *Ibid.*, C3$^v$, p. 14 below.

[88] See *cmts* on *HV*, E1$^r$–E3$^r$, pp. 388–9 below.

[89] *HV*, E3$^r$, p. 32 below.

[90] *Ibid.*, E1$^r$–E3$^r$, pp. 28–32 below. Also see Joëlle Ducos, 'Entre latin et langues vernaculaires, le lexique météorologique', in *Lexiques et glossaires philosophiques de la Renaissance*, ed. Jacqueline Hamesse and Marta Fattori, Fédération internationale des instituts d'études médiévales: Louvain-la-Neuve, 2003, pp. 55–71.

[91] *HV*, E3$^v$–F7$^r$, pp. 32–44 below.

about general winds, and in particular about the wind that blows from east to west in the tropics. His interest in this is no surprise for it represents an incursion (detached from its theoretical context) of his cosmological thinking into natural history, and so no doubt a reason why he chose the history of the winds as one of the six that he aimed to write to illustrate the natural-historical programme.[92] On Bacon's cosmological theories, the heavens moved from east to west about a central Earth. For physico-chemical reasons this cosmic motion reached down into the sublunar region to embrace the air and the waters of the oceans and produced the east-to-west wind and the sexhorary motion of the tides, both of which phenomena Bacon used as consilient crucial instances to support his geocentric, geostatic conception of the universe.[93]

In *Historia ventorum* Bacon avoids exposing in full this theoretical context. He notes the phenomena—that the general east wind is detectable mainly in the tropics where it has more room to move, that in Europe it appears to blow very faintly when there is no other wind, and that it can be detected on high ground on calm evenings. He then invokes the 'indirect' or consilient phenomena, i.e. that if weathercocks point steadily west, and if such a motion can be detected in the tides, as well as in the heavens, 'it would not be unlikely that the air which lies in between them would join in the same motion'. The *fact* that this general wind exists is, Bacon says, indisputable; only its cause is doubtful. It blows either because the heat in the tropics expands the air strongly or because it is part of the universal motion already mentioned.[94] We know of course from other works that Bacon believed that the second of these explanations was true,[95] but here in the spirit of natural history he does not deliver a verdict on the matter. Suffice it to say that the fact that he broached the subject at all may be regarded as evidence that his speculative theories silently determined some of the content of the history and, as I have already suggested, his choice of natural-historical titles. In the *Historia vitæ* and (for that matter) the *Historia densi* the influence is a lot less silent, and much more of a shaping force on the character and content of those works.

[92] *HV*, E3ᵛ–E6ᵛ, pp. 32–6. For the reasons for Bacon's choice of natural-historical topics see Graham Rees, 'Francis Bacon's semi-Paracelsian cosmology and the *Great Instauration*', *Ambix*, **22**, Part 3, 1975, pp. 161–73.

[93] *OFB*, VI, pp. li, 78, 379; XI, pp. 316, 324–6, 561–2.

[94] *HV*, E5ᵛ–E6ʳ, pp. 34–6 below.　　[95] See, for instance, n. 93 above.

As for recurrent and prevailing winds, the most important of the former in Europe are these: 'the north winds at the solstice, and they occur both before and after the rising of the Dog-star; west winds from the autumn equinox, and easterlies from the spring one; for the winter solstice needs less attention paid to it because of the variability of winter weather'. Prevailing winds (*Venti Asseclæ*) is a designation that Bacon says he has invented. These are winds which blow in all but one of up to five parts of any given country, with the remaining part alone being subject to a contrary wind. Lastly Bacon discusses marine winds, though not, as the particular topics at the beginning of *Historia ventorum* lead us to expect, under a separate heading.[96] Moister and more violent than land winds, these are the prevailing winds of all countries especially in coastal areas. They are not recurrent because they are generated from vapours which are randomly distributed over the surfaces of the oceans.[97]

Having dealt with the species of winds, Bacon turns to the qualities and powers of winds. Very little need be said of this except to observe that it draws very heavily on the text of the (pseudo)-Aristotelian *Problems*—a text of which Bacon had a good opinion[98]—, on Pliny's *Historia*, and to a lesser extent on modern authorities such as Gilbert, Acosta, Knolles, and Severinus.[99] Interestingly, Aristotle is nowhere mentioned by name in the *Historia ventorum*, a fact in line with Bacon's policy of avoiding great parades of authorities. Acosta and Gilbert on the other hand are named several times in the text, and Pliny half a dozen times.[100] This warns us that while some of Bacon's data is highly original, a great deal of it is perforce, as we shall see in a moment, highly derivative.

Bacon now turns to the local origins of winds. There are three such manifested in winds which originate from (a) the earth, (b) from above, and (c) within the body of the air around us. He presents evidence in favour of the proposition that all three sources produce winds. An appreciable portion of the evidence is derived from Acosta, Gilbert, and Pliny, but an exception to this is the discussion of the third source. Here Bacon introduces an entirely novel theory of the origin of

---

[96] *HV*, D1ʳ⁻ᵛ, p. 20 below.    [97] *Ibid.*, F2ʳ–F7ʳ, pp. 38–44 below.

[98] For Bacon on the *Problems* see *OFB*, XI, p. 517.

[99] *HV*, F7ʳ–G8ʳ, pp. 44–54 below.

[100] For Acosta see *HV*, F4ᵛ, H4ʳ, I7ʳ (pp. 42, 58, 70 below); for Gilbert see *ibid.*, H3ᵛ, P2ᵛ (pp. 58, 114 below); for Pliny see *ibid.*, G3ʳ, G3ᵛ, K6ʳ, L1ᵛ, P7ʳ, R3ʳ (pp. 48, 76, 80, 120, 130 below).

winds.[101] This theory is based on a crucial idea—an idea canvassed, defended, and supported by experiment in a number of his works— namely that the Aristotelian notion that one part of water could be converted into ten parts of air was false. The truth (in Bacon's view) was that a volume of water would convert into eighty, a hundred, or even more volumes of air. The consequence of this would be that the water converted into air would overburden the air already present, and that this in turn would produce winds.[102] In fact, this cause of winds is the one that Bacon adopts as the most important of all.[103]

After a brief history of accidental generations of winds, Bacon comes to extraordinary winds and sudden blasts: typhoons, hurricanes, whirlwinds, and the like. This is pretty much lifted en bloc from Pliny's *Historia naturalis*. The material from Pliny can be found in the commentary on the *Historia ventorum* presented below.[104] In this respect the treatment of extraordinary winds foreshadows the later discussion of prognostics of winds, which draws still more extensively on the Roman natural historian. Pliny, and a sprinkling of materials from other writers drawn upon elsewhere in the *Historia ventorum*, also crop up in the next and rather miscellaneous section on factors contributing to starting winds. Here Bacon considers a range of factors that help generate winds. He discounts the Aristotelian exhalation theory of the origin of winds, considers the roles of Sun, Moon, and stars, and examines ways of calming winds. In the course of this he introduces a couple of characteristically Baconian artificial experiments—with a calendar glass, and an experiment on wind in a 'tower closed on all sides'. The former evinces his fascination with the sense-extending possibilities of a relatively new instrument; and both exhibit his profound awareness that human beings could not hope to acquire reliable knowledge of nature if they assumed (in the Aristotelian manner) that it was possible to approach the world with senses and intellect in their native fallibility.[105]

Bacon then proceeds to the limits, successions, and motions of winds. Winds are limited as to height, latitude, and duration and, in the matter of successions (i.e. how different winds are followed by different

[101] *HV*, G8ᵛ–I1ᵛ, pp. 54–64 below.
[102] See *PhU* in *OFB*, VI, pp. 48–52, 372, 373; *NO*, in *OFB*, XI, pp. 82, 510, 569; *HDR* in *OFB*, XIII, pp. 38, 70, 270, 277.
[103] See *HV*, L4ʳ, pp. 82–4 below.
[104] *Ibid.*, I3ᵛ–I5ʳ, pp. 66–8 below, also see *cmts* pp. 398–9 below.
[105] *Ibid.*, I8ᵛ–K1ʳ, pp. 70–2 below; also *cmts* thereon p. 400 below. For experimentalism and the senses in Bacon's philosophy, see *OFB*, XI, pp. xl–xlii, lxxxix–xc.

meteorological conditions) we are offered a brief, six-item history of which the final and longer three items are derived from Aristotle's *Problems* and Pliny's *Historia*.[106] In other words, here we have yet more evidence of heavy reliance on these two sources. As for the motions of winds, Bacon's analysis rests on two things principally: an exploration of the similarities of and differences between the motion of water in rivers, and on variations on the closed-tower experiment which support the fundamental theory that many winds follow from overcharging of the air caused by the conversion of relatively small volumes of vapour into very large volumes of air.[107]

Having dealt with the motion of winds in nature, Bacon turns to one of the most arresting sections of the *Historia ventorum*—the motion of winds in machines and, in particular, ships and windmills. This, together with the material on the prognostics of winds which follows it, accounts for a quarter of the entire text of the *Historia ventorum*.[108] In other words, like the *Historia vitæ*, this history is replete with material aimed at *utility*. Of course this represents a much smaller proportion of text devoted to such matters in comparison with (say) Agricola's *De re metallica*, but here Bacon seems to have been lured, notwithstanding his theoretical devotion to his new conception of natural history, away from experiments of light and towards experiments of fruit. This is as true of the prognostics of winds as it is of the material on machines, for the prognostics are concerned in effect with weather forecasting in its widest sense, and that in turn has fundamental implications for agriculture, transport, and so on. Isn't Bacon allowing himself to be turned aside from the race by Atalanta's golden apples?

In his discussion of ships and windmills Bacon is pushing at the soft limits of the organic economy.[109] In an age which still relied enormously on human and animal muscle power, he is concerned with the potentialities of that great inorganic prime mover, the kinetic energy of the wind, with its promise of doing exactly what Bacon always had in mind,

---

[106] *HV*, K8ᵛ–L2ʳ, pp. 78–80 below, and *cmts* thereon pp. 401–2 below.

[107] See *cmts* on *HV*, H7ᵛ–H8ʳ, p. 397 below.

[108] *HV*, M2ᵛ–P7ᵛ, pp. 90–120.

[109] For the organic economy see, for instance, the recent and closely argued book by John Landers, *The field and the forge: population, production, and power in the pre-industrial West*, Oxford University Press: Oxford, 2003. The topics covered by Landers can be seen as a new set of contexts—hard, practical ones—against which the early modernity of Bacon's philosophy can be gauged. Landers (pp. 161–4) notes that 'innovation failure' in organic economies, and the tendency to innovate by reasoning by analogy with existing technologies, often had *literally* lethal effects.

enlarging human power over nature. Taking ships first, we are concerned with the largest early modern ocean-going British vessels, ones larger than that depicted on the engraved title of *Instauratio magna*.[110] That vessel has just one mizzen mast whereas Bacon is thinking of ships with two: such, for instance, as the merchantman Trade's Increase,[111] or the royal-naval monster, the Prince Royal. This latter, the English flagship, was built between 1608 and 1610 by the first Master of the Company of Shipwrights, Phineas Pett (1570–1647). It carried fifty-five guns, it was the greatest naval vessel built in the reign of James, and was almost beyond doubt the one that Bacon took as his example.[112]

What is remarkable is the amount of detail in Bacon's description of his large ship: he describes the masts; he specifies the sails each mast carries; he describes the sizes of the sails (in minute detail); and much more besides. It is almost impossible to believe that he was not working directly from a manuscript, or a printed or oral source, though I have not been able to identify it. Yet it would have been easy for Bacon, while in office, to have obtained his figures from reliable sources. He himself had sat on the 1608 Commission of Enquiry into the Navy which took evidence from shipwrights who accused Pett of shoddy workmanship in the building of the Prince Royal.[113] Alternatively he could, for instance, have acquired his data from members or witnesses of the 1618 Commission of Enquiry into the Navy, or perhaps from Phineas Pett himself, who claimed (if he is to be believed) to have been on social terms with Bacon from at least 1613.[114]

---

[110] For a reproduction of this engraving see *OFB*, XI, facing p. xxxii.

[111] Wrecked on her first voyage in 1613 and burnt by the Javanese; see BL manuscript log L/MAR/A/IX–XII.

[112] The dimensions of the Prince Royal, as given in contemporary records, agree too closely with those given by Bacon to think otherwise; see pp. 90–102 below, and *cmts* thereon pp. 402–4 below. This ship was represented in several near-contemporary paintings. There is a broadside-to-port view in Adam Willaerts's Embarkation of the Elector Palatine in the Prince Royal at Dover, 25 Apr. 1613. Willaerts also produced a painting of the Elector's arrival at Flushing on 29 Apr., and in that work the vessel appears broadside to starboard. Finally there is Hendrick Cornelisz Vroom's painting, *The Return of Prince Charles from Spain*, 5 Oct. 1623, in which the ship is seen at the head of the fleet, broadside to starboard, on the port tack. These paintings belong to the collections of the National Maritime Museum, London. Their repro IDs are respectively BHC0266, BHC4176, and BHC0710.

[113] For the 1608 Commission, see *The Jacobean commissions of enquiry 1608 and 1618*, ed. A. P. McGowan, The Navy Records Society: London, 1971, esp. pp. xv, 2–4.

[114] For Pett's prominence in Jacobean affairs, and his acquaintanceship with Bacon, see *The autobiography of Phineas Pett*, ed. W. G. Perrin (Publications of the Navy Records Society, 51), Navy Records Society: London, 1918, pp. lxxxiii, 103, and *passim*.

This is where the new natural history emerges—up to a point. It is true that the material on ship and sail finds Bacon feeling around for immediate practical results; but it is also true that here technological data—*quantitative* technological data[115]—of the most precise kind is being introduced into a *natural* history, alongside data relating to purely natural phenomena. He never wanted his readers to forget that his natural history was to be a history of generations, pretergenerations, and the arts, and especially (as we have seen) of that last. It was, as we know, to have been a history of nature vexed by experiment or wrought by the hand of man, and here in the *Historia ventorum* is perhaps Bacon's nearest approach to his own ideal in action. Accordingly, it is perhaps a little bit strange that his discussion of his second wind-powered device, the windmill, represents, for all the improving experimentalism that suffuses it, a lost opportunity. Here too was a chance to display the potentialities of quantitative data, but it was a chance that Bacon allowed to pass by, and we are left with a rather brief, sketchy account of a mill's sails, an account which does not even allow us to deduce whether Bacon had in mind a post-mill or a tower mill. We are told very little about the sails, and nothing at all about their dimensions or the machinery which they drove.[116]

From mills, ships, and all the sometimes novel data accompanying Bacon's treatments of those subjects, we turn to something which while firmly utilitarian in thrust, is completely different in its historical character, namely a long historical section on the prognostics of winds. Like the materials on the powers and qualities of winds and on the local origins of winds, the section is firmly rooted in ancient sources. Indeed, it is remarkable in being rooted in a single source, Pliny's *Historia naturalis*, passages of which Bacon rearranged and reworded. In fact the section is almost certainly the most extensive uninterrupted sequence of direct borrowings from an ancient text in the whole Bacon corpus.[117] This speaks volumes about the poverty of fresh data available to him. He is in effect treading natural-historical water at this point, for he can do nothing else. The sharp contrast between, on the one hand, the numerical data concerning the large ship and, on the other, the

---

[115] For the importance of such data in Bacon's work, see, for instance, *OFB*, XI, pp. xc–xcii; XIII, pp. 40–6, 52–4. Also see Graham Rees, 'Quantitative reasoning in Francis Bacon's natural philosophy', *Nouvelles de la république des lettres*, 1985, pp. 27–48.

[116] For a brief account of post- and tower mills, see Donald Hill, *A history of engineering in classical and medieval times*, Routledge: London and New York, 1996, pp. 172–7.

[117] See *cmts* on *HV*, O2$^r$–P7$^r$, pp. 404–13 below.

wholesale appropriation of Plinian material illuminates with some clarity the nature of the dilemmas that Bacon the practitioner of the new natural history faced.

Next comes a brief section on imitations of winds—particularly interesting for what Bacon has to say about gunpowder, mercury fulminate, and fulminating gold—after which the *Historia ventorum* comes to an end with the last two items promised in the Norma: (i) a set of provisional rules or 'imperfect axioms', and (ii) a list of things of use to human beings, things 'which are closest and most akin to those things deemed impossible and undiscovered'. Little need be said of these items. The provisional rules are very brief—especially when compared with those of the *Historia vitæ*—and at their heart lies the theory, developed in the body of the text, that the primary cause of winds is an overburdening of the air wrought by the conversion of very small volumes of water into very large volumes of air. As for the list of things akin to things judged impossible, it offers no surprises, given Bacon's preoccupations in the rest of the text: what he has in mind is anything that might radically improve ships and mills or enable humans to raise and allay winds, and accurately forecast not just the weather but harvests and the onset of epidemics.[118]

The *Historia ventorum* may end with these matters, but the *Historia naturalis* does not. The 1622 volume ends with a natural-historical Pisgah sight in the shape of the prefaces to the five histories which Bacon intended to publish at the rate of one a month for the next five months. The preface to *Historia vitæ* is discussed briefly below. The discussion of the preface to the *Historia densi* appears in volume XIII of the *Oxford Francis Bacon*.[119] As for the prefaces to the *Historia grauis & leuis*, *Historia sympathiæ, & antipathiæ rerum*, and *Historia sulphuris, mercurij, & salis*, remarks on these can be found in several volumes of *The Oxford Francis Bacon*, as well as in the commentaries on the edited texts presented in this one.[120]

All in all, the *Historia ventorum* accomplishes the very considerable feat of making the weather boring. It contains little or nothing in the way of prodigies or weather marvels. It presents a mass of data relating to the meteorological department of what Bacon called history of

---

[118] See *HV*, Q3ᵛ–R3ᵛ, pp. 124–30 below. For fulminating gold, etc., see *cmts* on *HV*, Q1ᵛ–Q2ʳ, p. 414 below.

[119] *OFB*, XIII, pp. 36–9, 269–70.

[120] See *OFB*, VI, p. lxxxiv; XIII, pp. xx–xxi, xxxvii, 307, 316. Also see pp. 415–17, cf. pp. 132–8 below.

nature in her ordinary course. It contains none of the old erudition, no philology, and precious little jocundity. It also strays beyond the limits of history of generations and into history of the mechanical arts, and beyond that into the domain of quantified data. In these senses the *Historia ventorum* is an exemplary *simulation* of the new model, functional natural history, which Bacon was seeking to propagate. But a simulation it remains, for he was well aware (and wished his readers to be aware) of how far his specimen histories fell short of his ideal. He knew he did not have enough data or enough good data. He relied to a remarkable degree on unverified materials derived from Pliny and Aristotle among the ancients, and, to a lesser extent, from Acosta and Gilbert among the moderns. But to say this is perhaps to miss the point. Bacon knew perfectly well that he simply did not have the means to collect the kind of rigorously tested data that his programme for the instauration of the sciences required. So he did what he could: he poured existing data, old and recent, derived from books or informants, into a new mould, and trusted that he had done enough to convey a radically new conception of natural history to his successors. His trust in himself seems not to have been entirely misplaced.[121]

### (e) *Historia vitæ & mortis*

As I have written in another volume of this edition, the aim of prolonging life epitomizes the aims of Bacon's programme as a whole. Certain that he lived in an age ordained by Providence for the advancement of knowledge, he believed that philosophy should improve material conditions of the human race, and so in part restore prelapsarian felicity. He marked out the prolongation of life as the first and highest objective of the new philosophy. Realization of that ancient dream would fulfil a programme proposing a material soteriology for this world.[122]

Accordingly it is no wonder that the preliminaries and endmatter to the *Historia naturalis et experimentalis* are recalled straight away at the beginning of the *Historia vitæ & mortis*. At the start of the former he had promised to bring out the latter as the last of his six promised specimen natural histories. But now, in the brief exordium to the *Historia vitæ*, he tells us that because of the extreme importance of the subject he has now promoted it to second place. This exordium is followed

---

[121] For a sketch of the reception of Bacon's ideas, including his ideas on natural history, see *OFB*, XI, pp. xxii–xxxviii.

[122] *OFB*, VI, p. lxv.

immediately by a preface which is in every substantive detail identical to the one presented at the end of the *Historia naturalis* volume.

The preface, employing language and especially imagery first used in the *De vijs mortis* (*c*.1611–19),[123] begins with a brief vindication of the subject under inquiry, a summary attack on chemical medicines and (at greater length) the radical moisture and natural heat theory of ageing and death,[124] a theory to which Bacon returns later. The rest of the preface presents a summary of the Baconian alternative, i.e. that the true causes of ageing and death are that the body's reparable parts are undermined by the gradual failure of the less reparable; and that the root of this failure is the conspiracy between, on the one hand, the body's internal vital and inanimate spirits and, on the other, the external air. He also asserts (again echoing *De vijs mortis*)[125] that to prolong life investigators must adopt a twin-track approach, and see the body as an inanimate substance prone to decay like any material object of any kind, and as an entity possessing an animate spirit which gives living beings their distinctive faculties but which requires them to take on nourishment. This talk of inanimate and animate spirits reminds us that the *Historia vitæ* is—strangely for a natural-historical work—one of the most explicitly theory-laden of all Bacon's writings.

Like *Historia ventorum*, the *Historia vitæ* begins with a set of topics of inquiry which take up what Bacon believed were his key doctrines, namely the twin-track approach and the theory of spirits. The latter theory is absolutely central to so many sections of the *Historia vitæ* that I give an account of it now. The account will be very brief and very summary, for a great deal has been written about it in other volumes of *The Oxford Francis Bacon*.[126]

The theory rests on the assumption that there are two mutually convertible states of matter in the universe: tangible and pneumatic. Tangible matter is heavy, gross, inactive, and resistant to change or motion. Spirit matter is weightless, tenuous or vaporous, highly active, and the principle of motion and change in almost the whole universe. The universe is a finite, geocentric plenum in which the region from the

---

[123] For instance, the imagery of the vestal fire and the torture of Mezentius; see *HVM*, A6ʳ, A7ᵛ, pp. 144–6 below; cf. *DVM* in *OFB*, VI, pp. 270–4, 352, 354.

[124] For the centrality of this theory in medieval and Renaissance thought, see *OFB*, VI, pp. lxv–lxix.

[125] *OFB*, VI, pp. 274–6.

[126] See, for instance, (*OFB*, VI, pp. xliii–lxv; XI, lxxxiii–lxxxvi, 436, 584; XIII, pp. 60–6, 80–2, 188.

Moon upwards consists entirely of free pneumatic substances, and in which the Earth's core is made up of tangible matter alone. Only in the region below the Moon and above the Earth's core do tangible matter and spirit meet and interact. In fact, in the sublunar world there exist not just free pneumatic bodies—air and flame—but two classes of pneumatic substances (both air–flame compounds) locked into tangible matter which effect change within it.

The two kinds of spirit, both thoroughly corporeal, are inanimate and animate (or vital) spirit. The former exists in all tangible bodies, living and non-living, at or near the surface of the Earth; the latter are to be found in living bodies alone. The two kinds differ in that in all varieties of inanimate spirit the airy component predominates whereas in the vital spirits the flamy has the upper hand.[127] Inanimate bodies possess only inanimate spirit distributed through their mass in discontinuous portions. Vegetable bodies have these spirits *and* vital spirits organized in a network of branching channels. Animals have discontinuous inanimate and branching vital spirits but have their branching spirits connected to a cell or cerebral concentration of spirit which endows animals with their sensory-motor functions. Vital spirits are always reluctant to leave the organisms enclosing them for nothing outside is akin to them; but inanimate spirits long to escape from tangible bodies for their dominant airy component draws them to the ambient air with which they conspire.[128] Vital spirits restrain the inanimate, but eventually the latter prevail and destroy their hosts.[129] Possession of vital spirit entails consumption of the body and therefore the need for nourishment.[130]

All disintegration arises from an *actio triplex* of inanimate spirit. The stages of the *actio* are attenuation, escape, and contraction. Attenuation happens when the spirit attacks the matter imprisoning it and converts some of it into itself. This weakens the object's structure and, since the spirit is weightless and tangible matter is not, makes it lighter. After attenuation comes escape for, once the spirit has increased its volume and weakened the object, it can decamp into the air. Escape is followed by contraction and desiccation. Inanimate spirits differ in their intrinsic qualities and their distribution within tangible matter. Distribution

[127] *HVM*, 2D3ʳ–2D5ʳ, pp. 350–2 below.

[128] *HVM*, A6ʳ–B2ᵛ, D8ʳ–E6ʳ, 2D6ʳ–2D8ʳ, pp. 144–8, 172–6, 354–6 below; *DVM* in *OFB*, VI, pp. 272–82; also see *SS*, M3ᵛ–M4ʳ, Z2ᵛ–Z4ʳ (*SEH*, II, pp. 451–2, 557–61).

[129] *DVM* in *OFB*, VI, p. 274.

[130] *HVM*, A6ʳ–B2ᵛ, Y2ᵛ–Y6ʳ, pp. 144–8, 310–14 below.

varies with the size of the spirit's particles and the evenness of its dispersion through tangible matter. Spirit finely divided favours durability; spirit collected in large portions accelerates dissolution. As for evenness, the greater it is the more durable the body. Otherwise the inanimate spirit, like a demented prisoner, hurls itself against the walls of its tangible prison, and slowly but surely breaks them down. This process is aided by the *actio* of attenuation and self-multiplication which converts susceptible parts of the body into more spirit. The second *actio* is to escape and unite with cognate bodies—especially the air. This impulse is inimical to durability in inanimate bodies and longevity in living ones. Consequently factors hindering escape are very important. A compact body with narrow pores resists escape. Likewise, if the spirit is enclosed in friendly bodies it will be less vulnerable to the seductions of cognate external bodies, and the urge to escape will be muted yet more if there is plenty of suitable matter which the spirit can multiply itself upon.

For Bacon, then, living organisms had to be considered as cradles of the vital spirit *and* as entities undergoing the same processes of decay as lifeless things. Accordingly, they had to be considered as if they were *inorganic* things to the extent that their tangible parts embodied inanimate spirits.[131] Hence the first natural-historical section of the *Historia vitæ* is given over to the 'nature of durable', a section devoted to data relating to what causes all substances, living and non-living, to last.[132] At this point Bacon's natural-historical enthusiasms meet his theoretical-speculative convictions. And this is why *Historia vitæ* differs from the earlier essay on the causes of ageing, the *De vijs mortis*. The *De vijs mortis* is the *Historia vitæ* without data. The former is not a history but an exercise in theory formation; the latter an imperfect model of an ideal Baconian history, with the historical material shaped and to a degree directed by theoretical preferences developed in *De vijs*.

The history concerning the nature of durable leads Bacon to conclude that the most durable substances are hard or oily because these detain the inanimate spirit by reducing the spirit's capacity to escape into the air; and that is the reason why, for instance, stones last, and why cooked food lasts longer than uncooked, the watery part of the food having been driven off, and the oily remaining. These conclusions are

---

[131] *HVM*, A6ʳ–B2ᵛ, 2D3ʳ–2D5ʳ, pp. 144–8, 350–2 below; cf. *DVM* in *OFB*, VI, pp. 274, 344, 350–2; *DAS*, 2D4ʳ⁻ᵛ (*SEH*, I, p. 600).

[132] *HVM*, B7ᵛ–C2ᵛ, pp. 154–8 below.

followed by a brief history of durability in all kinds of plants, and then by data concerned with the causes and prevention of that great enemy of durability, desiccation. The conclusions drawn from (or rather associated with) this body of data amount to nothing less than a classic statement of the inanimate spirit's *actio triplex*.[133]

The same *actio* is not mentioned in the next section of the *Historia vitæ*—on length and shortness of life in animals—but is effectively a substrate of the entire discussion. Like the discussions of the nature of durable and desiccation, this one begins with historical data and ends with major observations or inferences from the data.[134] Once again Bacon is perfectly frank about the lack of good data, and about the difficulties of correlating longevity with factors which appear to promote long life in some animals but not in others. Foxes seem to be well set up for long life but are nonetheless short-lived. They are 'very well clad, carnivorous, and live in earths'.[135] These factors should favour long life because (though Bacon does not say so) they all resist the *actio triplex*—especially the harmful conspiracy between inanimate spirit and the external air. The goat too seems a promising candidate for long life, but its advantages in respect of the unspoken *actio triplex* are cancelled out by its enthusiastic sexuality, which for sure abbreviates its existence.[136]

After the investigation of longevity in animals Bacon takes up the other track of his argument: that living bodies differ from inanimate ones in that they need nourishment. As he promised earlier (in the topics of inquiry) he deals with this very briefly, and he confines himself to three points: (i) that nourishment should be simpler than the body nourished and not closely akin to the body nourished; (ii) that preparation is the key to proper nourishment for that allows the nourishment to be assimilated more efficiently; and (iii) that if nourishment could be accomplished from the outside of the body such a practice might compensate for the weak digestions of the old.[137] We shall see that this was not the last time that Bacon proposed this notion.

Bacon returns to data with an assemblage of material on longevity in human beings. This is probably the most original and arresting section in the work, and accounts for some six thousand words or 17.5 per cent of the text. Although its paragraphs are numbered consecutively and without a break, the section has two parts: the first (paragraphs 1 to 22)

---

[133] *HVM*, C2ᵛ–E6ʳ, pp. 158–76 below.  [134] *Ibid.*, E6ᵛ–G6ᵛ, pp. 176–92 below.
[135] *Ibid.*, E8ʳ, p. 178 below.  [136] *Ibid.*, F2ʳ, p. 180 below.
[137] *Ibid.*, G7ʳ–H2ᵛ, pp. 192–6 below.

presents cases of longevity; the second (paragraphs 23 to 50) tries to identify correlates of longevity.[138]

Taking paragraphs 1 to 22 first, Bacon presents one hundred named instances of longevity, and another 267 unnamed derived via Pliny's *Historia naturalis* from Vespasian's census.[139] Naturally, Bacon's data is derived from printed sources—the Old and New Testaments, Greek and Roman sources, and various kinds of post-classical literature.[140] To modern eyes the limitations of Bacon's data (in this pre-eminently 'inductive' form of inquiry) are obvious. For instance, his data were too 'literary', i.e. they were not collected especially for the purpose, and of course Bacon had no prospect of sending out what the *New Atlantis* called 'Merchants *of* Light' to scour the world for suitable first-hand data, or even of hiring 'Depredatours' to collect '*the* Experiments *which are in all* Bookes'.[141] Again, Bacon seems to have assumed that lumping data together from very different times and social conditions was unproblematic, i.e. that biblical, Greek, Roman, or medieval conditions were similar enough to those of his day to be taken without serious reservation. His biblical data are assembled on the assumption that the ages of the patriarchs could be taken not just as the Word of God but as the *literal* Word of God. Lastly, Bacon's sample was, of course, minute: modern actuaries, demographers, and health statisticians characteristically deal with samples running not to hundreds of instances but to millions. No doubt Bacon was to a degree aware of some at least of these limitations, and, if he had no inkling of what might count as a good sample from a modern demographer's point of view, he certainly knew that his sample was too small and the instances it comprised not always trustworthy.[142]

But for all that, the collection of data on human longevity is startlingly original, as original perhaps as the nautical data presented in *Historia ventorum*. Who before Bacon had made such a collection? And who before Bacon had used that data in so convincing and proto-statistical a way to rebut early seventeenth-century arguments that human strength, stature, health, and life expectancy had declined in the ages since Noah's flood, arguments which if true or widely accepted might have made Bacon's entire programme for rebuilding the sciences difficult to sustain or downright pointless?[143]

[138] *HVM*, H2$^v$–M7$^v$, pp. 196–232 below.   [139] *Ibid.*, I4$^v$, p. 206 below.
[140] See *cmts* pp. 429–40 below.   [141] *NA*, f4$^{r-v}$ (*SEH*, III, p. 164).
[142] See *HVM*, H6$^v$, p. 200 below.   [143] See *cmts* pp. 437–8 below.

li

That apart, the data assembled in paragraphs 1 to 22 would have had little value had Bacon's instances not been correlated with variables which (he believed) affected longevity. These variables are touched on in the potted biographies that he attached to those instances for which biographical material was available. As Bacon puts it himself, 'I have adjoined to each a true and very concise character sketch or biography such, in my judgement, as bears on longevity (which is governed to no small degree by habit and fortune); and this in two ways; for they are either generally long-lived, or are not so predisposed, but sometimes could be'.[144]

Now the information collected in the potted biographies comes into its own in the second part (paragraphs 23–50) of the history of longevity in humans. Using the biographical data Bacon attempts to specify groups of interrelated variables which he supposed had a bearing on the dependent variable, i.e. lifespan. Among the interrelated variables men live longer in cooler, island, and upland places, because their inanimate spirits are less predatory in these conditions. Factors such as the physical and mental states of the parents when the child is conceived, and the state of the mother during gestation and nursing are also influential in determining the child's life expectancy. Diet and exercise are also important, as are the individual's career and mode of life, which, like other factors, affect the behaviour of the spirits be they animate or inanimate. Physiognomical signs also indicate longevity or lack of it, though here it is not at all clear how Bacon derives the conclusions from the data presented.[145]

Much of this is, of course, fragile even from Bacon's own perspective, never mind a modern one which might insist, for instance, on starting with mortality rates, i.e. the percentage change in the number of people of a specific age in a given sample who die in a given year. And for one who set so much store in *Novum organum* on negative instances, it is perhaps odd that he did not look for cases of individuals whose lives were short although they possessed many of the principal factors indicating long life. Likewise, one might have expected some attempt to look for evidence of concomitant variations in the relations between, on the one hand, possible factors influencing length of life and, on the

---

[144] *HVM*, I5$^v$, p. 206 below.
[145] See para. 36 (*HVM*, M1$^r$, p. 226 below) on hairiness, for example; the remarks here are not supported by data on hair given earlier; see *HVM*, K7$^r$, p. 216 below.

other, actual length of life. All the same, Bacon's was one of the very first attempts at a proto-statistical treatment of some very difficult intellectual and practical issues, and one which struck a chord with his successors. It also delivered, as *Historia densi* did with the decuple proportion theory,[146] a knock-out empirical blow to the then-popular notion that human stature, strength, health, and longevity had all diminished during the centuries since the Noachian flood.

Next comes a relatively short excursus on medicines for prolonging life, mainly 'those they call cordials'. This is rather more favourable to inorganic remedies and popular nostrums than the preface to *Historia vitæ* might lead one to expect; it is also a prelude to a substantial section of the work, a section which presents a great deal of materia medica for the would-be Methuselah.

This section—by far the longest section of the work, amounting as it does to some 13,500 words or almost 40 per cent of the text—is concerned with practical procedures and measures for prolonging life, and to that extent is not natural history in the strict Baconian sense at all.[147] But certainly practicality is Bacon's watchword, for not only must the measures recommended be safe but they should involve no serious curtailment of the business of life—so forget about living in caves, on top of pillars, painting the body, or adopting Cornaro's mode of life: all of these may promote longevity but they will not help you plough a field, plead in court, or set up colonies in the Americas.[148] In addition, Bacon exhorts readers once again not to fall for the radical moisture theory of ageing, or for the superstitious nostrums usually peddled by writers on the subject.[149] He also assures readers that although what he is about to propose is safe, not all his recommendations have been tested by experiment (*Experimento*), though all are based on principles and presuppositions 'cut and dug from the very rock and veins of nature itself'.[150] He is, in other words, not going to deceive anyone that this natural history is a perfect model of the kind of history that the programme of the *Instauratio magna* requires.

---

[146] See *HDR* in *OFB*, XIII, pp. 38, 70.     [147] *HVM*, N4ʳ–2A3ᵛ, pp. 236–326 below.

[148] *Ibid.*, N7ʳ–N8ᵛ, pp. 238–40 below. Living in caves might be reserved for the hermits in Bensalem alone; see *NA*, e2ᵛ (*SEH*, III, p. 157).

[149] Bacon had read the standard authorities—Aristotle, Galen, Avicenna, Roger Bacon, Arnaldus de Villanova, Ficino, Paracelsus, Fernel, Cornaro, and Telesio—but set little store by any of them, see *DAS*, 2D3ᵛ (*SEH*, I, pp. 598–9); *HVM*, C6ᵛ, K8ʳ⁻ᵛ, N4ʳ–N8ʳ, R4ʳ, Z4ᵛ, pp. 162, 218, 236–40, 268, 320 below.

[150] *HVM*, O1ʳ–O2ʳ, pp. 240–2 below.

Preliminaries over, Bacon organizes the section in terms of three *intentions*, a word he uses in the classical medical sense (*intentio curationis*) to mean the aim or purpose of a curative procedure and hence a plan of treatment.[151] The intentions are: 'the prohibition of consumption, the accomplishing of repair, and the renovation of what has grown old'. To these intentions Bacon attaches ten *operations*, i.e. sets of procedures, treatments, and routines designed to accomplish the intentions. To the first intention he attaches four operations; to the second another four; and to the third, two.

The first intention, 'the prohibition of consumption', has as its first operation that which works on the spirits 'to keep them young and help them recover their strength'.[152] The discussion of this operation, running to ninety-nine numbered paragraphs, is far and away the longest of the ten that comprise this part of the text. Here Bacon immediately reintroduces the spirits, the primary hypothetical entities of his theory of matter. When dealing with the spirits the *Historia vitæ* is about as explicit as any of Bacon's works which touch on the subject. We have seen that already, and we will see it again when we look at the provisional rules with which the *Historia vitæ* ends. Anyone wishing to look at this in detail may see above, the commentaries below or, indeed, consult other *Oxford Francis Bacon* volumes on the subject.[153] Suffice it to say at this point that Bacon speaks mainly (though not exclusively) of the animate spirits, and in particular of the need to 'condense' them. There are four ways to do that: by concentrating them (mainly with opiates and their subordinates); by cooling them (mainly with nitre and its subordinates); by soothing them (with various organic remedies); and by curbing their motions (with sleep principally). Lastly come some general observations on the spirits.[154]

The other three operations associated with the first intention need not detain us for long. They comprise measures to prevent the external air encouraging the body's inanimate spirits to become predatory, to make the blood become less liable to predation,[155] and to induce 'a dewiness spread through, and (if you like) radical, in the very substance of the body'. The language associated with this last is especially telling ('dewiness', 'radical'): for even in a discussion of ageing predicated on the pneumatic theory of matter, the terms of a pre-Baconian orthodoxy,

---

[151] See *OED*, **Intention**, II, 10.  [152] *HVM*, O3$^r$–O5$^r$, pp. 242–4 below.
[153] See p. xlvii n. 126 above.  [154] *HVM*, R4$^v$–R6$^v$, pp. 270–2 below.
[155] See *ibid.*, R6$^v$–V1$^v$, pp. 272–92 below.

the radical moisture theory, break through. As I have argued elsewhere, Bacon was not immune to a species of intellectual slippage through which received ideas could exact their tribute from the most independent of minds.[156]

As for the four operations connected with the second intention—the intention aimed at 'the accomplishing of repair'—here too Bacon adapts the existing pharmacopoeia and therapeutic practice to his purposes. Many of his recommendations for the repair of the body—diet, preparation of food, exercise, massage, maintence of the principal organs, and so on—would have raised no early seventeenth-century physician's eyebrows. Bacon's only departure here into relatively new territory may have been his recommendations regarding methods of keeping the liver in good condition, recommendations aimed at preventing the senescent liver's progressive arefaction, parching, and salination.[157] These evils are exactly those which Bernardino Telesio (1509–88) had identified as the primary causes of ageing and death.[158] Clearly no stone was to be left unturned in work of prolonging life and so, with this acknowledgement of Telesio's contribution, we see again the kind of practical, bet-hedging eclecticism that typifies Bacon's approach to his subject.

The third and last intention, 'the renovation of what has grown old', is again an exercise in diet and polypharmacy but with an accent on the thoroughly Baconian idea, mentioned earlier, that nutrition from the outside (i.e. through the skin) might help compensate for the weakness of digestion in old people. Among the means considered prominence is given to bathing in blood, especially the blood of kittens and infants.[159] These procedures are dismissed in the end as disgusting and loathsome, but that did not stop Bacon looking for substitutes—the application of fatty meats, butter, and egg yolks to the body. Prolonging life, as he so often remarked, was never going to be easy. An elixir, potable gold, a single nostrum—none of these was going to do the trick, for once the causes of ageing (especially the actions of the spirits) had been acknowledged, the achieving of long life was going to be a hard business involving all ten of the operations that Bacon's three intentions required.[160]

[156] See *HVM*, T7$^r$, p. 288 below. Also see *OFB*, VI, pp. lxv–lxix.
[157] *HVM*, V6$^{r-v}$, p. 198 below.
[158] See *cmts* p. 450 below, and *OFB*, VI, pp. lxvi–lxvii, 270–2, 437–8.
[159] *HVM*, Z4$^v$–Z6$^r$, pp. 320–2 below.
[160] *Ibid.*, 2A3$^{r-v}$, p. 326.

After all the particulars relating to longevity, after all the operations needful for procuring long life, and after all the speculative reflections as to the causes of decay, what is the common end of all pathways to death? The answer supplied will come as no surprise: if the flame-like living spirit is deprived of three things death will ensue. Here Bacon simply rehearses points implicitly or explicitly made earlier for the three things needful: suitable motion, proper cooling, and aliment. It needs suitable motion because inability to move, or the wrong motion (as in discharge of the spirits), will cause sudden death, as in strokes or severe head injuries. Want of coolness by respiration causes the spirit to be destroyed by its own heat, as in strangulation and suffocation. The living spirit is also destroyed by lack of aliment for, although the spirit exists in identity (i.e. it does not require fuel or renovation), it still causes the body to require aliment. Lack of any or all of these requirements shows as the precursors of death: convulsions and labouring of heart, and all the other appearances (pallor, mental confusion, cold sweats, etc.) of imminent death.[161]

Bacon rounds off the historical sections of *Historia vitæ* with a short account of the differences between youth and age. He runs through the physical differences, marking the decline from bilious, sanguine, and robust to phlegmatic, melancholy, and feeble. Then he considers differences regarding the 'affections of the mind', differences he tackles by recalling a French friend of his youth who seems to have shared with Hamlet a taste for reflections unfavourable to the elderly: he argued 'that the vices of their minds had in a way some agreement or parallel with the defects of their bodies'. He aligned their dry skin with their effrontery; their tough guts with their pitilessness, and so on. But Bacon's brief diversion into jocundity is short-lived: there are serious points to be made for 'as with the body so with the mind: old men do get better in one or two respects, unless they are completely superannuated. For instance, though they are not so quick at thinking things up, they have sounder judgement and settle rather on safer, more reasonable conclusions than outwardly impressive ones. They also become more garrulous and ostentatious, since being less a match for things, they seek satisfaction in words, so that the poets' fiction that Tithonus was turned into a cricket is not completely absurd.'[162]

The *Historia vitæ* ends with a set of provisional rules. In this it resembles the earlier *Historia ventorum* and the later *Historia densi*,

[161] *HVM*, 2A5ʳ–2C1ʳ, pp. 328–40 below.     [162] *Ibid.*, 2C1ᵛ–2C5ᵛ, pp. 340–6 below.

albeit with two differences. First, in the other two histories the provisional rules are followed by a brief list of desiderata (with their closest approximations) of potential benefit to human life, desiderata arising from subject matter of the histories in question. There is no such list in the *Historia vitæ*. Secondly, in the other two histories the provisional rules amount to no more than a list of bare propositions the whole of which runs to a few hundred words in each case.[163] In the *Historia vitæ* the rules are much more extensive and developed. Each rule is accompanied by an *explicatio* and, taken together, the rules and *explicationes* run to some 3,700 words of text. This makes the work rather more like the *De vijs mortis* (which often uses the rule–*explicatio* structure) than like the other Latin histories. The provisional rules are among the most explicit and extensive, concerted accounts (amounting to just over a tenth of the entire text) of Bacon's theoretical positions in relation to spirits and types of spirit in humans. The first nineteen provisional rules deal with the fundamental principles of the pneumatic theory of matter, and the remaining thirteen deal with assimilation, alimentation, and the nature of animate spirit.[164]

The sheer volume of such material is, if the *Historia vitæ* is approached with Bacon's natural-historical ideals in mind, rather disconcerting. The extent and weight from first to last of causal discourse, of talk about what makes ageing and death happen, of talk that is as unequivocal in its self-confidence as it is in its apparent denial of the procedural prescriptions of *Novum organum*, all such causal talk goes well beyond anything allowed for in any of his programmatic writings on natural history. In this respect the *Historia vitæ* is, in the end, a generic hybrid. It is a natural history *and* an anticipation of the new philosophy which Bacon looked forward to. It is an anticipation of what natural history might be, but one compromised by an anticipation of what natural philosophy might become. The history is Bacon's imperfect simulation of his own generic characterization of natural history crossed with an 'illegitimate' anticipation of his characterization of the genre natural philosophy.[165]

Of course Bacon could (and did) use the weaknesses of the work to his own advantage to point out that the work itself justified his belief that natural history had to be placed on a properly financed, concerted,

---

[163] *HV*, Q3ᵛ–Q8ᵛ, pp. 124–8 below; also see *HDR* in *OFB*, XIII, pp. 162–6.

[164] *HVM*, 2C7ʳ–2F8ᵛ, pp. 346–76 below.

[165] On the notion of 'anticipation' in Bacon's work, see *OFB*, VI, pp. xix, xxv, xxxvi–xxxvii; XI, pp. 74–6, 505–6; XIII, p. xx.

and cooperative footing to have any chance of fulfilling its promise. In addition, the *Historia vitæ*, like Bacon's other natural histories, is to a degree a sound expression of his new model natural history: it steers clear of all traditional erudition, displays little jocundity, and shows no interest in philological and textual questions whatever.

The *Historia vitæ* is to that extent a new model natural history, but it is not new enough. In the first place, the history has a profoundly practical bias, and many of the practical recommendations were underpinned by Bacon's theoretical preferences. But the practical recommendations are not of the kind that he believed would be produced by the powerful operational knowledge flowing from the full implementation of his programme. Rather they are of a kind that he repeatedly warned against, i.e. results of a search for short-term practical benefits which might actually get in the way of the greater benefits that would come through the deferred gratification entailed by the routines of *Novum organum*. To use Bacon's own language, the *Historia vitæ* offers Atalanta's golden apples instead of the strenuous effort needed to win the race. In fact the *Historia vitæ* is not a history in the strict Baconian sense at all. It is history as enumerative recital of particulars relating to durability, to longevity of plants, animals, and above all humans. It is also a history in one of that word's medical senses, i.e. not a case history, but a history of particular means (dietary, medicinal, etc.) designed to have a single outcome, i.e. long life. We see this meaning of 'history' predominating in the discussion of intentions, for instance. On the other hand, the *Historia vitæ* presents too many untested particulars, and (like the *Historia ventorum*) too few particulars whether tested or not. In the end the two Latin histories published before Bacon's death are hybrid creatures. They resist simple categorization—even in terms of his own natural-historical prescriptions.

2

## The Texts and their Transmission

### (*a*) Title-pages, Collational Formulae, and Imprints

The two texts presented in this volume were both printed in common octavo by John Haviland's printing house. The Haviland octavos were the only editions of the two texts published in Bacon's lifetime, and (almost) the only ones with any claim to authority.[166] Their title-pages (also see Plates I and II in Appendix I), and collational formulae are as follows:

I. *Historia naturalis et experimentalis.*

FRANCISCI | BARONIS | de | VERVLAMIO, | VICE-COMITIS | Sancti Albani, | HISTORIA NATVRALIS | et experimentalis | ad condendam | Philosophiam: | Sive, | PHÆNOMENA VNIVERSI: | Quæ eſt Inſtaurationis Magnæ | pars tertia. | [rule] | [fleuron] | [rule] | Londini, | In Officina Io. Haviland, impenſis | *Matthæi Lownes* & *Guilielmi Barret.* | 1622.

*coll:* common 8°: *A*⁴ B⁸–T⁸ [$4 signed (–*A*1, *A*2, *A*3, *A*4)] 148 leaves present [*i–viii*] 1–17 *18* 19–285 [*286–8*]

II. *Historia vitæ & mortis.*

FRANCISCI | BARONIS | de | VERVLAMIO, | VICE-COMITIS | Sancti Albani, | Hiſtoria *Vitæ* & *Mortis.* | Sive, | TITVLVS SECVNDVS | in Hiſtoriâ Naturali & Experimentali | ad condendam Philoſophiam: | Quæ eſt | *INSTAVRATIONIS MAGNÆ* | pars tertia. | [rule] | [fleuron] | [rule] | Londini, | In Officina Io. Haviland, impenſis | Matthaei Lownes. 1623.

*coll:* common 8°: *A*⁸–2F⁸ [$4 signed (–*A*1, 2*A*4)] 232 leaves present, pp. [*i–vi*] 1–191 *192* 193–368 *369–70* 371–404 *405–6* 407–9 410 (410 corrected from 406) 407–54 (407–54 uncorrected sequence in all copies)

The title-page of the *Historia vitæ* is clearly modelled on the title-page of the *Historia naturalis*, and that, together with their common octavo formats, is but one of a number of features which tell us that the two natural histories were designed to be in broad terms uniform. The two histories also share to a very large extent a common mis-en-page, a

---

[166] For the 'almost', see pp. lxxv–lxxix below.

common *staging* of the texts—especially in their typographical conventions, which go far beyond the mere use of common fonts,[167] and page layout. But more of this in a moment.

As for the imprints, the individuals who took responsibility for producing the two natural histories were all prominent participants in the early seventeenth-century London book trade. William Barrett (d. 1624) and Matthew Lownes (d. 1625), who financed the publication of the *Historia naturalis,* and Lownes who financed the *Historia vitæ,* were very substantial figures. Barrett had been apprenticed (1597–1605) to Bonham Norton, who was later to be associated with the printing of Bacon's *Novum organum.*[168] Barrett was elected member of the Livery of the Stationers' Company in 1610; he was a partner (1608–13) of Edward Blount, and in 1622 had an interest in the Stationers' Company's unprosperous Latin Stock.[169] In addition to his interest in the *Historia naturalis,* he was also associated with the printing of Bacon's *Henry the seventh* in the same year. Matthew Lownes was a member of a veritable tribe of Lownses involved in the early seventeenth-century book trade. He was elected to membership of the Livery of the Stationers' Company in 1602. At the time of the printing of the Haviland octavos he had sufficient standing to act with John Bill in the 1622 sequestration of Bonham Norton's moiety of the King's Printing House.[170] In fact, like Barrett, Lownes seems to have been part of an informal network that connected them to Bacon and the affairs of the King's Printers.[171]

The two Latin natural histories were physically produced by John Haviland (d. 1638). Among other things Haviland was one of the successors of Richard Tottell in the patent for printing common-law books, which he issued not under his own imprint but always 'for the Company of Stationers'. He was a partner in the Eliot's Court Press, and under his own imprint was a prolific publisher of a large number and wide range of titles in the period when he produced the Bacon

---

[167] See pp. lxiii–lxiv below.

[168] See *Dictionaries of the printers and booksellers who were at work in England, Scotland and Ireland 1557–1775,* ed. H. R. Plomer, H. G. Aldis, et al., The Bibliographical Society: London, 1977, p. 24; *OFB,* XI, pp. xcviii–ciii.

[169] Michael G. Brennan, 'The literature of travel', in *The Cambridge History of the Book in Britain,* iv: *1557–1695,* ed. John Barnard and D. F. McKenzie with the assistance of Maureen Bell, Cambridge University Press: Cambridge, 2002, pp. 246–73 at p. 256. Also see Maria Wakely, 'Printing and double-dealing in Jacobean England: Robert Barker, John Bill, and Bonham Norton', *The Library,* 7th series, vol. 8, no. 2, 2007, pp. 119–53, 137.

[170] *STC,* III, p. 109. On the sequestration see National Archives C22/601/28.

[171] See National Archives C2/JASI/N7/44.

histories.[172] More will be said of Haviland and his associates in the *Oxford Francis Bacon* edition of *De augmentis scientiarum*, a work which Haviland printed in 1623.

## (*b*) Skeletons, Box Rules, and Typographical Conventions

With one or two exceptions considered below, the type-pages of *Historia naturalis* and *Historia vitæ* were imposed such that one skeleton imposed the eight type-pages of one forme, and a second the eight type-pages of the other. On almost every page the typographical parts of the skeleton comprise the headline and seven rules.[173] The rules are disposed as in Plate III in Appendix I, such that there are three vertical rules, two set 12 mm apart adjacent to the fore-edge, and the third set 60 mm from the inner of the two other verticals and adjacent to the spine. The vertical closest to the spine and the one nearest the fore-edge are each about 133 mm long; the other vertical is about 2 mm shorter,[174] and is topped and tailed by two 81 mm horizontal rules which terminate inside the head and tail ends of the outermost verticals. Two further horizontals, 60 mm and 12 mm long, lie approximately 9 mm below the top horizontal. The disposition of the rules is such that the ones on a verso page are the mirror image of those on the facing recto.

There are therefore seven rules to each page and fifty-six to each skeleton forme. Each of the four rectangles produced by the seven rules has a particular function. The larger of the two boxes at the head of the page accommodates the headline with the exception of the page number, which is lodged in the smaller. The smaller of the other two boxes is for marginalia (section headings, paragraph numbers, etc.), while the larger rectangle contains the body of the text and the direction line. In their dimensions and functions the boxes are pretty much identical in our two natural histories, and their quite deliberate

[172] See J. H. Baker, 'English law books and legal publishing', in *The Cambridge History of the Book in Britain*, iv, pp. 474–503 at pp. 481–4. Also see *STC*, III, pp. 79–80.

[173] A skeleton has three essential components (chase, furniture, and quoins), and two accidental (headline and rules). The chase, a rigid rectangular iron frame, has eight type-pages placed in it and these, together with the accidental components, are locked into the chase with the quoins and other wooden furniture. The accidental or typographical components (headline and rules) show up on the pages of the edition and, since skeletons can be used time and time again, their recurrences can be tracked, and that allows us to draw inferences about the text's transmission through the press. For another account of skeletons and what can be learnt from them, see *OFB*, XI, pp. ciii–cxiv.

[174] The dimensions vary very slightly from forme to forme and type-page to type-page. This is due to differential shrinkage of the paper, bending stresses on the rules, and so forth.

uniformity in this and other respects[175] no doubt encouraged their owners to have them bound together.[176]

Haviland and others used this configuration of rules for books in all the common formats.[177] Indeed, the dimensions of the *Historia naturalis* and *Historia vitæ* rules seem to have had a very particular history. In the same years that the histories were printed Haviland also produced two parts of a series by the prolific Joseph Hall, namely *Contemplations . . . the sixth volume* (*STC* 12657a) and *Contemplations . . . the seventh volume* (*STC* 12658). The first of these was entered in the Stationers' Register 11 May 1622, i.e. over four months before *Historia naturalis*, and was no doubt printed before the latter. As for the seventh volume of *Contemplations*, that was entered on 20 June 1623, i.e. some four months after *Historia vitæ* saw the light of day. Now the pages of the sixth and seventh *Contemplations* are ruled in the same manner and with the same dimensions. But more to the point, both have the same arrangement of rules in almost the exact same sizes as the *Historia naturalis* and *Historia vitæ* and, like the histories, have a distinctive 20-line font,[178] save that from Sheet E onwards in the seventh *Contemplations* that font is used for a 21-line page which, as a result, has longer vertical rules than the histories. More remarkably still, Haviland's sixth and seventh volume of *Contemplations* follow very closely the typography, disposition and dimensions of rules of the fifth volume of *Contemplations* (*STC* 12657) printed by E. Griffin for N. Butter in 1620. Indeed, the disposition and dimensions can be traced as a model right back to the first volume (*STC* 12650) printed by Bradwood in 1612.[179] In short, the page layout of the Bacon histories seems to conform to conventions characteristic of octavo editions which belonged to a series—in fact editions which

---

[175] For typography, see below pp. lxiii–lxiv.

[176] See the Corpus Christi, Oxford volume (ST. N4. 16(1), and ST. N4. 16(2) ). This, the gift of a fellow of the college, Samuel Byfield, dated 1652, is a perfect instance of how the two works go together in a single binding.

[177] Examples are Haviland's folio edition (the first) of *De augmentis* (1623) (Gibson, nos. 129a and b); and his quarto edition of the *Essayes* (1632) (Gibson, no. 16).

[178] On this font see p. lxiii below.

[179] The fifth *Contemplations* has spine- and fore-edge vertical rules measuring 132 mm, with a vertical of 126–9 mm at 16 mm from the fore-edge vertical. The head and tail horizontals are 80 mm, with two further horizontals (62 mm and 12 mm) 9 mm beneath the head horizontal. This edition also has the 20-line type typical of parts of the Bacon-Haviland octavos. The rules of the first *Contemplations* have spine- and fore-edge vertical rules measuring 139 mm each, with a vertical of 135 mm at 17 mm from the fore-edge vertical. The head and tail horizontals are 85 mm, with two further horizontals (64 mm and 12 mm) 8 mm beneath the head horizontal. The extra height of the vertical rules is to accommodate 21 lines of type per type-page.

may have been *signalling* that they belonged to a series. A speculative conclusion might be that the staging of the text of the histories is the typographical equivalent of Bacon's promise to produce a succession of specimen histories at monthly intervals.

Together with the uniformity imposed by box rules, the *Historia naturalis* and *Historia vitæ* share other typographical conventions. The body of both texts shares the same roman and italic types. The italic and roman fonts shared by and used for the body of *both* editions came in two sizes such that (direction line excluded) a typical type-page might be set either in a larger font yielding sixteen lines per page or in a smaller giving twenty lines of type to the page.[180] Many type-pages have roman and italic mixed; some have some lines in 16-line and others in 20-line type (see Plates III and IV), though the two sizes, of course, never occur in the same line. The standard type for most of the text is 20-line; the 16-line serving for special purposes. In fact the histories embody an elaborate set of typographical distinctions.

To take the *Historia vitæ* as an example,[181] some materials are set in 16-line italic with a few words in roman; others are set in 20-line roman with a sprinkling of italic. Cutting across this distinction is another, viz. some materials have their headings set within the box-ruled outer margins; others have headings centred in the text above the passages they identify. The Aditus (preface), Obseruationes majores (major observations), Intentiones (intentions), list of Operationes (operations), a Commentatio (speculation), and the Canones mobiles (provisional rules) are all set in 16-line italic, with headings centred in the text. However, the Explicationes (explanations) attached to the Canones are set in 20-line italic with a small admixture of roman. All the Connexiones (connections) are also in 16-line italic with a few words in roman, but in every case the heading Connexio is set in the outer margin. All Mandata (directions) are so headed in the outer margin and the substance of each is set in 20-line roman with a few italics, a mix which is also used for every Historia (history), while the heading of each Historia is centred in the text. All Monita (pieces of advice) are exceptional for, while each is so headed in the margin, its substance is set in 20-line *italic* with a few words here and there set in roman. More exceptional still are the Topica particularia (particular topics) which are

---

[180] The vertical height of twenty lines of 20-line type is 117 mm. Twenty lines of 16-line set solid would have a vertical height of 143 mm.

[181] *HV* is less elaborate in these respects than *HVM*.

set in the 20-line, predominantly roman mix but with transitions (without headings) set in 20-line type with italic predominant and (uniquely) indented from the vertical rules enclosing the type-pages. Where items are numbered, the numbers stand within the box rules of the outer margins. One might also add that the same or similar types were used in another Bacon–Haviland production, the *De augmentis scientiarum* (1623), a work which was box-ruled in the same way as the two natural histories though on the grander scale appropriate to folio format.[182]

This kind of signposting of structure and organization of the text is typical of Bacon. He always wanted to tell the reader how to read his text (*Novum organum* is a classic example). Here the typographical features of the Haviland octavos discriminate between different kinds of material, and so reflect the pragmatics of new natural-historical discourse. For instance, *obseruationes* with their generally high theoretical content are quite distinctly marked off from the work's purely historical sections, and the theoretically charged *canones* are likewise distinguished from their *explicationes*. Intriguingly some of these formal features recur in the *Historia densi*, a work published long after Bacon's death, which suggests that the manuscript from which it was printed was marked up in much the same ways that printer's copy of our two histories must have been, or that William Rawley, who saw the *Historia densi* through the press, so marked it up.[183]

## (*c*) The Production of the Haviland Octavos: The Skeletal Pas de Deux

Production of the two Haviland natural histories seems to have been a pretty routine affair, with little to entertain professional analysts on the look out for bibliographical surprises. In the case of these editions we could study orthography, punctuation, recurrences of distinctive types, and so on, but reap few rewards for our trouble. So let us start by looking at some of the higher-level features of their production and see if they tell us what we need to know of the transmission of the texts through the press.

There is no evidence that the quires of the two octavos were set other than seriatim, i.e. in reading order; rather than by formes. If the editions had been in folio or quarto format setting by formes would have been

---

[182] *DAS* also uses the 16-line type of *HV and HVM* for its chapter headings.
[183] For the typographical features of *HDR*, see *OFB*, XIII, pp. xxx–xxxii.

a real possibility worthy of close investigation.[184] But in the case of octavo format seriatim setting is what we expect, and in the case of the Haviland histories what we get. Octavo does not lend itself to accurate casting off—the more so when two different type sizes are in play. But there is clear and positive evidence that seriatim setting was the rule in an anomalous quire of one copy of the *Historia vitæ*, a copy which will be examined later.[185]

As for imposition, the box rules and headlines show that two skeleton formes were used in the production of almost all of the sheets of the *Historia naturalis*, and another (and different) two for the sheets of the *Historia vitæ*. To take the *Historia naturalis* first, with one or two important exceptions to be discussed later, each of the seven rules found on almost every type-page retained its position relative to the other six and to the other forty-nine rules of its skeleton forme for most of the time that the sheets of the edition were in production. Yet while both skeletons had a high degree of internal stability, their relations with each other were anything but steady. They liked to move about, one skeleton sometimes imposing type-pages of an inner forme of one sheet before changing places with its partner to impose outer-forme type-pages of the next sheet, and in the process sometimes turning through 180°. In describing the dance of the skeletons I follow the order of sheets as they occur in the *Historia naturalis*, an order in which most of the type-pages of the edition were, as will be seen later, set, imposed, and machined.

Now we begin with an exception since the first of the nineteen quires of *Historia naturalis*, the unsigned quire A, is a half-sheet imposed for work and turn which was *possibly* set, imposed, and machined after some or all of the other sheets had been printed. The half-sheet carries the title-page, the dedication to Prince Charles, and the list of histories that Bacon was promising to publish, i.e. some of the preliminaries to the *Historia naturalis* in general rather than to the *Historia ventorum* in particular. Moving on to the next quire (B), the first in full common octavo, this was imposed using just one skeleton for both formes (and, as we shall see, in this respect it was like the last quire, T). The B skeleton then appears as the one that is imposed with the type-pages for quire C(i), i.e. C inner forme, which perhaps suggests that the skeleton for C(o), i.e. the outer forme, was machined first, for if (as seems likely)

---

[184] The folio *IM* affords a spectacular instance of setting by formes: as many as eight skeleton formes were in play at once during the edition's production; see *OFB*, XI, pp. ciii–cxiv.

[185] See pp. lxvii–lxx below.

the sheets were wrought off in signature order,[186] the skeleton which was imposed for B may still have been tied up in the press. For C(o) a new skeleton was assembled and this was also used for D(o) and E(o), while the one first used for quire B was now used for D(i) and E(i). With sheet F the skeletons swapped over such that the inner and outer of E(i) and E(o) were put to use, respectively, for the outer and inner formes of F, G, and H, such that the parts of the skeleton that were imposed with type-pages $1^r$, $8^v$, $7^r$, and so on of E, were now imposed with pages $2^r$, $7^v$, $8^r$, etc. of F, G, and H, while the parts of the skeleton that accompanied type-pages $2^r$, $7^v$, $8^r$, etc. of E now accompanied pages $1^r$, $8^v$, $7^r$, and so on.

With Sheet I the skeletons changed places again in such a way as to resume the positions they had occupied in relation to quires C, D, and E. Sheets K and L see the skeletons swapping again but this time only the shift of quire I(i) to K(o) and L(o) is a straight swap. The shift from I(o) outer to K(i) and L(i) is accompanied by a 180° turn such that the parts of the skeleton that had imposed $3^r$, $6^v$, $5^r$, and so on, now carry type-pages $2^r$, $7^v$, $8^r$, and so on. As for quires M and N, the L(o) skeleton reappears and is to be found at M(i) and N(i); L(i) becomes M(o) and N(o). Again L(i) and L(o) were rotated through 180° such that what had been imposed with L$1^r$, L$8^v$, L$7^r$, etc. now imposed M and N $4^r$, $5^v$, $6^r$, and so on. With quires O, P, Q, R, and S the skeletons swap again and resume the positions and orientations in which they had stood in K and L. As for quire T, that was produced using one skeleton forme only, the one that had carried the outer forme of S through the press. All this seems to suggest normal, uninterrupted, and fluent production.

Compared with the tergiversations of the skeleton formes used in the production of the *Historia naturalis*, the skeleton formes of the *Historia vitæ* exhibit a degree of tranquillity which may indicate a still more fluent impositional history, and smoother passage through the press generally. Two skeletons per quire were used throughout, and, with one exception in one copy, the format of *Historia vitæ* is uniformly common octavo, i.e. unlike the *Historia naturalis* it does not begin with a half sheet.

The two skeletons were established quickly. Quires A and B carry some features which recur in later quires (later alphabetically, though

---

[186] See pp. lxx–lxxii below for cumulative damage to rules as evidence for this.

not necessarily in order of production) and some which do not,[187] but by quire C the skeletons have settled down to a condition of stability. From quire C to quire Z the outer skeleton forme never changed places with the inner. However, when the outer skeleton forme left quire C it was rotated through 180°, and remained in its new orientation from quire D to quire Z. As for the final quires, 2A to 2F, they have two new skeletons but ones made up from typographical elements of the old pair in an entirely logical way. The inner skeleton of 2A–2F retains 3$^v$ and 2$^r$ elements exactly as they were in earlier quires. The remaining typographical elements of the inner skeleton of 2A–2F come from the earlier quires' *outer* skeleton and were transferred from this to the inner in regular order. If we proceed clockwise, 1$^r$ of the earlier quires goes to 4$^r$, 8$^v$ to 1$^v$, 7$^r$ to 8$^r$, 6$^v$ to 7$^v$, 5$^r$ to 6$^r$, and 4$^v$ to 5$^v$. These changes almost exactly parallel those which produce the 2A–2F outer skeleton forme: 3$^r$ and 2$^v$ of the earlier quires stand in the same places in 2A–2F, and 6$^r$ skeletal elements move to 7$^r$, 5$^v$ moves to 6$^v$, 4$^r$ to 5$^v$, 1$^r$ to 4$^v$, 8$^r$ to 1$^r$, 7$^v$ to 8$^v$. I do not know what the re-forming of the skeletons that takes place at 2A signifies. Is it a reflection of the disposition of the typographical elements of the skeletons on the imposing stone after the type-pages of quire Z were stripped from the skeletons? Does it signal a break in production? There is no evidence to substantiate or falsify any such hypotheses.

It is worth adding that there exists at least one copy of the *Historia vitæ* in which at one point the usual pattern of imposition breaks down. In the British Library copy (shelfmark 1250.a.53) all seems well until we reach the type-pages of quire S. The type-pages are all present and correct: the box rules, running-titles, and text of each are bibliographically pretty well identical to the corresponding type-pages of all other copies collated.[188] Yet the type-pages in the BL copy are disposed in a very strange way. First comes S1$^r$ followed by two blank pages, then appear what in normal circumstances would be S1$^v$ and S2$^r$. These are followed by another two blanks succeeded by S2$^v$ and S3$^r$ and two further blanks, beyond which lie S3$^v$ and S4$^r$, another two blanks and,

---

[187] All the typographical elements (with spatial relations unaltered) of the skeleton which frames A1$^v$, A2$^r$, B3$^v$ (A3$^v$ is blank), A4$^r$, A5$^r$, and A5$^v$ persist in those positions in subsequent quires. The typographical elements of A2$^v$ (recurring in B2$^v$) occur in C and later quires at 7$^v$; of A6$^v$ (recurring in B6$^v$) occur in later quires at 3$^v$; of A8$^r$ (recurring in B8$^r$) occur later at 1$^r$. The remaining typographical elements of A and B do not appear to crop up again in any later quires.

[188] There is a single press variant in quire S; see *tns* to S3$^v$ below, but it is not unique to the BL copy.

finally, S4$^v$. Four type-pages (S1$^r$, S2$^r$, S3$^r$, and S4$^r$) are signed in exactly the same way that they are in any normal instance of S. Now between this anomalous quire (hereafter S$^i$) and quire T stands another quire (hereafter S$^{ii}$) which, among the copies inspected, occurs in the BL copy alone. Like the final eight type-pages of any normal gathering of the *Historia vitæ*, none belonging to S$^{ii}$ is signed but otherwise the gathering follows the pattern present in S$^i$: what would normally have been S5$^r$ is followed by a couple of blank pages, then by S5$^v$ and S6$^r$, two more blanks, then by S6$^v$ and S7$^r$, two more blanks, then by S7$^v$ and S8$^r$, yet another pair of blanks and, lastly, by S8$^v$.

What has happened here? I believe there can be but one answer: the first eight type-pages were imposed as if for a half-sheet of octavo set up for work and turn (octavo in 4s, half-sheet imposition).[189] Thus imposed, and after correction, the forme went to the press and an impression or impressions were wrought off. But the work-and-turn cycle was not completed: S$^i$ was never turned and perfected, it was left machined on one side alone, it was folded in that state and in the manner appropriate to half-sheet octavo—save that the sheet was not cut in half parallel to its shorter sides as perfected half-sheet octavo would have been. Exactly the same fate overtook the remaining eight type-pages of what would ordinarily have been quire S, i.e. the type-pages of S$^{ii}$: they too were imposed as if for work and turn, and impressions taken from that forme were folded without being cut in half, and eventually paired up with S$^i$ for binding.

It is easier to describe this oddity than to explain it. Establishing *what* occurred is a matter of mere logic; understanding *why* it occurred less straightforward. But I think there are two likely explanations for the anomaly. The first is relatively easy to state. The paper of S$^i$ and S$^{ii}$ is appreciably thinner than the paper used for printing any other sheet either in the BL copy or in any other copy that I have seen, except for my own copy, which shows some thinning to sheet S, which does give some indirect support to one explanation of the BL anomaly, namely that Haviland's shop adopted the anomalous imposition because it allowed a stock of abnormally thin sheets to be used up without the show-through that would have been a consequence of the use of normal octavo imposition with sheets so thin. Clearly the stock of thin sheets would have had to have been relatively large for it to have been worth

---

[189] For a format diagram of a sheet imposed thus see Philip Gaskell, *A new introduction to bibliography*, Clarendon: Oxford, 1972, repr. 1985, fig. 53, p. 95.

Haviland's while to adopt the normal imposition for most copies of quire S but the anomalous one for the rest. There are problems with this hypothesis: (i) I have not yet come across another copy with the anomalous imposition, and (ii) I know of no other case (Baconian or otherwise) where two different impositions were used for the printing of copies of a single sheet of a single edition.

Alternatively, the anomalous imposition may have been a mistake. Maybe a compositor setting seriatim produced all the type-pages destined for S;[190] he found the skeleton elements from whichever type-pages had been machined just before $S^i$ and $S^{ii}$ on the imposing stone and married them up with the corresponding type-pages of the anomalous sheets.[191] Maybe the compositor had several jobs on the go at the same time and, when he returned to this one, he did so with the half-sheet octavo format of some other edition he was working on still fresh in his mind, found sixteen type-pages each with their proper skeleton elements laid out before him on the imposing stone but not laid out in a way that screamed common octavo; they were laid out perhaps in a row or rows seriatim. He then shuffled them into two work-and-turn configurations, adjusted them to their chases, locked them up, and went on to his next task. He may have been betrayed by his own skill and experience; he may have been so used to imposing type-pages of all formats that he simply did not notice that one of the two formes which he was now manipulating with such dexterity was unsigned, and allowed himself to be guided by pagination or catch-words. Yet even though the shop had already produced as many as sixteen sheets of the *Historia vitæ* in common octavo without, as far as we know, making a similar mistake, the impositional lapse still occurred. This may suggest that people were being switched from one unfinished job to another, and that when the shop was busy (not the norm for shops in this period) confusion exacted its tribute. All this assumes that the anomalous imposition existed before the correct one, an assumption which is perhaps as speculative as much else to do with this issue.[192]

---

[190] Seriatim composition is of course implicit in the impositions of $S^i$ and $S^{ii}$; the compositor could not have imposed $S^{ii}$ without having the last eight type-pages destined for a normal inner forme, and the last four of a normal outer, available simultaneously.

[191] The formes machined before $S^i$ and $S^{ii}$ were probably those of R. Sheets for the edition were probably produced in signature order, as we shall see imminently.

[192] Further observations on this question can be found in my 'Francis Bacon: some bibliographical remarks', *Nouvelles de la république des lettres*, 1996, pp. 107–13.

Incidentally, the anomaly does not mean that work-and-turn imposition was *generally* used in production of our Haviland octavos. That the type-pages were overwhelmingly imposed in anything but common octavo is obviously quite inconsistent with the skeleton dances—especially when 180° rotations of skeleton formes take place at the same time as the swapping of inner and outer skeletons, or when rotations take place and the typographical elements of one skeleton turn up in new but predictable positions in the next quire. Such features are consistent only with common octavo imposition, and not with half-sheet imposition. In the latter a 180° rotation would make the typographical parts of the skeleton for $1^r$ move to $2^v$ and not to $3^r$ as in common octavo.

### (*d*) How the Sheets Went through the Press

I said earlier that the sheets of the two Latin histories went through the press (with the possible exception of the first quire of the *Historia naturalis*) seriatim or in reading order. The regularities of the behaviour of skeleton formes, and their smooth and routine progression through the press support this view, a view further corroborated by evidence of damage to the box rules, damage which took place while our editions were going through the press. It is not necessary to present a biography of every bit of this evidence—of every nick, bend, and bulge—but it would be useful to give some brief indication of how this kind of evidence can be used to establish the order in which the sheets went through the press.

I begin with the *Historia naturalis*. Many sheets have a large kink to the inner of the outer two vertical rules. This kink does not appear anywhere in the formes of sheets A, B, or C. They occur on sheets D, E, and F not far from the foot of the page, and are turned inwards (see $D1^r$, $E1^r$, $F2^r$). On sheets G to T the kink is now nearer the top and turned outwards (i.e. inverted relative to the D, E, and F position). Accordingly we may assume that D, E, and F were imposed and machined one after the other, and that sheets G to T were also imposed and machined one after the other, though at this stage we cannot assume on that evidence alone that they were imposed and machined in alphabetical order. Next, there is a small kink which appears on sheets A to I on the inner of the two outer vertical rules ($A2^r$, $B2^r$, and its subsequent rotations and transpositions). Insofar as this kink overlaps D, E, and F, and G to I, it reinforces the evidence that these latter two runs of sheets were imposed and machined not just as separate groups but in alphabetical order. And the more such overlaps occur the more that conclusion grows stronger.

Take, for instance, the small kink to the inner of two outer vertical box rules, a kink turned inward, and located not far from the head of the page. This runs from sheet C to T ($C4^v$ and its subsequent rotations and transpositions). Taken with evidence already presented this means that sheets C to I must have been imposed continuously, uninterrupted by the imposition of the formes of any other sheets, for C to I all have several kinks in the same positions on each of their appropriate type-pages, and no other sheets share *all* these kinks. Again, at $M2^r$ the horizontal rule immediately below the running title is a turning through 180°. This change persists until at least sheet Q. Likewise with $M7^v$ we find that the outermost vertical rule acquires a nick two-thirds of the way down its length. This nick appears on $M7^v$, $N7^v$, $O8^v$, $P8^v$, $Q8^v$, and $S8^v$. These pieces of evidence are, as it were, subsets of evidence already presented, and show that smaller runs of nicks indicate seriatim subgroups with larger sets of features indicating the same thing, i.e. that it would be perverse to assume that the type-pages of the work were set, imposed, and machined in anything other than reading order.

Only two further points need to be made in relation to the *Historia naturalis*. The first is the single exception to the point just made. The formes of A (a half sheet) could well have been set, imposed, and machined after all the others, a procedure far from unprecedented in early-modern London printing. It is a half sheet, it carries a discrete set of preliminaries—the title-page, dedication, and list of titles Bacon planned to produce in the coming months—, it lacks a kink common to sheets B and C, and it shares no kinked or damaged rules found on any of the sheets from D to T. As for the second anomaly, the last sheet, T, has only one skeleton (the one last used for the inner forme of S), which was used twice—once for the inner and once for the outer forme.

As for the *Historia vitæ*, the same sorts of evidence lead to the same conclusion, namely that, as with the *Historia naturalis*, the sheets comprising the edition were printed in reading order. Consider, for instance, the vertical rule nearest the fore-edge on $1^r$: from quires D to M this is undamaged; from N to 2A a nick appears towards the rule's foot; from 2B to 2F, the damaged rule is superseded by a new and undamaged one. Likewise, after quire T a bulge appears in the inner of the two vertical rules nearest to the fore-edge, and there it remains up to and including the final quire, 2F. Again, after quire G the vertical rule next to the spine acquires a kink level with line 15 of pages with 20-line type. This rule is replaced by a new one after 2C. The rule beneath the headline gains a kink below the *t* of *Vitæ* with quire V, and this kink persists in all quires up to and including the last. One could adduce more evidence of this kind but enough has been presented to make the point that each piece of evidence, taken *in relation* to all the other pieces, points emphatically to seriatim production.

(*e*) Press Correction

Only twenty-seven corrections have been detected in collation of copies of the *Historia naturalis* on a McLeod Portable Collator, and some of these are extremely doubtful (registered as a result of show-through or other marks on individual copies). Of these twenty-seven, two involve replacement of broken letters.[193] and there are changes to diacriticals on H4ʳ and O1ʳ.[194] Most corrections are minor changes to punctuation, and one or two of these may be phantom (i.e. inking variations),[195] and only one change makes much difference, i.e. where on K8ʳ (line 8) a closing bracket has been inserted. The same forme of the same sheet also had two other changes: on K2ᵛ line 9 'Aeri,' becomes 'Aeri' (without a comma); and there is minor relining on K1ʳ line 7 with 'ſu-| ſpendimus' becoming 'ſuſ-| pendimus'.[196]

More significant alterations take place on O3ᵛ and P5ᵛ. On the former (lines 10–11), in some copies we read 'leuitèr excauatus, *Ventos;* ſi ca-|uus in profundo, *Imbres.*' ('*Imbres.*' ending a paragraph);[197] in others, 'leuitèr excauatus, *Ventos;* ſi ca-| in profundo, *Imbres.*'.[198] In other words some copies omit 'uus' from the start of line 11, an omission rectified in others. As for one short paragraph on P5ᵛ (lines 4–8), some copies read 'At *terrestres volucres* contrà, a-| quam pete*n*tes, eámqu*e* alis percu-| tientes, & clangores dantes, & | ſe perfundentes; ac præcipuè | *Cornix,* tempeſtates portendunt.' In other copies we read (with the suppression of a contraction), 'At *terrestres volucres* contrà, | aquam petentes, eámqu*e* alis per-| cutientes, & clangores dantes, | & ſe per-

---

[193] Compare Huntington 12569 and University of Minnesota copies, at C2ᵛ, where the T of 'Titulos' is replaced in its two occurrences (lines 2 and 16).

[194] H4ʳ, line 17: BNF '*calidúm*' versus Pierpont Morgan '*calidúm*'; O1ʳ, line 15: '*motûs*' of BNF versus '*motûs*' of Huntington 12569.

[195] See *tns* to F8ᵛ, I5ʳ, K2ᵛ, K8ʳ, L7ʳ, L8ʳ, M1ʳ, R2ᵛ, S4ʳ below.

[196] BNF copy versus Library of Congress QC 931.B3 1638b Fabyan Coll; and BNF copy versus Linda Hall Library copy.

[197] See for instance Bodl., 8° V 16 Art. Seld.

[198] See for instance the Oxford Philosophy Library copy.

fundentes; ac præcipuè | *Cornix*, tempeſtates portendunt.'[199] And that is all for *Historia naturalis* in the way of correction—except for one pen correction on Q2ᵛ, line 8 where we find that 'Amne,' has been changed in all copies I have seen to 'Amnes', the original comma having been overwritten by hand with the final 's'. Even after the presses had finished their work a sharp lookout for errors was still being kept. But only a couple of substantive changes detectable in the final stages of printing were made to the entire text. In short, correction had been so thorough that very few traces of the early stages of proofreading remain.

With the *Historia vitæ* the story is different—there is a scattering of relatively insignificant press corrections into which little can be read, but the scatter is accompanied by a couple of rather surprising features connected with correction: the overwhelming majority of the fifty-five press corrections picked up in collation occur in quires O–Z (with thirty-two corrections in quires V, X, and Z alone), with relatively few corrections identified in A to I, none at all in K to N, and only five in the whole of quires 2A to 2F. Let's take quires V, X, and Z first, with the Bibliothèque nationale copy as our control. The inner and outer formes of V are remarkable in that both exist in no fewer than four states. The outer forme exists in an uncorrected state represented by the Biblioteca comunale copy. A first-state correction is exemplified in the Bibliothèque nationale copy on V4ᵛ (line 3) where the 'infuſi,' of, for instance, Folger STC 1156 (copy 2) is corrected to 'infuſi;'. Second-state corrections appear on V7ʳ (line 11) where '*Epati*, modò' becomes '*Hepati*,modò'; on V1ʳ (line 12) where 'Similitèr' becomes 'Similitèr,' (represented in this writer's copy). A third-state correction occurs in the Bibliothèque nationale copy where a comma (V1ʳ, line 17) disappears between '*Radices*' and '*Potado*'. Lastly there is a substantive first-state correction to the outer forme (V1ʳ, lines 19–20) such that 'Eſculentas' becomes 'Eſculentæ', a correction that would perhaps have required reference to printer's copy. In these last two cases the uncorrected forme is represented in Folger STC 1156 (copy 2).

As for the inner forme of quire V, 'Alimenti V.' on V2ʳ (line 4) becomes 'Alimenti. V.' (Newberry Library versus BNF); 'cum' on V3ᵛ (line 4) becomes 'cùm' (Folger copy 3 versus BNF); 'fortaſſe' on V5ᵛ (line 1) becomes 'fortaſsè' (Biblioteca comunale versus BNF); and (line 5) '*Matutinus*,' becomes '*Matutinus*' (Newberry Library versus BNF). On V7ᵛ (line 7) '*chalibeatum*' becomes '*chalybeatum*' (Newberry versus

---

[199] Cf. e.g. Bodl., 8° E 105(1) Linc, and BNF R2908A.

BNF); on V8$^r$ (line 3) '*Fructibus*' becomes '*Fructibus,*' (Newberry versus BNF). The corrections to the inner forme of V represent no fewer than four states, states represented in the Biblioteca comunale (uncorrected), the Newberry (first-state correction), Folger copy 3 (second-state) and Bibliothèque nationale (third-state) copies. The exactitude of the inner-forme corrections has the same character as the nice, almost finicky outer-forme ones, a fact that reassures us that a great deal of care was taken to get the printed text just right. The outer-forme substantive correction may have been the stimulus for the other corrections to V's formes.

The next quire, X, features corrections to the inner forme only. On X4$^r$ (line 3) '*Fuliginofi*' becomes '*Fuliginofi,*'; '*Kermes*' (line 9) becomes '*Kermes,*' and '*Fragrantia*' (line 12) acquires a comma '*Fragrantia,*'. Then on X5$^v$ '*Calida,*' (line 10) becomes '*Calida;*'. These are all second-state corrections apparent when the corrected BNF forme is compared with the uncorrected Folger STC 1156 (copy 2). A first-state correction is evident in the BNF copy when compared to Huntington 56659 whose '*Nitra*' is corrected to '*Nitro*' (X5$^v$, penultimate line).[200] On X6$^r$ (line 11) '*Viuacitatem*' is followed in some copies with a comma. On X8$^r$ '*Fricatio,*' (line 4) is followed by a semicolon (BNF has the heavier punctuation; Folger STC 1156 (copy 2) the lighter). As for quire Y, on Y6$^v$ a title has been reset with consequential and slight positional changes to other items on that page.[201]

Quire Z features several corrections to the inner forme. On Z2$^r$ (line 3) 'arifieri' becomes 'arefieri'; and 'In teriùs' (with space and diacritical) becomes 'Interius' (line 8). On Z5$^v$ (line 5) 'vehementrèr' is corrected to 'vehementèr', and 'Multum' (line 8) becomes 'Multùm'; and on Z6$^r$ (line 18) '*Vnctuofa,* fcilicet' becomes '*Vnctuofa* fcilicet,'.[202] There are first- and second-state corrections here,[203] another indication of lengths to which the printing house went to ensure accurate transmission of the text. The fact that so few such corrections appear in quires other than those from V to Z rather suggests that the earliest stages of proof-correction on other formes were very thorough, so thorough that

---

[200] For the same error see P7$^r$ below.

[201] For instance, '**Connexio**' is set marginally higher in some copies (e.g. CUL Keynes E.1.20 and Emmanuel 512.6 (1)) than in others (e.g. CUL LE. 7. 59). The section title 'OPERATIO' has been reset.

[202] See *tns* to pp. 318, 320, 322 below.

[203] An uncorrected state is represented by Folger STC 1156 (copy 2); the second state represented by BNF; the first by Folger STC 1156 (copy 1) where 'arifiere' and 'vehementrèr' are corrected.

few traces of the process remain. In fact, on these other formes we find mainly sporadic corrections to accidentals[204] with, in addition, several substantive changes. Of the latter, one appears on D4$^v$ (line 1), and another on P6$^r$ (line 20). The first of these consists in the removal of 'pro' from 'profundo' as the 'pro' (followed by a hyphen) had already appeared at the very end of the previous page (followed by the catchword 'fundo').[205] The other involves the insertion of 'illis' in the final line of the page, and would certainly have required reference to printer's copy.[206] A further substantive alteration occurs on S5$^r$ (line 4) where 'ſit *Diæta*' becomes 'ſi *Diæta*'.[207] There are further press corrections to quire S to the punctuation.[208] As for quires 2A to 2F, there is only a tiny scattering (five in all) of insignificant, non-substantive corrections.[209] In addition there is a correction to a page number: in some copies 2C8$^v$ is numbered 406 and in others 410. The former is incorrect and it is an error which, coming at the end of a quire, caused all subsequent pages in the next and later quires in all copies to be numbered (2D1$^r$ onwards) from 407 and not from 411.[210]

## (ƒ) The Genesis of the Texts, and Printer's Copy

We do not know much about the genesis of the texts to which the two Haviland octavos are overwhelmingly the principal witnesses, or about the manuscript copy from which the compositors set them. However, there is no reason to believe that the *Historia naturalis* and *Historia vitæ* did not come into being by a process of incremental revision of the kind which we see taking place in the *De vijs mortis*, and which apparently took place in the genesis of *Instauratio magna*.[211] Nor is there any reason to believe that Bacon did not have a fair copy or copies of his two texts made by a professional scribe who would have acted as a copy-editor—as the scribe who worked on the *De vijs mortis* had done, or as the scribe who prepared the manuscript of *Henry VII* must have done at about the time when the *Historia naturalis* was in the final stages of preparation.[212]

---

[204] For these see *tns* to the edited text, p. 168 below.

[205] The BNF copy represents the corrected state; the Newberry the uncorrected.

[206] The BNF copy lacks the correction which is represented in Folger STC 1156 (copy 1).

[207] See *tns* to p. 278 below.

[208] See p. 276 below: S2$^r$ ll. 14–15 'habuerunt-*Lintea*' loses the hyphen; S3$^v$ 'peſſimum,' (S3$^v$ penultimate line) becoming 'peſſimum;'.

[209] See *tns* to 2A5$^r$, 2B1$^r$, 2D2$^r$, 2E6$^v$, and 2F7$^v$ below.

[210] The BNF copy has '410'; Folger STC 1156 (copy 1) has '406'.

[211] On the genesis of *DVM* and *IM*, see *OFB*, VI, pp. xcv–cvi; *OFB*, XI, pp. cxvii–cxxiii.

[212] BL Add. MS 7084 (*IELM*, BcF 215) is a manuscript (with autograph revisions) of *Henry VII*. For the *DVM* scribe as copy-editor, see *OFB*, VI, pp. 460–6.

Such fair copies would themselves have become printer's copy or the immediate ancestors of printer's copy.

This line of inquiry can be carried a little but not much further, and here I shall take the two works in reverse chronological order for some of what may be said about the *Historia vitæ* can be applied to the *Historia naturalis*, a work for which less evidence is available. In connection with the *Historia vitæ* three pieces of evidence come to our aid: (i) the version of the *Historia vitæ* aditus at the end of the *Historia naturalis*; (ii) a holograph manuscript version of two paragraphs of *Historia vitæ*; and (iii) evidence afforded by *De vijs mortis*.

(i) The two versions of the aditus coincide in all substantive respects, and almost all accidental ones too—down to minutiae of punctuation, orthography, paragraphing, italics, and diacriticals. This at least assures us that Haviland's men could produce text accurately from copy. In fact it seems probable that the copy-text for the aditus was the printed *Historia naturalis* version, and not manuscript. But even so a good text was still further tightened up and polished—with the suppression of a typo, expansion of contractions, addition of diacriticals, and so on.[213] The first of these could have been a result of authorial intervention or the intervention of someone acting for the author. If it does no more, comparison of the two versions of the aditus at least reassures us that care was being taken in the transmission of the text through the press.

(ii) The only two extant holograph paragraphs of *Historia vitæ* crop up in a sheaf of papers which consist in the main of drafts of material in William Rawley's hand, drafts which eventually appeared in the *Sylva sylvarum*.[214] They are endorsed by Rawley with the words 'Hist. of Life & Death p. 3'. The page number refers to the 1651 'official' translation of the *Historia vitæ* and so must have been added to the text after that date.[215] The two paragraphs exhibit holograph revisions of precisely the kind that Bacon introduced into his text when he was drafting new material or revising a scribal fair copy. The revisions—deletions, interlinear insertions, etc.—bring these paragraphs very close to the form in

---

[213] For such minute differences as there are, see *tns* pp. 144–8 below.

[214] For a photographic reproduction of these paragraphs (BL Add. MS 38693, fo. 49ʳ), see Graham Rees, 'An unpublished manuscript by Francis Bacon: *Sylva sylvarum* drafts and other working notes', *Annals of Science*, **38**, 1981, pp. 377–412 at p. 409.

[215] Rawley's endorsements are all much later than the drafts. For the 'official' translation of *HVM*, first published by Rawley in 1638, and published again with the 1651 *SS* (Gibson, no. 176), see *OFB*, XIII, p. lxxvii. The manuscript endorsements to other drafts, drafts that eventually found their way into *SS*, also seem to refer to the 1651 *SS*.

which they appear in the printed text but they lack the copy-editing refinement which they would have had if they had been copied by a professional scribe. The printed version of the two paragraphs differs from the holograph in ways absolutely typical of those which distinguish holograph drafts of other works from, on the one hand, scribal practice authorized by Bacon and, on the other, printed material in a form that Bacon would have authorized.[216] In short, the two manuscript paragraphs represent a late stage in Bacon's work on the text and, in their closeness to the published text, suggest that Haviland's compositors followed their copy faithfully in respect of the rest of the text too. It must be more than mere coincidence that the only holograph material we possess for the *Historia vitæ* happens to approximate very closely to the published version.

(iii) The *De vijs mortis* is an unfinished manuscript work, parts of which were drafted by a scribe and corrected by Bacon, other parts of which were drafted as continuous, and separate holograph additions and revisions of the scribal draft. The manuscript very probably reached its present form at some point (or points) between *c.*1611 and not later than 1619. *De vijs* is in effect the *Historia vitæ* without (among other things) the *historia*. It is, in other words, an account of the causes of ageing and death and not at all a set of structured assemblages of data which support (or appear to suggest) the theoretical/causal thinking that the history also comprises. Now the theoretical and causal speculations presented in *De vijs* recur in the natural history, and sometimes recur word for word. The manuscript therefore represents a stage or, more accurately stages, in the forming of materials which were later to find their way by undocumented intermediate stages into a work whose purposes were rather different: the *De vijs* was perhaps intended for Part V of the *Instauratio magna*, whereas the *Historia vitæ* was certainly meant for Part III. However, differences of purpose notwithstanding, verbal similarities and identities between the two texts are such that the

---

[216] For the notion of 'authorized' style see *OFB*, VI, p. 460. Each of the two paragraphs is approximately seventy words long. The printed version of the first paragraph lacks the four contractions of the manuscript; it adds four extra points, twelve initial capitals, nine italicized words, and one diacritical; it suppresses one word ('paruis') and transposes two words. The drafting and correction of this paragraph brought it close to the printed form. As for the other paragraph, there are thirty-five differences between the manuscript and the printed version: the latter adds italics and diacriticals; contractions are expanded, punctuation and capitals are added, and every 'et' is turned into '&'. The draft has two important interlinear insertions of substance which subsequently turned up in the printed *HVM*. Lastly, two grammatical lapses are corrected in the printed version; these two corrections were perhaps introduced before printer's copy was prepared. For these alterations see *tns* to pp. 156, 158 below.

*De vijs* intermittently validates or authenticates the parts of the *Historia vitæ* to which it corresponds, and so constitutes another test of the accuracy of the printer's transmission of Bacon's text.[217]

As for the history of *Historia naturalis* / *Historia ventorum*, we have no manuscript material, scribal or holograph, bearing on the genesis of the text directly. All the same, the phrasing of parts of the text is duplicated or closely similar to wording in other texts. Most notably the wording of the norma of the *Abecedarium* echoes that of the norma of the *Historia naturalis*, and both echo the *Parasceue*[218]—facts which provide independent though limited grounds for confidence that the octavo *Historia naturalis* transmitted Bacon's text accurately. From the point of view of Bacon's theoretical preoccupations the octavo text is thoroughly in accord with positions developed in earlier works, in the same manner as the doctrines of the *Historia vitæ* stand four-square with those of the *De vijs*. Indeed the *De vijs* happens to have belonged to the same manuscript volume as a draft of *De fluxu et refluxu maris*, and this latter presents ideas about the general, 'cosmological' wind, ideas which recur unchanged in *Historia ventorum*.[219]

To end our consideration of the transmission of the texts, two further points are worth raising: one relating to *Historia naturalis* alone, the other to both the Haviland octavos. On the first point, the aditus to the *Historia Sulphuris, Mercurij, & Salis* ends with a couple of substantive errors so gross as quite to confound the ideas being elaborated at that juncture. The absurd 'Salis' appears instead of the correct 'Mercurialis' (S7$^r$), and twenty-three words later the equally preposterous 'Aqua' occurs instead of 'Flamma' (S8$^r$). In fact, 'Aqua' is doubly wrong for it also appears as the catchword on the preceding page.[220] It is impossible to pin these blunders on Haviland's compositors: as there is no possibility of eyeskip, these particular faults cannot possibly be ascribed to misreading of printer's copy. Accordingly they must have been present in printer's copy and in whatever manuscript genealogy preceded it—right back to whatever Bacon holograph represented his final draft. Except that that too is impossible—Bacon himself cannot have committed these textual sins: he knew the theoretical framework to which the correct terms belonged, and knew it so intimately that to write

---

[217] See, for instance, *cmts* on *DVM* in *OFB*, VI, pp. 437, 438–9, 442, 449.
[218] See *OFB*, XIII, pp. 325–6. Also see pp. 12–16 above, and *cmts* pp. 384–5 below.
[219] *DFRM* in *OFB*, VI, pp. li, 78, 379.
[220] See p. 138 below.

'Salis' for 'Mercurialis' would have been like inadvertently writing 'atom' instead of 'molecule'. But if one is at a loss to explain the origin of the faults, the fact of their existence means that Bacon either did not read proofs of these pages or read them too lazily. It also means that Bacon, or his secretary William Rawley, spotted the errors after publication, for the latter had the true readings incorporated in the edition of *Historia naturalis* published in the *Operum moralium et ciuilium tomus* of 1638.[221]

Finally, another unsolved problem: how were the special typographical conventions of the Haviland octavos—16-line type versus 20-line, italic preponderances versus roman, and so on—signalled to the compositors? The fact that the same rather systematic conventions were used in both texts with a high degree of consistency surely means that they were authorized by Bacon himself, and signalled to the compositors with great precision. How otherwise would they have known when to switch from one convention to another? Did Bacon mark up printer's copy or an antecedent of printer's copy? If so, what form did the mark-up take? Or were the conventions actually represented literally in printer's copy with a larger hand for 16-line type, a smaller for 20-line, with the differences used in addition to represent what became the octavos' varying concentrations of italic and roman? There are, alas, currently no answers to these questions.

---

[221] See *tns* to *HNE*, S7ᵛ–S8ʳ, p. 138 below; also see *OFB*, XIII, p. lxxvii for Rawley and the 1638 *Operum . . . tomus.*

# THIS EDITION: PRINCIPLES, CONVENTIONS, AND A NOTE ON THE TRANSLATIONS

## (i) Principles and Conventions

Editorial intervention in the texts has been kept to a minimum, but the edited texts are neither literal transcripts nor quasi-facsimiles of the copy-text (*c-t*). *C-t* ornaments and initials with their following small capitals have not been recorded or reproduced. Turned letters and wrong-font letters have been recorded only if bibliographically significant. The running headlines in this edition do not match those of the printed *c-ts*.

Materials in the box-ruled margins of the copy-texts (i.e. paragraph numbers, titles, and so on) have been incorporated into the edited text, and have been distinguished in **bold type**.

The edited texts are accompanied by two banks of textual footnotes (*tns*): the first relates to substantives, the second to accidentals. After the edited texts stand the commentaries (*cmts*). In the main the *cmts* elucidate difficult passages and indicate (especially by quotation) sources, analogues, and parallels. In addition they supplement the *tns* and cross-reference passages of the edited texts with passages in other Bacon writings.

Since *c-t* signatures and folio numbers are the means by which texts in the edition are cross-referenced, superscript vertical bars ( $^|$ ) have been set in the edited text to distinguish each *c-t* page from the next. In the outer margin next to every line of text containing such a bar, the appropriate *c-t* signature or folio number is set in square brackets ( [ ] ).

Emendation of substantives has been confined to instances where *c-ts* appear to be deficient or corrupt. Emendations stand in the established text, and are recorded in the *tns* where the emended word(s) appear as a lemma (or lemmata) preceded by a line number and followed by (i) a closing square bracket ( ] ), and then (ii) the *c-t* reading and/or editorial remarks. Unemended words requiring editorial comment are lemmatized in the same manner. All editorial remarks in the *tns* are distinguished from the lemma and record of the *c-t* reading by a forward slash ( / ).

Where a substantive reading has been adopted from a witness other than the *c-ts*, the adopted reading becomes the *tn* lemma and the *c-t* reading is placed after the square bracket followed by a forward slash and a brief history of the adopted reading. See, for instance, a *tn* in *HNE* (S7ᵛ): Mercurialis] Salis / emended thus in *SEH* (II, p. 83) following the 1638 edition (Gibson nos. 196–7), i.e. the one presented in *Operum . . . tomus*, 4I4ʳ.

Emendation of accidentals: in general, *c-t* accidentals have been preserved. Where accidentals have been emended (generally for the sake of clarity), *c-t* readings have been scrupulously preserved in the *tns* and keyed to the text in the same manner as for substantives. A swung dash to the right of the closing square bracket stands for the lemma; thus, for example, *lemma,*] ~; means that the comma following the lemma in the edited text replaces a semicolon in the *c-t*. Where the edited text has punctuation but the *c-t* has none, the ~ is followed by a caret mark ( ∧ ); for brackets inserted by the editor, ~ is followed by ∧*lemma*∧. In cases where *c-t* has stop-press alterations to accidentals, the *tn* to X4ʳ of *HVM* might read, for instance, '*Kermes,*] / uncorrected inner forme (e.g. Folger STC 1156 (copy 2)) without a comma', meaning that the corrected inner forme has no comma while the uncorrected (Bibliothèque nationale) copy has.

1. Punctuation: texts have their own regularities or eccentricities which it would be unwise to disturb more than is absolutely necessary. Punctuation has been emended only for clarity or to avoid ambiguity. Emendations are noted in *tns* in the manner indicated in the previous paragraph. The punctuation of the *c-ts* is roman throughout, and this feature has been retained in the edited texts.

2. *C-t* paragraphing has been retained without exception. Beginnings of paragraphs are indented, outdented, or neither in- nor outdented in accordance with *c-t* practice.

3. *C-t* orthography has been retained for the most part. It makes no sense in what aims to be an honest edition to normalize printers' practices on some arbitrary 'modern' standard. Apparent orthographical errors (typographical in some cases) have been corrected and *c-t* forms noted in the *tns*. Modern practice in the cases of i vs. j and u vs. v has been (quite properly) ignored. The long s (be it roman or italic, or doubled as ss, ſs or ß) has been silently altered to the modern form in the edited texts, but has been retained in editorial matter where bibliographical issues are under discussion. In the edition *c-t* digraphs have been retained.

4. Abbreviations: all *c-t* contractions have been expanded (in italics where the *c-t* has roman, and in roman where the *c-t* has italic). Where the *c-ts* omit a contraction sign (generally a tilde or a horizontal straight line) normally used in the *c-ts*, the missing letters have been supplied in italics enclosed in square brackets. Caudate *e* has been represented as an italicized *æ*.

5. Diacriticals: all have been retained or omitted according as the *c-ts* retain or omit them.

6. *C-t* initial capitals and lower-case letters: these have been emended now and then for clarity's sake. Emendations have been recorded in textual notes thus: *lemma*] l.c. where editorial capitals have been introduced, and *lemma*] *Lemma* where editorial lower-case initials have displaced *c-t* capitals. Small-caps following display initials have not been recorded or reproduced.

7. *C-t* roman and italic have been retained in the edited texts. Contractions are represented by italic in the roman of edited text or roman in the italic of edited texts. *C-t* words which begin with capitals but end with lower-case letters (e.g. 'CVm') are given only their initial capital (e.g. 'Cvm') in the edited texts.

8. In the small number of cases in which Bacon holograph manuscript has been quoted, the following conventions (used elsewhere in *OFB*) have been observed: all matter deleted by Bacon has been restored; deletions have been delimited by double angle brackets (<< . . . >>); interlineations have been delimited by normal and reverse primes (' . . . '); contractions have been expanded and set in italics; but superscript contractions have been retained, e.g. 'apponunt'.

## (ii) A Note on the Translations

I have spoken about the translations presented in *The Oxford Francis Bacon* in other volumes of the edition,[222] and so I have little to add here. However it is worth repeating that I aim to translate Bacon's Latin into modern English, achieve consistent translations of specialist, semi-technical, and technical terms, and avoid anachronism. I have tried to carry over and represent in English Bacon's distinctive Latin lexis. The effort to reproduce the meanings of such terms *and* simultaneously alert readers to them may very occasionally result in a degree of strain which is perhaps unavoidable when modern English is being used to convey early seventeenth-century Latin distinctions and differences.

---

[222] See *OFB*, VI, pp. cxv–cxvi; XI, pp. cxxvi–cxxviii; XIII, pp. xciv–xcvi.

The translations presented below represent a fairly sharp break with the past for unlike (say) the *Novum organum*, the Latin natural histories have been translated into English very infrequently, and not at all since the nineteenth century,[223] and a glance at these is quite enough to demonstrate the evanescence of idiom. An edition of Bacon's original Latin may last for a very long time as the standard for scholars, and afterwards enjoy a shadowy immortality as an object of reference for the editions which supersede it. Translations, on the other hand, soon start to look dated for one is not just trying to mould one semantic system into another, but trying to graft a modern system onto one in another, and seventeenth-century, form of an ancient language, knowledge of which even among the most highly educated has declined steeply in the last half-century. The best that can be said for a facing-page translation is that it acts as a gloss on the original, while the original acts as a check on the translation.

---

[223] See, for instance, Gibson, nos. 115, 153, 154. Also see *SEH*, V, pp. 125–206, 213–335 for the anonymous nineteenth-century translations of *HNE* and *HVM*.

# THE TEXTS AND TRANSLATIONS

# FRANCISCI BARONIS
## DE VERVLAMIO,
## VICE-COMITIS SANCTI ALBANI,

## HISTORIA NATVRALIS ET EXPERIMENTALIS
### AD CONDENDAM PHILOSOPHIAM:

### SIVE,
## PHÆNOMENA VNIVERSI:
Quæ est Instaurationis Magnæ
#### PARS TERTIA.

<sup>|</sup> FRANCIS
LORD VERULAM,
VISCOUNT ST ALBANS

HIS NATURAL AND EXPERIMENTAL HISTORY
FOR THE BUILDING UP OF PHILOSOPHY

OR

PHENOMENA OF THE UNIVERSE
Which is the Great Instauration's
THIRD PART

[A1$^v$: blank]    |

[A2$^r$]    | ILLVSTRISSIMO, ET EXCELLEN-
tissimo Principi, CAROLO,
Serenissimi Regis IACOBI
Filio, & Hæredi.

5    *Illustrissime, & Excellentissime*
*Princeps,*

PRIMITIAS *Historiæ* nostræ *Naturalis* Celsitudini tuæ humillimè
[A2$^v$]    offero. Re*m* | mole perpusillam, veluti *Granum Sinapis*; Sed tamen
pignus eorum, quæ Deo volente sequentur. Obstrinximus enim
10    nosipsos tanquam voto, singulis nos mensibus, ad quos Dei
bonitas (cuius agitur Gloria tanquam in Cantico nouo) vitam
nostram produxerit, vnam, aut plures eius partes, prout fuerint
magis aut minùs arduæ, aut copiosæ, confecturos, & edituros.
[A3$^r$]    Moti etiam | fortassè erunt alij, nostro exemplo, ad similem indu-
15    striam, præsertim postquàm penitùs perspexerint, quid agatur.
Nam in *Historiâ Naturali* bonâ, & benè institutâ, claues sunt &
Scientiarum, & Operum. Deus Celsitudinem tuam diu seruet
incolumem.

*Celsitudinis tuæ Seruus*
*humilis & deuotus,*
FR. S$^t$. ALBAN.

8 *Sinapis*,] ~:    11 cuius] / *nld* in *SEH* (II, p. 9) as cujus    12 eius] / *nld* in *SEH*
(II, p. 9) as ejus    14 alij] / *nld* in *SEH* (II, p. 9) as alii    14–15 industriam,] ~:

4

|

# ᐟ TO THE MOST ILLUSTRIOUS AND EXCELLENT
## Prince CHARLES,
### Son and Heir to his Most Serene Majesty
## JAMES

Most Illustrious and Excellent
Prince,

Most humbly do I offer your Highness these the first fruits of my Natural History. Like a grain ᐟ of mustard-seed it is a tiny thing but nevertheless a token of those things which, God willing, are to follow. For I have pledged myself, as if by a vow, to finish and publish one or more parts of it, according as they are more or less arduous or extensive, for every month that the goodness of God (whose glory is celebrated as in a new song) prolongs my life. Perhaps others too ᐟ will be moved by my example, especially when they have fully appreciated what I am up to. For in a good and solidly constructed *Natural History* lie the keys both to knowledge and to works. May God keep your Highness in health.

*Your Highness's humble*
*and devoted servant,*
Francis St Albans

[A3ᵛ: blank] |

[A4ʳ]    | TITVLI
Historiarum & Inquisitionum
in primos sex menses destinatarum.

   *Historia Ventorum.*
5 *Historia Densi & Rari, nec-non Coitionis, & Expansionis Materiæ*
   *per spatia.*
   *Historia Grauis & Leuis.*
   *Historia Sympathiæ, & Antipathiæ Rerum.*
   *Historia Sulphuris, Mercurij, & Salis.*
10 *Historia Vitæ & Mortis.*

[A4ᵛ: blank] |

[B1ʳ]    | HISTORIA NATVRALIS ET
*Experimentalis, ad condendam*
*Philosophiam:*

SIVE,
15 *Phænomena Vniversi: quæ* est Instaurationis Magnæ
pars tertia.

   *Monendi vtique sunt Homines, & per Fortunas suas rogandi, atque*
[B1ᵛ] *obsecrandi, vt animos submittant, & sci* |*entias in Mundo Maiore*
   *quærant; quinetiam de Philosophiâ vel cogitationem abjiciant, vel*
20 *modicos saltem, & tenues fructus ex illâ sperent, vsque dum Historia*
   *Naturalis, & Experimentalis, diligens & probata, comparata sit, &*
   *confecta. Quid enim sibi volunt ista Cerebella Hominum, & potentes*
   *Nugæ? Fuerunt apud antiquos, placita Philosophorum valdè nume-*
[B2ʳ] *rosa;* Pythagoræ, Philolai, Xenophanis, | Heracliti, Empedoclis,
25 Parmenidis, Anaxagoræ, Leucippi, Democriti, Platonis, Aristotelis, Theophrasti, Zenonis, *aliorum. Hi omnes Mundorum argu-*
*menta, tanquam Fabularum, pro arbitrio confinxerunt, easq́ue*

18 *Maiore*] / *nld* in *SEH* (II, p. 13) as *Majore*   19 *quærant;*] ~:   25 Anaxagoræ,] ~;

| TITLES
of the Histories and Investigations
destined for the first six months

History of the Winds.
History of Dense and Rare; or of the Coition and Expansion of
    Matter in Space.
History of Heavy and Light.
History of the Sympathy and Antipathy of Things.
History of Sulphur, Mercury, and Salt.
History of Life and Death.

|

| A HISTORY NATURAL AND
Experimental for the building up of
Philosophy:

OR,

Phenomena of the Universe: which is the Great Instauration's
third part.

Men are to be advised and indeed for the sake of their fortunes
begged and beseeched to moderate their pride and seek | for
the sciences in the greater world, and cast aside thoughts of
philosophy or at least hope for few and trivial fruit from it until
a tried and tested natural history has been collected and constructed. For what is the point of men's feeble cerebrations
and proud nothings? Among the ancients there were no end of
philosophical systems, the systems of Pythagoras, Philolaus,
Xenophanes, | Heraclitus, Empedocles, Parmenides, Anaxagoras,
Leucippus, Democritus, Plato, Aristotle, Theophrastus, Zeno,
and the rest. All of them got up fictions of worlds at will, worlds
like tales, tales which they told and made public, some more neat

*Fabulas suas recitarunt, publicarunt, alias magis concinnas certè, &*
*probabiles, alias duriores.* At nostris sæculis, propter instituta Scho-
[B2ᵛ] *larum, & Collegiorum, cohibentur ingenia magis, neque* ¹ *proptereà*
*omninò cessatum est;* Patricius, Telesius, Brunus, Seuerinus *Danus,*
5　Gilbertus *Anglus,* Campanella, *scænam tentârunt, & nouas Fabulas*
*egerunt, nec plausu celebres, nec argumento elegantes. Num hæc*
*miramur? Quasi verò non possint infinita oriri huiusmodi placita, &*
*Sectæ, omnibus sæculis? Neque enim est, aut erit, huiusce rei finis*
9　*aliquis, aut modus.* Alius aliud arripit, alijs alia placent, nihil est
[B3ʳ] *luminis sic* ¹ *ci, & aperti; quisque ex Phantasiæ suæ cellulis, tanquam*
*ex specu* Platonis *Philosophatur; Ingenia sublimiora, acutiùs,*
*fœliciùs, tardiora, minore successu, sed æquâ pertinaciâ. Quin non ita*
*pridem, ex quorundam Virorum doctorum, & prout nunc sunt res,*
*excellentium disciplinâ, scientiæ* (credo propter Varietatis, *& Licentiæ*
15　*tædia) intra certos, & descriptos Authores coercentur, atque ita*
[B3ᵛ] *cohibitæ, senioribus imponun* ¹ *tur, adolescentibus instillantur, vt iam*
(*quod cauillatus est* Cicero *in Cæsaris annum*) Stella Lyræ ex edicto
oriatur, *& Authoritas pro Veritate, non Veritas pro Authoritate sit.*
*Quod genus Institutionis, & Disciplinæ, ad vsum præsentem egregiè*
20　*valet, sed idem meliorum indicit exilium. Nimirum primorum*
*Parentum Peccatum, & luimus, & imitamur. Illi Dei similes esse*
[B4ʳ] *voluerunt, Posteri eorum adhùc magis. Ete* ¹ *nim Mundos creamus,*
*Naturæ præimus, & dominamur, omnia ita se habere volumus, prout*
*nostræ Fatuitati consentaneum fore videtur, non prout Diuinæ*
25　*Sapientiæ, nec qualia inueniuntur in rebus ipsis; nec scio an Res, aut*
*Ingenia magis torqueamus; sed planè Sigilla Imaginis nostræ,*
*Creaturis, & Operibus Dei imprimimus, non Creatoris Sigilla cum*
*curâ inspicimus, & agnoscimus. Itaque non immeritò iterum de*
[B4ᵛ] *Imperio* ¹ *in Creaturas decidimus, & cùm post lapsum Hominis,*
30　*nihilominùs dominatio nonnulla in Creaturas reluctantes relicta*
*fuerit, vt per veras, & solidas artes, subigi & flecti possint, id ipsum*
*ex insolentiâ nostrâ, & quia Dei similes esse volumus, & propriæ*
*Rationis dictamina sequi, maximâ ex parte amittimus. Quamobrem,*
*si qua est erga Creatorem Humilitas, si qua Operum eius Reuerentia*

---

15 *coercentur*] / *nld* in *SEH* (II, p. 14) as coërcentur　　16 *instillantur*] / thus in *c-t* but
*SEH* (II, p. 14) reads installantur　　25 *ipsis;*] ~,

and probable, others harder to swallow. But in our time, despite the fact that wits are less extravagant (restrained as they are by the establishment of schools and colleges), these ' habits have not entirely ceased; Patrizi, Telesio, Bruno, Severinus the Dane, Gilbert the Englishman, and Campanella have held the stage with fresh tales, neither noted for applause nor attractive in argument. Are we to be astonished by this, as if innumerable doctrines and sects of this kind could not have sprung up in all ages? For there is not nor ever will be any end of any kind to this sort of thing. One seizes upon one thing, another on another; of dry ' and open light there is none; everyone philosophizes out of the cells of his own fantasy, as if from Plato's cave—the more sublime wits more acutely and successfully, the slower-witted with less success but the same tenacity. And so not long since, by direction of certain learned and, by current standards, first-rate men (bored, I believe, by the lack of uniformity and moderation) the sciences are pent up in certain set books, and thus confined they are, imposed ' on the old, and instilled into the young; so that now (as Cicero joked of Cæsar's year) the star Lyra rises by decree, and authority is taken for truth, not truth for authority. Now this practice and discipline is splendid for present use but does not point the way to a better future. Evidently we imitate our first parents' sin and we suffer for it. They wanted to be like God, but their descendants want more still. For ' we conjure up worlds, and dictate to nature like despots; we want to have things our own way and in accordance not with the Divine Wisdom, or how we find the actual facts, but with depths of our own folly. Indeed, I do not know whether we more abuse the things themselves or our own wits but we plainly set the seal of our own image on the creatures and works of God rather than carefully examining and recognizing the seal that the Creator has set upon them. Thus again do we deservedly ' lose our power over created things; and while after the fall of man some degree of control over the recalcitrance of creatures still remained—so that they could be subdued and steered by true and solid arts—yet this too we for the most part forfeit through our insolence, and because we want to be like God, and follow the dictates of our own reason. For which reason if there be

[B5ʳ] *& Magnificatio, si qua Charitas in | Homines, & erga necessitates, &*
*ærumnas humanas releuandas Studium, si quis Amor Veritatis in*
*Naturalibus, & Odium Tenebrarum, & Intellectûs purificandi*
*Desiderium; orandi sunt Homines iterùm atque iterùm, vt missis*
5  *paulisper, aut saltem sepositis Philosophijs istis volaticis, & præpo-*
*steris, quæ Theses Hypothesibus anteposuerunt, & Experientiam*
*captiuam duxerunt, atque de Operibus Dei triumphârunt, summissè,*
[B5ᵛ]  | *& cum veneratione quâdam, ad Volumen Creaturarum euoluendum*
*accedant; atque in eo moram faciant, meditentur, & ab opinionibus*
10  *abluti, & mundi, castè, & integrè versentur.* Hic est ille Sermo, &
*Lingua, qui exiuit in omnes fines terræ, nec confusionem Babyloni-*
*cam passus est; Hunc perdiscant Homines, & repuerascentes, atque*
*iterùm Infantes facti, Abecedaria eiusdem in manibus habere dignen-*
[B6ʳ] *tur. In Interpretati | one autem eius eruendâ, atque enucleandâ, nulli*
15  *operæ parcant, sed strenuè procedant, persistant, immoriantur. Cum*
*igitur in* Instauratione *nostrâ, Historiam Naturalem, qualis sit in*
*ordine ad finem nostrum, in tertiâ Operis parte collocauerimus, hanc*
*rem præuertere, & statìm aggredi visum est. Etsi enim haud pauca,*
19  *eáque ex præcipuis, supersint in* Organo *nostro absoluenda, tamen*
[B6ᵛ] *consilium est, vniuersum Opus* Instau | rationis, *potius promouere*
*in multis, quàm perficere in paucis, hoc perpetuò, maximo cum*
*ardore (qualem Deus Mentibus, vt planè confidimus, addere solet)*
*appetentes, vt quod adhuc nunquam tentatum sit, id ne iam frustrà*
*tentetur. Simul subijt animum illa cogitatio: Spargi proculdubiò per*
25  Europam, *complura Ingenia, capacia, libera, excelsa, subtilia, so-*
*lida, constantia. Quid si quis, tali Ingenio præditus, rationem, &*
[B7ʳ] *v | sum* Organi *nostri capiat, probet? Tamen non habet, quid agat, nec*
*quomodo se ad Philosophiam comparet, aut accingat. Si esset res, quæ*
*lectione Librorum Philosophicorum, aut disputatione, aut medita-*
30  *tione perfici posset, sufficeret fortasse ille, quisquis sit, & abundè illud*
*præstaret. Quod si ad Historiam Naturalem, & Experimenta*
*Artium, illum remittimus (id quod facimus) hæret, non est Instituti*
[B7ᵛ] *eius, non Otij, non | Impensæ. Atqui non est postulandum nobis, vt*
*quis vetera dimittat, antequam in possessionem Meliorum inducatur.*
35  *Postquam autem Naturæ, & Artium Historia, fidelis & copiosa,*

---

17  *collocauerimus,*] ~;      22  *ardore*] ~,

humility towards the Creator, if there be reverence or willingness to magnify his works, if there be charity in | men and eagerness to relieve human necessities and afflictions, if there be any love of truth in nature, hatred of shadows, and desire to purify the intellect, we should beg men again and again to set aside for a while or at least discard these fickle and wrong-headed philosophies, which have put theses before hypotheses, led experience captive, and exulted over God's works; | and to read through with due humility and reverence the volume of creatures, and dwell and reflect on it, and, purged of opinions, to study it with a pure and honest mind. This is that speech and language which went out to the ends of the earth, and did not suffer the confusion of Babel; let men learn this thoroughly and, becoming childlike, return to infancy again and deign to take its abecedaria into their hands. They should | spare no effort in interpreting and unravelling it, but advance energetically, and stick at it until death. Now since in my *Instauration* I have placed natural history, history of the kind fit for my purpose, in the third part of the work, I have thought it good to turn my attention to this business first and deal with it at once. For although not a few things, and of those some of the principal, are still to be finished in my *Organum*, my resolution is nevertheless rather to advance the whole work of the | *Instauration* in many things than to perfect it in a few; and this is always my fervent desire (a desire of the kind that I believe God is accustomed to vouchsafe) that this which has so far never been tried may not now be tried in vain. At the same time this thought comes to mind: that there are without doubt many capacious, candid, sublime, subtle, solid, and steadfast intellects scattered the length and breadth of Europe. And what if one such intellect were to appropriate the plan and purpose | of my *Organum* and put it to the test? He still does not know how to proceed, nor how to get ready to re-equip himself for philosophy. If it were something which could be achieved by poring over books of philosophy, by disputation, or by meditation, he, whoever he may be, might be up to the job, and do it well. But if I refer him (as I do) to natural history, and the experiments of the arts, he is at a loss: it is not what he is used to, and he has neither the time nor | the money for

*collecta, & digesta fuerit, atque veluti ante oculos Hominum posita,*
*& explicata, non tenuis est spes, Ingenia, de quibus diximus, grandia*
*(qualia & in antiquis Philosophis viguerunt, & adhùc non rarò*
[B8$^r$] *reperiuntur) cum tantæ ante*$^|$*hàc fuerint efficaciæ, vt veluti ex*
5 Scalmo, *aut* Conchâ (*rarâ scilicet Experientiâ, & friuolâ*) Nauiculas *quasdam Philosophiæ, admirabili structurâ, quoad opificium,*
*ædificauerint; multò magis postquam Syluam, & Materiem nacta*
*sint, solidiores structuras excitaturæ; idque licet viâ veteri pergere*
9 *malint, nec viâ nostri* Organi (*quæ vt nobis videtur, aut vnica est,*
[B8$^v$] *aut optima) vti. Itaque hùc res redit, vt* Organum *no*$^|$*strum; etiamsi*
*fuerit absolutum, absq*ue Historiâ Naturali, *non multùm;* Historia
Naturalis *absq*ue Organo, *non parùm, Instaurationem Scientiarum*
*sit prouectura. Quare omninò, & ante omnia, in hoc incumbere,*
*satius, & consultius visum est. Deus Vniuersi Conditor, Conseruator,*
15 Instaurator, *hoc Opus, & in ascensione ad Gloriam suam, & in*
*descensione ad Bonum humanum, pro suâ erga Homines, Beneuo-*
[C1$^r$] *lentiâ, & Miseri*$^|$*cordiâ, protegat & regat, per Filium suum vnicum,*
Nobiscum Deum.
[C1$^v$: blank]   $^|$

[C2$^r$]          $^|$ Norma Historiæ præsentis.

20 *Qvamuis sub finem eius partis* Organi *nostri, quæ edita est, præcepta*
*de* Historiâ Naturali *&* Experimentali *conscripserimus; visum est*
*tamen, huius, quam nunc aggredimur,* Historiæ Normam, *& Fi-*
*guram, & accuratiùs describere, & succinctiùs.* Titulis *in* Catalogo
[C2$^v$] *comprehen*$^|$*sis, qui pertinent ad* Concreta, Titulos, *de* Naturis
25 Abstractis (*quarum ibidem, vt* Historiæ Reseruatæ, *mentionem*
*fecimus), superaddimus. Hi sunt,* Materiæ Schematismi *diuersi,*
*siue* Formæ primæ Classis, Motus simplices, Summæ Motuum,
Mensuræ Motuum, *alia quædam. De his* Abecedarium Nouum
*confecimus, & sub finem huius voluminis collocauimus.*

5 Conchâ] ~,     10 *no*$^|$*strum;*] ~,     25 Abstractis] ~,     26 *fecimus*),] ~,)

12

it. And yet I do not lay it down that anyone should give up what he already has before he has taken possession of something better. For once a faithful and abundant history of nature and the arts has been collected and arranged, and once it has been unfolded and placed as it were before men's eyes, there will be no mean hope that those great intellects of whom I have spoken (such as flourished in the ancient philosophers, and are even now not unusual), who till now have built with ¦ such efficiency as far as the work goes certain philosophical skiffs of ingenious construction from a plank or shell (i.e. from slight and paltry experience), will, once the right timber and material have been obtained, raise much more solid constructions, and that too although they prefer to follow the old ways and not the way of my *Organum* (which seems to me to be either the only or the best way). And so it comes down to this, that my *Organum*, ¦ even if it were finished, would not carry forward the Instauration of the Sciences much without Natural History, whereas Natural History without the *Organum* would advance it not a little. And so I have thought it better and more prudent above and before all, to engage with this. May God, the Founder, Preserver, and Renewer of the universe, protect and govern this work, in his goodwill and ¦ mercy towards us humans, both in its ascension to His glory, and in its descent to the benefit of human kind through His only Son, God with us.
¦

¦ The Rule of the present History

Although at the end of the published part of my *Organum* I drew up precepts concerning the *Natural and Experimental History*, I nevertheless think it right to describe accurately and succinctly the rule and make-up of the history which I now attempt. To the titles of the *Catalogue* ¦ which deal with things concrete, I add on titles relating to abstract natures (which I have mentioned as a history kept back for myself). These are the various schematisms of matter or forms of the first class, simple motions, sums of motions, measures of motions, and some other things besides. I have drawn up a *New Abecedarium* of these which I have located at the end of this volume.

[C3ʳ]    Titulos (*cum ad omnes* | *nullo modo sufficiamus*) *non ex ordine, sed ex delectu sumpsimus; quorum scilicet Inquisitio, aut propter vsum erat grauissima, aut propter copiam Experimentorum maximè commoda, aut propter obscuritatem rei, maximè difficilis,* & *nobilis,*
5    *aut propter discrepantiam* Titulorum *inter se, latissimè patens ad Exempla.*

    *In* Titulis *singulis, post* Aditum *quendam, aut præfationem, sta-*
[C3ᵛ]    *tim* Topica particularia, *siue Articulos* | *Inquisitionis proponimus, tum ad Lumen Inquisitionis præsentis, tum ad Prouocationem*
10    *futuræ. Domini enim Quæstionum sumus, Rerum non item. Neque tamen Quæstionum Ordinem, in* Historiâ *ipsâ præcisè obseruamus, ne impedimento sit, quod pro auxilio adhibetur.*

    Historia & Experimenta *omninò primas partes tenent. Ea si*
[C4ʳ]    *Enumerationem,* & *seriem Rerum particularium exhibeant, in* |
15    Tabulas *conficiuntur; aliter seorsùm excipiuntur.*

    *Cum* Historia & Experimenta *sæpissimè nos deserant, præsertim* Lucifera *illa,* & Instantiæ Crucis, *per quas, de veris Rerum causis, Intellectui constare possit,* Mandata *damus de Experimentis nouis, quantùm prospicere animo possumus, aptis ad id, quod quæritur.*
20    Hæc Mandata *tanquam* Historia designata *sunt. Quid enim aliud*
[C4ᵛ]    *nobis, primò viam ingredi* | *entibus, relinquitur?*

    Experimenti *alicuius subtilioris* Modum, *quo vsi sumus, expli-camus, ne* Error *subsit; vtque alios ad meliores,* & *magis exactos modos excogitandos, excitemus.*
25    Monita, & Cautiones, *de Rerum Fallacijs,* & *qui in inquirendo,* & *inueniendo possint occurrere,* Erroribus & *scrupulis, aspergimus; vt Phantasmata omnia, quantum fieri potest, tanquam* Exorcismo *fugemus.*

[C5ʳ]    Obseruationes *nostras,* | *super* Historiam & Experimenta, *sub-*
30    *teximus, vt Interpretatio Naturæ magis sit in procinctu.*

    Commentationes, & *tanquam* Rudimenta quædam, Interpre-tationis *de* Causis, *parcè,* & *magis suggerendo quid esse possit, quam definiendo quid sit, interponimus.*

---

18 *possit,*] ~;    26 *Erroribus*] ~,    30 *Naturæ*] ~,    32 *suggerendo*] ~,
33 *definiendo*] ~,

I have not taken the titles (since I am not up to ˡ dealing with them all) in order but I have picked some out, which are most weighty in respect of use, handiest on account of the abundance of experiments, most difficult and noble on account of the obscurity of the thing, or, on account of the differences between the titles, the ones which present the widest range by way of example.

In the particular titles, after a preamble or preface, I at once lay down particular topics or articles ˡ of inquiry both as light to the present and stimulus to future inquiry. For we command questions where we cannot command things. Nevertheless, in the history itself I do not follow the order of the questions slavishly, in case what was meant to be a help becomes a hindrance.

History and experiments take the first place above all. These, if they display an enumeration or series of particular things, are ˡ organized in tables; otherwise they are taken by themselves.

Since I very often lack history and experiments, especially experiments of light and crucial instances which can inform the mind about the true causes of things, I give directions for new experiments suitable, as far as I can tell at present, for the subject under inquiry. These directions are like history in embryo, for what other alternative is left to me who is just setting out ˡ on the road?

I explain the ways of performing any subtler experiment in case it is flawed, and also so as to prompt others to work out better ways.

I intersperse advice and cautions about the fallacies of things, and the errors and snags which may crop up in the course of inquiring and discovering, so that all spectres can as far as possible be driven off as if by exorcism.

I append my observations ˡ on the history and experiments so as to make the interpretation of nature readier.

I put forward speculations and, as it were, certain imperfect attempts at the interpretation of causes; I do this sparingly, more to hint at what might be the case than to present it cut and dried.

I outline and establish rules (yet only provisional ones), or imperfect axioms which crop up in the course of inquiry, and not

Canones, *sed tamen* Mobiles, *siue Axiomata inchoata, quæ nobis*
[C5ᵛ] *inquirentibus, non pronunciantibus se offerunt, præscribimus,* | *&*
*constituimus. Vtiles enim sunt, si non prorsus veri.*

　　*Vtilitatis humanæ nunquam obliti* (*licet Lux ipsa dignior sit ijs,*
5　*quæ à Luce monstrantur*) Vellicationes de Practicâ, *attentioni &*
*memoriæ Hominum subjicimus; cum nobis constet, talem, & tam*
*infœlicem esse Hominum Stuporem, vt quandóque, res ante pedes*
*positas, nisi moniti, non videant, sed prætereant.*

[C6ʳ]　　Opera, *&* Res impossibiles, *aut saltem adhuc* | *non inuentas, quæ*
10　*sub singulis Titulis cadunt, proponimus; atque vnà ea, quæ iam*
*inuenta sunt, & in Hominum potestate, atque* Impossibilibus *illis,*
*& non Inuentis, sunt* Proxima, *& maximè cognata, subiungimus; vt*
*simul & Industria humana excitetur, atque animi addantur.*

　　*Patet ex antedictis,* Historiam *præsentem, non tantum* Tertiæ
15　Partis Instaurationis *vices supplere, sed præparationem*
[C6ᵛ]　　　*esse non* | *contemnendam ad* Quartam, *propter*
Titulos ex Abecedario, *&* Topica; *& ad*
Sextam, *propter* Obseruationes
Maiores, Commentationes,
20　　　　*&* Canones.

with the intention <sup>|</sup> of laying down the law. For they are useful if not altogether true.

Never unmindful of utility for mankind (though light itself is more noble than what it discloses), I subjoin incentives to practice, for men's attention and memory, for I well know that such and so unfortunate is their stupidity that sometimes they do not see what is in front of their noses, but fail to notice it.

I set out works and things deemed impossible, or at least so far <sup>|</sup> undiscovered which fall under the individual titles; and together with them I subjoin things already discovered and lying within human power, which are closest and most akin to those things deemed impossible and undiscovered, so that human industry may be stimulated and souls fired.

It is clear from the foregoing that the present history not
only supplies the requirements of the third part of the
*Instauration*, but it is (because of the titles from
the *Abecedarium* and of the Topics) no <sup>|</sup>
negligible preparation for the fourth
part; and (because of the major
observations, speculations,
and rules) for the sixth.

[C7$^r$]
# Historia Ventorum.
## Aditus, siue Præfatio.

Venti *humanæ Genti alas addiderunt. Eorum enim dono, feruntur Homines, & volant, non per Aerem certè, sed per Maria; atque*
5 *ingens patet ianua Commercij, & fit Mundus peruius. Terræ autem*
[C7$^v$] *(quæ Gentis Humanæ Sedes est, & Do|micilium) Scopæ sunt. Eamque, atque simul Aerem ipsum, euerrunt & mundant. Attamen & Mare infamant, alioqui tranquillum, & innoxium; neque aliàs sine maleficio sunt. Motum, absque operâ humanâ, cient magnum,*
10 *& vehementem; vnde & ad Nauigandum, & ad Molendum, veluti Operarij conducti sunt, & ad multò plura adhiberi possunt, si*
[C8$^r$] *Humana non cesset Diligentia. Natura ipsorum inter se|creta, & abdita reponi solet; Nec mirum, cum nec Aeris Natura, & Potestas, cognita quoquo modo sit; cui famulantur, & parasitantur Venti, vt*
15 *(apud Poetas) Æolus Iunoni. Primariæ Creaturæ non sunt, nec ex Operibus sex Dierum; quemadmodum nec reliqua Meteôra quoad Actum, sed Post-nati, ex Ordine Creationis.*

[C8$^v$]
# Topica Particularia;
## Siue,
20
### *Articuli Inquisitionis de Ventis.*

1. **Nomina Ventorum.** Describito Ventos, ex diligentiâ Nauticâ, & imponito nomina ipsis, siue vetera, siue noua, modo constantia.

*Venti vel Generales sunt, vel Stati, vel Asseclæ, vel Liberi. Generales*
25 *voco, qui semper flant; Statos, qui certis temporibus; Asseclas, qui frequentiùs; Liberos, qui indifferentèr.*

[D1$^r$] 2. **Venti generales.** An sint *Venti* aliqui *Genera|les*, atque ipsissimi Motus Aeris; & si sint, in quâ consecutione Motûs, & in quibus locis spirent?

---

4 *volant*,] ~;    *Aerem*] / *nld* as aërem in *SEH* (II, p. 19)    9 *sunt*.] ~;    13 *Aeris*] / *nld* as aëris in *SEH* (II, p. 19)    15 *Poetas*] / *nld* as poëtas in *SEH* (II, p. 19)    24 et seqq. *SEH* (II, p. 20 et seqq.) has outdented words at the start of passages which in the original are indented and italic. In the Topica Particularia *SEH* has outdenting where *c-t* has indenting

<sup> </sup>History of the Winds
Preamble or Preface

The winds lend wings to the human race. For their gift is to carry men and let them fly, not of course through the air, but by sea; and a vast gateway of trade is opened, and the world becomes passable. To the earth (which is the seat and <sup>|</sup> dwelling-place of the human race) they are brooms which sweep clean both it and the air itself. And yet they give the sea a bad name, which apart from them would be calm and inoffensive; nor are they devoid of bad behaviour in other respects. Without human intervention they cause great and violent motion, as a result of which they are brought in as workers to drive shipping and mills, and could be used for many other things, provided that human industry does not sleep. Their nature is usually assigned to the category of the se<sup>|</sup>cret, and hidden; which is no wonder since the nature and power of the air, which the winds serve and batten on (according to the poets) as Æolus on Juno, is quite unknown anywhere. They are not original creations, nor are they among the works of the Six Days; just as the other meteors in their actualization are not original but born later in the order of creation.

<sup>|</sup> Particular Topics;
Or,
Articles of Inquiry
concerning the Winds

1. *Names of Winds.* Describe the winds with nautical precision and give them names old or new, but be consistent.

Winds are either General, Recurrent, Prevailing or Free. The ones I call General always blow; Recurrent ones blow at particular times; Prevailing winds blow more often; Free ones blow indiscriminately.

2. *General Winds.* Whether there be any general winds, <sup>|</sup> and motions of the very same air; and if there are, in what direction and in what places they blow?

**3. *Venti Stati.*** Qui *Venti Anniuersarij* sint, aut redeuntes per vices, & in quibus Regionibus? An inueniatur *Ventus* aliquis ita præcisè *Status*, vt redeat regularitèr ad dies certos, & horas, instar Æstûs Maris?

**4. *Venti Asseclæ.*** Qui *Venti* sint *Asseclæ*, & familiares *Regionum*; qui
5 *Temporum*, in ijsdem *Regionibus*; qui *Verni*; qui *Æstiuales*; qui *Autumnales*; qui *Brumales*; qui *Æquinoctiales*; qui *Solstitiales*; qui *Matutini, Meridiani, Vespertini, Nocturni?*

[Dɪʳ]    **5. [*Venti Marini.*]** Quales sint *Venti Marini*; ˡ quales qui spirant à *Continente?* Differentias autem *Marinorum*, & *Terrestrium* diligentèr exci-
10 pito, tam eorum qui in Terrâ, & Mari; quàm eorum qui à Terrâ, & Mari.

**6. *Venti Liberi.*** An non spirent *Venti* ex omni *Plagâ Cœli?*

> *Venti non multò plus Plagis Cœli, quam Qualitatibus variant. Alij Vehementes, alij Lenes; alij Constantes, alij Mutabiles; alij Calidi, alij Frigidi; alij humectant magis, & soluunt, alij desiccant, & constipant;*
> 15 *alij congregant nubes, & sunt pluuiosi, vel etiam procellosi, alij dissipant, & sunt sereni.*

**7. *Qualitates Ventorum diuersæ.*** Inquirito & narrato, qui sint *Venti*
[D2ʳ] vniuscuiusque speciei ex ˡ prædictis, & quomodo varient, secundum Regiones, & Loca?

20   > *Origines locales Ventorum triplices. Aut deijciuntur ex alto; aut emanant à terrâ; aut conflantur in ipso Corpore Aeris.*

**8. *Origines locales Ventorum.*** Secundùm has tres *Origines*, de *Ventis* inquirito. Qui scilicet ex ipsis, deijciantur de *Mediâ* (quam vocant) *Regione Aeris;* Qui vero expirent è *Cauis Terræ;* siue illi erumpant con-
25 fertìm, siue efflent insensibilitèr, & sparsìm; & postea glomerent, vt riuuli in fluuiu*m;* Qui deniq*ue* generentur passim ex tumoribus, siue expansionibus *Aeris* proximi?

> *Neque generationes Ventoru*m *Originales tantùm; Sunt &*
[D2ᵛ] > *Accidentales, ex compressioni*ˡ*bus Aeris scilicet, & percussionibus, &*
30 > *repercussionibus eius.*

**9. *Generationes Accidentales Ventorum.*** De huiusmodi *Ventorum Generationibus Accidentalibus* inquirito. *Generationes Ventorum* propriè

---

8 [*Venti Marini.*]] / marginalium missing in *c-t*, inserted on model of marginalia before and after; not so emended in *SEH* (II, p. 20)      10 eorum] ~,        28 *tantùm*;] ~:

3. *Recurrent Winds.* What winds are annual and come back at regular intervals, and in what regions? And are winds found which return so exactly as to come back on certain days or hours like the tides?

4. *Prevailing Winds.* What winds are prevailing in, and habitual to, particular regions; and when do they blow there—which in spring, summer, autumn, and winter; which at the equinoxes, and which at the solstices, and which in the morning, at midday, in the evening, and by night?

5. [*Maritime Winds.*] What are marine winds like; ' and those which blow from land masses? Identify the differences between marine and land winds, as well as those which blow land- and seaward, and those which blow from land and sea.

6. *Free Winds.* Whether there are not winds which blow from every quarter of the sky?

Winds do not vary much more in the quarters they blow from than in their qualities. Some are strong, others gentle; some are steady, others changeable; some are hot, others cold; some moisten more and loosen, others dry and consolidate; some gather clouds and bring rain, others disperse them and bring clear weather.

7. *The Various Qualities of Winds.* Inquire into and recount to which of the ' above-mentioned classes each wind belongs, and how they vary by region and place.

The local origins of winds are three: either they are cast down from above; or they come from the earth; or they are hatched in the body of the air itself.

8. *The Local Origins of Winds.* Investigate the winds in relation to these three origins; namely which of them are cast down from what they call the middle region of the air; which exhale from the hollows of the earth—be they ones which burst out as a body or flow out imperceptibly and dispersedly, and afterwards come together like tributaries to a river; and which are generated by the way from swellings or expansions of air close by.

Generations of winds are not only original, they are also accidental and caused by compressions, ' percussions, and repercussions of the air.

9. *Accidental Generations of Winds.* Investigate accidental generations of winds of this kind. They are not strictly generations of winds,

non sunt; etenim augent, & fortificant Ventos potiùs, quàm producunt & excitant.

*De Communitate Ventorum hactenus. Reperiuntur autem Venti Rari, & Prodigiosi, quales sunt Præster, Turbo, Ecnephias: Hi super Terram.*
5 *At sunt & subterranei, quorum alij sunt Vaporosi, & Mercuriales: ij percipiuntur in Mineris; Alij Sulphurei: illi emittuntur, nacti exitum in*
[D3ʳ] *Terræ Motibus, aut etiam efferuescunt ex Montibus ardentibus.* |

**10. Venti extraordinarij & Flatus repentini.** De huiusmodi *Ventis, Raris & Prodigiosis,* atque adeò de omnibus *Ventorum Miraculis,*
10 inquirito.

*A speciebus Ventorum transeat Inquisitio ad Confacientia ad Ventos (ita enim loqui volumus, quia vocabulum Efficientis plus significat, Vocabulum Concomitantis minus quam intelligimus) atque ad ea, quæ Ventos putantur excitare, aut sedare.*

15 **11. Confacientia ad Ventos, & excitantia, & sedantia ipsos.** Circa *Astrologica* de *Ventis* inquirito parcè, nec de accuratis *Schematibus Cœli* curato; tantummodo obseruationes manifestiores, de Ventis ingruentibus circa exortus aliquorum Astrorum; aut circa Eclipses Luminarium;
[D3ᵛ] aut Coniunctiones | Planetarum, ne negligito; nec minùs, quatenus
20 pendent ex vijs Solis, aut Lunæ?

**12.** Quid confaciant *Meteôra* diuersorum generum ad *Ventos?* Quid *Terræ Motus,* quid *Imbres,* quid *Concursus Ventorum* ad inuicem? Concatenata enim sunt ista, & alterum alterum trahit.

**13.** Quid confaciant ad *Ventos, Vaporum* & *Exhalationum* diuersitas;
25 & quæ ex ipsis sint magis generatiua *Ventorum*; & quatenus Natura *Ventorum* sequatur huiusmodi *Materias* suas?

**14.** Quid confaciant ea, quæ hic in *Terrâ* sunt, aut fiunt, ad *Ventos*; Quid Montes, & solutiones Niuium in ipsis; Quid Moles glaciales,
[D4ʳ] quæ in Mari innatant, | & deferuntur alicubi; Quid differentiæ Soli,
30 aut Terræ (modo hoc fuerit per Tractus Maiores) veluti Paludes, Arenæ, Syluæ, Campestria; Quid ea, quæ hic apud Homines aguntur, veluti Incensiones Ericæ, & similium, ad culturam agrorum; Incensiones Segetum, aut villarum in Bellis; Desiccationes Paludum; Displosiones continuæ Bombardarum; Sonitus Campanarum simul in magnis Vrbibus;
35 & similia? Festucæ certè sunt res nostræ, sed tamen aliquid possunt.

**15.** De omnimodis *Excitationibus,* aut *Sedationibus Ventorum*
[D4ᵛ] inquirito, sed parcè de Fabulosis, aut superstitiosis. |

---

4 *Terram.*] ~:     5 *Mercuriales.*] ~,     6 *Sulphurei:*] ~;     30 Terræ] ~,

but they increase and strengthen them rather than produce or stimulate them.

So much then for winds in the common way. But there are winds rare and prodigious, of which kind are the hot wind, whirlwinds, and hurricanes, which happen above ground. But there are underground winds, of which some are vaporous and mercurial which are detectable in mines; others are sulphurous and escape in earthquakes, or erupt from burning mountains. |

10. *Extraordinary Winds and Sudden Blasts.* Investigate this kind of rare and prodigious winds; as well as all miracles of winds.

Let the inquiry move on from the species of winds to factors contributing to winds (I do not want to use the term efficients for that is too strong; and for my purposes concomitants is too weak); and to those things which are thought to stimulate winds or damp them down.

11. *Factors Contributing to Winds and to Stimulating and Damping them down.* Investigate astrological doctrine about the winds sparingly, and take no notice of the finer configurations of the heavens, only do not neglect the more obvious observations concerning winds increasing when certain stars come up, or at the eclipses of luminaries, or at conjunctions | of planets; and no less the extent to which they depend on the approaches of the Sun and Moon.

12. What do the various meteors contribute to the winds, and what do earthquakes, rains, and the rushing together of the winds do? For these factors are linked together, and one influences another.

13. What do the various vapours and exhalations contribute, and which of them generates more winds; and how far does the nature of the winds follow from suchlike materials?

14. What do things here in the earth contribute to the winds: what do mountains or their melting snows do; what do the masses of ice which float on the seas, | and get carried everywhere contribute; what about differences of soil or earth (if they cover large areas), as swamps, sands, woods, flatlands; and what about things to do with man, as heather burning and the like for clearing the land, the burning of crops or villages in wars, draining of swamps, continual cannon fire, bells ringing together in big cities, and so on? Human deeds are certainly insignificant, but they still have some influence.

15. Look into all ways of raising or calming winds, but inquire sparingly into ways which are fabulous or superstitious. |

*A Confacientibus ad Ventos transeat Inquisitio ad inquirendos*
*Limites Ventorum: de Altitudine, Extensione, Duratione eorum.*

**16. Limites Ventorum.** Inquirito diligentèr de *Altitudine*, siue *Eleua-*
*tione Ventorum*; atque si sint Fastigia Montium, ad quæ Venti non
aspirent; aut si conspiciantur Nubes quandoque stare, & non mouere,
flantibus eodem tempore Ventis fortitèr, hîc in Terrâ?

**17.** Inquirito diligentèr de *spatijs*, quæ *Venti* deprehensi sunt simul
occupare, & ad quos terminos? Exempli gratiâ, si *Auster* flauerit tali
loco, an constet, quod eodem tempore, *Aquilo* flauerit ab illinc mil-
liaribus decem? | Contrà, in quantas angustias *Venti* redigi possint, ita ut
fluant *Venti* (id quod fieri videtur in *Turbinibus* nonnullis) tanquam per
canales.

**18.** Inquirito, ad quod tempus, vel maximum, vel medium, vel
minimum, continuari soleant *Venti*, & deinde flaccescere, & tanquam
expirare, qualis etiam esse soleat Ortus, & inceptio *Ventorum*, qualis
languor & cessatio; subitò, gradatìm, quoquo modo?

*A limitibus Ventorum, transeat Inquisitio ad successiones Vento-*
*rum, vel inter se, vel respectu Pluuiæ, & Imbrium. Cùm enim choreas*
*ducant, ordinem Saltationis nosse iucundum fuerit.* |

**19. Successiones Ventorum.** An sit aliqua Regula, aut obseruatio
paulò certior, de *successionibus Ventorum* ad inuicem, siue ea sit in
ordine ad Motum Solis, siue aliàs; & si sit, qualis sit illa?

**20.** Circa *Successionem* & *Alternationem Ventorum*, & *Pluuiæ*
inquirito; cum illud familiare, & frequens sit, vt *pluuia* sedet *Ventos*,
*Venti* compescant, & dissipent *Pluuiam*.

**21.** An post certam *Periodum* Annorum, redintegretur *Successio*
*Ventorum*, & si ita sit, quæ sit ea *Periodus*?

*A Successionibus Ventorum, transeat Inquisitio ad Motus ipsorum.*
*Motus Ventorum, septem Inquisitionibus absol|uuntur. Quorum tres,*
*superioribus Articulis continentur, Quatuor adhùc manent intactæ.*
*Nam de Motu Ventorum dispertito per Plagas Cœli, inquisitum*
*est. Etiam de Motu trium Linearum sursùm, deorsùm, lateralitèr.*
*Etiam de Accidentali Motu Compressionum. Restant Motus quartus*

1 Ventos] ~,   Inquisitio] ~,   2 Altitudine,] ~;   Extensione,] ~;   18 Imbrium.] ~:
21 ad inuicem] adinuicem

24

From factors contributing to winds let the inquiry pass on to investigating the limits of winds: i.e. their height, extension, and duration.

**16.** *The Limits of Winds.* Look carefully into the height or elevation of the winds, and see if there are mountain tops to which the winds do not reach; or if one can see clouds standing still and motionless while down here on the ground stiff winds are blowing.

**17.** Look carefully into the spaces which the winds are understood to occupy at the same time, and into what the limits of these spaces are. For instance, if a southerly is blowing at such and such a place, will it happen that a north wind will be blowing ten miles away from there? ⏐ On the other hand, how narrow are the spaces into which winds can be squeezed, such that the winds (as seems to happen in some whirlwinds) flow as it were in channels.

**18.** Look into the greatest, average, and shortest time that winds generally persist before they ease off and die out; also look into the nature of their rising and beginning, and their weakening and stopping, be it sudden, gradual, or otherwise.

From the limits of winds the inquiry passes on to their successions, either among themselves, or in relation to rains and showers. For seeing that they perform dances, it would be delightful to know the steps. ⏐

**19.** *Successions of Winds.* Is there any rule or relatively certain observation regarding the successions of winds in relation to each other, be it associated with the motion of the Sun, or something else? And if there is such a rule, what is its nature?

**20.** Look into the successions and alternations of wind and rain; for the usual view is that rain calms the winds, and that winds suppress and disperse rain.

**21.** After a certain period of years does the succession of winds repeat itself, and if that is the case what is that period?

From successions of winds the investigation passes to their motions. The motions of the winds comprise seven inquiries, ⏐ of which three are contained in the articles above, and four still remain to be tackled. For I have already inquired into the distribution of the winds' motion according to their quarters, and the three directions of motion, upwards, downwards, and sideways, as well as into the accidental motion of compressions. The fourth or

*Progressiuus; quintus Vndulationis; Sextus Conflictûs; Septimus in Organis, & Machinis humanis.*

**22. Motus diuersi Ventorum.** Cum *Progressus* sit semper à termino;
4 de loco primi Ortûs, & tanquam Fontibus alicuius *Venti*, quantum fieri
[D6ʳ] potest, diligentèr inquirito. Siquidem videntur *Venti Famæ* similes. |
Nam licet tumultuentur, & percurrant, tamen *Caput inter nubila condunt.* Item de *Progressu* ipso. Exempli gratiâ, Si *Boreas* vehemens, qui flauerit *Eboraci,* ad talem diem, aut horam; flauerit *Londini* biduo post?

**23.** De *Vndulatione Ventorum* Inquisitionem ne omittito: *Vndula-*
10 *tionem* vocamus eum motum, quo *Ventus* ad parua interualla intenditur,
& remittitur, tanquam vnd*æ* Aquarum: quarum vices optimè percipiuntur, ex auditu in ædibus: Eò autem magis notato diligentèr differentias *Vndulationis, siue Sulcationis* inter Aerem & Aquam; quia in
[D7ʳ] *Aere, & Ventis,* deest Motus grauitatis, qui pars mag|na est *Vndulationis*
15 in Aquis.

**24.** De *Conflictu,* & Concursu Ventorum, flantium ad idem tempus, diligentèr inquirito; primò vtrum flent simul plures *Venti Originales,* non dicimus *Reuerberantes?* Et si hoc ita sit, quales *Euripos* in Motu, quales rursus *Condensationes,* & *Alterationes* in Corpore Aeris,
20 gignant?

**25.** An *Venti* alij eodem tempore flent superiùs, alij hic apud nos in imo; quandoquidem obseruatum est à nonnullis, interdum nubes ferri in contrarium versionis Pinnaculi: Etiam nubes ferri forti Aurâ, cum hic
24 apud nos fuerit summa Tranquillitas?
[D7ᵛ] **26.** Fiat descriptio diligens admodùm, & particularis *Motûs* | *Ventorum,* in impulsu *Nauium,* per vela.

**27.** Fiat Descriptio *Motûs Ventorum* in velis *Molendinorum,* ad *Ventum;* in volatu Accipitrum & Auium; etiam in vulgaribus, & ludicris, veluti Signorum explicatorum, Draconum volantium, Duellorum ad
30 ventum, &c.

*A Motibus Ventorum transeat Inquisitio ad vim, & Potestates ipsorum.*

**28. Potestates Ventorum.** Quid possint, & agant *Venti,* circa *Currentes,* & Æstus Aquarum, & circa Detentiones, Immissiones, &
35 Inundationes ipsarum?

progressive motion, the fifth or fluctuating motion, the sixth or
motion of conflict, and the seventh or motion in man-made
instruments and machines remain to be considered.

**22. *The Various Motions of Winds.*** Since progress always comes
from a beginning, inquire as carefully as possible into the place where
the winds first rise, and into their sources. For the winds appear to be
like Fame, ¹ for although they swell and surge hither and thither, *they
hide their heads in the clouds.* Inquire too into the progress itself; for
instance, if a strong northerly blew at York on this or that day or hour,
did it blow (say) two days later in London?

**23.** Do not forget to inquire into the fluctuation of winds. By fluctuation I mean that motion by which the winds rise and fall in a short
time, like waves in the waters, and which can be best sensed by listening
in buildings. But the differences between the fluctuation and furrowing
of the air as against the water should be carefully noted because, in the
air and winds, motion of gravity is absent, a motion which has ¹ a large
part in wave motion in water.

**24.** Inquire carefully into the conflict and rushing together of the
winds blowing at the same time; and first whether many original winds
(not reverberating ones) can blow at the same time, and if that is the
case, what channels they develop in their motion, and what condensa-
tions and alterations in the body of the air?

**25.** Do some winds blow above at the same time as others blow
down here with us? For some people have observed that sometimes
clouds are carried in a direction opposite to that indicated by the
weathervane, and are also carried by a strong breeze when down here
with us there is a dead calm.

**26.** Let a very careful and particular description be made of wind ¹
motion in propelling ships with sails.

**27.** Let a description be made of wind motion in windmill sails
turned into the wind; in the flight of hawks and birds; and also in
common effects and amusements such as banners unfurled, flying kites,
battledore and shuttlecock, etc.

From the motions of winds the inquiry passes to their force and
powers.

**28. *The Powers of Winds.*** What can the winds do and accomplish
with respect to currents and tides, and to holding them back, pushing
them on, and making them flood?

**29.** Quid circa *Plantas,* & *Insecta,* inducendo Locustas, Erucas, [D8$^r$] malos Rores? |

**30.** Quid circa *Purgationem,* & *Infectionem* Aeris, & circa *Pestilentias,* morbos, & affectus Animalium?

5 **31.** Quid circa *delationem specierum* (quas vocant) *spiritalium,* vt *Sonorum, Radiorum,* & similium?

*A Potestatibus Ventorum, transeat Inquisitio ad Prognostica Ventorum, non solum propter vsum Prædictionum, sed quia manu ducunt ad Causas. Prognostica enim, aut præparationes rerum monstrant,* 10 *antequam perducantur ad Actum; aut Inchoationes, antequam perducantur ad sensum.*

**32. Prognostica Ventorum.** Colligantur, cum diligentiâ bonâ, *Prog-* [D8$^v$] *nostica* Ventorum omnigena (præter Astrologica, | de quibus superiùs diximus, quatenus sint inquirenda) siue petantur ex *Meteoricis,* siue ex 15 *Aquis,* siue ex *Instinctu Animalium,* aut quouis alio modo.

*Postremò Inquisitionem claudito, inquirendo de Imitamentis Ventorum, siue in Naturalibus, siue in Artificialibus.*

**33. Imitamenta Ventorum.** Inquirito de *Imitamentis Ventorum,* in *Naturalibus,* qualia sunt *Flatus in Corporibus Animalium, Flatus in recep-* 20 *taculis Distillationum?* &c.

**[34.]** Inquirito de Auris factis, & *Ventis Artificialibus,* vt Follibus, Refrigeratorijs in cœnaculis, &c.

[E1$^r$] Articuli tales sint. Neque nobis dubium est, quin ad non|nullos horum, responderi non possit, secundum copiam Experientiæ, quam 25 habemus. Verum quemadmodum in Causis Ciuilibus; quid causa postulet, vt interrogetur, nouerit Iurisconsultus bonus; quid Testes respondere possint, non nôrit: Idem nobis circa *Historiam Naturæ* accidit. Posteri cætera viderint.

HISTORIA.
30 *Nomina Ventorum.*

**Ad Artic. 1.** Nomina *Ventis,* potiùs ex ordine & gradibus, numerata, [E1$^v$] quàm ex Antiquitate propria, imponimus: | hoc perspicuitatis, &

---

21 [34.] ] / not emended thus in *SEH* (II, p. 25)
26 Iurisconsultus] Iureconsultus / silently emended thus in *SEH* (II, p. 25)

9 *Causas.*] ~:    31 ordine] ~,

29. What can they do with plants and insects, by inducing locusts, caterpillars, and bad damps? <sup>|</sup>

30. What can they do in relation to the purgation and infection of the air, and to pestilences, illnesses, and the affections of animals?

31. What do they do in carrying spiritual species (as they call them), as of sounds, radiations, and the like?

From the powers of winds the inquiry passes on to the prognostics of winds, not only for the usefulness of the forecasts but because they lead straight to causes. For forecasts show the preparations of things before they are put into effect, or their beginnings before they affect the sense.

32. *Prognostics of Winds.* Let all kinds of prognostics of winds be collected with satisfactory diligence (except for astrological ones <sup>|</sup> of which, as far as they should be looked into, I have spoken above), be they sought from meteors, the water, or instincts of animals, or by any other means.

Lastly finish off the inquiry by looking into simulations of winds be they in things natural or artificial.

33. *Simulations of Winds.* Inquire carefully into simulations of winds, of which kind are winds in the bodies of animals, winds in distilling vessels, etc.

[34.] Inquire into manufactured breezes and artificial winds, as bellows, ventilators in rooms, &c.

These then are the articles. But I have no doubt but that <sup>|</sup> some of them cannot be answered from the stock of experience currently available. For just as in civil cases, a good lawyer knows what case to put to the question but yet does not know what the witnesses may answer, so it falls out with me in natural history. Let posterity see to the rest.

# THE HISTORY
## Wind Names

*To Article* 1. For the sake of clarity and memorability <sup>|</sup> I assign names to the winds, and not the ones sanctioned by antiquity but rather ones assigned according to their order and degree. But I also add ancient names in recognition of the views of the old authors, from whom I have

memoriæ gratiâ. Sed Vocabula antiqua adjicimus quoque, propter suffragia Authorum veterum, ex quibus cum haud pauca (licet anxio quodam iudicio) exceperimus, non agnoscentur ferè illa, nisi sub nominibus, quibus illi vsi sunt. Partitio autem generalis ea esto: vt sint
5 Venti *Cardinales*, qui spirant à Cardinibus Mundi; *Semicardinales* qui in dimidijs; *Mediani* qui in intermedijs. Etiam ex Intermedijs, *Mediani Maiores* vocentur, qui in quadris; *Minores* reliqui. Particularis autem
[E2$^r$] diuisio ea est, quæ sequitur. |

| | |
|---|---|
| Cardin. | *Boreas.* |
10 | | *Bor.* 1. *ad Eurum.* |
| Med. Maj. | *Bor.* 2. *ad Eur. siue Aquilo* |
| | *Bor.* 3. *ad Eur. siue Meses.* |
| Semica. | *Euroboreas.* |
| | *Eurus.* 1. *à Boreâ.* |
15 Med. Maj. | | *Eurus.* 2. *à Boreâ siue Cœcias.* |
| | *Eur.* 3. *à Boreâ.* |

| | |
|---|---|
| Cardin. | *Eurus, siue Subsolanus.* |
| | *Eur.* 1. *ad Austrum.* |
| Med. Maj. | *Eur.* 2. *ad Aust. siue Vulturnus.* |
20 | | *Eur.* 3. *ad Aust.* |
| Semica. | *Euro-Auster.* |
| | *Auster.* 1. *ab Euro.* |
| Med. Maj. | *Aust.* 2. *ab Euro, siue Phœnicias.* |
| | *Aust.* 3. *ab Euro.* |

| | |
|---|---|
| [E2$^v$] | Cardin. | *Auster, siue Notus.* |
26 | | *Aust.* 1. *ad Zephyrum.* |
| Med. Maj. | *Aust.* 2. *ad Zeph. siue Lybonotus.* |
| | *Aust.* 3. *ad Zephyrum.* |
| Semica. | *Zephyro-aust. siue Lybs.* |
30 | | *Zephyr.* 1. *ab Austro.* |
| Med. Maj. | *Zeph.* 2. *ab Aust. siue Africus.* |
| | *Zeph.* 3. *ab Austro.* |

---

6 intermedijs.] ~:        9 ff. Cardin.] / the table is set out here as in *c-t*; but it is set out differently in *SEH* (II, p. 25)        15 Maj.] ~ₐ    *Cœcias*] / *nld* as Cæcias in *SEH* (II, p. 25)
25 *Auster,*] ~:        27 *Lybonotus*] / *nld* in *SEH* (II, p. 25) as Libonotus

selected (not without careful reflection) a fair number of terms, since the matters in hand might barely be made out except by the names that they used. Let the general distribution be this: that there are Cardinal winds which blow from the cardinal points of the world; Semi-cardinal halfway between these; and Median halfway between these again. And let these intermediate winds between the Cardinal and Semi-cardinal be called Greater Medians, the rest the Lesser. The particular division of the winds is set out below. |

| | |
|---|---|
| Cardinal | North |
| | 1. North by east |
| Greater Median | 2. North-north-east, or Aquilo |
| | 3. North-east and by north, or Meses |
| Semi-cardinal | North-east: |
| | 1. North-east and by east |
| Greater Median | 2. East-north-east, or Cœcias |
| | 3. East and by north |
| Cardinal | East, or Subsolanus |
| | 1. East and by south |
| Greater Median | 2. East-south-east, or Vulturnus |
| | 3. South-east and by east |
| Semi-cardinal | South-east: |
| | 1. South-east and by east |
| Greater Median | 2. South-south-east, or Phœnicias |
| | 3. South and by east |
| | Cardinal | South, or Notus |
| | 1. South and by west |
| Greater Median | 2. South-south-west, or Libonotus |
| | 3. South-west and by south |
| Semi-cardinal | South-west, or Libs |
| | 1. South-west and by west |
| Greater Median | 2. West-south-west, or Africus |
| | 3. West and by south |

| | |
|---|---|
| Cardin. | *Zephyrus, siue Fauonius.* |
| | *Zeph.* 1. *ad Boream.* |
| Med. Maj. | *Zep.* 2. *ad Bor. siue Corus.* |
| | *Zeph.* 3. *ad Boream.* |
| 5 Semica. | *Zephyro-boreas.* |
| | *Boreas.* 1. *à Zephyro, siue Thrascias.* |
| Med. Maj. | *Bor.* 2. *à Zeph. siue Circias.* |
| [E3<sup>r</sup>] | *Boreas.* 3. *à Zephyro.* | |

Sunt & alia *Ventorum* nomina, *Apeliotes, Argestes, Olympias, Scyron,*
10 *Hellespontius, Iapyx,* Ea nil moramur. Satis sit Nomina *Ventorum* ex
ordine, & distributione plagarum cœli, fixa imposuisse. In Interpre-
tatione Authorum non multum ponimus, cum in ipsis Authoribus
parum sit.

## *Venti Liberi.*

15 ***Ad Artic.* 6.** [1.]   Non est plaga cœli, vnde *Ventus* non spiret. Quin
si plagas cœli, in tot partes diuidas, quot sunt gradus in *Horizonte,*
[E3<sup>v</sup>] inuenias *Ventos* aliqua*n*do, alicubi à singulis flantes. |
   **2.** Sunt Regiones totæ, in quibus non pluit, aut rarò admodum: At
non sunt Regiones, vbi non flent *Venti,* & sæpiùs.

20 ## *Venti Generales.*

***Ad Artic.* 2.** De *Ventis Generalibus* Phœnomena rara. Nil mirum, cum
intra *Tropicos* præcipuè perspiciantur, loca damnata, apud Antiquos.
   [1.]   Constat nauigantibus inter *Tropicos,* libero æquore, flare *Ventum*
Constantem, & iugem (*Brizam* vocant Nautæ) ab Oriente in
25 Occidentem. Is non ita segnis est, quin partìm flatu proprio, partìm
[E4<sup>r</sup>] regendo Cur|rentem Maris, id efficiat, vt nequeant Nauigantes versus
*Peruuiam,* eâdem redire, quâ proficiscuntur, viâ.
   **2.** In nostris Maribus *Europæ,* percipitur cœlo sereno, & sudo, &
cessantibus *Ventis* particularibus, Aura quædam lenis ab Oriente,
30 solisequa.
   **3.** Recipit obseruatio vulgaris, nubes sublimiores ferri plerunque ab
Oriente in Occidentem, idque cùm iisdem temporibus, circà terram aut

15 [1.] ] / silently emended thus in *SEH* (II, p. 26)   spiret.] ~:      21 De *Ventis* . . . apud
Antiquos] / this para. in 20-line type in *c-t* but represented in *SEH* (II, p. 26) as if it were
16-line type   Phœnomena] / here as elsewhere *nld* as phænomena in *SEH* (II, p. 26)
23 [1.] ] / silently emended thus in *SEH* (II, p. 26)

| | |
|---|---|
| Cardinal | West, or Favonius |
| | 1. West and by north |
| Greater Median | 2. West-north-west, or Corus |
| | 3. North-west and by west |
| Semi-cardinal | North-west: |
| | 1. North-west and by north or Thrascias |
| Greater Median | 2. North-north-west, or Circias |
| | 3. North and by west ¹ |

There are also other names of winds: Apeliotes, Argestes, Olympias, Scyron, Hellespontius, and Iapyx. But I do not bother with these, for it is enough to have imposed set names on the winds according to the order and distribution of the quarters of the heavens. I set little store by the interpretation of authors; for there is not much of substance in these same individuals.

## Free Winds

*To Article 6.* [1.] There is no quarter of the heavens from which a wind may not blow. Indeed, if the quarters of the heavens were divided into as many parts as there are degrees in the horizon, you will find winds blowing from each at some time. ¹

2. There are whole regions where it does not rain, or rains very rarely; but there are no places where winds do not blow, and blow often.

## General Winds

*To Article 2.* Phenomena to do with general winds are uncommon, which is no surprise since they are perceived mainly within the tropics, a region thought by the ancients to be blighted.

[1.] Voyagers sailing within the tropics in the open sea and a flat calm find that a steady and constant wind (sailors call it a breeze) blows from east to west. Now this wind is not so sluggish but that partly by its own force, and partly by directing the ¹ sea's current, it so arranges matters that it prevents voyagers to Peru from returning by the same route.

2. Here in our European waters when the skies are calm and clear and particular winds have stopped blowing, we feel a certain light breeze from the east which follows the Sun.

3. That the higher clouds generally drift from east to west is a common observation, and this happens at the same time as there is a calm or a different wind down here. The fact that this does not always happen

tranquillitas sit, aut *Ventus* diuersus. Id si non semper faciant, poterit in causâ esse, quod *Venti* particulares quandoque flant in sublimi, qui [E4ᵛ] *Ventum* istum Generalem obruunt. |

    *Monitum. Si quis sit talis* Ventus Generalis, *ex ordine Motûs cœli, is* 5 *non adeo firmus est, quin* Ventis Particularibus *cedat. Manifestior autem est intra* Tropicos, *propter Circulos, quos conficit, maiores; Etiam in sublimi, propter eandem causam, & propter cursum liberum. Quamobrem, si hic extra* Tropicos, *& iuxta terram (vbi mollis admodum, & segnis est) eum deprehendere voles, Fiat* Experimentum *in Aere libero, & in summâ* 10 *tranquillitate, & in locis altis; & in corpore valdè mobili, & tempore pomeridiano, quia per id Tempus,* Ventus *Orientalis particularis, parciùs flat.*

    *Mandatum. Fiat diligens obseruatio, circa Pinnacula, & eiusmodi* [E5ʳ] *flabella in fastigijs turrium & templorum, annon* | *in maximis tranquil-* 15 *litatibus, stent perpetuò versus Occidentem?*

    **4.** *Phœnomenon obliquum. Constat* Eurum, *in Europâ nostrâ, esse* Ventum *desiccantem & acrem,* Zephyrum *contra humectantem & almum. Annon hoc fit, quia (posito quod Aer moueat ab Oriente in Occidentem) necesse est, vt* Eurus *qui moueat in eâdem Consecutione, Aerem dissipet, &* 20 *attenuet, vnde fit Aer mordax, & siccus,* Zephyrus *autem, qui in contrariâ, Aerem in se vertat, & condenset, vnde fit obtusior, & demum humidus?*

    **5.** *Phœnomenon obliquum. Consulito Inquisitionem de Motu, & fluxu Aquarum, vtrum illæ moueant, ab Oriente ad Occidentem. Nam si* 24 *extrema hoc motu gaudeant, Cœlum & Aquæ, parùm abest, quin Aer,* [E5ᵛ] *qui intermedius est, ex eodem participet.* |

    *Monitum.* Phœnomena *duo, proximè posita,* Obliqua *appellamus, quia rem designatam non rectâ monstrant, sed per consequens; id quod (cum deest copia* Phœnomenorum *rectorum) etiam auidè recipimus.*

    *Mandatum. Quod* Briza *illa, inter* Tropicos *luculentèr spiret, res certa,* 30 *Causa ambigua. Possit ea esse, quia Aer, more cœli, mouetur; Sed extra* Tropicos, *quasi imperceptibilitèr, propter circulos minores, intra, mani-festò, propter Circulos maiores, quos conficit. Possit alia esse, quia calor omnem Aerem dilatat, nec se priore loco contineri patitur. Ex dilatatione* 34 *autem Aeris, necessariò fit impulsio Aeris contigui, quæ Brizam istam pariat* [E6ʳ] *prout progreditur Sol. Sed illa intra* | Tropicos, *vbi Sol est ardentior, insig-nior est; extra, ferè latet. Videtur esse* Instantia Crucis, *ad ambiguitatem istam tollendam, si inquiratur, vtrum Briza noctù flet annon? Rotatio enim Aeris etiam noctù manet, at calor* Solis *non item.*

---

8 *terram*] ~,    14 *annon*] *Annon*    24 *Aquæ,*] ~;    37 *annon*] / *SEH* (II, p. 27) has an non

may be because particular winds sometimes blow aloft and smother this general wind. |

*Advice.* If there is such a general wind following the motion of the heavens it is not powerful enough not to give way to particular winds. It is more evident in the tropics on account of the greater circles which it makes there, and also at altitude for the same reason, and because it gets a free run there. Accordingly, if you want to detect it outside the tropics and near the ground (where it is very light and slow-moving, perform the experiment in the open air, in a dead calm, in high places, with a body very sensitive to motion, and in the afternoon, for then the particular east wind blows less briskly.

*Direction.* Observe carefully to see whether weathercocks and vanes of that kind on the tops of towers and churches | do not always face west in perfectly calm weather.

4. *An indirect Phenomenon.* It is plain that with us in Europe the east wind is drying and bitter; the west wind, on the other hand, is moistening and kindly. Does not this happen because (allowing that the air moves from east to west) the east wind, moving of necessity in the same direction, thins and dissipates the air and so makes it biting and dry, whereas the west wind, blowing in the opposite direction, condenses and folds the air into itself and so makes it thicker, and at last moist?

5. *An indirect Phenomenon.* Seek guidance from the inquiry into the motion and flow of the waters to see whether they move from east to west. For if the extremes, the heavens and waters, rejoice in this motion, it would not be unlikely that the air which lies in between them would join in the same motion. |

*Advice.* I call these last two phenomena *indirect* because they do not demonstrate the matter in hand directly but by consequence; and this (when I lack a stock of direct phenomena) I eagerly embrace.

*Direction.* It is certain that this breeze blows spendidly in the tropics, but its cause is undecided. It could be because the air moves in the way the heavens do, but that outside the tropics it is barely perceptible because it makes smaller circles, but within the tropics it is obvious because its circles are larger. Or perhaps it is because heat dilates all the air so that it cannot be contained in its original place, and its dilatation necessarily gives a push to the air next to it, and so produces this breeze as the Sun advances. But within | the tropics where the Sun is stronger this is more striking, yet outside it is practically undetectable. There appears to be a crucial instance to remove this uncertainty, if we ask whether this breeze blows at night or not. For the rotation of the air carries on but the Sun's heat does not.

6. At certum est illam noctù non flare, sed manè, aut etiam aurorâ adultâ. Nihilominus non determinat illa *Instantia* quæstionem. Nam condensatio Aeris nocturna, præsertim in illis Regionibus, vbi nox & dies non magis pares sunt spatijs, quam differentes calore & frigore, possit Motum illum naturalem Aeris (qui lenis est) hebetare, & confundere.

[E6ᵛ]

7. Si Aer participet ex Motu cœli, sequitur non tantùm, quod *Eurus* cum motu Aeris concurrat, *Zephyrus* concertet; verum etiam quod *Boreas* tanquam ab alto spiret, *Auster* tanquam ab imo, in hemisphærio nostro, vbi *Polus Antarcticus* sub terrâ est, *Arcticus* eleuatur; idque etiam ab antiquis notatum est, sed titubantèr & obscurè; optimè autem conuenit cum experientiâ modernâ, quia *Briza* (quæ possit esse Motus Aeris) non est *Eurus* integer, sed *Euro-aquilo.*

[E7ʳ]

Venti Stati.

**Ad Artic. 3. Connexio.** *Vt in Inquisitione, de* Ventis Generalibus, *homines* Scototomiam *passi sunt; Ita in illâ, de* Ventis Statis, *vertiginem. De illâ silent, de hâc sursùm, & deorsùm, sermones faciunt inconditos. Ignoscendum hoc magis, quod varia res est; quia* Stati Venti *cum locis permutantur, vt non ijdem in* Ægypto, Græciâ,

[E7ᵛ]

Italiâ, *spirent.*

1. Esse alicubi *Statos Ventos,* etiam Nomen impositum declarat; vt & Nomen alterum *Etesiarum,* quod *Anniuersarios* sonat.

2. Apud Antiquos, inter causas Inundationis *Nili,* ascripta est, quòd eo Anni tempore, *Venti Etesiæ* (*Aquilones* scilicet) flarent, qui cursum fluuij in Mare inhibebant, & retrorsùm voluebant.

3. Inueniuntur in Mari Currentes, qui nec naturali motui Oceani, nec decursui ex locis magis eleuatis, nec Angustiis ex litoribus aduersis, aut Promontoriis excurrentibus attribui possint; sed planè reguntur *à* Ventis Statis.

[E8ʳ]

4. *Columbum,* qui nolunt à relatione *Naucleri Hispani,* & leuius putant, ab obscuris Antiquitatis vestigiis, & auris, tam certam, & fixam, de *Indijs Occidentalibus,* opinionem concepisse, huc se conuertunt, quòd à *Statis Ventis,* ad litora *Lusitaniæ,* coniecerit, *Continentem* esse à parte Occidentis: Res dubia, nec admodum probabilis, cùm *Ventorum*

---

21 1. Esse alicubi] / this is not preceded by a blank line in the *c-t* but by a page break; but on the evidence of the text below (see e.g. F2ᵛ below), a *connexio* should be followed by a blank line

6. But it is certain that it does not blow at night but at dawn and even at full sunrise. Nevertheless this instance does not decide the question. For the condensation of air at night, especially in those regions where the lengths of day and night are no more equal than their differences in heat and cold, can weaken and confuse this natural motion (gentle as it is) of the air. [¹]

7. If the air shares in the motion of the heavens, it follows not only that the east wind goes with the motion of the air and the west wind goes against it, but that the north wind blows as if from above and the south wind as if from below in our hemisphere where the antarctic pole lies beneath the Earth and the arctic one above it. And this was noticed even by the ancients, though darkly and uncertainly, but it agrees very well with modern experience because the breeze (which could be a motion of the air) is not a plain easterly but a north-easterly.

## [¹] Recurrent Winds

*To Article* 3. *Connection.* As in the inquiry about general winds, men have been afflicted with bad eyesight, so in the inquiry into recurrent winds they have suffered from dizzy spells. Of the former they say nothing, of the latter they write stuff that will not stand up. For we must allow more that this is a variable business, for recurrent winds change with their places, and the same ones do not blow in Egypt, Greece, and Italy. [¹]

1. The name given to them states that there are recurrent winds in some places, as too does the other name of Etesian, which denotes Anniversary winds.

2. Among the causes of the Nile's flooding the ancients set down that the Etesian winds (i.e. north winds) blew at that time of the year and checked the river's running into the sea, and drove it backwards.

3. We find currents in the sea which cannot be ascribed to the natural motion of the ocean, to downrush from places higher up, to straits formed by opposite shores, or to protruding headlands, but are plainly governed by recurrent winds.

4. Those who deny that Columbus conceived so firm and fixed opinion of the West Indies [¹] from the report of a Spanish captain, and think it improbable that he got the idea from obscure hints and rumours of antiquity, fall back on the notion that from the recurrent winds blowing to the Portuguese coast he guessed that there was a

Itinerarium, ad tam longos tractus vix attingat: Magnus interim honos huic Inquisitioni, si vni Axiomati, aut Obseruationi, ex iis quas multas complectitur, Inuentio Noui orbis debeatur.

4    **5.** Vbicunque siti sunt Montes alti & niuales, ab eâ parte, flant *Venti*
[E8ᵛ] *Stati* ad tempus, quo niues soluuntur. |

   **6.** Arbitror & à paludibus magnis, qu*æ* aquis cooperiuntur hieme, spirare *Ventos Statos*, sub tempora, quibus à calore Solis siccari cœperint; sed de hoc mihi compertum non est.

   **7.** Vbicunque generationes vaporum fiunt in abundantiâ, idque
10 certis temporibus; ibi scias *Ventos Statos* iisdem temporibus orituros.

   **8.** Si *Venti Stati* flent alicubi, nec causa eorum reperiatur in pro-pinquo; scias *Ventos* huiusmodi *statos* peregrinos esse, & à longè venire.

   **9.** Notatum est, *Ventos Statos* noctu non flare, sed tertiâ, ab ortu
[F1ʳ] Solis, horâ, insurgere; sunt certè huiusmodi *Venti*, veluti ex | longo
15 itinere, defessi, vt condensationem aeris nocturnam vix perfringant, at post exortum Solis excitati, paulispèr procedant.

   **10.** Omnes *Stati Venti* (præterquam ex locis propinquis) imbecilli sunt, & *Ventis* subitis se submittunt.

   **11.** Sunt complures *Venti Stati*, quos nos non percipimus, aut obser-
20 uamus, propter infirmitatem ipsorum, vnde à ventis liberis obruuntur. Ideò vix notantur hyeme, cum *Venti* liberi vagantur magis; sed potius versus æstatem, cum *Venti* illi Erratici magis deficiant.

[F1ᵛ]    **12.** In partibus *Europæ*, ex *Ventis Statis*, hi potissimi sunt. *Aqui*|*lones* à Solstitio; suntque exortûs *Caniculæ*, tum prodromi, tum sequaces;
25 *Zephyri*, ab Æquinoctio Autumnali, *Euri* à verno; Nam de Brumali Solstitio minus curandum, propter hyemis varietates.

   **13.** *Venti Ornithij*, siue *Auiarij*, qui nomen traxerunt, quod Aues à regionibus gelidis transmarinis, regionibus apricis immittant, nihil pertinent ad *Ventos Statos*; quia illi tempore sæpiùs fallunt. Aues autem
30 eorum Commoditatem, siue citiùs, siue tardiùs flent, expectant; etiam non rarò postquam flare paululum incœperint, & se subinde verterint,
[F2ʳ] destituuntur Aues, & merguntur in pelago, aliquan|do in naues decidunt.

   **14.** Præcisus reditus *Ventorum* ad diem & horam, instar Æstûs Maris,

---

5 *Stati*] ~,    14 insurgere;] ~:    20 obruuntur.] ~;    24 Solstitio;] ~:
sequaces;] ~:    25 verno;] ~:    29 fallunt.] ~:

continent out in the west. But this is doubtful and lacks plausibility since the winds could scarcely cover such vast distances. Meanwhile it lends great prestige to this inquiry if the discovery of the New World can be credited to one axiom or observation of the many the inquiry comprises.

5. Wherever there are high and snowy mountains, recurrent winds blow from that quarter at the time when the snows melt. |

6. I also reckon that from large marshes which are completely inundated in winter, recurrent winds blow at the time when the Sun's heat starts drying them, but I have no reliable information on this.

7. Wherever vapours are generated in abundance, and at particular times, you may understand that recurrent winds will arise there at those times.

8. If recurrent winds are blowing somewhere and you do not find their cause close by, you may understand that winds of this kind are foreigners come from afar.

9. It has been reported that recurrent winds do not blow at night but get up three hours after sunrise. Certainly winds of this kind are tired out | as from a long journey so that they can barely break through the air's nocturnal condensation, but after sunrise they are stimulated and move on for a short while.

10. All recurrent winds (save those coming from places nearby) are feeble, and surrender themselves to sudden winds.

11. There are many recurrent winds which we cannot feel or observe because of their weakness which causes them to be overwhelmed by free winds. Thus in winter when free winds range more widely they can hardly be detected, but rather they can be felt in summer when these wandering winds die out more.

12. In parts of Europe, these are the most important of the recurrent winds: the north winds | at the solstice, and they occur both before and after the rising of the Dog-star; west winds from the autumn equinox, and easterlies from the spring one; for the winter solstice needs less attention paid to it because of the variability of winter weather.

13. The Ornithian or Bird-winds (which get their name because they send birds from cold regions overseas to sunny lands) have nothing to do with prevailing winds, for they more often fail in point of time. But whether they blow earlier or later the birds wait to take advantage of them, but as often happens soon after they have begun to blow they change, and the birds are left in the lurch, and are sent to the bottom of the sea, or sometimes | fall on ships.

14. We do not find exact recurrences of the winds down to the day

non inuenitur. Designant quandoq*ue* Authores nonnulli diem; sed potius ex coniecturâ, quam ex obseruatione constante.

## Venti Asseclæ.

***Ad Artic. 4. & 5. Connexio.*** Ventorum Asseclarum *vocabulum*
5   *nostrum est, quod imponere visum est, ne aut pereat obseruatio circa*
[F2ᵛ]  *ipsos, aut confundatur.* Sensus talis est.  ᐠ *Diuide, si placet, annum in tres, quatuor, quinque partes, in aliquâ Regione.* Quod si Ventus *aliquis ibi flet, duas, tres, quatuor portiones ex ipsis,* Ventus *contrarius, vnam;* illum Ventum, *qui frequentiùs flat, eius regionis*
10  Asseclam *nominamus. Sic de temporibus.*

1.  *Auster* & *Boreas Asseclæ* Mundi sunt; Frequentiùs enim per vniuersum spirant illi, cum suis Sectionibus, quàm *Eurus,* & *Zephyrus* cum suis.

[F3ʳ]  2.  Omnes *Venti Liberi* (non *Stati*) magis Asseclæ hyemis ᐠ sunt, quàm
15  Æstatis, maximè autem Autumni, & Veris.

3.  Omnes *Venti* Liberi, potiùs Asseclæ sunt regionum extra Tropicos, atque etiam Circulos polares, quàm intra; In regionibus enim torridis, & conglaciatis, plerunq*ue* parciùs spirant, in medijs frequentiùs.

4.  Etiam omnes *Venti* Liberi, præsertim fortiores ex ipsis, flant
20  sæpiùs, & intensiùs, Mane & Vesperi, quàm meridie & noctù.

5.  *Venti* Liberi in regionibus fistulosis & cauernosis, frequentiùs spirant, quàm in firmis, & solidis.

***Mandatum.*** *Cessauit ferè humana diligentia, in obseruatione* Vento-
[F3ᵛ]  rum Asse|clarum, *in regionibus particularibus, quod tamen fieri debuit, &*
25  *ad multa vtilis foret. Memini me à Mercatore quôdam, prudenti viro, qui ad* Terram Piscationis, *coloniam duxerat, ibique hyemârat, causam quæsiuisse, cur regio illa tam impensè frigida haberetur, cùm clima satis benignum esset. Respondit, rem esse famâ aliquanto minorem, causam autem duplicem: Vnam, quod moles glaciales, à currenti maris Scythici,*
30  *iuxta ea litora deueherentur; Alteram (quam longe potiorem duxit) quod longè pluribus Anni partibus, spiraret apud eos* Zephyrus, *quàm* Eurus; *quod etiam facit apud nos (inquit) sed apud illos à Continenti, & gelidus;*
[F4ʳ]  *apud nos à Mari, & tepidus.* ᐠ *Quod si (addidit) tam frequentèr & diù spiraret in* Angliâ Eurus, *quàm apud eos* Zephyrus, *longè forent intensiora*
35  *frigora apud nos, & paria illis, quæ ibi fiunt.*

29 *duplicem:*] ~.      30 *deueherentur,*] ~.

40

and hour as we do with the tides. Some authors sometimes specify a day, but more by conjecture than settled observation.

## Prevailing Winds

*To Articles* 4 *and* 5. *Connection.* Prevailing Winds is my own expression; and I have imposed it in case observation of them gets lost or confounded. This is what I mean by it: | for any region divide the year into three, four, or five parts as you please; and if any wind blows there for two, three, or four of these intervals, and a contrary wind for just one of them, that wind which blows more frequently I call the prevailing wind of the region, and I do the same with the weather.

1. The south and north winds prevail throughout the world. For they with their subdivisions blow more often all over than do the east and west winds with theirs.

2. All free winds (not recurrent ones) prevail more in winter | than summer, but most of all in spring and autumn.

3. All free winds prevail more in regions outside the tropics, and even inside the polar circles, than inside the tropics. For in torrid and freezing regions they blow less often but in the places between more frequently.

4. Again, all free winds, especially the stronger ones, blow more often and stiffly in the morning and evening than at noon or night.

5. Free winds blow more often in regions full of cavities and hollows than in hard and solid districts.

*Direction.* Human diligence has been pretty well absent in the observation of prevailing | winds in particular regions, and yet it should have been pursued as it would have been useful in many ways. I remember that a certain merchant, a man of prudence, who led settlers to Fishing Land, and wintered there, answered, when I asked him why the country had a reputation for severe cold when its latitude was low, that its reputation was overblown, but that there were two reasons: that icy masses were carried down to those shores by the current of the Scythian Sea, and secondly (which he considered much more important) that the west rather than the east wind blew there with them for most months of the year, which (he said) happens here with us, but that with them the westerly came cold from the continent but with us warm from the sea. | But if (he continued) the east wind blew in England as often and long as the west wind did with them, here with us the cold would be much more severe and like what they experienced there.

**6.** *Zephyri* sunt *Asseclæ* horarum pomeridianarum. Declinante enim Sole frequentiùs spirant *Venti* ab Occidente, ab Oriente rariùs.

**7.** *Auster* noctis Assecla est; nam noctu & sæpiùs oritur, & flat vehementiùs. *Boreas* autem interdiu.

5 **8.** *Asseclarum* verò Maris, & continentis, multæ & magnæ sunt differentiæ. Ea præcipuè, quæ *Columbo* ansam præbuit inueniendi

[F4ᵛ] Noui Orbis: Quod *Venti* marini *Stati* non sunt, terrestres au|tem maximè. Cùm enim abundet vaporibus mare, qui vbique ferè indifferentèr adsunt, vbique etiam generantur Venti, & magnâ inconstantiâ

10 hùc illùc feruntur, cùm certas origines, & fontes non habeant. At terra, ad Materiam *Ventorum*, valdè inæqualitèr se habet; cum alia loca ad *Ventos* pariendos, & augendos, magis efficacia sint; alia magis destituta. Itaque flant ferè à parte fomitum suorum, & inde directionem sortiuntur.

15 **9.** Non satis constat sibi *Acosta*. Ait ad *Peruuiam*, & maritima *Maris Australis*, ferè per totum Annum, spirare *Austros*. Idem alibi ait, ad eas

[F5ʳ] oras, spirare potissimum *Ventos* Mari|nos. At *Auster* illis terrestris est, vt & *Boreas*, & *Eurus*, tantúmque *Zephyrus* est illis marinus. Sumendum quod certius ponit, hoc est, *Austrum* esse *Ventum Asseclam*, &

20 familiarem earum regionum; nisi fortè ex nomine *Maris Australis*, vel phantasiam, vel modum loquendi corrupit, intelligens *Zephyrum* per *Austrum*, quòd à Mari Australi spiret. At Mare, quod vocant, *Australe* propriè *Australe* non est, sed tanquam Oceanus secundus Occidentalis; quando simili, cum *Atlantico*, situ exporrigatur.

25 **10.** Marini *Venti* sunt proculdubiò terrestribus humidiores, sed

[F5ᵛ] tamen puriores, quíque faci|liùs, & æqualiùs, cum Aere puro incorporentur. Terrestres enim malè coagmentati, & fumei. Neque opponat quispiam, eos debere esse, propter salsuginem Maris, crassiores. Natura enim terrestris salis, non surgit in vaporibus.

30 **11.** Tepidi, vel gelidi sunt *Venti* marini, pro ratione qualitatum duarum prædictarum, humiditatis, & puritatis. Humiditate enim frigora mitigant (Siccitas siquidem vtrumque, & calorem, & frigus, intendit); at puritate refrigerant. Itaque extra Tropicos, tepidi, intra, gelidi.

35 **12.** Arbitror vbique *ventos* marinos, Asseclas esse regionum (præ-

[F6ʳ] sertim maritimarum) sin|gularum; frequentiùs scilicet spirare *ventos*, à parte, vbi collocatur mare; propter copiam longè vberiorem Materiæ, ad *ventos* in mari, quam in Terrâ; nisi fortè, sit aliquis *ventus* status, spirans

---

1 pomeridianarum.] ~:      32 mitigant] ~,      33 intendit);] ~:)      at] At
35 regionum] ~,

6. West winds prevail during the afternoon; for when the Sun sinks winds blow more often from the west and less so from the east.

7. The south wind prevails by night for it rises more often at night and blows more violently, whereas the north wind blows by day.

8. The differences between the prevailing winds of land and of sea are many and great. The main one is that which gave Columbus the chance to discover the New World, namely that sea winds are not recurrent as land winds <sup>|</sup> mostly are. For since the sea abounds with vapours, which are present everywhere and almost without distinction, winds are also generated everywhere, and since they have no fixed origins and sources, they are carried hither and thither with great fickleness. But the land has very little equality when it comes to matter for winds, for some places are better at generating and increasing winds, while others lack that ability. Thus they generally blow from their nurseries, and acquire their direction thence.

9. Acosta is at odds with himself when he says that southerlies blow towards Peru and on the coasts of the South Sea for the best part of the year; but elsewhere that sea winds generally blow towards those <sup>|</sup> coasts. But in those places the south wind is a land wind as are the north and east winds, with only the west wind being a sea wind. We should take up what he has laid down with more certainty, which is that the south wind is the prevailing one, and customary in those regions, unless perhaps from the name of South Sea his imagination or way of speaking misled him so that he took the west for the south wind because it blew from the South Sea. For the sea which they call South is not really southerly but like a second western ocean, for it spreads in the same direction as the Atlantic.

10. Sea winds are doubtless moister than land ones but yet purer, and more <sup>|</sup> easily and evenly embodied in pure air. For land winds are poorly combined and smoky. And let no one object that sea winds ought to be grosser on account of the saltiness of the sea. For the earthy nature of salt does not rise up in vapours.

11. Sea winds are warm or cold in proportion to those two qualities just mentioned, i.e. moistness and purity. For cold is moderated by moistness (for dryness intensifies both heat and cold), but increased by purity. Therefore the ones outside the tropics are warm, the ones within cold.

12. In my judgement sea winds are in each and every region (especially maritime ones) the prevailing ones; <sup>|</sup> for winds blow more often from where the sea lies because of the far greater supply of matter for

à terrâ, ex causâ peculiari. Nemo autem confundat *ventos* statos, cum *ventis* Asseclis, cum Asseclæ semper frequentiores sint; stati, sæpiùs rariores. Id tamen vtrisq*ue* commune est, quod *venti* spirent à parte fomitum suorum.

5   **13.** Vehementiores plerunq*ue* sunt *venti* marini, quàm terrestres, ita tamen, ut cum cessent, maior sit Malacia, in medio Mari, quàm ad

[F6ᵛ] litora; adeo vt Nautæ quandoq*ue* ament, potius li|torum obliquitates premere, quàm vrgere altum; ad euitandas Malacias.

  **14.** Spirant à Mari ad litora *Venti Tropæi* siue versarij, qui scilicet

10 postquam paulisper progressi sunt, subitò vertuntur. Omnino est quædam refractio inter Auras maris, & Auras terræ, & inæqualitas. Omnis autem inæqualitas Aeris, est inchoatio quædam *Venti.* Maximè autem fiunt *Tropæi, & Euripi Ventorum,* vbi mare sinuat.

  **15.** Spirant quædam Auræ plerunque circa omnes Aquas maiores;

15 potissimùm autem sentiuntur manè; at magis circa fluuios, quàm in

[F7ʳ] mari, propter dif|ferentiam auræ Terræ, & auræ Aquæ.

  **16.** In locis proximis iuxta mare, flectunt fere se arbores, & incuruant, quasi auersantes auras maris. Neque tamen malicia est; sed *Venti* maritimi, ob humiditatem, & densitatem, sunt tanquam ponderosiores.

20            Qualitates & Potestates Ventorum.

*Ad Artic.* 7. 28. 29. 30. 31. ***Connexio.*** *Circa* Qualitates *&* Potest-

[F7ᵛ] ates Ventorum, *obseruatum est ab hominibus, non diligentèr, &* | *variè. Nos certiora excerpimus, reliqua, vt leuia, ipsis* Ventis *permittemus.*

25   **1.** *Auster* Pluuiosus, *Boreas* Serenus apud nos sunt: Alter nubes congregat, & fouet; Alter dissipat, & discutit. Itaque *Poetæ,* cùm narrant de *Diluuio,* fingunt, eo tempore, *Boream* in carcere conclusum; *Austrum* cum Amplissimis mandatis emissum.

  **2.** *Zephyrus* apud nos pro *Aureæ Ætatis Vento* habitus est, qui comes

30 esset perpetui veris, & mulceret flores.

  **3.** *Paracelsi* Schola, cùm tribus suis principijs, etiam in *Templo*

[F8ʳ] *Iunonis* (Aere scilicet) locum | quærerent; tres collocarunt, *Euro* locum non repererunt.

---

21 7] / immediately after this *SEH* (II, p. 33) adds 27. This is not warranted by *c-t*

26 fouet;] ~,

winds out to sea than on land, unless perhaps there is some recurrent wind, brought on by an exceptional cause, blowing from the land. But no one should confuse recurrent with prevailing winds, since prevailing winds are always more frequent, recurrent ones always rarer. Yet the two have this much in common, that they blow from the direction of their nurseries.

13. Sea winds are generally more violent than land ones, but nevertheless such that when they stop the calm is greater out to sea than inshore, so that sailors like rather to hug ᐧ the ins and outs of the coast to avoid calms than to take to the high seas and risk them.

14. There blow from sea to shore changeable winds or ones which turn about, i.e. ones which once they have come on a little suddenly turn tail. This is altogether a result of a certain refraction or imbalance between the sea breezes and the ones from the land. In fact all inequality of air is an embryonic wind. These changeable and straitened winds happen mostly where the sea curves in and out.

15. Certain breezes generally blow about all major bodies of water; but more about rivers than seas on account of the ᐧ differences between breezes from land and water.

16. In places very close by the sea, the trees bend and curve as if avoiding the sea breezes. Nevertheless this is not harmful, it is just that their moistness and density make them as it were heavier.

## The Powers and Qualities of Winds

*To Articles* 27, 28, 29, 30, 31. **Connection.** Men have not investigated the qualities and powers of winds carefully, and ᐧ in different ways. I pick out the more certain instances; the rest, which are light, I shall leave to the winds themselves.

1. Here with us, the south wind is rainy, and the north clear. The former gathers and cherishes clouds, the latter disperses and breaks them up. Thus the poets, when talking of the deluge, pretend that the north wind lay in prison at that time, while the south wind was let out with plenary powers.

2. Here with us the west wind is taken for the Golden Age, the companion of endless spring, and fosterer of flowers.

3. The school of Paracelsus, with its three principles, sought a place for them, even in the Temple of Juno (i.e. in the air), ᐧ lodged them with three winds but left the east wind out:

*Tincturis liquidum qui* Mercurialibus *Austrum.*
*Diuitis & Zephyri rorantes* Sulphure *venas,*
*Et Boream tristi rigidum* sale.

4. At nobis, in *Britanniâ, Eurus* pro Malefico habetur, vt in prouer-
5 bio sit, *Eurum,* neque homini, neque bestiæ, propitium esse.

5. *Auster* à præsentiâ Solis, *Boreas* ab absentiâ spirat, in hemisphærio
nostro: *Eurus* in Consecutione Motûs Aeris, *Zephyrus* in contrarium
vbique. *Zephyrus* à Mari, *Eurus* à Continente; plerunque in *Europâ,* &
[F8ᵛ] *Asiâ* Occidentali. Hæ sunt differentiæ *Ventorum,* maximè radicales, |
10 vnde plurimæ ex qualitatibus, & potestatibus *Ventorum,* reuerâ
pendent.

6. *Auster* minùs Anniuersarius est, & status, quàm *Boreas,* sed magis
vagus, & liber; & quando est status, tam lenis est, vt vix percipiatur.

7. *Auster* magis humilis est, & lateralis, *Boreas* celsior, & spirans ex
15 alto; neque hoc de Eleuatione, & depressione polari dicimus, de qua
supra, sed quod origines suas habeat plerunque magis in vicino *Auster,*
magis in sublimi *Boreas.*

8. *Auster* nobis pluuiosus (vt iam dictum est) *Africæ* vero serenus, sed
[G1ʳ] magnos immittens feruores, non frigidus (vt alij | dixerunt). Est tamen
20 *Africæ* satis salubris; At nobis, si flauerit paulò diutiùs in sudo, absque
pluuiâ *Auster,* valdè pestilens est.

9. *Auster* & *Zephyrus* non generant vapores, sed spirant à partibus,
vbi maxima est copia ipsorum, propter auctum calorem Solis, qui
vapores elicit, ideóque sunt pluuiosi. Quod si spirauerint à locis
25 siccioribus, & ieiunis à vaporibus, sunt sereni; sed tamen aliquando
puri, aliquando æstuosi.

10. Videntur hic apud nos *Auster* & *Zephyrus* fœderati, suntque
tepidi & humidi; at ex alterâ parte Affines sunt *Boreas* & *Eurus,* suntque
29 frigidi, & sicci.

[G1ᵛ]  11. *Auster* & *Boreas* (quod & an|teà attigimus) frequentiùs spirant,
quàm *Eurus,* & *Zephyrus;* quia magna est inæqualitas vaporum, ex illis
partibus, propter absentiam, & præsentiam Solis; At Orienti, &
Occidenti, Sol tanquam adiaphorus est.

12. *Auster* Saluberrimus marinus, à continente magis morbidus;
35 contra *Boreas* à mari suspectus, à terrâ sanus; Etiam frugibus & stirpibus
*Auster* marinus valdè benignus, fugans rubigines, & alias pernicies.

---

8 vbique.] ~:      Continente;] ~:      9 Occidentali.] ~:      19 frigidus] ~, /(BNF
copy with comma; Pierpont Morgan copy with semicolon)      dixerunt.] ~.)

> The clear south wind streaked with mercurial tints,
> Veins of the rich west wind, moist with sulphur,
> And the wind from the bitter north, stiff with salt.

4. But for us in Britain, the east wind is seen as noxious, so that there is a proverb: When the wind is in the east, it bodes no good to man or beast.

5. In our hemisphere, the south wind blows from where the Sun is; the north wind from where it is not. The east wind everywhere follows the motion of the air, whereas the west wind opposes it. On the whole in Europe and Western Asia, the west wind blows from the sea, the east wind from the land mass. And these are the most fundamental differences between winds, ˡ and from them do many of the qualities and powers of the winds really follow.

6. The south wind is less of an anniversary or recurrent wind than the north, but more wandering and free; and when it is recurrent it is so soft that it can scarcely be felt.

7. The south wind blows low and more laterally, the north blows high, and more from above; and this is not to do with the polar elevation and depression mentioned above, but because the south wind has its origins nearer to the ground, while the north wind has its higher up.

8. As I said, with us the south wind is wet, but in Africa clear and brings great heatwaves, and (as others ˡ have said) it is not cold. In Africa it is nevertheless healthy, but with us, if it blows for a long while and stays rain-free, it is very pestilential.

9. The south and west winds do not generate vapours, but blow from places where the vapours' abundance is greatest by reason of the Sun's greater heat which draws them, and so these winds are rainy. But if they were to blow from dry places where vapours are scarce, they are clear, but are nevertheless fresh at times and sultry at others.

10. Here with us it seems that the south and west winds are allied, and are warm and moist, while on the other hand, the north and east winds are akin and are cold and dry.

11. The north and south winds (as I intimated ˡ earlier) blow more often than the east and west winds, because in those places the variability of the vapours is great on account of the comings and goings of the Sun. But in the east and west the Sun is more neutral so to speak.

12. The south wind from the sea is good for you, from the land mass bad; on the other hand, a northerly off the sea cannot be trusted, as it can from the land. A southerly off the sea is also very good for fruits and plants, driving off blights and other injuries.

**13.** *Auster* lenior non admodùm cogit nubes, sed sæpè serenus est, præsertìm si sit breuior; Sed flans commotiùs, aut diutiùs, facit cœlu*m* [G2ʳ] nubilum, & inducit pluuia*m*; sed potiùs cum desinat, | aut flaccescere incipiat, quam à principio, aut in ipso vigore.

5    **14.** Cum *Auster* aut oritur, aut desistit, fiunt ferè Mutationes tempestatum, à sereno ad nubilum, aut à calido ad frigidum, & è contrà; *Boreas* sæpè, & oritur, & desinit, priore tempestate manente, & continuatâ.

**15.** Post pruinas, atq*ue* etiam Niues paulò diuturniores, non alius ferè *Ventus*, quàm *Auster* spirat, tanquam factâ concoctione frigorum, quæ 10 tum demùm soluuntur, neque proptereà semper sequitur pluuia, sed fit hoc etiam in regelationibus serenis.

**16.** *Auster*, & frequentiùs oritur, & fortiùs spirat, noctù, quàm [G2ᵛ] interdiù, præsertìm noctibus hyᴵbernis. At *Boreas*, si noctù oriatur (quod contra suam consuetudinem est), non vltrà triduum ferè durat.

15    **17.** *Austro* flante, maiores voluuntur fluctus, quàm *Boreâ*, etiam quando pari, aut minore impetu, spirat.

**18.** Spirante *Austro*, fit Mare cœruleum, & magis lucidum; *Boreâ* contrà, atrius, & obscurius.

**19.** Cum Aer subitò fit tepidior, denotat interdum pluuiam; rursus 20 aliàs, cum Aura subitò fit gelidior, pluuiam præmonstrat. Sequitur vero hoc natura*m* *Ventoru*m; nam si flante *Austro* aut *Euro* intepescit Aer, pluuia in propinquo est, itidemq*ue* cùm flante *Septentrione*, aut *Zephyro*, [G3ʳ] refrigescit. |

**20.** *Auster* flat plerunque integer, & solitarius; At *Boreâ*, & præcipuè 25 *Cæciâ*, & *Coro* flantibus, sæpè contrarij, & alij diuersi *Venti*, simul spirant; vndè refringuntur, & turbantur.

**21.** *Boreas* sementi faciendæ, *Auster* Insitionibus, & Inoculationibus, cauendus.

**22.** A parte *Austri*, folia ex arboribus, citiùs decidunt; At palmites 30 vitium, ab eâ parte erumpunt, & eò ferè spectant.

**23.** In latis pascuis, videndum est pastoribus (vt ait *Plinius*) vt greges ouium, ad *Septentrionale* latus adducant, vt contra *Austrum* pascant. Nam si contra *Boream*, claudicant, & lippiunt, & aluo mouentur; Quin [G3ᵛ] etiam *Boreas* | coitum illis debilitat adèo vt si in hunc *Ventu*m spectantes 35 coeant oues, fœmellæ vt plurimu*m* gignantur. Sed in hoc *Plinius* (vtpote transcriptor) sibi non constat.

---

3 pluuia*m*,] ~:   flaccescere] flaccessere / silently emended thus in *SEH* (II, p. 34)
14 est,] ~,)   19 pluuiam;] ~:   21 *Ventoru*m;] ~:   33 mouentur;] ~:
35 gignantur.] ~:

13. A gentle southerly does not gather clouds much but is often clear, especially if it is brief. But, blowing longer and more roughly, it makes the sky cloudy, and brings on rain, though rather when the wind stops <sup>|</sup> or begins to slacken, than when it starts or blows strong.

14. When the southerly is getting up or dying down, then the weather begins to change: from clear to cloudy, or from hot to cold. On the other hand, the northerly often gets up and dies away, with the weather staying as it was.

15. After frosts and long-lasting snows, the southerly is about the only one that blows, as if the cold stuff had been worked up, and then dissolved, not that that is always followed by rain, for that also happens in calm thaws.

16. The south wind blows more often and strongly by night than by day, especially on winter <sup>|</sup> nights. But the north wind, if it departs from its usual behaviour, and gets up at night, hardly lasts longer than three days.

17. When the south wind blows, greater waves are whipped up than when the north does, even though the former blows with equal or less strength.

18. When the south wind blows, the sea becomes blue and brighter, but the north wind makes it blacker and duller.

19. When the air suddenly grows warmer, that sometimes betokens rain; but again the same thing is presaged by a sudden cooling. But this follows from the nature of the winds concerned, for if the air grows warmer with a southerly or easterly wind rain is imminent, as it is when the air is cooled by a northerly or westerly. <sup>|</sup>

20. The south wind generally blows entire and by itself. But north winds, especially when Cæcias and Corus are blowing, encounter contrary and other kinds of winds, and that breaks them up and makes them turbulent.

21. Beware of sowing seed in a north wind, and grafting in a southerly.

22. Leaves fall more rapidly from the south side of the tree; but vine shoots burst on that side, and practically bend towards that place.

23. In broad pastures, shepherds should see to it (as Pliny says) that they lead their flocks to the north side, so they can graze facing south; for if they face north they limp, and get bad eyes and diarrhoea; and if they face north <sup>|</sup> their sexual powers decline such that if they have the north wind in their eyes, they generally produce ewe lambs. But in this matter Pliny (as a transcriber) is inconsistent.

**24.** *Venti* tribus temporibus, frumento & segetibus nocent; in flore aperiente, & deflorescente, & sub maturitatem; tum enim exinaniunt aristas deiectis granis, at prioribus duobus temporibus, florem, aut in calamo constringunt, aut decutiunt.

5   **25.** Flante *Austro* anhelitus hominum magis fœtet, Appetitus animalium deijcitur magis, morbi pestilentes grassantur, grauedines incumbunt, homines magis pigri sunt, & hebetes; At flante *Boreâ*, magis [G4ᵛ] alacres, | sani, auidiores cibi. Phthisicis tamen nocet *Boreas*, & tussiculosis, et podagricis, & omni fluxui acuto.

10   **26.** *Eurus* Siccus, mordax, mortificans; *Zephyrus* humidus, clemens, Almus.

**27.** *Eurus* spirans vere adulto, calamitas fructuu*m*, inducendo Erucas & vermes, vt vix foliis parcatur; nec æquus admodu*m* segetib*us*. *Zephyrus* contrà, herbis, floribus, & omni vegetabili, maximè propitius, 15 & amicus. At *Eurus* quoque, circa æquinoctiu*m* autumnale, satis gratiosus.

**28.** *Venti* ab Occidente spirantes, sunt vehementiores, quàm illi ab [G4ᵛ] Oriente, & magis curuant, & contorquent arbores. |

**29.** Tempestas pluuiosa quæ incipit spirante *Euro*, longiùs durat, 20 quàm quæ spirante *Zephyro*, & ferè ad diem integrum extenditur.

**30.** *Eurus* ipse, & *Boreas*, postquam inceperint flare, constantiùs flant; Auster, & Zephyrus magis mutabiles.

**31.** Flante *Euro* visibilia omnia maiora apparent; At flante *Zephyro* Audibilia; Etiam longiùs deferuntur Soni.

25   **32.** *Cœciam nubes ad se trahere*, apud *Græcos* in prouerbium transijt, comparando ei Fœneratores, qui pecunias erogando, sorbent; vehemens est *Ventus*, & latus, vt non possit summouere nubes, tam citò, quàm illæ [G5ʳ] re|nitantur, & se vertant; quod fit etiam in maioribus incendijs, quæ contra *Ventum* inualescunt.

30   **33.** *Venti Cardinales*, aut etiam *Semicardinales*, non sunt tam procellosi, quàm *Mediani*.

**34.** *Mediani*, à *Boreâ* ad *Euro-boream*, magis sereni, ab *Euro-boreâ* ad *Eurum*, magis procellosi; Similitèr ab *Euro* ad *Euro-austrum*, magis sereni, ab *Euro-austro* ad *Austrum*, magis procellosi; Similitèr, ab *Austro*, 35 ad *Zephyro-austrum*, magis sereni, à *Zephyro-austro* ad *Zephyrum*, magis procellosi; Similitèr, à *Zephyro* ad *Zephyro-boream*, magis sereni, à

---

1 nocent;] ~:    2 maturitatem;] ~:    7 hebetes;] ~:    13 segetib*us*.] ~:
22 flant;] ~:    24 Audibilia;] ~:    25 *Cœciam*] / *nld* as Cæciam in *SEH* (II, p. 36)
26 sorbent;] ~:    33 procellosi;] ~:    34 procellosi;] ~:    35 *Zephyro-austrum*]
/ *c-t* lacks hyphen    36 procellosi;] ~:

24. Winds are harmful to corn and grain crops at three times of the year: when they blossom, when they blow, and before ripening. In the first two cases they empty the ears and scatter the grain; in the last case they either stifle the flower in the stalk or cast it down.

25. When the south wind blows, people's breath stinks more, animals' appetites are worse, pestilent diseases run riot, catarrhs take hold, and men become sluggish and feeble. But when the north wind blows, they get more active, |healthier, and hungrier. Yet the northerly is bad for consumptions, niggling coughs, gouts, and all acute fluxes.

26. The east wind is dry, biting, and fatal; the west wind moist, gentle, and life-giving.

27. The east wind in season is disastrous for fruits, for it induces caterpillars and worms which consume almost all the leaves, and is very unfavourable to crops. The west wind by contrast is very well disposed and friendly to herbaceous plants, flowers, and every vegetable thing. But around the autumnal equinox, the east wind is agreeable enough.

28. Winds blowing from the west are more violent than ones from the east, and they bend and lash the trees more. |

29. Wet weather which starts while an easterly is blowing lasts longer than when a westerly blows, and it goes on for almost the whole day.

30. The east wind and the north, once they have got going, blow more steadily; the south and west winds are more changeable.

31. When the east wind is blowing all things visible seem larger; when the west wind blows sounds seem longer and are carried further.

32. Among the Greeks it became proverbial that *Cæcias attracts clouds*, as moneylenders suck in money by disbursing it. It is a violent wind, and wide so that it cannot shift the clouds as quickly as they |turn back and resist it, a thing which also happens in big blazes which rise up against the wind.

33. Cardinal or even semi-cardinal winds are not so stormy as median winds.

34. Median winds from the north towards the north-east are calmer; from the north-east towards the east more stormy. Likewise, from the east to the north-east they are calmer; from the south-east to the south more stormy. Likewise, from the south to the south-west they are calmer; from the south-west to the west stormier. Likewise, from the west to the north-west they are calmer; from north-west to the north

[G5ᵛ] *Zephyro-boreâ* ad *Boream,* magis procellosi; Ita vt ¦ progrediendo secundum ordinem cœli, semper *Mediani* prioris Semicardinis, disponantur ad Serenitatem; posterioris, ad Tempestates.

**35.** Tonitrua, & fulgura, & Ecnephiæ fiunt, spirantibus *Ventis* fri-
5 gidis, quíque participant ex *Boreâ,* quales sunt *Corus, Thrascias, Circias, Meses, Cœcias;* Ideoque fulgura sæpiùs comitatur grando.

**36.** Etiam Niuales *Venti* à *Septentrione* veniunt, sed ab ijs *Medianis,* qui non sunt procellosi, veluti *Corus,* & *Meses.*

9 **37.** Omnino *Venti* quinque modis naturas suas & proprietates
[G6ʳ] nanciscuntur: vel ab absentiâ aut præsentiâ Solis; vel à con¦sensu & dissensu cum naturali motu Aeris; vel à diuersitate Materiæ fomitum suorum à quibus generantur, Maris, Niuis, paludum, &c.; vel à tincturâ regionum per quas pertranseunt; vel ab originibus localibus suis, in alto, sub terrâ, in medio; quæ omnia sequentes articuli meliùs explanabunt.

15 **38.** *Venti* omnes habent potestatem desiccandi, etiam magis quam ipse Sol; quia sol vapores elicit sed nisi admodum feruens fuerit, non dissipat; at *ventus* eos & elicit, & abducit. Attamen *Auster* minimè omnium hoc facit; quin etiam saxa, & trabes sudant, magis flante non-
[G6ᵛ] nihil *Austro,* quàm in tranquillo. ¦

20 **39.** Martij magis longè desiccant, quàm æstiui: adeo vt Artifices Instrumentoru*m* Musicoru*m Ventos* Martios expectent, ad Materiam Instrumentorum suorum desiccandam, eamque reddendam porosam, & sonoram.

**40.** *Venti* omnis generis purgant Aerem, eumque à putredine vindi-
25 cant, vt Anni in quibus *venti* frequentiùs spirent, sint maximè salubres.

**41.** Sol Principum fortunam subit; quibuscum ita sæpè agitur, vt præsides in Prouincijs remotis, magis obnoxios habeant subditos, & quibus obsequia præstentur magis, quàm principi ipsi. Certè *Venti* qui
[G7ʳ] pote¦statem, & originem habent à Sole, æquè, aut plus gubernant tem-
30 peraturas regionum, & affectus Aeris, quàm ipse Sol, in tantùm, vt *Peruuia* (quæ propter propinquitatem Oceani, vastitatem Amnium, & altissimos, & maximos montes niuales, maximam habet copiam *Ventorum,* & aurarum spirantium) cum *Europâ,* de temperamento, & clementiâ Aeris certet.

35 **42.** Nil mirum si *Ventorum* tantus sit impetus, quantus inuenitur, quandoquidem *Venti* vehementes, sint tanquam Inundationes, atque Torrentes, & Fluctus magni Aeris. Neq*ue* tamen, si attentiùs aduertas,

---

3 Serenitatem;] ~,    10 nanciscuntur:] ~.    vel] Vel    Solis;] ~:    11 Aeris;] ~:
12 generantur,] ~;    &c.;] ~ ‸    13 pertranseunt;] ~:    14 medio;] ~:
17 abducit.] ~:

stormier. Thus if we go ˡ round from east to west, median winds of the
first semi-cardinal incline to calm; winds of the last to storms.

**35.** Thunder, lightning, and tornadoes occur when cold winds from
the north blow, such as are the Corus, Thrascias, Circias, Meses, and
Cœcias; and so thunder often goes with hail.

**36.** Snowy winds also come from the north, but from Median ones,
such as Corus and Meses, which are not stormy.

**37.** The winds in general acquire their natures and properties in five
ways: from the Sun's absence or presence; or from their consent with ˡ
and dissent from the natural motion of the air; or from differences of
the material of the nurseries from which they are generated—sea, snow,
swamp, and so on; or from the colouring of the countries through
which they pass, or from the places from which they originate—from
beneath the ground, in the middle air; all of which will be more fully
explained in the articles following.

**38.** All winds have the power of drying, even more so than the Sun
itself; for the Sun draws out vapours but, unless it is extremely hot, it
does not dissipate them; the wind on the other hand draws them off
and carries them away. However, the south wind does this least of all,
and stones and beams sweat rather more when a southerly is blowing
than in calm weather. ˡ

**39.** March winds are much more drying than summer ones, so that
makers of musical instruments wait for the March winds to dry their
products and make them porous and tuneful.

**40.** Winds of all kinds clear the air and rescue it from putrefaction,
so that years when winds blow more often are the healthiest.

**41.** The Sun enjoys the fortune of princes, in that it often happens
that in distant provinces their deputies have more respect and deference
paid to them by their subjects than is paid to the prince himself. Certainly winds which ˡ take their power and origin from the Sun have
more authority over the climates of countries and states of the air than
the Sun itself, and to such a degree that Peru (which, by reason of its
closeness to the ocean, the colossal size of its rivers, the enormous height
of its snowy mountains, has an abundant supply of winds and breezes)
may rival Europe for the mildness and equability of its air.

**42.** It is no wonder that the winds have so much force seeing that
strong winds are like floods, torrents, or great tidal waves of air. Nevertheless, if you think about it more thoroughly, their power is in fact not
that great. They can ˡ knock trees down whose crowns are spread out
like sails to help topple them with their own weight. They can do the

[G7ᵛ] magnu*m* quiddam est eoru*m* potentia. Possunt | deijcere arbores, quæ cacuminum onere, tanquam velis expansis iisdem commoditatem præbent, & se ipsæ onerant; possunt etiam Ædificia infirmiora; sed structuras solidiores, nisi fiant cum terræ motibus, non subuertunt.

5 Niues quandóque tanquam integras deijciunt ex Montibus, vt planitiem subiacentem ferè sepeliant, quod accidit *Solymanno* in campis *Sultaniæ*; Etiam magnas quandóque immittunt Inundationes aquarum.

**43.** Amnes quandóque tanquam in sicco ponunt *Venti*, & fundos ipsorum discooperiunt; Si enim post magnam siccitatem, *Ventus* [G8ʳ] robustus in consecutione | fili aquæ, pluribus dieb*us* spirârit, ita vt aquas 11 amnis, tanqua*m* euerrendo, deuexerit in mare, Aquas marinas prohibuerit; fit Siccatio Amnis in multis locis insolitis.

*Monitum.* Verte Polos, *& verte simul Obseruationes, quâtenus ad* Austrum *& Boream. Cum enim Absentia, & Præsentia Solis in causâ sit,* 15 *variat pro ratione polorum. At illud, constans res esse possit, quod plus sit* Maris versus Austrum, *plus sit* Terræ versus Boream, *quod etiam ad* Ventos *non parum facit.*

*Monitum. Mille modis fiunt* Venti, *vt ex Inquisitione sequenti patebit. Itaque in re tam variâ, figere obseruationes, haud facile est. Attamen quæ à* 20 *nobis posita sunt, pro certo plerumque obtinent.*

[G8ᵛ] <sup> </sup>| Origines locales Ventorum.

*Ad Artic.* **8.** *Connexio.* Ventorum Origines locales *nosse, arduæ est Inquisitionis, cum illud* Vnde, *& Quò* Ventorum, *vt res abdita, etiam in* Scripturis *notata sit. Neque loquimur iam de Fontibus* 25 Ventorum *particularium (de quibus posteà), sed de Matricibus* Ventorum *in genere. Alij ex alto eas petunt, Alij in profundo riman-* [H1ʳ] *tur, in medio autem vbi* | *vt plurimum generantur, vix eas quærunt; vt est mos hominum, quæ ante pedes posita sunt, præterire, & obscuriora malle. Illud liquet,* Ventos *aut Indigenas, aut Aduenas esse;* 30 *sunt enim* Venti *tanquam* Mercatores Vaporum, *eosque in nubes collectos, & important in regiones, & exportant, vnde iterum* Venti, *tanquam per permutatione*m: *Sed inquiramus iam de Natiuis. Qui* [H1ᵛ] *enim aliundè* Aduenæ, *alibi* Natiui. Tres | *igitur* Origines locales: *Aut expirant & scaturiunt è terrâ; aut deijciuntur ex sublimi; aut* 35 *conflantur hic in Corpore Aeris. Qui autem deijciuntur ex alto,*

---

18 *patebit.*] ~: 20 *plerumque*] / *nld* as plerunque in *SEH* (II, p. 38)
25 *particularium*] ~, *posteà*),] ~) ∧

same damage to unsound houses, but they leave solid structures unscathed unless they are accompanied by earthquakes. Sometimes they cast down whole slews of snow from the mountains, so as to pretty well bury the plains below, a fate which overtook Solyman on the plains of Sultania. Sometimes they also let fly floods of water.

43. Winds sometimes dry up great rivers and expose their beds; for if after a great drought a strong wind blows ' for a number of days downriver such that the waters are swept off into the sea and the sea water is held back, the river dries up in unaccustomed places.

*Advice.* If you switch poles, you must also switch observations, as far as north and south are concerned, for when the Sun's absence or presence is at issue, that varies with the poles. But this can be taken as a constant, i.e. that there is more sea to the south and more land to the north, which affects the wind not a little.

*Advice.* Winds happen in a thousand ways, as the following investigation will make plain. Thus it is not easy to establish observations in so slippery a subject. Yet those that I have set down may be in general taken for certain.

## ' The Local Origins of Winds

*To Article* 8. **Connection.** Inquiring into the places winds come from is hard work, since, even as it is noted in Scripture, whence the wind comes and whither it goes is a mystery. But I am not speaking here of the sources of particular winds (more on these later), but of the wombs of winds in general. Some seek for these on high; others search for them in the depths; but few look for them in the middle region where ' most of them are generated. For as people usually do, they ignore what is in front of their noses, and prefer what lies in darkness. Now this is clear, that winds are either native or foreign, for they are in effect merchants of vapours which, gathered into clouds, they import or export from place to place, whence they get winds by way of exchange. But let me now investigate native winds, for those which are foreign in one place are native in another. The winds ' originate from three kinds of place: either they are exhaled or spring from the earth, or they are thrown down from above, or they are hatched in the body of the air. Those thrown down are generated in two ways: for they are either thrown down before they are

*duplicis generationis: Aut enim deijciuntur, antequam formentur in nubes, aut postea ex nubibus rarefactis, & dissipatis. Videamus quæ sit harum rerum historia.*

4
[H2ʳ]
**1.** Finxerunt *Poetæ, Regnum Æoli,* in Antris & Cauernis, sub terram fuisse collocatum, vbi *Carcer* esset *Ventorum,* qui subindè emittebantur. |

**2.** Etiam *Theologos* quosdam, eosdemque *Philosophos,* mouent *Scripturæ* verba: *Qui producit Ventos de thesauris suis;* tanquam *Venti* prodirent ex locis thesaurarijs, subterraneis scilicet, vbi sunt Mineræ: sed hoc nihil est. Nam loquitur etiam *Scriptura,* de Thesauris Niuis, &
10 grandinis, quas in sublimi generari, nemo dubitat.

**3.** In subterraneis proculdubiò magna existit Aeris Copia, eamque & expirare sensìm verisimile, & emitti confertìm aliquando, vrgentibus causis, necesse est.

**Phænomenon obliquum.** *In magnis siccitatibus, & mediâ æstate, cum*
15 *magis rimosa sit terra, solet erumpere in locis aridis, & arenosis, magna vis*
[H2ᵛ] *aquarum.* | *Quod si faciant Aquæ* (*Corpus Crassum*) *rarò; Aerem* (*Corpus tenue, & subtile*) *hoc frequentèr facere, probabile est.*

**4.** Si expirat Aer è terrâ, sensìm & sparsìm; parùm percipitur primò; sed postquam Aeris illius emanationes multæ minutæ confluxerint,
20 tum fit *Ventus;* vt ex scaturiginibus Aquarum Riuus. Hoc vero ita fieri videtur; quoniam notatum est ab Antiquis, *Ventos* complures in ortu suo, & in locis à quibus oriuntur, primò spirare exiguos, deinde in progressu inualescere prorsùs, more Fluuiorum.

24
**5.** Inueniuntur quædam loca in Mari, ac etiam lacus, qui nullis
[H3ʳ] flantibus ventis, maiorem in | modum tumescunt, vt hoc à subterraneo flatu fieri appareat.

**6.** Magna vis requiritur spiritûs subterranei, vt Terra concutiatur, aut scindatur; leuior, vt Aqua subleuetur. Itaque tremores terræ rari; tumores & subleuationes Aquarum, frequentiores.

30
**7.** Etiam vbique notatum est, nonnihil attolli, & tumescere Aquas, ante tempestates.

**8.** Spiritus subterraneus exilis qui sparsìm efflatur, non percipitur super terram, donec coierit in *Ventum,* ob porositatem terræ; sed exiens subter aquas, ob continuitatem aquæ, statìm percipitur ex tumore
35 nonnullo.

[H3ᵛ]
**9.** *Asseclas* esse *Ventos,* terrarum | cauernosarum anteà posuimus; vt prorsus videantur Venti illi, habere *origines* suas *locales* è terrâ.

7 verba:] ~.     15 *terra,*] ~;     16 *aquarum.*] ~:     20 Riuus.] ~:
28 scindatur;] ~:

formed into clouds, or afterwards when they have been rarefied or dissipated from clouds. Let us now see what their history is.

1. The poets have pretended that the kingdom of Æolus was located below ground in vaults and caverns where the winds' prison was situated, whence they were now and again discharged. [|]

2. Scripture too prompted some theologians who were also philosophers with these words: *Who brings forth winds out of his treasures*, which they took to mean that winds came from underground treasuries where the rich lodes lie. But this makes no sense; for Scripture also speaks of the treasures of snow and hail, which no one doubts are generated on high.

3. No doubt an abundance of air exists underground, which probably exhales gradually, but pressing causes sometimes make it go out in a body.

*An indirect Phenomenon.* In great droughts and in midsummer, when the ground has more cracks, large volumes of water can burst out in dry and sandy places. [|] But if water (a gross body) seldom does this, it is likely that air (a thin and subtle body) does it often.

4. If air exhales from the earth gradually and dispersedly it can barely be felt at first, but when many of these little streams of air come together, then a wind is produced, just as a river comes from springs. But this in fact seems to happen, for the ancients observed that many winds in their origins and in the places they spring from are feeble at first but then grow stronger as they go on, just as rivers do.

5. People find certain places in the sea, and in some lakes too, which develop a heavy swell without any winds; [|] and this seems to result from underground air currents.

6. Great pressure of subterranean spirit is needed to shake or split the earth, but a lesser pressure to raise the water. Accordingly earthquakes are rare but swellings and upliftings of the waters are more frequent.

7. It is also observed everywhere, that waters rise and swell somewhat before storms.

8. The thin subterranean spirit which is released dispersedly cannot be felt (as earth is porous) until it is concentrated as wind; but coming out underwater it can be felt at once (as water is continuous) by the swelling.

9. I said earlier that prevailing winds [|] are associated with cavernous territories, so that it certainly seems that they come from places in the ground.

10. In montibus magnis & saxeis, inueniuntur *Venti,* & citiùs spirare (antequam scilicet percipiantur in vallibus) & frequentiùs (cùm scilicet valles sint in tranquillo); At omnes Montes, & Rupes cauernosi sunt.

5  11. In Comitatu *Denbigh* in *Britanniâ,* montosâ regione & lapidosâ, ex cauernis quibusdam, tam vehementes (ait *Gilbertus*) sunt *Ventorum* eruptiones, vt iniecta vestimenta, pannique, rursus magnâ vi efflentur, & altiùs in Aerem efferantur.

[H4$^r$]  12. In *Aber Barry* iuxta *Sabrinam* | in *Walliâ,* in quodam cliuo saxoso, 10 in quo sunt foramina, si quis aurem apposuerit, Sonitus varios, & murmur flatuum sub terrâ exaudiet.

*Phœnomenon obliquum.* Notauit Acosta, *oppida* Platæ *&* Potosæ, *in* Peruuiâ, *non longè esse distantia, & vtrumque situm esse, in terrâ eleuatâ, aut montanâ, vt in hoc non differant; & nihilominus habere* Potosam 15 *temperaturam aeris frigidam, & hyemalem,* Platam *clementem, & vernam; id quod videtur Argenti fodinis, iuxta* Potosam, *attribui posse; quod demonstrat esse spiracula terræ, quatenus ad calidum, & frigidum.*

[H4$^v$]  13. Si terra sit primum frigidum, vt voluit *Parmenides* (non con|-temnendâ vsus sententiâ, cum frigus, & densitas, arcto copulentur vin-20 culo) non minùs probabile est, eijci halitus calidiores, à frigore centrali terræ, quam deijci à frigore Aeris sublimioris.

14. Sunt quidam putei in *Dalmatiâ,* & Regione *Cyrenaicâ,* vt qui-dam ex Antiquis memorant, in quibus si deijciatur lapis, excitantur paulò post tempestates, ac si lapis perfringeret operculum aliquod, in 25 loco, vbi vis *Ventorum* erat incarcerata.

*Phœnomenon obliquum. Flammas euomunt* Ætna, *& complures Mon-tes; similitèr & Aerem erumpere posse consentaneum est, præsertìm calore in* [H5$^r$] *subterraneis dilatatum, & in motu positum.* |

15. In terræ motibus, *Ventos* quosdam noxios, & peregrinos, & ante 30 eruptionem, & postea flare, obseruatum est; vt fumi quidam minores solent emitti, ante, & post Incendia magna.

*Monitum.* Aer in terrâ conclusus, erumpere ob varias causas compellitur. *Quandoque Massa terræ malè coagmentata in Cauum terræ decidit; quandoque Aquæ se ingurgitant; quandoque expanditur Aer per ignes*

---

5 *Denbigh*] / *SEH* (II, p. 40) notes that this should be *Derbiæ* but does not emend accordingly. It is not the duty of an editor always to correct mistakes of this kind

1 spirare] ~,      2 vallibus)] ~,)      frequentiùs] ~,      3 tranquillo);] ~:)
10 apposuerit,] ~;      15 *vernam;*] ~:      17 *calidum*] / thus BNF copy; Pierpont Mor-gan copy has *calidúm*      18 *Parmenides*] ~,      30 est;] ~:      33 *decidit;*] ~:
34 *ingurgitant;*] ~:

10. In great and rocky mountains we find winds which blow more quickly (i.e. before they are felt in the valleys) and more often (i.e. when the valleys are still); but then all mountains and rocks are cavernous.

11. In Denbighshire in Britain, a mountainous and stony place, there are (as Gilbert tells us) such violent eruptions of winds from certain caverns that clothing or rags put down them are blown back up again with great force, and carried high into the air.

12. In Aber Barry on the Severn $^|$ in Wales, there is a rocky cliff with holes which, if you put your ear to it, you will hear various sounds and rumblings of underground blasts.

*An indirect Phenomenon.* Acosta has noted that although the Peruvian towns of Plata and Potosi lie fairly close to each other, and are situated in similar highland or mountainous country, Potosi none the less has a cold and wintry climate, while Plata's is mild and vernal. And this seems to be due to the silver mines near Potosi, which shows that the earth has hot and cold air vents.

13. If the earth is the primal source of cold, as Parmenides held (and it is a doctrine not to be $^|$ scorned since cold and density are closely linked), it is not less likely that warmer exhalations should be thrown up from the central cold of the earth, than that they should be cast down by the cold of the air up above.

14. In Dalmatia and the region of Cyrenaica there are, according to certain of the ancients, certain wells which if you throw a stone into them, tempests are stirred up, as if the stone had broken through some seal in a place where a supply of winds was trapped.

*An indirect Phenomenon.* Ætna and many other mountains spew forth flames; and in line with this air can probably break out in the same way—especially when dilated and set going by underground heat. $^|$

15. In earthquakes certain noxious and alien winds are seen to blow, both before and after the shock; just as certain lesser fumes are usually given off before and after great conflagrations.

*Advice.* Various causes make air shut up in the earth break out. Sometimes a poorly attached mass of earth falls into a hollow; sometimes waters become engulfed; sometimes air is expanded by sub-terranean fires, so that it needs more room; sometimes earth, which was

*subterraneos, vt ampliorem locum quærat; quandoque Terra, quæ anteà solida erat, & concamerata, per ignes, in cineres versa, se amplius sustinere non potest, sed decidit: & complura id genus.*

[H5ᵛ]   Atque de primâ *Origine locali Ventorum*, videlicet è subterra|neis, hæc
5   inquisita sunt; Sequitur origo secunda, ex sublimi; nempè mediâ, quam appellant, regione Aeris.

**Monitum.** *At nemo tam malè, quæ dicta sunt, intelligat, quasi negemus, & reliquos* Ventos, *è terrâ, & mari, per vapores educi; Sed hoc prius genus erat,* Ventorum *qui exeunt è terrâ iam* Venti *formati.*

10   **16.** Increbescere murmur siluarum, antequam manifestò percipiantur *Venti*, notatum est; ex quo conijcitur *Ventum* à superiore loco descendere; quod etiam obseruatur in Montibus (vt dictum est) sed causâ magis ambiguâ, propter caua montium.

[H6ʳ]   **17.** Stellas sagittantes (vt loquimur) & vibratas, sequitur *Ven|tus;*
15   atque etiam ex eâ parte, ex quâ fit iaculatio; ex quo patet, aerem in alto commotum esse, antequàm ille motus perueniat ad nos.

**18.** Apertio cœli, & disgregatio nubium, præmonstrat *Ventos*, antequam flent in terrâ; quod itidem ostendit *Ventos* inchoari in alto.

**19.** Stellæ exiguæ, antequam oriatur *Ventus*, non cernuntur, licet
20   nocte serenâ; cum scilicet (vt videtur) densetur, & fiat minùs diaphanus Aer, propter materiam, quæ posteà soluitur in *Ventos*.

**20.** Circuli apparent circa corpus Lunæ; Sol quandoque Occidens
[H6ᵛ]   conspicitur sanguineus; Luna rubicundior est in ortu | quarto; & complura alia inueniuntur prognostica Ventorum in sublimi (de quibus suo
25   loco dicemus) quæ indicant materiam *Ventorum* ibi inchoari, & præparari.

**21.** In istis Phœnomenis, notabis illam, de quâ diximus, differentiam, de duplici generatione *Ventorum*, in sublimi; nimirum ante congregationem vaporum in Nubem, & post. Nam prognostica
30   Halonum, & colorum Solis, & Lunæ, habent aliquid ex nube; at Iaculatio illa, & Occultatio stellarum exiguarum, fiunt in sereno.

**22.** Cum *Ventus* prodit à nube formatâ, aut totalitèr dissipatur
[H7ʳ]   nubes, & vertitur in *Ventum;* | aut secernitur, partìm in pluuiam, partìm in *Ventum;* aut scinditur, & erumpit *Ventus*, vt in procellâ.

35   **23.** Plurima sunt *Phœnomena obliqua*, vbique in naturâ rerum, de repercussione per frigidum; Itaque cùm constet esse in mediâ regione Aeris, frigora valde intensa, planum fit, vapores maximâ ex parte, ea loca

20 fiat] fit / error noted in *SEH* (II, p. 42 n. 3) but not emended accordingly

1 *quærat;*] ~:    14 sagittantes] ~,    loquimur)] ~,)    24 sublimi] ~:
35 Phœnomena] Phœnomina

once solid and overarching, is reduced to ashes by fires, and can no longer hold itself up; and many other causes of this kind.

So much then for the inquiry concerning the first place whence the winds come, i.e. from beneath ' the earth. Next comes the second place, from on high, i.e. from the middle region (as they call it) of the air.

*Advice.* But let no one think that what I have said amounts to a denial that other winds may also be brought forth by vapours both from earth and sea. But that first kind comprised winds which come out of the earth as winds already formed.

16. People have noticed that the woods begin to rustle before winds are plainly felt; whence they guess that wind comes down from somewhere above; and this is also observed (as I have said) in mountains, but the cause is less clear-cut because of their hollows.

17. Wind follows shooting (as we say) and ' flashing stars, and comes from the same quarter as the shooting; whence it is clear that the air on high is stirred before the motion reaches us.

18. The clearing of the sky and the dispersal of clouds presage winds, before they blow at ground level; which also shows that the winds start up on high.

19. Before a wind gets up small stars cannot be seen, though the night be clear, since the air (as it seems) gets thicker, and less transparent, because of the matter which is afterwards resolved into winds.

20. We find halos around the body of the Moon, a blood-red setting Sun, the Moon getting redder in its fourth ' rising, and many other prognostics on winds high up (more of this later) which show that the matter of winds is set off and prepared up there.

21. In these phenomena you shall see the difference I have spoken of between the two ways in which the generation of winds takes place high up, namely before and after the gathering of vapours in clouds. For the prognostics of halos and of the colours of Sun and Moon have something of the cloud about them, but the shooting or obscuring of the smaller stars takes place in clear weather.

22. When the wind comes from a formate cloud, either the cloud is wholly dissipated, turned into wind, ' or separated partly into rain and partly into wind, or it is torn asunder, and the wind bursts out as in a storm.

23. There are many indirect phenomena everywhere in nature concerning repercussion by cold. Thus, since there are extremely bitter colds in the middle region of the air, it is obvious that for the most part

perfringere non posse, quin aut coagulentur, aut vibrentur; secundum opinionem veterum, in hac parte sanam.

Tertia *Origo localis ventorum*, est eorum, qui hîc in inferiore Aere 4 generantur, quos etiam tumores siue Super-onerationes Aeris appel-[H7ᵛ] lamus. Res maxime | familiaris; & tamen silentio transmissa.

*Commentatio. Horum* Ventorum, *qui conflantur in aere infimo, Generatio abstrusior aliqua res non est, quàm hæc ipsa. Quòd scilicet Aer nouitèr factus ex aquâ, & vaporibus attenuatis, & resolutis, coniunctus cum aere priore, non potest contineri ijsdem, quibus* 10 *anteà, spatijs, sed excrescit, & voluitur, & vlteriora loca occupat.* [H8ʳ] *Huius tamen rei duo sunt Assumpta: Vnum, quod gutta | Aquæ in aerem versa (quicquid de* decimâ proportione Elementorum *fabulentur) centuplò ad minus plus spatij desiderat, quàm prius; Alterum, quod parum Aeris noui, & moti, super-additum Aeri* 15 *veteri, totum concutit, & in Motu ponit; vt videre est, ex pusillo Vento, qui ex follibus, aut rimâ fenestræ efflat, qui tamen totum Aerem in cubiculo in motu ponere possit; vt ex flammis lucernarum* [H8ʳ] *facilè apparet.* |

**24.** Quemadmodum rores, & nebulæ, hîc in aere infimo generantur, 20 nunquam factæ nubes, nec ad mediam regionem penetrantes; eodem modo, & complures *Venti.*

**25.** Aura continua spirat circa maria, & aquas, quæ est *Ventus* pusillus, nouitèr factus.

**26.** Iris, quæ est ex Meteôris quasi humillima, & generatur in proxi-25 mo, quando non conspicitur integra, sed curtata, & quasi frusta eius tantùm in cornibus, soluitur in *Ventos;* æquè ac in pluuiam, & magis.

**27.** Notatum est, esse quosdam *Ventos* in regionibus, quæ di-[Iĩʳ] sterminantur, & separantur per Montes intermedios; qui ex al|terâ parte Montium spirant familiares, ad alteram non perueniunt, ex quo 30 manifestum, eos generari, infra altitudinem ipsorum Montium.

**28.** Infiniti sunt *Venti,* qui spirant diebus serenis, atque etiam in Regionibus, vbi nunquam pluit; qui generantur vbi flant, nec unquam erant nubes, aut in mediam regionem ascenderunt.

*Phænomena obliqua. Quicunque nôrit, quàm facilè vapor soluatur in* 35 *Aerem; & quàm ingens sit copia vaporum; & quantum spatium occupet*

---

11 *Assumpta:*] ~.    24–5 proximo,] ~;

the vapours cannot penetrate those places, but are solidified or forced to fall swiftly; and the view of the ancients was sound in this respect.

The third place that winds come from belongs to those generated in the lower air, which I call swellings or overburdenings of the air. This is a very familiar | thing but nevertheless one passed over in silence.

*Speculation.* The generation of those winds which blow in the lower air is no more mysterious a thing than this: namely, that air, newly made from water and attenuated and resolved vapours, and joined with the air already there, cannot be kept in the same bounds as before, but swelling and turning outwards, fills a larger volume. Yet in this connection we assume two things: first, that a drop | of water turned into air needs (whatever the fables people make up about the decimal proportion of the elements) at least a hundred times more space than before; and secondly, that a little new air in motion, when added to the old, agitates and sets the whole in motion, as is apparent from the draught of bellows or a crack in a window which can set in motion all the air in the room, as is obvious from lamp flames. |

24. Just as dews and mist are generated here in the air around us without making clouds or advancing into the middle region, so also that is the case with many winds.

25. A continual breeze blows about the seas and waters, which is a weak wind newly created.

26. The rainbow, the lowest of the meteors, and generated nearest the earth, when it does not appear whole but broken off and resembles a pair of horns, is dispersed in winds, and as much if not more in rain.

27. It has been observed, in regions marked off and separated by mountains in their midst, that there are certain winds which, common | on one side of the range, do not reach the other. From this it is obvious that they are generated at a lower altitude than the mountains themselves.

28. Numberless are the winds which blow on clear days and even in places where it never rains, and which, generated where they blow, were never clouds or climbed to the middle region.

*Indirect Phenomena.* Whoever knows how easily a vapour is resolved into air, and how large is the volume of vapours, and how much more space is (as I said before) filled by a drop of water after turning into air

*gutta Aquæ versa in Aerem, præ eo quod anteà occupabat (vt dictum est),*

[I1ᵛ] *& quam modicum sustineat se comprimi Aer, non dubitabit quin necesse* |
*sit, etiam à superficie terræ, vsque ad sublimia aeris, vbique generari* Ventos. *Neque enim fieri potest, vt magna copia vaporum, cum cœperint*
5 *expandi, ad mediam Aeris regionem attollantur, absque Super-oneratione*
*Aeris, & tumultu in viâ.*

## Accidentales generationes Ventorum.

***Ad Artic. 9. Connexio.*** Accidentales Generationes Ventorum *eas*
9 *vocamus, quæ non efficiunt, aut gignunt Motum impulsiuum*
[I2ʳ] Ventorum, *sed* | *eum compressione acuunt, repercussione vertunt,*
*Sinuatione agitant, & voluunt. quod fit per causas extrinsecas, &*
*posituram Corporum adiunctorum.*

1. In locis, vbi sunt colles minùs eleuati, & circa hos subsidunt valles,
& vltra ipsos, rursùs colles altiores, maior est agitatio Aeris, & sensus
15 *Ventorum,* quàm aut in montanis, aut in planis.

2. In vrbibus, si sit aliquis locus paulò latior, & exitus angustiores,
aut Angiportus, & Plateæ se invicem secantes; percipiuntur ibi Flatus,
[I2ᵛ] & Auræ. |

3. In Ædibus refrigeratoria per *Ventos* fiunt, aut occurrunt; vbi Aer
20 est perflatilis, & ex vnâ parte introit Aer, ex aduerso exit; Sed multò
magis, si Aer intrat ex diuersis partibus, & facit concursum auræ ad
Angulos, & habet exitum illi Angulo communem. Etiam Concameratio
cœnaculorum, & rotunditas, plurimum facit ad Auras; quia repercu-
titur Aer commotus ad omnes lineas. Etiam Sinuatio porticuum magis
25 iuuat, quàm si exporrigantur in recto; Flatus enim in recto, licet non
concludatur, sed liberum habeat exitum, tamen non reddit Aerem,
[I3ʳ] tam inæqualem, & voluminosum, & vndantem, | quam confluxus ad
Angulos, & anfractus, & glomerationes in rotundo, & huiusmodi.

4. Post magnas tempestates in Mari, continuatur *Ventus Accidentalis*
30 ad tempus, postquam *Originalis* resederit; factus ex collisione, & per-
cussione Aeris, per vndulationem fluctuum.

5. Reperitur vulgò in hortis, repercussio *Venti,* à parietibus, &
ædibus, & aggeribus; ita vt putaret quis, *Ventum* in contrariam partem
spirare eius, à quâ reuerâ spirat.

35 6. Si Montes regionem, aliquâ ex parte, cingant, & *Ventus,* paulò

---

1 *dictum est*),] ~)ₐ    22 communem.] ~:    24 lineas.] ~;

than it did before, and how little compression the air suffers, will not doubt but that winds <sup>|</sup> are necessarily generated everywhere, right from the surface of the Earth up to the highest reaches of the air. Nor can vast volumes of vapours, once they have started expanding, be drawn to the middle region of the air without overloading the air and causing commotion in the process.

## Accidental Generations of Winds

*To Article 9. Connection.* I call them accidental generations of winds when they do not produce or give rise to an impulsive motion of winds, but <sup>|</sup> stimulate it by compression, turn it back by repercussion, or agitate and deflect it with winding; and this springs from external causes, and the disposition of adjacent bodies.

1. The air is more agitated and easily detectable in places with low hills surrounded by valleys with higher hills beyond them than in mountainous or flat districts.

2. In cities, if there is some slightly wider place with narrower ways out, or alleys and streets which cross each other, you can feel winds and breezes there. <sup>|</sup>

3. Cooling down of buildings takes place and crops up where the air blows through, and comes in on one side and leaves on the other. But it happens much more if it comes in from different sides and causes draughts to come together in the corners, and leaves by a common outlet in such a corner. Vaulting and doming to upper rooms also does a lot for draughts because air in motion bounces off in all directions. Winding galleries also help more than straight ones, for a blast going straight, though unimpeded and with a clear way out, does not make the air so irregular, full, and fluctuating <sup>|</sup> as does confluence in corners, coiling, and concentration in domes, and the like.

4. After great storms at sea an accidental wind goes on for a while after the original one has subsided; and this is caused by the collision and percussion of the air by the billowing of the waves.

5. We find that in gardens that the wind bounces off walls, buildings, and banks, so that you might think that it was blowing in the opposite direction to its true one.

6. If mountains girdle one side of a region, and a wind blows for

diutiùs, ex plano contra montem spirauerit, fit vt ipsâ repercussione
[I3ᵛ] montis, aut con'trahatur *Ventus* in pluuiam, si fuerit humidior; aut
vertatur in *Ventum* contrarium, sed qui breui tempore duret.

7. In flexionibus Promontoriorum experiuntur Nautæ sæpiùs
5 mutationes *Ventorum.*

## Venti extraordinarij & Flatus repentini.

***Ad Artic.* 10. *Connexio.*** *De* Ventis extraordinarijs *sermocinantur*
*quidam, & causantur,* Ecnephiâ *siue* procellâ, Vortice, Typhone,
[I4ʳ] Prestere: *Sed rem non narrant,* | *quæ certè ex chronicis, & historiâ*
10 *sparsâ, peti debet.*

1. *Repentini flatus* nunquam cœlo sereno fiunt, sed semper nubilo, &
cum imbre; vt eruptionem quandam fieri, & flatum excuti, aquas con-
cuti, rectè putetur.

2. *Procellæ,* quæ fiunt cum nebulâ, aut caligine, quas Belluas vocant,
15 quæque se sustinent, instar Columnæ, vehementes admodum sunt, &
diræ nauigantibus.

3. *Typhones* maiores, qui per latitudinem aliquam notabilem cor-
ripiunt, & correpta sorbent in sursum, rarò fiunt; at *Vortices,* siue
[I4ᵛ] *Turbines* exigui, & | quasi ludicri, frequentèr.

20 4. Omnes *Procellæ,* & *Typhones,* & *Turbines* maiores, habent manifestum motum præcipitij, aut vibrationis deorsùm, magis quam alij
Venti; vt torrentum modo ruere videantur, & quasi per canales defluere,
& postea à terrâ reuerberari.

5. Fit in pratis, vt cumuli fœni, quandoquè in altum ferantur, & tum
25 instar Conopæi spargantur; etiam in agris, vt caules pisarum inuoluti, &
aristæ segetum demessæ, quin etiam lintea ad exsiccandum exposita,
attollantur à *Turbinibus,* vsque ad altitudinem Arborum, aut supra
[I5ʳ] fastigia Ædium; hæcque fiunt, absque aliquo maiore *Venti* | impetu, aut
vehementiâ.

30 6. At quandoque fiunt *Turbines* leues, & admodùm angusti, etiam in
sereno; ita vt Equitans, videat pulueres, vel paleas corripi, & verti propè
se, neque tamen ipse magnoperè *Ventum* sentiat; quæ proculdubiò fiunt
hic propè, ex auris contrarijs se mutuò repellentibus, & circulationem
aeris ex concussione facientibus.

25 inuoluti] inuolutæ / emended thus in *SEH* (II, p. 45)
1 fit] ~,

a fair time against the mountains from the plain, it comes about that, repelled by the mountain, the wind is either | contracted into rain if it be moist, or turned in the opposite direction though only for a short time.

7. Sailors often feel the winds change when rounding capes.

## Extraordinary Winds and Sudden Blasts

*To Article* 10. **Connection.** Some people rattle on about extra-ordinary winds such as the hurricane or storm, whirlwind, typhoon, and the hot wind. But their words lack the substance | which should certainly be sought from the chronicles and scattered history.

1. Sudden blasts never happen in a clear sky but always in the presence of cloud with rain, so that it is rightly thought to be a certain eruption, a blast thrown out and waters stirred up.

2. Storms which accompany cloud and fog, which they call 'monsters', which raise themselves up like columns, are extremely savage, and dangerous to sailors.

3. Large-scale typhoons, which have considerable breadth and carry things off into the blue yonder, seldom happen, but slight eddies or playful whirlwinds | occur often.

4. All storms, typhoons, and larger whirlwinds manifestly fall or dart downwards more than other winds, so that they appear to rush like torrents, to flow down as if channelled, and then to be beaten back from the ground.

5. It sometimes happens in meadows that haystacks are carried up into the air, and then spread out like a coverlet. In fields too, bundled-up pea stalks, corn sheafs, and linen put out to dry are carried right up to the treetops or up over the roofs by whirlwinds, and this happens without the wind being at all | strong or violent.

6. But sometimes these light and very restricted whirlwinds even occur in clear weather, so that a horseman may see dust and straws snatched up and turned about him, yet without him feeling any appreciable wind; and these things no doubt happen by contrary breezes driving each other back, and producing circular motions of the air by their collision.

7. Certum est, esse quosdam flatus qui manifesta vestigia relinquunt Adustionis, & Torrefactionis in plantis. At *presterem*, qui est tanquam fulgur cœcum, atque Aer feruens, sed sine flammâ, ad Inquisitionem de
4 Fulgure rejicimus.

[I5$^v$]
|Confacientia ad Ventos;
Originales scilicet,
nam de Accidentalibus, suprà inquisitum est.

*Ad Artic.* II. I2. I3. I4. I5. *Connexio. Qvæ à veteribus, de* Ventis *eorumque causis, dicta sunt, confusa planè sunt, & incerta, nec*
10 *maximâ ex parte, vera. Neque mirum si non cernant clarè, qui non spectant propè. Loquuntur, ac si* Ventus *aliud quippiam esset,*
[I6$^r$] *separatum ab Aere Moto; atque ac si Exhalationes generarent,* | *& conficerent corpus integrum* Ventorum; *atque ac si Materia* Vento-*rum esset Exhalatio tantum calida, & sicca; Atque ac si origo Motûs*
15 Ventorum, *esset tantummodo deiectio, & percussio à frigore mediæ regionis: omnia phantastica, & pro Arbitrio. Attamen ex huiusmodi filis, magnas conficiunt telas;* Operas *scilicet* Aranearum. *At omnis Impulsio Aeris, est* Ventus; *Et Exhalationes permistæ Aeri, plus con-*
[I6$^v$] *fe'runt ad Motum, quam ad Materiam; Et vapores humidi, ex calore*
20 *proportionato, etiam faciliùs soluuntur in* Ventum, *quàm exhala-tiones siccæ; & complures* Venti *generantur in Regione infimâ Aeris, & ex terrâ expirant, præter illos, qui dejiciuntur, & repercutiuntur. Videamus qualis sit sermo rerum ipsarum.*

I. Rotatio naturalis Aeris (vt dictum est in *Articulo* de *ventis Genera-*
25 *libus*) absque causâ aliâ externâ, gignit *ventum* perceptibilem intra
[I7$^r$] tropicos, vbi Aeris | conuersio fit per Circulos maiores.

2. Post Motum Aeris naturalem, antequam inquiramus de Sole, qui est Genitor *ventorum* præcipuus, videndum, num quid sit tribuendum Lunæ, & alijs Astris, ex experientiâ clarâ.
30 3. Excitantur *venti* magni, & fortes, nonnullis, ante *Eclipsin Lunæ,* horis; ita vt si *Luna* deficiat medio noctis, flent *venti* vesperi præcedente; si *Luna* deficiat manè, flent *venti* medio noctis præcedente.

---

3 flammâ,] ~;     17 *telas,*] ~:

7. It is clear that there are certain blasts which leave behind manifest traces of burning or scorching on plants. But I refer the hot wind which is like latent lightning, and a burning air but without flame, to the investigation into lightning.

<sup></sup>ᵗ Factors contributing to Winds,
i.e. to original ones;
for the accidental ones have been investigated above.

*To Articles* 11, 12, 13, 14, 15. *Connection.* What the ancients have told us about the winds and their causes is plainly confused, uncertain, and for the most part untrue. And that is no surprise, since those who do not look closely do not see clearly. They talk as if the wind were something separate from the motion of the air, and as if exhalations generated ᵗ and put together the whole body of the winds; and as if the material of winds were only a hot and dry exhalation, and as if the origin of the winds' motion were just a casting down or rebounding from the cold of the middle region—all of which ideas are fanciful and arbitrary. Yet from suchlike threads they weave great webs, as spiders do. But in fact all impulsion of the air is a wind; and exhalations mixed in with the air ᵗ contribute more to their motion than to their matter; and moist vapours, produced by an appropriate heat, are even more easily resolved into a wind than are dry exhalations; and many winds, besides the ones which are thrust down and repelled, are generated in the lowest region of the air, or suspire from the earth. Let us heed then the voice of the things themselves.

1. As I said in the article concerning general winds, the natural rotation of the air generates without any external cause a noticeable wind within the tropics, where the revolution ᵗ of the air takes place in more ample circles.

2. After the natural motion of air and before investigating the Sun, which is the winds' main progenitor, we must see whether, on plain experience, anything can be attributed to the Moon or other stars.

3. Great and strong winds are stirred up some hours before a lunar eclipse, such that if the Moon be eclipsed at midnight, the winds blow on the evening before; but if it is eclipsed in the morning, they blow on the previous midnight.

4. In *Peruviâ*, quæ regio est admodum flatilis, notat *Acosta*, maxime flare *ventos* in *Plenilunijs*.

[I7ᵛ]   *Mandatum.* *Dignum certè esset obseruatione,* | *quid possint super* Ventos, *Motus & tempora* Lunæ, *cum liquidò possint super Aquas; veluti,* vtrùm Venti *non sint paulo commotiores, in* Plenilunijs, *&* Nouilunijs, *quàm in dimidijs, quemadmodum fit in Æstibus Aquarum; licet enim quidam commodè fingant,* Imperium Lunæ *esse super* Aquas; Solis *vero, &* Astrorum, *super* Aerem; *tamen certum est, Aquam & Aerem esse corpora valdè homogenea; &* Lunam, *post* Solem, *plurimum hic apud nos posse, in omnibus.*

5. Circa *Coniunctiones Planetarum*, non fugit hominum obseruationem flare *Ventos* maiores.

[18ʳ]   6. Exortu *Orionis*, surgunt plerumque *venti*, & tempestates | variæ; sed videndum, annon hoc fiat, quia exortus eius sit, eo tempore Anni, quod ad generationem *ventorum* est maximè efficax; vt sit potius Concomitans quiddam, quam Causa; quod etiam de ortu *Hyadum*, & *Pleiadum*, quoad imbres; & *Arcturi*, quoad tempestates, similitèr meritò dubitari possit. De *Lunâ*, & *Stellis*, hactenus.

7. *Sol* proculdubiò est Efficiens primarius *ventorum* plurimorum, operans per calorem, in Materiam duplicem: Corpus scilicet Aeris, & Vapores, siue Exhalationes.

8. *Sol* cum est potentior, Aerem, licet purum, & absque immistione [18ᵛ] vllâ, dilatat fortassè ad | tertiam partem, quæ res haud parua est. Itaque per simplicem dilatationem, necesse est vt oriatur Aura aliqua, in vijs *Solis*; præsertìm in magnis feruoribus, ídque potius duas aut tres horas, post exortum eius, quàm ipso mane.

9. In *Europâ*, noctes sunt Æstuosiores; in *Peruviâ*, tres horæ Matutinæ; ob vnam eandemq*ue* causam; videlicet cessationem Aurarum, & *ventorum*, illis horis.

10. In *vitro calendari*, Aer dilatatus deprimit aquam, tanquam flatu. At in *vitro Pileato* Aere tantummodo impleto, Aer dilatatus inflat [K1ʳ] vesicam, vt *ventus* manifestus. |

11. *Experimentum* fecimus in turri rotundâ, vndique clausâ, huius generis *venti*. Nam foculum in medio eius locauimus, cum prunis penitùs ignitis, vt minus esset fumi. At à latere foculi in distantiâ nonnullâ, filum suspendimus, cum cruce ex plumis, vt facilè moueretur. Itàq*ue* post paruam moram, aucto calore, & dilatato Aere, agitabatur crux plumea cum filo suo, hinc inde, motu vario; Quin etiam facto

---

4 *Aquas;*] ~.     6 *Aquarum;*] ~:     13 variæ;] ~:     18 possit.] ~;     35 fumi.] ~;

4. In Peru, which is a very windy place, Acosta observes that the winds blow most when the Moon is full.

*Direction.* It would be a good idea to observe ¹ what the motions and changes of the Moon do to the winds, since they plainly affect the waters, and to see whether the winds, like the tides, are not a bit more unsettled when the Moon is full or new than at the quarters; for though certain people conveniently make out that the Moon's empire is over the waters, and that of the Sun and stars over the air, it is still certain that water and air are very closely akin, and that next to the Sun, the Moon has the greatest influence on all things here where we are.

5. It has not escaped human observation that winds blow stronger around planetary conjunctions.

6. When Orion rises more winds and storms spring up; ¹ but we should see whether this happens because its rising coincides with the time of the year which is most conducive to the generation of winds, so that it may be rather a concurrent event than a cause; and this is also a doubt that may be properly objected in the case of the rising of the Hyades and Pleiades in respect of rain, and of Arcturus in respect of storms. So much then for the Moon and stars.

7. The Sun is no doubt the main efficient cause of many winds, working as it does by heat on two kinds of matter, namely air and vapours or exhalations.

8. The Sun, when it is strong, dilates air, though it be pure and with no intermixture, by perhaps ¹ as much as a third, an amount by no means negligible. By simple dilatation, therefore, some current must arise when the Sun approaches, especially in heatwaves; and that happens two or three hours after sunrise rather than at dawn itself.

9. In Europe the nights are more sultry; in Peru, the three hours of morning; and both are caused by the cessation of breezes and winds during those hours.

10. In a calendar glass dilated air pushes the water down as if it were a breeze. But in a capped glass, filled only with air, the dilated air expands the bladder like an evident wind. ¹

11. I conducted an experiment on this kind of wind in a round tower closed on all sides. I placed a brazier in the middle of it with coals well alight so that there would be less smoke. But at some distance from a side of the brazier I hung a thread with a cross of feathers to make it move more easily. In a short time then, when the heat had increased and the air dilated, the cross and its thread were stirred back and forth with changeable motion; and when a hole was made in the window of the

foramine, in fenestrâ turris, exibat flatus calidus, neque ille continuus, sed per vices, & vndulans.

**12.** Etiam receptio Aeris per frigus, à dilatatione, creat eiusmodi
[Kₗᵛ] *ventum,* sed debiliorem, | ob minores vires frigoris; adeo vt in *Peruviâ,*
5 sub quauis paruâ vmbrâ, non solum maius percipiatur refrigerium, quam apud nos (per *Antiperistasin*) sed manifesta Aura ex receptione Aeris, quando subit vmbram.

Atque de *Vento,* per meram dilatationem, aut receptionem Aeris facto, hactenus.

10 **13.** *Venti* ex meris motibus Aeris, absque immistione Vaporum, lenes, & molles sunt. Videndum de *ventis Vaporarijs* (eos dicimus qui generantur à vaporibus) qui tantò illis alteris, possunt esse vehementiores, quantò dilatatio *guttæ* aquæ, versæ in Aerem, excedit aliquam dila-
[K₂ʳ] tationem Aeris, iam | facti: quod multis partibus facit, vt superiùs
15 monstrauimus.

**14.** *Ventorum vaporariorum* (qui sunt illi qui communitèr flant) Efficiens est *Sol,* & calor eius proportionatus; Materia, Vapores, & Exhalationes, qui vertuntur, & resoluutur, in Aerem; Aerem inquam (non aliud quippiam ab Aere) sed tamen ab initio minùs syncerum.

20 **15.** *Solis* calor exiguus non excitat vapores, itaque nec *ventum.*

**16.** *Solis* calor medius excitat vapores, nec tamen eos continuò dissipat. Itaque si magna fuerit ipsorum copia, coeunt in pluviam, aut
[K₂ᵛ] simplicem, aut cum *vento* coniunctam; si minor | vertuntur in *ventum* simplicem.

25 **17.** *Solis* calor in incremento, inclinat magis ad generationem *ventorum;* in decremento *Pluviarum.*

**18.** *Solis* calor intensus, & continuatus, attenuat, & dissipat vapores, eosque sublimat, atque interim Aeri æqualiter immiscet, & incorporat; vnde Aer quietus fit, & serenus.

30 **19.** Calor *Solis* magis æqualis, & continuus, minùs aptus ad generationem *Ventorum;* magis inæqualis & alternans, magis aptus. Itaque in nauigatione ad *Russiam,* minus afflictantur *Ventis,* quam in *Mari Bri-*
[K₃ʳ] *tannico,* propter longos dies; at in *Peruviâ* sub Æquinoctio, crebri |
*Venti,* ob magnam inæqualitatem caloris, alternantem noctù, &
35 interdiù.

**20.** In Vaporibus, & Copia spectatur, & Qualitas: Copia parua

11 *Vaporarijs*] ~,    12 vehementiores,] ~;    17 proportionatus;] ~:    Materia,] ~;
28 Aeri] / see Introduction, p. lxxii above    31 *Ventorum;*] ~:    33 dies;] ~:
34 Venti,] ~;

72

tower, a hot current of air went out, not steadily but intermittently and with fluctuating intensity.

12. The contraction of dilated air produces the same sort of wind but weaker | because of cold's lesser powers, such that in Peru, in any little shady spot, not only do you experience a greater cold than any here with us (because of antiperistasis) but there is a manifest breeze when the air contracts as it passes into the shade.

And so much for wind produced by mere dilatation or contraction of the air.

13. Winds caused merely by motions of the air, without intermixture of vapours, are soft and gentle. But we must look into vaporous winds (as I call those generated from vapours) which can surpass those others in violence, by as much as a drop of water turned into air exceeds any dilatation of air already | made; which it does many times over, as I have shown above.

14. The Sun with its proportionate heat is the efficient cause of vaporous winds (i.e. the ones that usually blow); the material cause is the vapours and exhalations which are turned and resolved into air, air I say (not anything apart from it), but air which from the start is still not quite pure.

15. Slight solar heat stirs up no vapours and so makes no wind.

16. Moderate solar heat stirs up vapours but does not dissipate them without more ado. Thus if their abundance be great, they come together as rain, either by itself or together with wind; but in smaller amounts | they turn into wind alone.

17. When it is increasing, solar heat inclines more to generating winds; when decreasing, to rains.

18. Intense and uninterrupted solar heat attenuates and dissipates vapours, and sublimes them, and at the same time mixes and incorporates them with air evenly; whence the air becomes quiet and still.

19. More even and steady solar heat is less apt for generating winds, but more uneven and variable heat makes it more apt. Thus the winds are less inconvenient on a voyage to Russia than they are in the British Sea, because the days are longer; but in Peru at the equinox, winds are | frequent because of the great variability of heat between day and night.

20. In vapours quantity and quality are significant. A small quantity produces gentle breezes; a moderate quantity stronger ones, and a large

gignit auras lenes; media *Ventos* fortiores; magna aggrauat Aerem, &
gignit pluuias, vel tranquillas, vel cum *Ventis*.

21. Vapores ex mari, & Amnibus, & paludibus inundatis, longè
maiorem copiam gignunt *Ventorum*, quam Halitus terrestres. Attamen,
5 qui à terrâ, & locis minùs humidis, gignuntur *Venti*, sunt magis obstinati, & diutiùs durant, & sunt illi ferè, qui deijciuntur ex alto, vt opinio
[K3$^v$] Veterum, in hâc parte, non fuerit omninò inutilis; nisi <sup>|</sup> quod placuit
illis, tanquam diuisâ hæreditate, assignare Vaporibus pluuias, & *Ventis*
solummodò Exhalationes; & huiusmodi pulchra dictu, re inania.

10 22. *Venti* ex resolutionibus niuium iacentium super montes, sunt
ferè medij inter *Ventos* Aquaticos, & Terrestres, sed magis inclinant ad
Aquaticos; sed tamen sunt acriores, & mobiliores.

23. Solutio niuium in montibus Niualibus (vt priùs notauimus)
semper inducit *Ventos Statos*, ex eâ parte.

15 24. Etiam *Anniuersarij Aquilones*, circa exortum Caniculæ, existi-
[K4$^r$] mantur venire à Mari glaciali, & partibus circa circulum Ar<sup>|</sup>cticum, vbi
seræ sunt solutiones glaciei, & niuium, æstate tùm valdè adultâ.

25. Moles, siue Montes glaciales, quæ deuehuntur versus *Canadam*
& *terram Piscationis*, magis gignunt Auras quasdam frigidas, quam *Ven-*
20 *tos* mobiles.

26. *Venti*, qui ex terris sabulosis, aut Cretaceis proueniunt, sunt pauci,
& sicci; ijdem in regionibus calidioribus, æstuosi, & fumei, & torridi.

27. *Venti* ex vaporibus marinis, faciliùs abeunt retrò in pluuiam,
aquâ ius suum repetente, & vindicante; aut si hoc non conceditur,
25 miscentur protinùs Aeri, & quietem agunt. At Halitus terrei, & fumei,
[K4$^v$] & vnctuosi, <sup>|</sup> & soluuntur ægriùs, & ascendunt altiùs, & magis irritati
sunt in suo motu, & sæpè penetrant mediam regionem Aeris, & sunt
aliqua materia meteororum ignitorum.

28. Traditur apud nos in *Angliâ*, temporibus, cum *Gasconia* esset
30 huius ditionis, exhibitum fuisse *Regi* libellum supplicem, per subditos
suos *Burdegaliæ*, & confinium; petendo vt prohiberetur Incensio
Ericæ in Agris *Sussexiæ*, & *Hamptoniæ*, quia gigneret *Ventum* circa finem
*Aprilis*, vineis suis exitiabilem.

34 29. Concursus *Ventorum*, ad inuicem, si fuerint fortes, gignunt
[K5$^r$] *Ventos* vehementes, & vorticosos; si lenes, & humidi, gig<sup>|</sup>nunt pluuiam,
& sedant *Ventos*.

---

2  tranquillas] tranquillos / emended thus in *SEH* (II, p. 49)

9  solummodò] solumodò / silently emended thus in *SEH* (II, p. 49)     18 *Canadam*] ~;

74

quantity burdens the air and produces rains either with or without winds.

21. Vapours from the seas, large rivers, and waterlogged fens generate much more abundant winds than do earthy exhalations. Yet winds generated from the earth and places less wet are more persistent and last longer, and these are pretty well the ones cast down from above, so that the opinion of the ancients would not have been absolutely useless in this connection, had it not pleased ⌐ them to split (so to speak) the inheritance by assigning the rains to vapours and the winds to exhalations alone—this and the like is pretty talk indeed and devoid of substance.

22. Winds from melting snows on mountains are practically inter-mediate between watery and earthy winds, but they lean more to the watery but are nevertheless more sharp and shifting.

23. As I noted earlier, the melting of snows on snowy mountains always gives rise for that reason to recurrent winds.

24. Northern anniversary winds, when the Dog-star rises, are also thought to come from the ice-bound sea, and places around the Arctic Circle, ⌐ where the ice and snow melt late and far into the summer season.

25. The masses or mountains of ice which are carried towards Canada and the Land of Fisheries produce certain cold breezes more than shifting winds.

26. Winds coming from sandy or chalky places are scanty and dry, but in hotter regions these same are sultry, murky, and parching.

27. Winds born of sea vapours more easily turn back into rain again, the water reclaiming and exacting its due; but if that be disallowed, they mix at once with the air, and stay quiet. But earthy, murky, and greasy exhalations ⌐ are less easily resolved, climb higher, their motion is more agitated, and they often penetrate into the middle region of the air and contribute to the matter of fiery meteors.

28. Here with us in England people say when Gascony was ours, the king's subjects in and around Bordeaux petitioned him to put a stop to the burning of heather in the fields of Sussex and Hampshire, because, around the end of April, it generated a wind pernicious to their vines.

29. Gatherings of winds, if they be strong, produce violent whirl-winds; but if they be gentle and moist, ⌐ they beget rains but calm the winds down.

**30.** Sedantur & coercentur *Venti*, quinque modis: cum aut Aer, vaporibus oneratus, & tumultuans, liberatur, vaporibus se contrahentibus in pluuiam; Aut cum vapores dissipantur, & fiunt subtiliores, vnde permiscentur Aeri, & bellè cum ipso conueniunt, & quietè degunt; Aut
5 cum Vapores, siue halitus exaltantur, & sublimantur in altum, adeò vt requies sit ab ipsis, donec à mediâ regione Aeris deijciantur, aut eam penetrent; Aut cum vapores, coacti in nubes, ab alijs *Ventis* in alto spirantibus, transuehuntur in alias regiones, vt pax sit ab ipsis, in
[K5ᵛ] regionibus, quas præter vo'lant; Aut denique, cum *Venti* à fomitibus suis
10 spirantes, longo itinere, nec succedente nouâ materiâ, languescunt, & impetu suo destituuntur, & quasi expirant.

**31.** Imbres plerunque *Ventos* sedant, præsertim procellosos, vt & *Venti* contrà sæpiùs detinent Imbrem.

**32.** Contrahunt se *Venti* in pluuiam (qui est primus ex quinque
15 sedandi modis, isq*ue* præcipuus); aut ipso onere grauati, cum vapores sunt copiosi; aut propter contrarios motus *Ventorum*, modò sint placidi; aut propter obices Montium, & Promontoriorum, quæ sistunt impe-
[K6ʳ] tum *Ventorum*, eosque paulatim in se | vertunt; aut per frigora intensiora, vnde condensantur.

20 **33.** Solent plerunque *Venti*, minores & leuiores, manè oriri, & cum Sole decumbere, sufficiente Condensatione Aeris nocturnâ, ad receptionem eorum. Aer enim nonnullam compressionem patitur, absq*ue* tumultu.

**34.** Sonitus Campanarum existimatur Tonitrua, & Fulgura dissipare.
25 De *Ventis* non venit in obseruationem.

*Monitum. Consule locum de* Prognosticis Ventorum; *est enim nonnulla Connexio Causarum, & Signorum.*

**35.** Narrat *Plinius Turbinis* vehementiam, aspersione Aceti in
29 occursum eius, compesci.

[K6ᵛ] | *Limites Ventorum.*

***Ad Artic.* 16. 17. 18. 1.** Traditur de Monte *Atho*, & similitèr de *Olympo*, consueuisse sacrificantes, in Aris, super fastigia ipsorum extructis, literas exarare in Cineribus sacrificiorum, & postea redeuntes, elapso anno (nam Anniuersaria erant sacrificia) easdem literas reperisse neutiquam
35 turbatas, aut confusas, etiamsi Aræ illæ non starent in Templo aliquo,

---

1 modis:] ~;          3 pluuiam:] ~:          4 degunt:] ~:          7 penetrent:] ~:
9 vo'lant:] ~:          15 præcipuus):] ~)ˏ          16 placidi:] ~:          24 dissipare.] ~,

30. Winds are calmed and curbed in five ways: either when the air, burdened and seething with vapours, is freed as the vapours contract themselves into rain; or when the vapours are dissipated and become more subtle, such that they mix in with the air, get on well with it, and stay quiet; or when the vapours or exhalations are raised up and sublimed on high, so that there is respite from them, until they are driven down from the middle region of the air, or cross over into it; or when the vapours, condensed into clouds, are carried into other regions by other winds blowing high up, so that they leave in peace the places they otherwise blow ' over; or lastly, when the winds blowing from their sources, after travelling a long way with no new matter to supply them, slacken, lose their force, and in a way give up the ghost.

31. Showers often calm winds down, especially if they are stormy, just as, on the other side, winds often hold back showers.

32. Winds contract themselves into rain (which is the first and most important of the five ways of calming them) either as they are weighed down when vapours are abundant; or because of the contrary motions of other winds, provided that they are gentle; or because of obstacles presented by mountains or promontories, which resist the winds' force, and gradually turn them back ' on themselves; or by condensing them with keener cold.

33. Slighter and lighter winds are wont to get up at dawn and go down with the Sun, when the nightly condensation of the air is enough to hold them back. For air will put up with some compression without a struggle.

34. The ringing of bells is supposed to dissipate thunder and lightning, but the like has not been observed with winds.

*Advice.* Look at the place concerned with the prognostics of winds; for there is some connection between causes and signs.

35. Pliny says that the violence of a whirlwind can be suppressed by sprinkling vinegar at its onset.

## ' The Limits of Winds

*To Articles* 16, 17, 18. 1. They say of Mount Athos and Olympus that those making sacrifices on altars built on their summits were given to drawing letters in the ashes of the animals sacrificed and that, coming back in a year's time (for the sacrifices were annual), they would find the letters not at all disturbed or confounded, and this even though the

sed sub dîo; vnde manifestum erat, in tantâ altitudine, neque cecidisse *Imbrem*, neque spirasse *ventum*.

[K7ʳ] 2. Referunt in fastigio *Pici* de *Tenariph*, atque etiam in *Andi* |*bus*, inter *Peruuia*m & *Chilem* niues subiacere per cliuos, & latera Montium; 5 at in ipsis cacuminibus, nil aliud esse quam Aerem quietum, vix spirabilem propter tenuitatem, qui etiam acrimoniâ quâdam, & os stomachi, & oculos pungat, inducendo illi nauseam, his suffusionem, & ruborem.

3. *Venti Vaporarij*, non videntur in aliquâ maiore altitudine flare; cum 10 tamen probabile sit, aliquos ipsorum altiùs ascendere, quam pleræque Nubes.

De *Altitudine* hactenus, de *Latitudine* videndum.

4. Certum est spatia, quæ occupant *Venti*, admodum varia esse, [K7ᵛ] interdùm amplissima, interdùm | pusilla, & angusta. Deprehensi sunt 15 *Venti* occupasse spatium Centenorum milliarium, cum paucarum horarum differentiâ.

5. Spatiosi *Venti* (si sint ex Liberis) plerunq*ue* vehementes sunt, non lenes: Sunt etiam diuturniores, & fere 24. horas durant. Sunt itidem minùs pluuiosi. Angusti contra, aut lenes sunt, aut procellosi; at semper 20 breues.

6. *Stati Venti* sunt itinerarij, & longissima spatia occupant.

7. *Venti procellosi* non extenduntur per larga spatia, licet semper euagentur vltra spatia ipsius procellæ.

24 8. Marini *Venti*, intra spatia angustiora multò quam Terrestres, [K8ʳ] spirant; in tantùm, vt in | Mari, aliquando conspicere detur, Auram satis alacrem, aliquam partem Aquarum occupare (id quod ex crispatione Aquæ facilè cernitur), cùm vndique sit malacia, & Aqua instar speculi plana.

9. Pusilli (vt dictum est) *Turbines* ludunt quandoquè coram Equi- 30 tantibus, instar ferè ventorum ex follibus.

De *latitudine* hactenus, de *Duratione* videndum.

10. Durationes *ventorum*, valde vehementium, in Mari longiores sunt, sufficiente copiâ vaporum; in Terrâ, vix vltra diem, & dimidiam, 34 extenduntur.

[K8ᵛ] 11. *Venti* valdè lenes, nec in Mari, nec in Terrâ, vltra triduum, | constantèr flant.

---

12 De *Altitudine* hactenus] / no new para. in *c-t*, but see K8ʳ l. 31 (De *latitudine* hactenus) 26 occupare] ~, 27 cernitur),] ~,) 29 dictum est)] ~ₐ / the closing bracket is a press correction: the BNF copy with a bracket, Huntington 12569 without

altars stood in the open and not in some temple. And from this it was clear that no rain fell or wind blew at such an altitude.

2. They say that on the summit of the Peak of Tenerife, and also in the Andes | between Peru and Chile, the snow lies on the sides and slopes of the mountains, but that at the very tops, there is nothing but still air so thin as to be scarcely breathable, and with a certain pungency that stings the mouth of the stomach and the eyes, provoking nausea in the former, and flushing and redness in the latter.

3. Vaporous winds do not seem to blow at any great altitude, though it is nevertheless likely that some of then climb higher than many clouds.

So much then for altitude; let us now investigate breadth.

4. It is certain that the spaces which winds cover are very variable, sometimes very wide, sometimes | narrow and constricted. People have found that winds have covered a space of one hundred miles in the space of a few hours.

5. Winds which cover large spaces (if they are of the free type) are often violent, not gentle. They are also longer lasting, and go on for up to twenty-four hours; they are also less rainy. Narrow winds are, on the contrary, either gentle or stormy but always brief.

6. Recurrent winds wander and cover the greatest distances.

7. Stormy winds do not cover extensive spaces, though they often stray beyond the area of the storm itself.

8. Sea winds blow within much narrower confines than land winds, to the extent that at | sea we sometimes see a brisk breeze covering some part of the sea (easily observable in the rippling of the water) while everywhere else it is as flat as a millpond.

9. As I have said, little whirlwinds play before horsemen, almost like puffs of wind from a bellows.

So much then for breadth; let us now investigate duration.

10. Very violent winds at sea last longest if they are supplied with enough vapours; on land they scarcely last a day and a half.

11. Very gentle winds at sea or on land do not blow | constantly for more than three days.

12. Non solum *Eurus Zephyro* magis est durabilis (quod alibi posuimus), sed etiam, quicunque ille *ventus* sit, qui manè spirare incipit, magis durabilis solet esse illo, qui surgit vesperi.

13. Certum est, *ventos* insurgere, & augeri gradatìm (nisi fuerint
5 meræ *Procellæ*); at decumbere celeriùs, interdum quasi subitò.

### Successiones Ventorum.

*Ad Artic.* **19. 20. 21. 1.** Si *Ventus* se mutet, conformitèr ad Motum Solis,
[L1ʳ] id est, ab *Euro* ad *Austrum*, ab *Austro* ad *Zephyrum*, à *Zephyro* ¹ ad
*Boream*, à *Boreâ* ad *Eurum*, non reuertitur plerunque; aut si hoc facit, fit
10 ad breue tempus. Si vero in contrarium Motûs Solis, scilicet ab *Euro* ad
*Boream*, à *Boreâ* ad *Zephyrum*, à *Zephyro* ad *Austrum*, ab *Austro* ad
*Eurum*; plerunque restituitur ad plagam priorem, saltem antequam confecerit Circulum integrum.

2. Si pluuia primum incœperit, & posteà cœperit flare *ventus*, *ventus*
15 ille pluuiæ superstes erit. Quod si primò flauerit *ventus*, posteà à pluuiâ occiderit, non reoritur plerunque *ventus*, & si facit, sequitur pluuia noua.

[L1ᵛ] 3. Si *venti* paucis horis varient, & tanquam experiantur, & de|inde cœperint constantèr flare, *ventus* ille durabit in dies plures.

20 4. Si *Auster* cœperit flare dies duos, vel tres, *Boreas* quandoque post eum subitò spirabit: Quod si *Boreas* spirauerit totidem dies, non spirabit *Auster*, donec *ventus* paulispèr ab *Euro* flârit.

5. Cum Annus inclinârit, & post Autumnum Hyems inceperit, si incipiente hyeme spirauerit *Auster*, & posteà *Boreas*, erit Hyems glacia-
25 lis; sin sub initijs hyemis spirauerit *Boreas*, posteà *Auster*, erit Hyems clemens, & tepidus.

6. *Plinius* citat *Eudoxum*, quod series *ventorum* redeat post qua-
[L2ʳ] driennium, quod verum mi|nimè videtur; neque enim tam celeres sunt reuolutiones. Illud ex aliquorum diligentiâ notatum est, tempestates
30 grandiores, & insigniores (feruorum, Niuium, Congelationum, Hyemum tepidarum, Æstatum gelidarum), redire plerunque ad Circuitum Annorum 35.

---

1 durabilis] ~,      2 posuimus),] ~)ᴧ      4 gradatìm] ~,      5 *Procellæ*);] ~;)
decumbere] / *SEH* (II, p. 52) has decumberre      8 *Austrum*,] ~;  *Zephyrum*,] ~;
9 *Boream*,] ~;      *Eurum*,] ~;      14 *ventus*,] ~; (1st occurrence)      30 insigniores] ~,
31 gelidarum),] ~)ᴧ      32 35] / *SEH* (II, p. 53) has triginta quinque

12. It is not just that (as I stated earlier) the east wind blows for longer than the west, but also that any wind that begins to blow in the morning generally lasts longer than one which gets up in the evening.

13. It is certain that winds get up and increase gradually (unless they are storms pure and simple); but they subside more quickly, and sometimes practically at once.

## Successions of Winds

*To Articles* 19, 20, 21. 1. If the wind changes in line with the motion of the Sun, i.e. from east to south, from south to west, from west ꞏ to north, and from north to east, it usually does not turn back again or, if it does, it is not for long. But if the winds change against the motion of the Sun, namely, from east to north, from north to west, from west to south, and from south to east, it usually returns to the previous quarter at least before it has gone full circle.

2. If it starts to rain before the wind gets up, the wind will last longer than the rain. But if the wind blows first and afterwards dies down because of rain, the wind usually does not get up again or, if it does, fresh rain follows it.

3. If winds change in a short time and as it were try things out, and then ꞏ start to blow steadily, that wind will last for many days.

4. If the south wind begins to blow for two or three days, the north wind will sometimes blow straight after it. But if the north wind has blown for many days, the south wind will not blow until the wind has blown for a brief while from the east.

5. As the year declines and autumn gives way to winter, if the south wind blows and afterwards the north, the winter will be freezing; but if the north wind is followed by the south at the beginning of winter, the season will be mild and warm.

6. Pliny cites Eudoxus to the effect that the series of winds comes back every four years, which seems ꞏ distinctly implausible, for the revolutions are not so quick. The diligence of others has recorded that great and outstanding periods of weather (heatwaves, snows, freezes, warm winters, and cold summers) generally come round every thirty-five years.

## Motus Ventorum.

*Ad Artic.* **22. 23. 24. 25. 26. 27.** *Connexio. Loquuntur homines, ac
si* Ventus *esset corpus aliquod per se, atque impetu suo, Aerem ante se*
[L2ᵛ] *ageret, & impelleret; Etiam cum* ¹ Ventus *locum mutet, loquuntur ac*
5 *si idem* Ventus, *se in alium locum transferret. Hæc vero cum loquuntur plebeij, tamen Philosophi ipsi, remedium huiusmodi opinionibus
non præbent, sed & illi quoque balbutiunt, neque erroribus istis
occurrunt.*

**1.** Inquirendum igitur, & de Excitatione Motûs in *ventis,* & de
10 Directione eius, cùm de *Originibus localibus* iam inquisitum sit. Atque
de ijs *ventis,* qui habent principium motûs, in suâ primâ Impulsione,
[L3ʳ] vt in ¹ ijs, qui dejiciuntur ex alto, aut efflant è terrâ, Excitatio motûs est
manifesta; alteri sub initijs suis descendunt, alteri ascendunt, & posteà
ex resistentiâ Aeris, fiunt voluminosi, maximè secundum angulos vio-
15 lentiæ suæ. At de illis, qui conflantur vbique in Aere inferiore (qui sunt
omnium *ventorum* frequentissimi) obscurior videtur inquisitio, cum
tamen res sit vulgaris, vt in Commentatione, sub Articulo octauo
declarauimus.

**2.** Etiam huius rei Imaginem, reperimus in illâ turri occlusâ, de quâ
20 paulò antè. Tribus enim modis, illud *Experimentum* variauimus. Primus
[L3ᵛ] erat is, de quo ¹ suprà diximus, Foculus ex prunis antè ignitis, & claris.
Secundus erat Lebes Aquæ feruentis, remoto illo Foculo; atque tum
erat Motus Crucis plumeæ, magis hebes, & piger, quam ex Foculo
prunarum, hærente in Aere, rore vaporis aquei, nec dissipato in mate-
25 riam *venti,* propter imbecillitatem caloris. At tertius erat ex vtrisque
simul, Foculo, & Lebete; Tum verò longè maxima erat Crucis plumeæ
agitatio, adeo vt quandoque illam in sursum verteret, instar pusilli
Turbinis; Aquâ scilicet præbente copiam vaporis, & Foculo, qui astabat,
29 eum dissipante.
[L4ʳ] **3.** Itaque *Excitationis* Motûs in ¹ *ventis,* causa est præcipua, super-
oneratio Aeris, ex nouâ accessione Aeris, facti ex vaporibus. Iam de
*Directione Motus* videndum, & de *Verticitate,* quæ est *Directionis*
Mutatio.

---

13 manifesta;] ~:     15 qui (first occurrence)] quæ / emended thus in *SEH* (II, p. 53)
22 atque] Atque

## The Motions of Winds

*To Articles* 22, 23, 24, 25, 26, 27. *Connection.* People talk as if the wind were some sort of body apart which by its own pressure pushed and shoved the air in front of it. They also ¹ talk as if when the same wind changed it shifted itself to another place. But when people utter these vulgar views, the philosophers themselves still offer no remedy for them, but also fall to babbling, and fail to tackle these errors head on.

1. We should, then, investigate what prompts and directs winds, seeing that we have already looked into their local origins. As for those winds whose principle of motion lies in their initial impulse, as in ¹ those cast down from above, or exhaled from the earth, it is obvious what sparks them off. At the start the former descend, and the latter rise up and then, because of the air's resistance, they grow fuller, principally in accordance with the way their violence is angled. But the investigation of those which blow everywhere in the lower air (the commonest winds of all) seems to be more obscure, even though, as I stated in the speculation subjoined to article 8, the thing itself is common.

2. Now I found a likeness of this thing in the enclosed tower that I mentioned earlier. For I varied that experiment in three ways. The first was the one ¹ I spoke of above with the brazier with coals well alight and smoke-free. The second was a kettle of boiling water which, when the brazier had been removed, moved the cross of feathers more weakly and lazily than the brazier did, since the feebler heat could not dissipate the dewy steam hanging in the air into windy matter. The third was with the brazier and kettle together, which stirred the cross of feathers most of all, such that it was sometimes rotated upwards as if by a tiny whirlwind, since the water supplied an abundance of vapour which the brazier was ready to dissipate.

3. Thus the main cause ¹ provoking wind motion is the overburdening of the air with added air made from vapour. But now let us consider the direction of motion or its verticity, which is change of direction.

4. *Directionem* Motûs progressiui *ventorum*, regunt fomites sui, qui sunt similes fontibus Amnium; loca scilicet, vbi magna reperitur copia vaporum, ibi enim est *Patria venti*. Postquam autem inuenerint currentem, vbi Aer minimè resistit (sicut Aqua invenit decliuitatem), 5 tum quicquid inueniunt similis materiæ in viâ, in consortium recipiunt, & suo currenti miscent; quemadmodum faciunt & Amnes; Itaque *venti* [L4ᵛ] ¦ spirant semper à parte Fomitum suorum.

5. Vbi non sunt Fomites insignes, in aliquo loco certo, vagantur admodum *venti*, & facilè currentem suum mutant; vt in medio Mari, & 10 Campestribus terræ latis.

6. Vbi magni sunt Fomites *ventorum*, in vno loco, sed in locis progressûs sui paruæ accessiones, ibi *venti* fortitèr flant, sub initijs, & paulatìm flaccescunt; vbi contrà, Fomites magis continui, leniores sunt sub initiis, & posteà augentur.

15 7. Sunt Fomites mobiles *ventorum*, scilicet in nubibus; qui sæpè [L5ʳ] à *ventis*, in alto spirantibus, transportantur in loca pro¦cul distantia, à Fomitibus vaporum, ex quibus generatæ sunt illæ Nubes; Tum verò incipit esse Fomes *venti*, ex parte, vbi Nubes incipiunt solui in *ventu*m.

8. At *Verticitas Ventorum*, non fit eò quod *Ventus* priùs flans, se trans-20 ferat; sed quod ille, aut occiderit, aut ab altero *vento* in ordinem redactus sit. Atque totum hoc negotium pendet ex varijs collocationibus Fomitum *Ventorum*, & varietate temporum, quando vapores ex huius-modi Fomitibus manantes soluuntur.

24 9. Si fuerint Fomites *Ventorum*, à partibus contrarijs, veluti alter [L5ᵛ] Fomes ab *Austro*, alter à *Boreâ*, præualebit scilicet *Ventus* forti¦or, neque erunt *Venti* contrarij, sed *Ventus* fortior continuò spirabit; ita tamen vt à *Vento* imbecilliore nonnihil hebetetur, & dometur; vt fit in Amnibus, accedente fluxu Maris; Nam Motus Maris præualet, & est vnicus, sed à Motu fluuij nonnihil frænatur. Quod si ita acciderit, vt alter ex illis 30 *Ventis* contrarijs, qui primùm fortior fuerat, succumbat, tum subitò spirabit *ventus* à parte contrariâ, vnde & antè spirabat, sed latitabat, sub potestate maioris.

10. Si Fomes (Exempli gratiâ) fuerit ad *Euro-boream*, spirabit scilicet 34 *Euro-boreas*. Quod si fuerint duo Fomites *Ventorum*, alter ad *Eurum*, [L6ʳ] alter ad *Boream*, ij *Venti* ¦ ad aliquem tractum spirabunt separatìm; At post Angulum confluentiæ, spirabunt ad *Euro-boream*, aut cum inclinatione, prout alter Fomes fuerit fortior.

---

4 resistit] ~,    decliuitatem),] ~,)    20–1 redactus sit.] ~;

4. The direction of the winds' progressive motion is governed by their nurseries, which are like the sources of great rivers. For where there is an abundance of vapours, there is the native land of the winds. Now once they have found a current where there is no air resistance (as water finds a slope), then whatever they find in the way of kindred matter in their journey they take it into partnership, and incorporate it in their flow in the same way that great rivers do. Thus do winds <sup>|</sup> always blow from where their nurseries lie.

5. When there are no notable nurseries in any precise place, the winds wander extremely, and readily alter their current, as they do in the midst of the sea and in broad plains on land.

6. When there are large nurseries of winds in one place, and the winds gather small reinforcements from the places through which they pass, they blow strong at the start but gradually slacken. But when, on the other hand, the nurseries are more spread out, the winds are gentler at the start but then strengthen.

7. There are wind nurseries which move, i.e. the clouds which are often carried by winds on high to places far <sup>|</sup> from the nurseries of vapours from which those clouds were bred. But then a nursery of wind begins to exist in that part where the clouds begin to dissolve into wind.

8. Wind verticity does not happen from a wind previously blowing that shifts itself, but from its sinking or being reduced to order by another wind. And this whole business hangs on the different spatial relations of the nurseries, and the different times when the vapours spreading from such nurseries are dissolved.

9. If the winds' nurseries are opposite one another, as when one is in the south and the other in the north, the stronger wind will <sup>|</sup> get the upper hand, and blow steadily without contrary winds, though in such a way that it is somewhat weakened and subdued by the feebler one. This is like what happens when great rivers run up against the tide; for the sea's motion prevails, and is the only one, but it is curbed somewhat by the river's flow. But if it happens that the stronger of the two winds should drop, then the opposing wind will blow from the direction it blew before when it lay hidden in the grip of the stronger.

10. If, for example, there is a nursery in the north-east, then a north-easterly will blow. But if there are two wind nurseries, one in the east and another in the north, <sup>|</sup> the winds will blow separately for some distance. But at the angle where they meet, they will blow north-east or lean in the direction of the stronger nursery.

**11.** Si sit Fomes *Venti* ex parte *Boreali*, qui distet ab aliquâ regione 20. Milliaribus, & sit fortior, alter ex parte *Orientali*, qui distet 10. Milliaribus, & sit debilior; spirabit tamen ad aliquas horas Eurus; paulò post (nimirùm post emensum iter) *Boreas.*

5 **12.** Si spiret *Boreas*, atque occurrat ab Occidente Mons aliquis, spirabit paulò post *Euro-boreas*, compositus scilicet ex *Vento* originali, & repercusso.

[L6ᵛ] **13.** Si sit Fomes *Ventorum* in terˈrâ, à parte *Boreæ*, halitus autem eius feratur rectâ sursùm, & inueniat nubem gelidam ab Occidente, quæ
10 eam in aduersum detrudat, spirabit *Euro-boreas.*

*Monitum. Fomites* Ventorum, *in Terrâ & Mari, sunt stabiles, ita vt fons & origo ipsorum, meliùs percipiatur.* at Fomites Ventorum in *Nubibus, sunt mobiles; adeò vt alibi suppeditetur Materia* Ventorum, *alibi vero ipsi formentur; id quod efficit Directionem Motûs in* Ventis *magis*
15 *confusam, & incertam.*

Hæc Exempli gratiâ adduximus; similia simili modo se habent. Atque de *Directione Motûs Ventorum*, hactenus. At de *Longitudine*, &
[L7ʳ] tanquam Itineˈrario *Ventorum*, videndum; licet de hoc ipso, paulò antè, sub nomine latitudinis *Ventorum*, inquisitum videri possit. Nam &
20 Latitudo, pro Longitudine, ab imperitis haberi possit, si maiora spatia *Venti* ex latere occupent, quam in longitudine progrediantur.

**14.** Si verum sit, *Columbum* ex oris *Lusitaniæ*, per *Ventos Statos* ab Occidente, de *Continente* in *Americâ*, iudicium fecisse, longo certè itinere possint commeare *Venti.*

25 **15.** Si verum sit Solutionem Niuium, circa *Mare glaciale*, & *Scandiam* excitare *Aquilones*, in *Italiâ* & *Græciâ*, &c. diebus Canicularibus, longa certè sunt spatia.

[L7ᵛ] **16.** Quantò citiùs in Consecutiˈone, in quâ *ventus* mouet (Exempli gratiâ, si sit *Eurus*) veniat tempestas, ad locum aliquem ab Oriente,
30 quantò verò tardiùs ab Occidente, non-dum venit in obseruationem.

De *Motu ventorum in progressu* hactenus; videndum iam de *vndulatione ventorum.*

**17.** *Vndulatio ventorum* ad parua momenta fit; adeò vt centies in horâ, ad minùs, *ventus* (licet fortis) se suscitet, & alternatìm remittat,
35 ex quo liquet inæqualem esse impetum *ventorum*. Nam nec Flumina,

---

5 aliquis,] ~;  5–6 spirabit] Spirabit  16 habent.] ~:  25 Niuium,] ~; / BNF copy with semicolon; Folger copy 1 with comma  28 mouet] ~,  31 De *Motu ventorum*] / no new para. in *c-t*, but evidence of crowding suggests that the compositor had to run on regardless; also see *tns* to K7ʳ  33 fit;] ~:  35 *ventorum.*] ~;

11. If there is a wind nursery to the north twenty miles away from a given region, and if it is stronger, and if there is a weaker one from the east which is ten miles off, the easterly will still blow for some hours, but soon after (when it has completed its journey) the northerly.

12. If the north wind is blowing, and runs up against a mountain to the west, it will soon turn north-east, made up as it is of the original wind and the rebound.

13. If there is a wind nursery in the earth | to the north, and its exhalation goes straight up and finds a cold cloud from the west which drives it east, a north-easterly will blow.

*Advice.* Wind nurseries by land and sea are stable, such that their source and origin can be more easily detected. But wind nurseries in the clouds are not fixed such that the matter of winds is supplied in one place but the winds themselves are formed in another; and that is what makes the direction of their motion more confused and unpredictable.

I have adduced these things by way of example; for similar things hold in similar cases. So much then for the winds' direction of motion. Now we must consider their longitude and, so to speak, their | journey, though it may seem that this very thing was looked into a little while ago under the heading of the latitude of winds. For people can confuse latitude with longitude if the winds take up space of greater breadth than the longitudinal distance that they move over.

14. If it be true that Columbus on the shore of Portugal guessed from the recurrence of westerly winds that the American continent existed, then the winds can certainly travel a great distance.

15. If it be true that snow melt in the Icy Sea and Scandinavia prompts northerlies in Italy and Greece during the Dog-days, then the distances involved are great indeed.

16. How rapidly a tempest comes | in the wake of a wind from a particular direction has not been observed; as for instance (if it be an easterly wind) how quickly a storm arrives from the east, or how slowly from the west.

So much then for the winds' progressive motion; now we must inquire into their fluctuation.

17. The winds' fluctuation happens in short order, for a wind (though strong) will alternately rise and fall at least a hundred times an hour, which shows that the winds' force is uneven. For neither rivers

licet rapida, nec Currentes in mari, licet robusti, vndulant nisi accedente flatu *ventorum*; Neque ipsa illa *vndulatio ventorum* aliquid æqualitatis [L8ʳ] habet in se; | Nam instar pulsûs manûs, aliquandò intercurrit, aliquandò intermittit.

5    **18.** *Vndulatio* Aeris, in eo differt, ab *vndulatione* Aquarum; quod in Aquis, postquam fluctus sublati fuerint in altum, sponte rursus decidant ad planum; ex quo fit, vt (quicquid dicant *Poetæ* exaggerando tempestates, quod *vndæ attollantur in Cœlum, & descendant in Tartarum*) tamen descensus vndarum, non multùm præcipitetur, vltra planum, & 10  superficiem Aquarum. At in *vndulatione* Aeris, vbi deest motus grauitatis; deprimitur, & attollitur Aer, ferè ex æquo.

[L8ᵛ]    De *vndulatione* hactenus: iam de *Motu Conflictûs* inquirendum est. |

**19.** De *Conflictu ventorum*, & compositis Currentibus, iam partìm inquisitum est. Planè constat, *Vbiquetarios* esse *ventos*, præsertim 15  leniores; id quod manifestum etiam ex hoc, quod pauci sunt dies, aut horæ, in quibus non spirent Auræ aliquæ lenes, in locis liberis, ídque satis inconstantèr, & variè. Nam venti, qui non proueniunt ex Fomitibus maioribus, vagabundi sunt, & volubiles, altero cum altero quasi ludente, modò impellente, modò fugiente.

20    **20.** Visum est nonnunquam in Mari, aduenisse duos ventos simul ex contrarijs partibus, id quod ex perturbatione superficiei Aquæ ab [Mrʳ] vtraque parte, atque | tranquillitate Aquæ in medio inter eos, facilè erat conspicere; postquam autem concurrissent illi *venti* contrarii, aliàs secutam esse tranquillitatem in Aquâ vndique, cum scilicet *venti* se ex 25  æquo fregissent, aliàs continuatam esse perturbationem Aquæ, cum scilicet fortior *ventus* præualuisset.

**21.** Certum est in Montibus *Peruvianis*, sæpe accidere, vt *venti*, eôdem tempore, super Montes ex vnâ parte spirent, in vallibus in contrarium.

30    **22.** Itidem certum apud nos, Nubes in vnam partem ferri, cùm *ventus* à contrariâ parte flet, hîc in proximo.

[Mrᵛ]    **23.** Quin & illud certum, ali|quando cerni Nubes altiores, superuolare Nubes humiliores; atque ita vt in diuersas, aut etiam in contrarias partes abeant, tanquam Currentibus aduersis.

---

27 21] 20 / this is the second item numbered 20 in *c-t*; accordingly all the remaining *c-t* numbers in this section have had to be corrected, as in *SEH* (II, p. 57), in the edited text

7 planum;] ~:    12 De *vndulatione*] / no new paragraph in *c-t*, but see *tns* to L7ᵛ 15 leniores;] ~:    23 conspicere;] ~:    31 flet,] / BNF copy with semicolon; Pierpont Morgan copy with comma    33 ita] ~,

though rapid, nor ocean currents though strong, fluctuate without winds blowing. And this fluctuation of the winds has no evenness to it, | for it is like the pulse at the wrist, sometimes running on, and sometimes slackening off.

18. The fluctuation of the air differs from that of the waters in this respect: that in the latter the waves rise and then fall flat spontaneously; whence it follows that (despite the poets' inflated talk of *waves that rise to the heavens and sink into hell*) they still do not fall much below the surface of the water when it is flat. But in the motion of the air, where motion of gravity is lacking, the air is raised and falls to pretty much the same degree.

So much then for fluctuation, and now for motion of conflict. |

19. We have already in part looked into wind conflict and mixed air currents. It is quite clear that winds are to be found everywhere, especially the gentler ones, as is also obvious from the fact that there are days or hours on which some light winds are not blowing in open spaces, and that inconstantly, and variably enough. For winds which do not come from the larger nurseries are wayward and unstable, sometimes coming on and sometimes running off, as if in play.

20. We sometimes see winds coming from opposite quarters at sea, as can easily be seen from the disturbance of the water's surface | on both sides and from the calm water in between. But after these contrary winds have come together then follows a calm sea everywhere, when the winds temper each other equally; but if a stronger wind prevails, then the disturbance of the water carries on.

21. It is certain that in the Peruvian mountains that at the same time winds often blow in one direction up the mountains, but in the opposite one in the valleys.

22. It is also certain that here with us the clouds are carried in one direction, while the winds down here blow in the opposite one.

23. This too is certain, that | we sometimes see higher clouds flying above lower ones, such that they go in different or even opposite directions, like contrary currents.

**24.** Itidem certum, quandoque in superiore Aere, *ventos* nec distrahi, nec promoueri; cum hic infra ad Semi-milliare insano ferantur impetu.

**25.** Certum etiam è contrà, esse aliquando tranquillitatem infrà, cum
5 supernè Nubes ferantur satis alacritèr; sed id rarius est.

*Phœnomenon obliquum. Etiam in fluctibus, quandoque supernatans Aqua, quandoque demersa, incitatior est; quinetiam fiunt (sed raro) varij*
[M2ʳ] *Currentes Aquæ, quæ voluitur suprà, & quæ labitur in imo.* |

**26.** Neque prorsus contemnenda illa testimonia *Virgilij*, cùm
10 Naturalis Philosophiæ non fuerit ipse omninò imperitus:

> *Vna* Eurus Notusq*ue ruunt, creberq*ue *procellis,*
> Africus.————
> Et rursus:
> *Omnia* ventorum *concurrere prælia vidi.*

15 De *Motibus Ventorum*, in Naturâ rerum, inquisitum est: Videndum de *Motibus* eorum in *Machinis humanis*; Ante omnia in *velis Nauium.*

| *Motus Ventorum in velis nauium.*

**1.** In *Nauibus* maioribus *Britannicis* (eas enim ad exemplum delegimus) quatuor sunt *Mali*, aliquando quinque; omnes in lineâ rectâ per
20 mediu*m* Nauis ductâ, alteri post alteros, erecti. Eos sic nominabimus:

**2.** *Malum principem*, qui in medio Nauis est; *Malum Proræ*; *Malum Puppis* (qui aliquando est geminus); & *Malum Rostri.*

**3.** Habent singuli *Mali* plures portiones; quæ sustolli, & per certos
24 *Nodos*, aut articulos figi, & similitèr auferri possunt; alij tres, alij duas
[M3ʳ] tantum. |

**4.** *Malus Rostri* stat ab inferiori *Nodo*, inclinatus versus mare, à superiori, rectus; reliqui omnes *Mali* stant recti.

**5.** His *Malis* superimpendent *Vela* decem, & quando *Malus Puppis* geminatur, duodecim. *Malus Princeps*, & *Malus Proræ* tres habent
30 *Ordines velorum.* Eos sic nominabimus: *Velum ab infrà, velum à suprà,* & *velum à summo.* Reliqui habent duos tantum, carentes *velo à summo.*

**6.** *Vela* extenduntur in transuersum, iuxta verticem cuiusque *Nodi Mali*, per ligna quæ *Antennas*, vel virgas dicimus, quibus suprema

2 Semi-milliare] ~;   10 imperitus:] ~.   13 rursus:] ~.   18 delegimus)] ~,)
20 nominabimus:] ~.   21 est;] ~:   *Proræ*,] ~:   22 geminus);] ~)ₓ
30 nominabimus:] ~.

90

**24.** It is also certain that in the upper air the winds are sometimes not subject to conflicting impulses, while here, half a mile below, they rage furiously.

**25.** The opposite is also certain: that sometimes when it is calm below the clouds above are carried along smartly enough; but this happens less often.

*An Indirect Phenomenon.* In waves too sometimes the water floating above, sometimes that sunk below, moves faster; and sometimes (though seldom) there are different currents which move rapidly above, but loiter below. |

**26.** Nor should we altogether despise the testimony of Virgil, since he was by no means a stranger to natural philosophy:

> East and South winds together, and the South-wester, thick
> with tempest.————
> And again:
> My own eyes have seen all the winds clash in battle.

So much then for the motions of winds; now let us examine their motions in human contrivances; and above all in the sails of ships.

## | Wind Motion in the Sails of Ships

**1.** On the larger British ships (and I take these as my example) there are four and sometime five masts, all standing upright one behind the next in a line down the ship's middle. This is what we shall call them:—

**2.** The mainmast amidships; the foremast, the mizzenmast (sometimes doubled up), and the bowsprit.

**3.** The individual masts have several parts, two or three in all, which can be raised and secured by certain knots or joints, and taken down likewise. |

**4.** From its lower fastening the bowsprit leans toward the sea; from the upper it is upright. All the other masts stand upright.

**5.** Ten sails hang from these masts, and twelve when the mizzenmast is doubled up. The mainmast and foremast have three ranks of sails, which we call the mainsail, topsail, and topgallant sail. This last is lacking on the other masts which carry only two sails.

**6.** The sails are stretched out sideways, near the top of each joint of the mast, on beams which we call yards or spars, to which the tops of the

*velorum* assuuntur, ima ligantur funibus ad angulos tantùm; *vela* scilicet [M3ʳ] | *ab infrà*, ad latera Nauis, *vela à suprà, aut à summo*, ad *antennas* contiguas. Trahuntur etiam aut vertuntur ijsdem funibus, in alterutrum latus, ad placitum.

5    7. *Antenna* siue virga cuiusque *Mali* in transversum porrigitur. Sed in *Malis puppis*, ex obliquo, altero fine eius eleuato, altero depresso; in cœteris in recto, ad similitudinem literæ *Tau*.

   8. *Vela ab infrà*, quatenus ad *vela Principis, Proræ*, & *Rostri*, sunt figuræ Quadrangularis, Parallelogrammæ; *vela à suprà*, *& à summo*, 10 nonnihil acuminata, siue surgentia in arctum; at ex *velis puppis*, quod à *suprà*, acuminatum, quod *ab infrà*, triangulare.

[M4ʳ]    9. In *Naui*, quæ erat mille & | centum amphorarum, atque habebat in longitudine, in carinâ, pedes 112, in latitudine, in alueo 40, *velum ab infrà, Mali Principis*, continebat in altitudine, pedes 42, in latitudine, 15 pedes 87.

   10. *Velum à suprà* eiusdem *Mali*, habebat in altitudine pedes 50, in latitudine pedes 84, ad basim; pedes 42, ad fastigium.

   11. *Velum à summo*, in altitudine pedes 27, in latitudine, pedes 42, ad basim; 21, ad fastigium.

20    12. In *Malo Proræ, velum ab infrà*, habebat in altitudine, pedes 40, cum dimidio; in latitudine pedes, 72.

[M4ᵛ]    13. *Velum à suprà* in altitudine pedes 46, cum dimidio; in latitu|dine pedes 69, ad basim; 36, ad fastigium.

   14. *Velum à summo*, in altitudine pedes 24, in latitudine, pedes 36, ad 25 basim; 18, ad fastigium.

   15. In *Malo Puppis, velum ab infrà*, habebat in altitudine, à parte *antennæ* eleuatâ, pedes 51; in latitudine, quâ iungitur *antennæ*, pedes 72, reliquo desinente in acutum.

   16. *Velum à suprà*, in altitudine pedes 30; in latitudine, pedes 57, ad 30 basim; 30, ad cacumen.

   17. Si geminetur *Malus puppis*, in posteriore, vela minuuntur ab anteriore, ad partem circiter quintam.

[M5ʳ]    18. In *Malo Rostri, velum ab infra*, habebat in altitudine, pedes 28, | cum dimidio; in latitudine, pedes 60.

35    19. *Velum à suprà*, in altitudine, pedes 25, cum dimidio; in latitudine, pedes 60, ad basim; 30, ad fastigium.

9 Parallelogrammæ;] ~:    18 27,] ~:    21 dimidio;] ~,    25 basim;] ~,
27 51;] ~.    29 30;] ~,    30 basim;] ~,

sails are attached, while the lower edges are tied with ropes at the corners only; so that the bottom of the ᴵ mainsail is secured to the sides of the ship, and the topsail or topgallant sail to the yards adjacent to them. The sails are pulled or turned by these ropes to either side at will.

7. The yard or spar of each mast is stretched out horizontally, but the one on the mizzenmast is set slantwise with one end higher and the other lower; the rest are set at right angles to the mast, like the top of the letter T.

8. The mainsails of the mainmast, foremast, and bowsprit are oblong or parallelogram in form; the topsail and topgallant sail are somewhat pointed and rise to a peak; but of the sails on the mizzenmast the higher rises to a point, while the lower is triangular.

9. In a ship of 1100 ᴵ tons, 112 feet at the keel, and 40 feet wide in the hold, the mainsail of the mainmast was 42 feet high and 87 feet wide.

10. The topsail of the same was 50 feet high, and 84 feet wide at the bottom and 42 at the top.

11. The topgallant sail was 27 feet high, and 42 feet wide at the bottom and 21 at the top.

12. The mainsail of the foremast was 40 and a half feet high, by 72 feet wide.

13. The topsail was 46 and a half feet high, and 69 feet wide ᴵ at the bottom and 36 at the top.

14. The topgallant sail was 24 feet high, and 36 feet wide at the bottom, 18 at the top.

15. The mainsail of the mizzenmast was 51 feet high from the higher end of the yard, and 72 feet wide where it joined the yard, with its width narrowing to a point.

16. The topsail was 30 feet high, and 57 feet wide at the bottom and 30 at the top.

17. If there are twin mizzenmasts, the sails of the rear one are about a fifth smaller than those of the front.

18. The mainsail of the bowsprit was 28 and ᴵ a half feet high, and 60 wide.

19. The topsail was 25 and a half feet high, and 60 feet wide at the bottom, 30 at the top.

20. Variant proportiones *Malorum* & *velorum*, non tantum, pro magnitudine *Nauium*, verum etiam pro varijs earum vsibus, ad quos ædificantur: ad pugnam, ad mercaturam, ad velocitatem, & cætera. Verum nullo modo conuenit proportio dimensionis *velorum* ad
5 numerum amphorarum, cum *Nauis* quingentarum Amphorarum, aut circiter, portet *velum ab infrâ principis Mali,* paucos pedes minus
[M5ᵛ] vndique, quam illa altera, ¹ quæ erat duplicis magnitudinis. Vnde fit, vt minores *Naues* longè *præstent* celeritate maioribus, non tantum propter leuitatem, sed etiam propter amplitudinem *velorum,* habito respectu
10 ad corpus *Nauis;* nam proportionem illam continuare in *Nauibus* maioribus, nimis vasta res esset, & inhabilis.

21. Cum singula *vela* per summa extendantur, per ima ligentur tantùm ad angulos; *ventus* necessariò facit *vela* intumescere, præsertim versus ima, vbi sunt laxiora.

15 22. Longè autem maior est tumor *veli,* in *velis* ab infrà, quam in
[M6ʳ] cæteris, quia non solu*m* parallelogramma sunt, cætera acumiʲnata; verùm etiam, quia latitudo *antennæ* tantò excedit latitudinem laterum *nauis,* ad quæ alligantur; vnde necesse est, propter laxitatem, magnum dari receptum *ventis;* adeo vt, in illâ magnâ, quam exempli loco
20 sumpsimus, *naui,* tumor in *Vento recto* possit esse ad 9. aut 10. pedes introrsùm.

23. Fit etiam, ob eandem causam, quod *vela* omnia à *Vento* tume-facta, ad imum colligant se in arcus, adeo vt multum *Venti,* præterlabi, necesse sit; in tantum, vt in illâ, quam diximus, *naui,* arcus ille ad
25 staturam hominis accedat.
[M6ᵛ] 24. At in *Velo puppis* illo triangulari, necesse est vt minor sit tuʲmor, quam in quadrangulari; tum propter figuram minùs capacem, tum quia in quadrangulari tria latera laxa sunt, in triangulari duo tantùm, vnde sequitur quod *Ventus* excipiatur magis rigidè.

30 25. Motus *Ventorum* in *velis,* quo magis accedat ad *Rostrum Nauis,* est fortior, & promouet magis; tum quia fit in loco, vbi vndæ, propter acumen *proræ,* facillimè secantur; tum maximè, quia Motus à *prorâ,* trahit nauem; Motus à *puppi,* trudit.

26. Motus *Ventorum* in *velis superiorum ordinum,* promouet magis,
35 quam in *velis ordinis inferioris;* quia Motus violentus maximè efficax est,
[M7ʳ] vbi plurimum ¹ remouetur à resistentiâ, vt in vectibus & *velis molendino-rum.* Sed periculum est demersionis, aut euersionis *nauis;* itaque &

---

3 ædificantur:] ~;     pugnam,] ~;     mercaturam,] ~;     10 *Nauis*,] ~:
18 alligantur;] ~,     19 *ventis*,] ~:     24 sit;] ~:

20. The proportions of masts and sails vary not just with the size of the ship but with the purposes for which they are built: for fighting, trade, speed, or whatever. But in no way is the amount of canvas proportional to tonnage, for a vessel of 500 tons or thereabouts will carry a mainmast mainsail of just a few feet less than a ship of ¹ twice the size. And this is the reason why smaller ships go much faster than large ones; they are not just lighter but they carry more sail relative to their size, and if the same proportion held for larger ships the sails would be enormous and impossible to handle.

21. Since each sail is stretched out at the top, and tied only at the corners at the bottom, the wind necessarily makes them fill out, especially towards the bottom where they are slacker.

22. The mainsails fill out much more than the others, not only because they are shaped like parallelograms, and the others pointed, ¹ but because the width of the yard so much exceeds the width of the ship's sides to which they are attached. Because of this the sails are necessarily slack and take up a great deal of wind, such that in the big ship that I have taken as an example the sail swells inwards by 9 or 10 feet in a following wind.

23. For the same reason it also happens that all sails filled out by the wind arch out at the bottom so that much wind necessarily slips past them, such that in the ship I have mentioned the arching comes close to a man's height.

24. But in the mizzenmast triangular sail, the swelling is necessarily ¹ slighter than in an oblong one, both because its shape is less capacious and because in an oblong one three sides are slack but only two are so in a triangular; whence it follows that it is stiffer for catching the wind.

25. The nearer the wind in sails gets to the ship's bow the stronger it is and the more it drives it on, both because it happens at a place where the waves are very easily sundered by sharpness of the bow, and most of all because motion from the stem pulls the ship, while motion from the stern pushes it.

26. The winds' motion in the upper sails drives the ship on more than motion in the lower ones because violent motion is most effective where it is most ¹ remote from resistance, as in levers and the sails of windmills. But as there is a risk of sinking or capsizing the ship, these

acuminata sunt illa, ne *Ventos* nimios excipiant, & in vsu præcipuè, cum spirent *Venti* leniores.

**27.** Cum *vela* collocentur in rectâ lineâ, altera post altera, necesse est, vt quæ posteriùs constituantur, suffurentur *Ventum* à prioribus, cum
5 *Ventus* flet rectâ; itaque si omnia simul fuerint erecta, tamen vis *Venti* ferè tantùm locum habet, in *velis Mali principis,* cum paruo auxilio, *veli ab infrà,* in *Malo Rostri.*

[M7ᵛ] **28.** Fœlicissima & commodissima dispositio *velorum,* in *Vento* ⟨recto,⟩ ea est; vt vela duo inferiora *mali proræ* erigantur; ibi enim (vt dictum est)
10 Motus est maxime efficax; erigatur etiam *velum à suprà Mali principis,* relinquitur enim spatium tantum subtèr, vt *Ventus* sufficere possit, *velis* prædictis *Proræ,* absque suffuratione notabili.

**29.** Propter illam, quam diximus, suffurationem *Ventorum,* celerior est nauigatio, cum *Vento laterali,* quam cum *recto. Laterali* enim flante,
15 omnia *vela* in opere poni possunt; quia latera sibi inuicem obuertunt, nec altera altera impediunt, neque fit furtum.

[M8ʳ] **30.** Etiam flante *vento laterali, vela* rigidiùs in aduersum *Venti* ex⟨-⟩tenduntur, quod *Ventum* comprimit nonnihil, & immittit in eam partem, vbi flare debet: vnde nonnihil fortitudinis acquirit. *Ventus* autem
20 maximè propitius est, qui flat in quadrâ, inter rectum & lateralem.

**31.** *Velum ab infrà, Mali Rostri* vix vnquam posset esse inutile; neque enim patitur furtum, quando colligat *ventum* qui flat vndequaque, circa latera nauis, & subter *vela* cætera.

**32.** Spectatur in *Motu Ventorum* in *nauibus* tum Impulsio, tum
25 Directio. At Directio illa quæ fit per *clauum,* non multum pertinet, ad Inquisitionem præsentem, nisi quatenus habeat connexionem, cum
[M8ᵛ] *Motu Ventorum in velis.* ⟨

*Connexio. Vt Motus* Impulsionis *in vigore est in* Prorâ, *ita motus* Directionis *in* puppi; *itaque ad eum,* velum ab infrà Mali puppis,
30 *est maximi momenti; & quasi copiam præbet auxiliarem* clauo.

**33.** Cum Pyxis nautica in plagas 32. distribuatur, adeo vt semicirculi eius sint plagæ sedecim, potest fieri Nauigatio progressiua (non angulata, quæ fieri solet in *ventis* planè contrarijs) etiamsi ex illis sedecim partibus, decem fuerint aduersæ, & sex tantum fauorabiles; at ea

---

5 rectâ;] ~:      9 enim] ~,      10 efficax;] ~:      *principis,*] ~:
21 inutile;] ~,      28 **Connexio.**] / the blank line preceding this para. is editorial. In *c-t*
the para. begins on a new page      33 contrarijs)] ~,)

sails rise to a point so that they do not catch too much wind, and they are used mostly when light winds blow.

27. Since sails are placed in a line one after another it is necessarily the case that in a following wind the sails behind rob the ones in front of wind; therefore if all the sails were raised at the same time the force of the wind would be almost entirely taken by sails of the mainmast, with little assistance from the bowsprit mainsail.

28. In a following wind the most propitious and convenient setting of ¹ sails is to hoist the two lower sails of the foremast for there (as I have said) the motion is most effective. Also set is the topsail of the mainmast, for then there remains enough space below for the wind to swell the aforementioned bowsprit sails without appreciable robbery.

29. On account of the robbery of winds which I have spoken of, you sail faster with a side than with a following wind. For when a side wind blows all sails can be put into action because they all turn their sides to each other, and do not obstruct one another, and no robbery occurs.

30. Indeed when a side wind is blowing the sails are stiffer against the wind, ¹ which compresses it and drives it to that quarter where it ought to blow; whence it gathers some strength. But the most favourable wind is one which blows from one of the aft quarters, i.e. between a following and a side wind.

31. The bowsprit mainsail can hardly ever be useless; for it is not a victim of robbery when it collects wind that blows everywhere round the ship's sides and from under the other sails.

32. We take account of impulse and direction in wind motion in ships. But direction by the rudder does not have much to do with the present inquiry, except insofar as it has a connection with the motion of winds in sails. ¹

*Connection.* As the motion of impulsion is strongest in the prow, so is motion of direction at the stern; therefore for the latter the mizzenmast mainsail is the most important, and lends substantial help to the rudder.

33. Since the mariner's compass is distributed into thirty-two points, such that sixteen make up a semicircle, a voyage can carry on ahead (and not by tacking as commonly happens with contrary winds) even if ten of the sixteen are contrary and six favourable. But this same voyage

[N1ʳ] Nauigatio, multum pendet ex | *velo ab infrà mali puppis*; Cùm enim *venti* partes contrariæ itineri, quia sunt præpotentes, & *Clauo* solo regi non possunt; alia vela obversuræ forent, vnà cum *Naui* ipsâ, in partem contrariam itineris, illud *velum* rigidè extensum, ex opposito fauens
5 *Clauo*, & eius motum fortificans, vertit & quasi circumfert *Proram* in viam itineris.

**34.** Omnis *ventus* in *velis* nonnihil aggrauat, & deprimit *Nauem*; tantóque magis, quò flauerit magis desupèr. Itaque tempestatibus maioribus, primò devolvunt Antennas, & auferunt *vela* superiora,
10 deinde, si opus fuerit, omnia; Etiam *Malos* ipsos incîdunt; quin &
[N1ᵛ] proji|ciunt onera mercium, tormentorum &c. vt alleuent *Nauem*, ad supernatandum, & præstandum obsequia vndis.

**35.** Potest fieri per motum istum *ventorum* in *velis nauium* (si ventus fuerit alacris, & secundus) progressus in itinere, 120 Milliarium *Italico-*
15 *rum* intra spatium 24. horarum; ídque in *Naui mercatoriâ*; sunt enim *Naues* quædam *Nunciæ*, quæ ad Officium celeritatis appositè extructæ sunt (quas *Caruvellas* vocant) quæ etiam maiora spatia vincere possunt. At cum *venti* planè contrarij sint, remedio ad iter promouendum
19 vtuntur hoc vltimo, & pusillo; vt procedant lateralitèr, prout ventus
[N2ʳ] permit|tit extra viam itineris, deinde flectant se versus iter, atque angulares istos progressus repetant; Ex quo genere progressûs (quod est minus quam ipsum serpere, nam serpentes sinuant, at illi angulos faciunt) poterint fortassè intra 24. horas, vincere milliaria 15.

Obseruationes maiores.

25 **1.** Motus *iste* Ventorum, *in* Velis nauium, *habet Impulsionis suæ*
[N2ᵛ] *tria præcipua Capita, & Fontes, vnde fluit; vnde* | *etiam præcepta sumi possint, ad eum augendum, & fortificandum.*

**2.** *Primus Fons est ex* Quanto venti, *qui excipitur. Nam nemini dubium esse possit, quin plus* Venti *magis conferat, quàm minus.*
30 *Itaque* Quantum *ipsum* Venti *procurandum diligentèr. Id fiet, si instar patrumfamiliâs prudentiorum, & frugi simus, & à furto caue-amus. Quare quantum fieri potest, nil* Venti *disperdatur, aut*
[N3ʳ] *effundatur;* | *nil etiam surripiatur.*

**3.** Ventus *aut supra latera* Nauium *flat, aut infra, vsque ad*
35 Aream Maris. Atque vt homines prouidi, solent etiam circa minima

---

1 *puppis*;] ~:     14 secundus)] ~,)     15 sunt] ~,

depends heavily on | the mizzenmast mainsail. For since the parts of the wind contrary to the journey are stronger, and cannot be countered by the rudder alone, they would turn the other sails and the ship itself in the opposite direction if that sail, stiffly stretched, did not favour the rudder the other way, and, strengthening its motion, turned and brought the bow round on to the correct course.

34. All wind in sails makes the ship somewhat heavier and weighs it down, and the more so the more the wind blows from above. In severe storms therefore they first lower the yards and furl the topsails, then if need be they furl all, and cut down the very masts themselves, and | throw cargo, guns, etc. over the side to lighten the ship so that it will float and go with the flow.

35. It can so happen by means of this motion of the winds in sails (if the wind be a following one and keen) that distance of 120 Italian miles can be covered in twenty-four hours; and that in a merchantman. But there are other ships (called caravels) which carry mail which have been built especially for speed, and they can cover an even greater distance. But when straight headwinds are blowing, they have this one final and feeble method of making progress: that as much as the wind allow, they go sideways | off their course and then turn back on to it again, and they repeat this tacking (which is slower than serpentine motion, for snakes wind while ships zigzag) and by this method can perhaps make just fifteen miles in twenty-four hours.

## Major Observations

1. The winds' motion in ships' sails flows from three springs and sources whence it draws its strength, | and from which directions can also be derived for increasing and strengthening it.

2. The first source springs from the quantity of the wind taken up. For no one doubts but that more wind is better than less. Accordingly wind in quantity should be carefully acquired. This can be done if, like prudent heads of families, we are frugal and watch out for theft. Therefore ensure as far as possible that no wind is lost, or wasted, | and none is carried off by stealth.

3. The wind blows either above ships' sides, or below right down to the sea's surface. And just as provident men generally look after the smallest matter (for no one neglects the larger

*quæque magis curare (quia maiora nemo non curare potest) ita
de istis inferioribus* ventis *(qui proculdubiò non tantùm possunt,
qua*ntum *superiores) primò videndum.*

**4.** *Ad* ventos, *qui circum latera* Nauium, *& subter* vela *ipsarum*
[N3ᵛ] *potissimùm* | *flant; planè est officium* veli ab infrà Mali rostri, *quæ*
6 *inclinata est, & depressa, vt excipiantur, ne fiat dispendium, &
iactura* venti. *Idque & per se prodest, & ventis, qui reliquis velis
ministrant, nil obest. Circa hoc, non video quid vlterius per diligentiam humanam fieri possit, nisi fortè etiam ex medio* Nauis, *similia*
10 vela *humilia adhibeantur, instar Pinnarum, aut Alarum, ex vtroque*
[N4ʳ] *latere gemina, cum* ventus *est* rectus. |

**5.** *At quod ad cauendum de furto attinet, quod fit, cum* vela
*posteriora* ventum *ab anterioribus surripiant, in* vento *recto (nam
in* Laterali *omnia* vela *cooperantur) non video quid addi possit dili-*
15 *gentiæ humanæ; nisi fortè, vt flante* vento recto, *fiat scala quædam*
velorum, *vt posteriora* vela *à* Malo puppis *sint humillima, media à*
Malo principis *mediocria, anteriora à* Malo proræ *celsissima; vt*
[N4ᵛ] *alterum* ve|lum *alterum non impediat, sed potiùs adiuuet, &*
ventum *tradat, & transmittat. Atque de primo Fonte Impulsionis,*
20 *hæc obseruata sint.*

**6.** *Secundus fons Impulsionis, est ex Modo* percussionis veli, *per*
ventum; *quæ si propter* ventum *contractum sit acuta, & rapida,
mouebit magis; si obtusa, & languida, minùs.*

**7.** *Quod ad hoc attinet, plurimùm interest vt* vela *mediocrem*
[N5ʳ] *extensionem, & tu* | *morem recipiant; nam si extendantur rigidè,*
26 *instar Parietis* ventum *repercutiunt; si laxè, debilis fit Impulsio.*

**8.** *Circa hoc, benè se expediuit in aliquibus Industria humana,
licet magis ex casu, quam ex iudicio. Nam in* Vento Laterali, *contrahunt part*em veli, *quæ* Vento *opponitur, quantum possunt; atque hoc*
30 *modo* Ventum *immittunt in eam partem, quâ flare debet. Atque hoc*
[N5ᵛ] *agunt, & volunt. Sed interim hoc sequitur (quod for* | *tassè non vident)
vt* Ventus *sit contractior, & reddat* percussionem *magis acutam.*

**9.** *Quid addi possit industriæ humanæ, in hâc parte, non video;*

---

1 *curare] ~, potest)] ~,)*   6 *inclinata] inlinata / nld* thus in *SEH* (II, p. 62)
8 *obest.] ~:*   12 *attinet,] ~,, / sic*   13 *recto] ~;*   14 *cooperantur)] ~;)*
25 *recipiant;] ~:*

ones), so we must first look to these lower winds (though they cannot do as much as the higher).

4. As for the winds which blow about ships' sides and especially beneath the sails, ꞁ it is clearly the function of the sail beneath the bowsprit, slanting and bent downwards as it is, to catch them, and prevent loss or expenditure of wind. And that is both useful in itself, and does not get in the way of the winds which serve the other sails. On this matter, I do not see what else can be accomplished by human effort, unless perhaps similar lower sails could be introduced amidships like feathers or wings, set two to each side, when the wind is right. ꞁ

5. As for guarding against the theft which happens when the foresails snatch wind from the back ones in a headwind (for in a sidewind the sails work together), I do not see what else human effort can add, except perhaps when a headwind is blowing to make a kind of ladder of sails, so that the foresails from the mizzenmast be set lowest, the sails of the mainmast at middling height, and those of the foremast highest, so that one sail ꞁ won't get in the way of the next, but rather help it and hand on and transmit the wind. And so much for the first source of force.

6. The second source of force is the way that the sails are struck by the wind, which if because of the wind's contraction the impact is sharp and quick, it will move more, but if obtuse and feeble less.

7. In this connection, it is important that the sails get a moderate stretching and swelling; ꞁ for if they are stretched tight, they behave like walls to knock back the wind; if they are slack, the wind's force is weaker.

8. In this regard, human industry has in some things given a good account of itself, though more by accident than design. For in a side wind they shorten as much as they can the part of the sail opposite to the wind; and in that way they drive the wind to that quarter from which it ought to blow. They do this quite deliberately. But meanwhile a consequence (which perhaps ꞁ they do not recognize) is that the wind is more compressed and its impact keener.

9. I do not see what human industry can add in this matter,

*nisi mutetur Figura in* velis, *& fiant aliqua* vela *non tumentia in rotundo, sed instar Calcaris, aut Trianguli cum* Malo, *aut ligno, in illo angulo verticis, vt &* Ventum *magis contrahant in acutum, &*

[N6ʳ] *secent Aerem externum potentiùs. Ille autem Angulus (vt ar|bitra-*

5 *mur) non debet esse omninò acutus, sed tanquam Triangulus curta-tus, vt habeat latitudinem. Neque etiam nouimus, quid profuturum foret, si fiat tanquam* velum *in* velo; *hoc est, si in medio* veli *alicuius maioris, sit bursa quædam, non omninò laxa ex carbaso, sed cum costis, ex lignis, quæ* Ventum *in medio* veli *excipiat, & cogat in*

10 *acutum.*

10. *Tertius fons Impulsionis, est ex loco, vbi fit percussio; isque*

[N6ᵛ] *duplex. Nam ex an|teriore parte* Nauis *facilior, & fortior est Impulsio, qua*m *ex posteriore; & ex superiore parte* Mali, *& veli, quàm ab inferiore.*

15 11. *Neque hoc ignorasse visa est industria humana, cùm & flante* vento *recto, plurimam in velis* Mali Proræ *spem ponant; & in malacijs, & tranquillitatibus, vela à summo erigere non negligant. Neque nobis in præsentiâ occurrit, quid humanæ, ex hâc parte,*

[N7ʳ] *industriæ addi possit; nisi fortè | quoad primum, vt constituantur*

20 *duo, aut tres* Mali *in* Prorâ *(medius rectus, reliqui inclinati) quorum* vela *propendeant; & quoad secundum, vt amplientur* vela proræ *in summo, & sint minus, quam solent esse, acuminata. Sed in vtroque cauendum incommodo periculi, ex nimiâ depressione* Nauis.

[N7ᵛ] | *Motus Ventorum in alijs Machinis humanis.*

25 1. Motus *Molendinorum* ad *Ventum,* nihil habet subtilitatis, & nihilo-minùs non benè demonstrari, & explicari solet. *Vela* constituuntur, rectâ in oppositum *Venti* flantis. Prostat autem in *Ventum* vnum latus veli, alterum latus paulatìm flectit se, & subducit à *Vento.* Conuersio autem, siue consecutio Motûs, fit semper à latere inferiore, hoc est eo,

30 quod remotius est à *Vento.* At *Ventus* superfundens se, in aduersum

[N8ʳ] Machinæ, à quatuor *velis* arctatur, & in quatuor inter|uallis, viam suam inire cogitur. Eam Compressionem non benè tolerat *ventus;* Itaque

---

21 *propendeant;*] ~:     *amplientur*] *amplientnr* / turned letter in *c-t*

except to alter the shape of the sails, and make some of them swell not in the round but in the shape of a spur or triangle, with a mast or spar in the angle where it comes to a point, so that they compress the wind more to a point, and cut the air on the other side more powerfully. Now that angle should not ⌐ (in my judgement) be altogether pointed but like a triangle blunted so that it has some breadth at the top. I also do not know what benefit may arise if one put one sail within another, i.e. if in the middle of any larger sail a kind of purse were set up not just of canvas and slack but with ribs of wood which would take the wind in the middle of the sail, and concentrate it to a point.

10. A third source of impulsion comes from the place where percussion happens, and this is twofold. For the impulsion ⌐ is stronger and easier from the front of the ship than from the rear, and from the upper part of mast and sail than from the lower.

11. Nor does human industry seem to have overlooked this: that when a following wind is blowing, sailors place great faith in the sails of the foremast; and in calms and quiet weather they do not neglect to hoist their topgallant sails. And at the moment I cannot think of anything which could in this case be added by human industry, except perhaps ⌐ in the first case to set up two or three foremasts (the middle one upright, the other two at an angle) with the sails hanging down; and in the second case to bulk out the foremast's topgallant sails, and to make them less pointed than usual. But in both cases care must be taken to beware of the inconvenient and dangerous effect of the ship settling too low in the water.

### ⌐ The Motion of Winds in other Human Contrivances

1. The motion of windmills has nothing subtle to it, but it is still not generally well demonstrated or explained. The sails stand facing straight into the wind, but with one side of the sail turned to the wind, while the other side gradually inclines and pulls itself away from the wind. Now the turning or rotary motion always starts from the lower side, i.e. the one furthest from the wind. But the wind, pouring itself up against the machine, is squeezed by the four sails, and forced to find its way through the gaps between them. ⌐ The wind does not suffer this

necesse est, vt tanquam cubito percutiat latera *velorum*, & proindè vertat, quemadmodum ludicra vertibula digito impelli, & verti solent.

2. Quod si *vela* ex æquo expansa essent, dubia res esset, ex quâ parte foret Inclinatio, vt in casu baculi. Cum autem proximum latus, quod occurrit *vento*, impetum eius deijciat in latus inferius, atque illinc in spatia, cumque latus inferiùs *ventum* excipiat, tanquam palma manûs, aut instar *veli* Scaphæ, fit protinùs conuersio ab eâ parte. Notandum autem est, Ori|ginem motus esse non à primâ Impulsione, quæ fit in fronte; sed à laterali Impulsione, post compressionem.

3. Probationes quasdam, & experimenta circa hoc, pro augendo hoc motu fecimus, tum ad pignus causæ rectè inuentæ, tum ad vsum; Imitamenta huius Motûs effingentes in *velis*, ex cartis, & *vento*, ex follibus. Igitur addidimus lateri *veli* inferiori, plicam inuersam à *vento*, vt haberet *ventus*, lateralis iam factus, campliùs quiddam, quod percuteret, nec profuit; plicâ illâ, non tam percussionem *venti* adiuuante, quam Sectionem Aeris in Consequentiâ impediente. Locauimus post ve|la, ad nonnullam distantiam, obstacula, in latitudinem diametri omnium *velorum*, vt *ventus* magis compressus, fortiùs percuteret; at hoc obfuit potiùs, repercussione motum primarium hebetante. At *vela* fecimus latiora in duplum, vt *ventus* arctaretur magis, & fieret percussio lateralis fortior. Hoc tandem magnoperè successit; vt & longè mitiore flatu fieret Conuersio, & longe magis pernicitèr volueretur.

*Mandatum.* Fortassè hoc augmentum motûs commodiùs fiet per octo vela, *quàm per* vela *quatuor, latitudine duplicatâ, nisi forte nimia moles aggrauauerit Motum. De hoc fiat Experimentum.*

*Mandatum. Etiam longitudo* velorum *facit ad Motum. Nam in rotationibus, leuis violentia versus circumferentiam, æquiparatur longè maiori versus centrum. Sed tamen hoc coniungitur incommodum; quod quo longiora sunt* vela, *eò plùs distant in summo, & minùs arctatur* ventus. Res *non malè fortassè se habeat, si* vela *sint paulò longiora, sed crescentia in Latum circa summitatem, vt palma Remi, sed de hoc nobis compertum non est.*

*Monitum. In his Experimentis, si ponantur in vsu ad* Molendina,

---

compression gladly, and it perforce elbows away at the sides of the sails and turns them accordingly, in the same way as toy pinwheels usually work when driven and turned with a finger.

2. But if the sails were spread out equally it would be uncertain as to which way they would incline, as in the case of a falling stick. But as the near side which encounters the wind, throws off the force of the wind to the lower side, and thence into the open air, and since the lower side picks up the wind like the palm of a hand or the sail of a skiff, the turning effect at once proceeds from that part. But it should be noted that | the origin of the motion does not arise from the initial impulse which comes from the front, but from the lateral impulse that follows compression.

3. I have performed certain tests and experiments on this to increase this motion, both to guarantee that the cause has been properly discovered, and for use. Simulations of this motion can be contrived with paper for sails and bellows for wind. Thus to the lower side of the sail I added a fold turned away from the wind, so that the wind now coming from the side would have a larger surface to strike against without flowing away. But this fold did not so much help the wind's percussion as get in the way of the cutting of the air. After the sails, and | to the whole width of their diameter, I placed at a certain distance obstacles so that the wind being compressed more would strike more forcibly. But this was more a hindrance than a help, as the repercussion weakened the primary motion. But I made the sails double the width to compress the wind more, and so that it would make the lateral percussion stronger. In the end this went very well, such that the sails were turned with a very much gentler current of air, and revolved more promptly.

*Direction.* This increase of motion will perhaps be accomplished by eight rather than four sails of double width, unless by chance the extra weight got in the way of the motion. Let an experiment be performed on this. |

*Direction.* The length too of the sails contributes to motion. For in rotary motions, a little pressure at the circumference is equal to a great deal more towards the centre. Nevertheless there is a disadvantage attached to this, i.e. that the longer the sails, the further they stand apart at the top, and the less the wind is compressed. It could turn out well if the sails were longer, but increasing in width towards the top like the blade of an oar. But this is something of which I have no personal knowledge.

*Advice.* These experiments, if they were applied to windmills, the

*robori totius Machinæ, præcipuè Fundamentis eius, subueniendum. Nam quanto magis arctatur* Ventus, *tanto magis* (*licet motu*m veloru*m incitet*)
[O2<sup>r</sup>] *tamen Machinam ipsam concutit.* |

4. Traditur alicubi esse *Rhedas* mouentes ad *Ventum*; de hoc diligen-
5 tiùs inquiratur.

**Mandatum.** Rhedæ *mouentes ad* Ventum *non poterint esse operæ pretium, nisi in locis apertis, & planitiebus. Prætereà, quid fiet, si decubuerit* Ventus? *Magis sobria esset cogitatio, de facilitando Motu Curruum, & plaustrorum, per* vela *mobilia, vt Equi, vel Boues minoribus viribus ea*
10 *traherent, quam de creando Motu, per* Ventum *solum.*

## Prognostica Ventorum.

[O2<sup>v</sup>] **Ad Artic. 32. Connexio.** Diuinatio *quò magis pollui solet vanitate,* |
*& superstitione, eò purior pars eius magis recipienda, & colenda.*
*Naturalis verò* Diuinatio *aliquandò certior est, aliquandò magis in*
15 *lubrico, prout subiectum se habet, circa quod versatur. Quod si fuerit*
*Naturæ Constantis, & regularis, certam efficit Prædictionem; si*
*variæ, & compositæ tanquam ex naturâ, & casu, fallacem. Attamen*
*etiam in subiecto vario, si diligentèr canonizetur, tenebit Prædictio*
[O3<sup>r</sup>] *vt plurimùm;* | *temporis fortè momenta non assequetur, à re non*
20 *multùm errabit. Quin etiam quoad tempora euentûs, & complementi, nonnullæ Prædictiones satis certò collimabunt, eæ videlicet,*
*quæ sumuntur non à causis, verùm ab ipsâ re iam inchoatâ, sed citiùs*
*se prodente in Materiâ procliui, & aptiùs dispositâ quàm in aliâ; vt*
*in* Topicis *circa hunc* 32 Articulu*m superiùs diximus.* Prognostica
[O3<sup>v</sup>] *igitur* ventorum *iam proponemus, miscentes* | *nonnihil necessariò*
26 *de* prognosticis pluuiarum, *& serenitatis, quæ bene distrahi non*
*poterant; sed iustam de illis inquisitionem proprijs titulis remittentes.*

1. *Sol* si oriens cernatur concauus, dabit eo ipso die *Ventos,* aut
*imbres.* Si appareat tanquam leuitèr excauatus, *Ventos;* si cauus in
30 profundo, *Imbres.*

2. Si *Sol* oriatur pallidus, & (vt nos loquimur) aqueus, denotat *pluuiam;* si occidat pallidus, *Ventum.*

---

6 *poterint*] / *SEH* (II, p. 65) reads poterunt

15 *versatur.*] ~:    24 32] ~.    29 cauus] / see Introduction, p. lxxii above.

strength of the whole machine, and especially of its footings, should be increased. For the more the wind is compressed (though it accelerate the sails' motion), the more does it shake the machine itself. |

4. They say that somewhere there exist carriages driven by the wind; look into this more thoroughly.

*Direction.* Wind-driven carriages cannot be much use except in the open and on the flat. Besides what can you do when the wind drops? It would be a better plan to help the motion of light vehicles and carts with movable sails so that horses or oxen of lesser strength could pull them, than to rely on the wind alone to produce the motion.

## Prognostics of Winds

*To Article* 32. *Connection.* Divination, to the extent that it is generally more polluted with vanity | and superstition, so much the more should its purer part be accepted and cultivated. But natural divination is sometimes more certain, and sometimes more slippery according to the subject under consideration. But if that subject be of a constant and regular nature, it makes for certain prediction; but if it be variable and a mixture as it were of the natural and accidental, the prediction may let you down. Nevertheless even in a variable subject, if it be carefully reduced to rules, a prediction will generally | hold good, and if it does not hit on the right time, it will not wander off the point by much. Again some predictions will certainly indicate the time of their accomplishment, namely those which are derived not from the causes but from the thing itself while it is beginning but showing itself more quickly in suitable matter better disposed than in other matter, as I have said before in the topics concerning this thirty-second article. So I will now set forth prognostics of winds, | mixing in some prognostics of rain and good weather which could not well be separated from them, though I give over a proper inquiry into these to their own proper titles.

1. If the rising Sun seems hollow, it will produce that very day winds or rain; winds if the Sun be just a little hollow, rain if it be deeply so.

2. If the Sun rises pale, and (as we call it) watery, that presages rain; if it sets pale, wind.

3. Si corpus ipsum *Solis,* in occasu cernatur, tanquam sanguineum,
[O4ʲ] præmonstrat magnos | *Ventos,* in plures dies.

4. Si in exortu *Solis,* Radij eius spectantur rutili, non flaui, denotat
*pluuias* potiùs quàm *Ventos;* idemque, si tales appareant in occasu.

5   5. Si in ortu, aut occasu *Solis,* spectantur radij eius tanquam contracti, aut curtati, neque eminent illustres, licet nubes absint, significat
*Imbres,* potiùs quàm *Ventos.*

6. Si ante Ortum *Solis,* ostendent se Radij præcursores, & *Ventum*
denotat, & *Imbres.*

10   7. Si in Exortu *Solis,* porrigat *Sol* Radios è nubibus, medio *Solis*
manente cooperto nubibus, significabit *pluuiam;* maximè si erumpant
[O4ʲ] Radij illi deor|sùm, vt *Sol* cernatur tanquam barbatus: Quod si Radij
erumpant è medio, aut sparsìm, orbe exteriore cooperto nubibus,
magnas dabit tempestates, & *Ventorum,* & *Imbrium.*

15   8. Si *Sol* oriens cingitur circulo, à quâ parte is circulus se aperuerit,
expectetur *ventus;* Sin totus circulus æqualitèr defluxerit, dabit
*Serenitatem.*

9. Si sub Occasum *Solis,* appareat circa eum circulus candidus,
leuem denotat *tempestatem,* eâdem nocte; Si ater, aut subfuscus, *ventum*
20 magnum in diem sequentem.

10. Si Nubes rubescant exoriente *Sole,* prædicunt *ventum;* Si
[O5ʲ] occidente, *Serenum* in posterum. |

11. Si sub Exortum *Solis,* globabunt se Nubes prope *Solem,* denunciant eodem die *tempestatem* asperam; quod si ab Ortu repellantur, & ad
25 Occasum abibunt, *serenitatem.*

12. Si in Exortu *Solis,* dispergantur Nubes à lateribus *Solis,* aliæ
petentes *Austrum,* aliæ *Septentrionem,* licet sit cœlum Serenum circa
ipsum *Solem,* præmonstrat *ventos.*

13. Si *Sol* sub Nube condatur occidens, *pluuiam* denotat in posterum
30 diem; quod si planè pluet occidente *Sole, ventos* potiùs. Sin Nubes
videantur quasi trahi versus Solem, & *ventos,* & *tempestatem.*

[O5ʲ]   14. Si *Nubes,* exoriente *Sole,* vi|deantur non ambire Solem, sed
incumbere ei desupèr, tanquam Eclipsim facturæ, portendunt *ventos,* ex
eâ parte orituros, quâ illæ *nubes* inclinauerint. Quod si hoc faciant
35 meridie, & *venti* fient, & *imbres.*

15. Si *Nubes,* Solem circuncluserint, quantò minus luminis relin-
quetur, & magis pusillus apparebit Orbis Solis, tantò turbidior erit

---

15 aperuerit] aperierit / emended thus in *SEH* (II, p. 67)

15 aperuerit,] ~;    16 *ventus,*] ~:    dabit] Dabit    27 *Septentrionem,*] ~;

3. If the very body of the Sun looks blood-red when it sets, it pre-figures strong | winds for some days.

4. If when the Sun rises its rays seem ruddy and not yellow, that presages rain rather than winds; the same is true if the Sun is sinking.

5. If on rising or setting, the Sun's rays look contracted or cut short, and do not shine out bright, and though the sky is cloudless, that signifies rain rather than wind.

6. If before sunrise we see rays forerunning it, that presages both wind and rain.

7. If at sunrise, the Sun spreads its rays from the clouds, with its centre staying covered with cloud, that will be a sign of rain; the more so if those rays shoot downwards, | making the Sun seem bearded. But if the rays shoot from the centre, or dispersedly, with the circumference covered with clouds, that will yield great storms of both wind and rain.

8. If there be a circle around the rising Sun, look for wind from the point at which the circle has opened out; but if the whole circle has melted away evenly, there will be fine weather.

9. If at the setting of the Sun a bright circle appears around it, that presages a moderate storm that night; but if the circle be dark and dingy, there will be strong wind next day.

10. Clouds reddening at sunrise foretell wind; but at sunset, fair weather the following day. |

11. If at sunrise the clouds accumulate near the Sun, that is a sign of a violent storm that day; but if they are driven from the east and go off to the west, the weather will be fine.

12. If at sunrise the clouds are dispersed from the Sun's flanks, some going north, others south, but with a clear sky about the Sun itself, that presages winds.

13. If the Sun sets behind a cloud, that signifies rain next day; but rain at sunset is rather a sign of wind. But if the clouds seem as it were to be pulled towards the Sun, look out for winds and a storm too.

14. If the clouds at sunset | do not seem to surround the Sun, but burden him from above, as if an eclipse were about to happen, they portend winds from the point at which the clouds leaned down. But if this happen at midday, the winds will come along with rain.

15. If the clouds shall have hedged the Sun in, the less light remains and the smaller the Sun's orb appears, so much the more violent will

*tempestas.* Si vero duplex, aut triplex Orbis erit, vt appareant tanquam duo, aut tres Soles, tantò erit tempestas atrocior, per plures dies.

**16.** *Nouilunia* dispositionum Aeris significatiua sunt; sed magis [O6ʳ] adhùc Ortus quartus, tanquam *Nouilunium confirmatum.* *Plenilúnia*
5 autem ipsa præsagiunt magis, quàm dies aliqui ab ipsis.

**17.** Diuturnâ obseruatione, *Quinta Lunæ* suspecta est Nautis, ob tempestates.

**18.** Si *Luna* à Nouilunio, ante diem quartum non apparuerit, turbidum aerem per totum Mensem prædicit.

10 **19.** Si *Luna* nascens, aut intra primos dies, cornu habuerit inferius, magis obscurum, aut fuscum, aut quouis modo non purum, dies turbidos, & tempestates dabit, ante *plenilunium*; si circa medium fuerit decolor, circa ipsum *plenilunium* sequentur *tempestates*; si cornu superius hoc patiatur, circa *Lunam decrescentem.*
14

[O6ᵛ] **20.** Si Ortu in quarto, pura ibit ¦ *Luna* per cœlum, nec cornibus obtusis, neque prorsùs iacens, neque prorsùs recta, sed mediocris, *Serenitatem* promittit maiore ex parte, vsque ad *Nouilunium.*

**21.** Si in Ortu illo rubicunda fuerit, *ventos* portendit; si rubiginosa, aut obatra, *Pluuias*; sed nil horum significat vltra *plenilunium.*

20 **22.** Recta *Luna* semper ferè minax est, & infesta, potissimum autem denunciat *ventos*; At si appareat cornibus obtusis, & curtatis, *Imbres* potius.

**23.** Si alterum Cornu *Lunæ* magis acuminatum fuerit, & rigidum, altero magis obtuso, *ventos* potiùs significat; si vtrunque, *pluuiam.*

25 **24.** Si circulus, aut *Halo* circa *lunam* appareat, *pluuiam* potius [O7ʳ] significat, quàm *ventos*; nisi stet recta *Luna* intra eum Circulum, tum verò vtrumque.

**25.** Circuli circa *lunam, ventos* sempèr denotant, ex parte quâ ruperint; Etiam splendor illustris circuli in aliquâ parte, *ventos* ex eâ
30 parte, quâ splendet.

**26.** Circuli circa *Lunam,* si fuerint duplices, aut triplices, præmonstrant horridas, & asperas tempestates; at multò magis, si illi Circuli non fuerint integri, sed maculosi, & interstincti.

**27.** *Plenilunia,* quoad colores, & Halones, eadem fortè denotant,
35 quæ Ortus quartus; sed magis præsentia, nec tam procrastinata.

[O7ᵛ] **28.** *Plenilunia* solent esse magis serena, quàm cæteræ ætates *Lunæ*, ¦ sed eadem, hyeme, quandoque intensiora dant frigora.

---

1 *tempestas.*] ~:    11 purum,] ~;    12 *plenilunium*;] ~:    16 mediocris,] ~;
35 quartus;] ~:

the storm be. But if there be a double or triple orb, so that there seem to be two or three suns, the storm will be even more terrible and last for days.

16. New Moons indicate the dispositions of the air, but more on its fourth rising, as if its newness were assured. But full Moons ' are themselves better forecasters than any days after it.

17. Long observation has caused sailors to look out for storms on the Moon's fifth day.

18. If the Moon has not appeared before the fourth day after it was new, that predicts turbulent air for the whole month.

19. If the Moon rising, or in its first days, has its lower horn obscure and dingy or any way indistinct, that means turbulent days and storms before the full; but if it is discoloured in the middle, there will be storms about the full; but if the upper horn is affected in that way, there will be storms about the wane.

20. If on its fourth rising the Moon ' is clear, with horns undulled, and neither quite prone nor completely upright but in a middling aspect, that promises good weather for the most part right up to the new Moon.

21. If it rise rubicund on that day, that portends wind; if rusty or dark, rain; but after the full these two signify nothing.

22. An upright Moon is almost always menacing and harmful, and presages winds in particular; but if it appear with blunt and shortened horns, it rather signifies rain.

23. If one horn of the Moon be sharp and stiff, and the other more blunt, it rather signifies wind, but if both are the same, rain.

24. If a circle or halo appear about the Moon, that ' rather signifies rain than wind; unless the Moon stands upright in the circle, when it presages both.

25. Circles about the Moon always forebode winds from the point where they break; and outstanding brilliance in any part of the circle signifies winds from the point which is brilliant.

26. Double or triple circles about the Moon presage wild and violent storms; but much more, if those circles are not complete but patchy and broken.

27. Full Moons, as far as colours and haloes go, perhaps predict the weather in the same way as the fourth rising, but more promptly and with less delay.

28. Full Moons are generally finer than the Moon's other ages, ' but they often presage sharper weather in winter.

**29.** *Luna* sub occasum solis ampliata, & tamen luminosa, nec subfusca, *serenitatem* portat in plures dies.

**30.** *Eclipses Lunæ*, quasi sempèr comitantur *venti*; Solis, *Serenitas*; *pluuiæ* rarò alterutrum.

**31.** A *Coniunctionibus* reliquis *planetarum*, præter Solem, expectabis *ventos*, & ante, & post; à Coniunctionibus cum *Sole, serenitatem*.

**32.** In Exortu *Pleiadum*, & *Hyadum*, sequuntur *Imbres*, & Pluuiæ, sed tranquillæ; In exortu *Orionis*, & *Arcturi*, Tempestates.

[O8ʳ] **33.** *Stellæ* (vt loquimur) discurrentes, & sagittantes, protinùs ' *ventos* indicant, ex eâ parte, vndè vibrantur. Quod si ex varijs, aut etiam contrarijs partibus volitent, magnas tempestates, & *Ventorum*, & *Imbrium*.

**34.** Cum non conspiciantur *Stellæ* minusculæ, quales sunt, quas vocant *Asellos*, ídque fit vbique per totum cœlum, magnas præmonstrat *Tempestates*, & *Imbres* intra aliquot dies; quod si alicubi *Stellæ* minutæ obscurentur, alicubi sint claræ, *ventos* tantùm, sed citiùs.

**35.** Cœlum æqualitèr splendens, in Noui-lunijs, aut Ortu quarto, *Serenitatem* dabit, per plures dies; æqualitèr obscurum, *Imbres*; inæqua-[O8ᵛ] litèr, *ventos*, ab ea parte, quâ cernitur Obscuratio. ' Quod si subitò fiat obscuratio, sine nube, aut caligine, quæ fulgorem Stellarum perstringat, graues & asperæ instant Tempestates.

**36.** Si Planetarum, aut Stellarum maiorum aliquam, incluserit Circulus integer, *imbres* prædicit; si fractus, *ventos* ad eas partes, vbi circulus deficit.

**37.** Cum tonat vehementiùs, quam fulgurat, *ventos* dabit magnos; sin crebrò inter tonandum fulserit, *Imbres* confertos, & grandibus guttis.

**38.** Tonitrua *Matutina ventos* significant; Meridiana *Imbres*.

**39.** Tonitrua mugientia, & veluti transeuntia, *ventos* significant; at [P1ʳ] quæ inæquales habent frago'res, & acutos, procellas, tam *ventorum*, quam *Imbrium*.

**40.** Cum cœlo sereno fulgurauerit, non longè absunt *venti*, & *Imbres* ab ea parte, quâ fulgurat; quod si ex diuersis partibus cœli fulgurauerit, sequentur atroces, & horridæ Tempestates.

**41.** Si fulgurauerit à plagis cœli gelidioribus, *Septentrione* & *Aquilone*, sequentur grandines; si à tepidioribus, *Austro* & *Zephyro*, *Imbres*, cum cœlo æstuoso.

**42.** Magni feruores, post solstitium Æstiuale, desinunt plerunque in tonitru, & fulgura; quæ si non sequantur, desinunt in *ventos*, & *pluuias* per plures dies.

6 *ventos*,] ~;    14 dies;] ~:    20 Tempestates.] ~,

29. The Moon enlarged at sunset, and yet luminous and not dingy, foretells fair weather for several days.

30. Lunar eclipses almost always coincide with winds; solar eclipses with good weather; but neither often coincides with rain.

31. You can expect winds before and after all planetary conjunctions, except for the Sun; from conjunctions with the Sun expect fair weather.

32. Rains and downpours in still weather follow the rising of the Pleiades and Hyades; storms follow the rising of Orion and Arcturus.

33. Discurrent or shooting stars (as I call them) presage [1] immediate winds from the point from which they dart. But if they fly from different or even opposite points, that means great storms of wind and rain.

34. When tiny stars such as those called Aselli cannot be seen in any part of the heavens, that indicates great storms and rains within a few days; but if these little stars are hidden in some places but clearly visible in others, expect only winds but sooner.

35. A sky evenly bright at new Moon or the fourth rising means fair weather for many days; sky evenly dull, rain; unevenly dull, winds from the direction where the dullness lies. [1] But if it suddenly gets dull without cloud or fog, so that it cuts down the brightness of the stars, that means rough and violent storms are at hand.

36. If a complete circle embraces any planet or larger star, that foretells rain; if the circle be broken, that means winds from the direction of the break.

37. When it thunders more than it produces lightning, that means great winds; but if lightning often blazes between the thunder-claps there will be heavy rain with large drops.

38. Morning thunder means wind; afternoon thunder, rain.

39. Rolling thunder, which seems to be passing over, means winds; but sharp, intermittent crashes [1] mean storms of both wind and rain.

40. When it thunders in clear weather, wind and rain come not far behind from the direction of the thunder; but if lightning flashes from different parts of the sky, wait for violent and severe storms.

41. If lightning flashes from the colder points of the heavens, from the north or north-east, there will be hail; but if from the warmer, from the south and west, rain with sweltering heat.

42. Great heats after the summer solstice often end in thunder and lightning; if not they end in wind and rain for many days.

[P1ᵛ]   **43.** Globus flammæ, quem *Castorem* vocabant Antiqui, qui cerᴵnitur nauigantibus in mari, si fuerit vnicus, atrocem Tempestatem prænunciat (*Castor* scilicet est Frater inter-mortuus) at multò magis, si non hæserit *Malo*, sed volvatur, aut saltet; Quod si fuerint gemini (præsente
5 scilicet *Polluce* Fratre viuo), ídque tempestate adultâ, salutare signum habetur. Sin fuerint tres (superveniente scilicet *Helenâ*, peste rerum) magis dira incumbet *Tempestas*. Videtur sanè vnicus, crudam significare materiam tempestatis; duplex, quasi coctam, & maturam; triplex, vel
9 multiplex, copiam ægrè dissipabilem.

[P2ʳ]   **44.** Si conspiciantur Nubes ferri incitatiùs, cœlo sereno, expeᴵctentur *venti*, ab eâ parte, à quâ feruntur Nubes. Quod si globabuntur, & glomerabunt simul, cum Sol appropinquauerit ad eam partem, in qua globantur, incipient discuti; quod si discutientur magis versus *Boream*, significat *ventum*, si versus *Austrum*, *pluvias*.

15   **45.** Si occidente Sole, Nubes orientur atræ, aut fuscæ, *Imbrem* significant; Si aduersus Solem, in Oriente scilicet, eâdem nocte; si iuxta Solem ab Occidente, in posterum diem, cum *ventis*.

   **46.** Liquidatio, siue disserenatio cœli nubili, incipiens in con-
19 trarium *venti*, qui flat, *Serenitatem* significat; sed à parte *venti*, nihil
[P2ᵛ] indicat, sed incerta res est. ᴵ

   **47.** Conspiciuntur quandoque plures, veluti Cameræ, aut contignationes Nubium, alteræ super alteras (vt aliquando quinque simul se vidisse, & notasse affirmet *Gilbertus*) & sempèr atriores sunt infimæ, licet quandóque secus appareat, quia candidiores visum magis lacessunt.
25 Duplex contignatio, si sit spissior, *pluvias* denotat instantes (præsertim si nubes inferior cernatur quasi grauida) plures contignationes perendinant *pluvias*.

   **48.** *Nubes*, si vt vellera lanæ spargantur, hinc inde, *tempestates*
29 denotant; Quod si instar squammarum, aut testarum, altera alteri
[P3ʳ] incumbat, siccitatem, & *serenitatem*. ᴵ

   **49.** *Nubes* plumatæ, & similes ramis palmæ, aut floribus Iridis, *Imbres* protinùs, non ita multò post, denunciant.

   **50.** Cùm *Montes*, & *Colles* conspiciantur veluti pileati, incumbentibus in illis Nubibus, eósque circumplectentibus, *Tempestates* præmon-
35 strat imminentes.

---

22 alteræ] / *SEH* (II, p. 70) emends to altera
28 spargantur] spargentur / silently emended thus in *SEH* (II, p. 70)

4 gemini] ~,   5 viuo),] ~)ᴧ   6 habetur.] ~:   Sin] lc   7 *Tempestas.*] ~:
Videtur] lc   15–16 significant;] ~:   16 nocte;] ~,   24 lacessunt.] ~;

43. The ball of fire which the ancients called *Castor*, that ¹ mariners see at sea, presages a terrible storm if it be single (*Castor* being the dead brother), but much worse if it does not stick to the mast, but spins and jumps about. But if there be two of them (i.e. if Pollux the living brother is there as well), and that too when the storm has grown, that is taken as a favourable sign. But if there be three of them (i.e. *Helen* the ruin of all appear on the scene), the storm will become even more terrible. It seems in fact that one ball by itself indicates that the tempest's material is crude; two that it is worked up and mature; and three or more that the material is so abundant that it can scarcely be dissipated.

44. If the clouds scud swiftly in calm weather, look ¹ out for wind from the direction from which the clouds are carried. But if they bunch up and gather together, they will begin to disperse when the Sun gets near to that part; but if they disperse more northwards look out for winds, if to the south rain.

45. If at sunset black or brown clouds rise up, that means rain—on the same night if they rise in the east opposite from the Sun; on the following day, and with winds, if they rise near the Sun in the west.

46. If a cloudy sky clears up beginning with the quarter opposite the wind that blows, that means fine weather; but if from where the wind blows, that means nothing, and the forecast is uncertain. ¹

47. Sometimes we see clouds stacked up like floors or layers, one above another (so that *Gilbert* affirms that he has sometimes seen and noticed five together at once) with the lowest always the blackest, although sometimes it seems different because the whiter ones are more striking. Two layers if they be thicker presage immediate rain (especially if the lower ones look pregnant); many layers mean rain later on.

48. Clouds scattered like fleece indicate storms to follow. But if they lie one on another like scales or tiles, that means fine, dry weather. ¹

49. Feathery clouds like palm fronds or iris petals foretell immediate or imminent rain.

50. When mountains or hills appear cloud-capped, the clouds lying and gathering about them, that forewarns us of imminent storms.

**51.** *Nubes* electrinæ, & aureæ, ante Occasum Solis, & tanquam cum fimbrijs deauratis, postquam Sol magis condi cœperit, *serenitates* præmonstrant.

**52.** *Nubes* luteæ, & tanquam cœnosæ, significant *Imbrem* cum *vento* instare.

**53.** *Nubecula* aliqua, non antè visa, subitò se monstrans, cœlo circùm [P3ᵛ] sereno, præsertìm ab Oc|cidente, aut circa Meridiem, *Tempestatem* indicat ingruentem.

**54.** *Nebulæ,* & caligines ascendentes, & sursùm se recipientes, *pluuias,* & si subitò hoc fiat, vt tanquam sorbeantur, *ventos* prædicunt; at cadentes, & in vallibus residentes, *Serenitatem.*

**55.** *Nube* grauidâ candicante, quam vocant Antiqui *Tempestatem albam,* sequitur æstate, Grando minutus, instar confituræ; hyeme, Nix.

**56.** Autumnus serenus ventosam Hyemem præmonstrat; ventosa Hyems, Ver pluuiosum; Ver pluuiosum, Æstatem serenam; serena Æstas, Autumnum ventosum; ita vt Annus (vt prouerbio dicitur) [P4ʳ] sibi debitor rarò | sit; neque eadem series tempestatum redeat, per duos Annos simul.

**57.** *Ignes* in focis, pallidiores solito, atque intra se murmurantes, *Tempestates* nunciant. Quòd si Flamma flexuosè volitet, & sinuet, *ventum* præcipuè; at Fungi, siue Tuberes in lucernis *pluuias* potiùs.

**58.** *Carbones* clariùs perlucentes, *ventum* significant; etiam cum fauillas ex se citiùs discutiunt, & deponunt.

**59.** *Mare* cum conspicitur in portu tranquillum in superficie, & nihilominùs intra se murmurauerit, licet non intumuerit, *ventum* prædicit.

[P4ᵛ] **60.** *Littora* in tranquillo resonan|tia, Marísque ipsius sonitus cum plangore, aut quâdam echô, clariùs, & longiùs solito auditus, *ventos* prænunciant.

**61.** Si in tranquillo, & planâ superficie *Maris,* conspiciantur Spumæ hinc inde, aut coronæ albæ, aut Aquarum bullæ, *ventos* prædicunt: Et si hæc signa fuerint insigniora, asperas tempestates.

**62.** In Mari fluctibus agitato, si appareant spumæ coruscantes (quas *Pulmones Marinos* vocant) prænunciat duraturam Tempestatem, in plures dies.

**63.** Si *Mare* silentio intumescat, & intra portum altiùs solito insur-[P5ʳ] gat, aut Æstus ad littora ce|leriùs solito accedat, *ventos* prænunciat.

---

14 ventosa] ventosus / emended thus in *SEH* (II, p. 71)

14 Hyemem] / *nld* as hiemem in *SEH* (II, p. 71)   præmonstrat;] ~:

15 pluuiosum;] ~:   serenam;] ~:   16 ventosum;] ~.   ita] Ita   21 præcipuè;] ~:

32 coruscantes] ~,

51. Golden or amber clouds before sunset, and with gilt edges after the Sun has gone down further, presage fine weather.

52. Dirty, muddy clouds signify rain and wind.

53. If a cloudlet not spotted before suddenly shows itself in a clear sky, and especially from the west, | that means a storm is brewing.

54. If mists and fogs lift and take themselves up, it will rain; if this happens suddenly, as if they were swallowed up, that presages winds; but if they fall and stop in the valleys, good weather.

55. White, heavily charged cloud which the ancients called a white storm, mean light, granular hail in summer, but snow in winter.

56. A fine autumn presages a windy winter; a windy winter a wet spring; a wet spring a fine summer; a fine summer a windy autumn; so that, as the saying goes, a year is rarely its own | debtor, and the seasons are not the same for two years running.

57. Domestic fires paler than usual, and murmuring inside themselves mean storms. But if the flames fly up twisting and curling as they go, that mainly means wind; but mildew or swellings on wicks rather mean rain.

58. Coals, when they burn bright, mean wind, as does rapid dispersal or dropping of their ashes.

59. The sea, when its surface looks calm in a harbour, and yet it rumbles within without any swell, foretells wind.

60. Shores resounding in calm weather, | and the sea itself sounding with a lowing echo clearer and longer than normal, foretell winds.

61. If on a calm and flat sea you discern foam here and there, white crests, or bubbles of water, expect winds. And if these signs are more marked, severe storms.

62. If there appear glittering foam (called sea-lungs) in a heavy sea, that presages a continuous storm for many days.

63. If the sea swells silently, and rises higher than usual in a harbour, or the tide rises quicker than | usual on the shore, expect winds.

**64.** Sonitus à *Montibus*, nemorúmq*ue* murmur increbrescens, atque fragor etiam nonnullus in campestribus, *ventos* portendit. Cœli quoque murmur prodigiosum, absque tonitru, ad *ventos* maximè spectat.

**65.** *Folia* & paleæ ludentes, sine aurâ, quæ sentiatur, & Lanugines
5 plantarum volitantes, plumæque in Aquis innatantes, & colludentes, *ventos* adesse nunciant.

**66.** *Aues aquaticæ* concursantes, & gregatìm volantes, *Mergíque* præcipuè, & *Fulicæ*, à Mari, aut stagnis fugientes, & ad littora, aut ripas
[P5ᵛ] properantes, præsertìm | cum clangore, & ludentes in sicco, *ventos*
10 prænunciant, maxime si hoc faciant manè.

**67.** At *terrestres volucres* contrà, aquam petentes, eámq*ue* alis percutientes, & clangores dantes, & se perfundentes, ac præcipuè *Cornix*, tempestates portendunt.

**68.** *Mergi, Anatésque*, ante *ventum* pennas rostro purgant; at *Anseres*,
15 clangore suo importuno, *pluuiam* inuocant.

**69.** *Ardea* petens excelsa, adeò vt nubem quandoque humilem, superuolare conspiciatur, *ventum* significat; At *Milui* contrà, in sublimi volantes, *serenitatem*.

19   **70.** *Corui* singultu quodam latrantes, si continuabunt, *ventos* deno-
[P6ʳ] tant; si vero carptìm vo|cem resorbebunt, aut per interualla longiora crocitabunt, *Imbres*.

**71.** *Noctua* garrula, putabatur ab antiquis, mutationem Tempestatis præmonstrare: si in sereno, *Imbres*, si in nubilo, *Serenitatem*; at apud nos, *Noctua* clarè, & libentèr vlulans, *Serenitates* plerunq*ue* indicat,
25 præcipuè Hyeme.

**72.** *Aues* in arboribus habitantes, si in nidos suos, sedulò fugitent, & à pabulo citiùs recedant, *Tempestates* præmonstrant; *Ardea* vero in arenâ stans tristis, aut *Coruus* spatians, *Imbres* tantùm.

29   **73.** *Delphini* tranquillo Mari lasciuientes, Flatum existimantur
[P6ᵛ] prædicere, ex quâ veniunt parte; | at turbato ludentes, & aquam spargentes, contrà, *Serenitatem*. At plerique Piscium, in summo natantes, aut quandóque exilientes, *pluuiam* significant.

**74.** Ingruente *vento*, Sues ita terrentur, & turbantur, & incompositè agunt, vt Rustici dicant illud solum Animal videre *Ventum*, specie
35 scilicet horrendum.

**75.** Paulò ante *Ventum*, *Araneæ* sedulò laborant, & nent, ac si prouidè præoccuparent, quia *Vento* flante nere nequeunt.

**76.** Ante pluuiam, *Campanarum* sonitus auditur magis ex longinquo;

---

12 perfundentes,] ~;    17 significat;] ~.    19–20 denotant;] ~,    23 *Serenitatem*;] ~:
31 plerique] plærique

64. A sound from the mountains, and a growing murmur in the woods, and a certain roaring on the plains portends wind. A great murmuring from the skies, without thunder, points principally to winds.

65. Leaves and straw dancing in still air, down flying from plants, and feathers drifting and playing on water tell us that wind is nigh.

66. Water birds coming together and flying as a flock, especially seagulls and coots when they hasten clamorously from the sea or ponds to the shores or banks | and play on dry land, presage wind; and this is especially true if it happens in the morning.

67. It is different with land birds, especially the crow, for when they seek out the waters they splash it with their wings, throw it over themselves, and cry out; and that foretells rain.

68. Before wind sea birds and ducks preen their feathers with their beaks; but geese with imploring cackles call up the rain.

69. A heron climbing high so as sometimes to overfly low clouds, signifies winds. But kites flying high signify fair weather.

70. Ravens foretell wind with their continuous barking cries; but if they silence their cawing | for longer or shorter intervals, that means rain.

71. The cries of the owl were thought by the ancients to presage a change in the weather, from fair to rain, or cloud to fine. But here with us clear and free owl cries generally mean fine weather, particularly in winter.

72. Birds living in trees, if they flee busily to their nests, and more quickly leave off feeding, that portends storms. But a heron standing gloomy on the sand, or a raven wandering around, only denote rain.

73. Dolphins frisking in a calm sea are believed to forecast wind from the direction whence they come; but if | they dance in a heaving sea and splash the water about, that on the other hand heralds fine weather. But most other fish swimming near the surface, or sometimes leaping about it, foretell rain.

74. When the wind gets up, pigs are so frightened, disturbed, and upset that country people say that this animal alone sees the wind, and that it must be dreadful to behold.

75. A little before the wind blows, spiders work hard, and spin hard, as if they were wisely anticipating it, because they cannot spin when the wind blows.

76. Before rain, bells can be heard further off; but before wind they

At ante *Ventum*, auditur magis inæqualitèr accedens, & recedens, quem-
[P7ʳ] admodum fit *vento* manifestò flante. ¹

77. *Trifolium* inhorrescere, & folia contra tempestatem subrigere,
pro certo ponit *Plinius*.

5   78. Idem ait, *vasa*, in quibus esculenta reponuntur, quandóque
sudorem in repositorijs relinquere, idque diras tempestates prænunciare.

**Monitum.** Cum Pluuia, *& Venti*, *habeant materiam ferè communem;*
*cumque* Ventum *semper præcedat nonnulla Condensatio Aeris, ex Aere*
*nouitèr facto, intra veterem recepto, vt ex plangoribus littorum, & excelso*
10   *volatu* Ardeæ, *& alijs patet; Cumque* pluuiam *similitèr præcedat Aeris*
*Condensatio* (*sed Aer in* Pluuiâ *posteà contrahitur magis, in* Ventis *contrà*
[P7ᵛ]   *excrescit*), *necesse est, vt* Pluuiæ *habeant com'plura* Prognostica, *cum*
Ventis *communia. De ijs consule* Prognostica Pluuiarum, *sub titulo suo.*

## Imitamenta Ventorum.

15   *Ad Artic.* **33.** *Connexio.* Si animum homines inducere possent, vt
*Contemplationes suas, in subiecto sibi proposito, non nimiùm*
*figerent, & cætera tanquam parerga rejicerent; nec circa ipsum*
*subiectum in infinitùm, & plerunque inutilitèr subtilizarent,*
[P8ʳ]   *haudquaquam talis, qualis solet,* ¹ *occuparet ipsos stupor, sed trans-*
20   *ferendo cogitationes suas, & discurrendo, plurima inuenirent in*
*longinquo, quæ propè latent. Itaque vt in* Iure Ciuili, *ita in* Iure
Naturæ, *procedendum animo sagaci, ad Similia, & Conformia.*

1.   *Folles* apud homines *Æoli vtres* sunt, vnde *ventum* quis promere
possit, pro modulo nostro. Etiam Interstitia, & Fauces Montium, &
25   Ædificiorum Anfractus, non alia sunt, quàm Folles maiores. In vsu
[P8ᵛ]   autem sunt Folles præcipuè, aut ad ex'citationem Flammarum, aut
ad Organa Musica. *Follium* autem ratio est, vt sugant Aerem propter
rationem *Vacui* (vt loquuntur), & emittant per Compressionem.

2.   Etiam *Flabellis* vtimur manualibus, ad faciendum Ventum, &
30   refrigeria, impellendo solummodo Aerem lenitèr.

3.   De cœnaculorum æstiuorum refrigerijs, quædam posuimus in
*Responso* ad *Artic.* 9. Possunt inueniri alij modi magis accurati,
præsertìm si Follium modo alicubi attrahatur Aer, alicubi emittatur. Sed
ea, quæ iam in vsu sunt, ad simplicem compressionem, tantum
35   referuntur.

---

12 *excrescit), necesse*] *excrescit) Necesse*     23 sunt,] ~:     28 *Vacui*] ~,     loquuntur),] ~,)

sound more uneven, the ringing rising and falling just as the wind does
as it evidently blows. ⌐

77. Pliny says with confidence that the hairs of the trefoil and its
leaves stand up against a storm.

78. He also says that vessels in which foods are stored sometimes
leave a sweat behind them in their repositories; and that this presages
terrible storms.

*Advice.* Since rain and winds share pretty much the same matter, and
since some condensation of the air always comes before wind because of
the newly created air being taken into the old—as appears from the
wailing of the shores, the heron's high flight, and other things—, and
since condensation of the air likewise precedes rain (but the air is contracted more in rain, while in winds it expands), it necessarily follows
that rains have many ⌐ prognostics in common. Concerning these consult the prognostics of rains under their own proper title.

## Imitations of Winds

*To Article* 33. *Connection.* If men could only persuade themselves
not to fix their contemplations exclusively on the subject in hand,
and not to put aside other things as incidentals, and not to split
the subject into infinite and mainly useless subtleties, they would
scarcely be as stupid ⌐ as they usually are, but by transferring their
thoughts and allowing them to branch outwards, they would find
out many things a good way off which are hidden close by. Thus
as it is in the civil law so it is in the natural; one must go forward
with a keen mind towards cases in the same mould.

1. With men bellows are the bags of *Æolus* whence one can bring out
wind on a human scale. The gaps too and passes of mountains, and the
twists and turns of buildings, are nothing other than bellows on a larger
scale. Bellows are mainly used either for ⌐ arousing flames or for running
musical organs. Bellows work on the principle that they suck in air to
prevent (as they say) a vacuum forming, and then force it out by
compression.

2. We also use hand-held fans to make wind, and to cool ourselves
down just by gently driving the air along.

3. I have already made some points about cooling down rooms in
summer in my reponse to article 9. Yet other, more precise ways can be
discovered, especially if air is sucked in one place, and expelled in

[Q1$^r$]  **4.** *Flatus* in *Microcosmo*, & Ani|malibus, cum *ventis* in Mundo maiore, optimè conueniunt. Nam & ex humore gignuntur, & cum humore alternant, vt faciunt *venti* & *pluuiæ*; & à calore fortiore dissipantur, & perspirant. Ab illis autem, transferenda est certe ea Obser-
5  uatio ad *ventos*; quod scilicet gignantur *Flatus*, ex materiâ, quæ dat vaporem tenacem, nec facilè resolubilem, vt Fabæ, & Legumina, & Fructus; quod etiam, eodem modo se habet, in *ventis* maioribus.

**5.** In destillatione vitrioli, & aliorum Fossilium, quæ sunt magis
9  flatuosa, opus est receptaculis valdè capacibus, & amplis, alioqui
[Q1$^v$]  effringentur.|

**6.** *Ventus* factus ex *Nitro*, commisto in *Puluere pyrio*, erumpens, & inflans flammam, *ventos* in Vniuerso (exceptis fulminosis) non tantum imitatur, sed exuperat.

**7.** Huius autem vires premuntur, in Machinis humanis, vt in
15  Bombardis, & Cuniculis, & domibus puluerarijs incensis; vtrum autem, si in Aere aperto, magna *pulueris pyrij* moles, incensa esset, *ventum* ex Aeris commotione, etiam ad plures horas excitatura esset, nondùm venit in Experimentum.

19  **8.** Latet spiritus flatuosus, & expansiuus, in *Argento viuo*, adeò vt
[Q2$^r$]  puluerem pyriu*m* (vt quidam volunt), imitetur, & parum ex | eo, pulueri pyrio admistum, eum reddat fortiorem. Etiam de *Auro* loquuntur *Chymistæ*, quod periculosè, & ferè tonitrui modo, in quibusdam præparationibus erumpat; sed de his mihi non compertum est.

## Obseruatio maior.

25  Motus Ventorum *tanquam in speculo spectatur, in* Motibus Aquarum *quoad plurima.*

Venti magni *sunt Inundationes Aeris, quales conspiciuntur*
[Q2$^v$]  *Inundationes A*|*quarum; vtræque ex aucto* Quanto. *Quemadmodùm Aquæ aut descendunt ex Alto, aut emanant è Terrâ; ita &* ventorum
30  *nonnulli sunt deiecti, nonnulli exurgunt. Quemadmodùm nonnunquam intra Amnes, sunt contrarij Motus; vnus fluxûs Maris, alter Cursûs Amnis; & nihilominùs vnicus efficitur Motus, præualente fluxu Maris; ita &* flantibus *ventis contrarijs, Maior in*
[Q3$^r$]  *ordinem redigit Minorem. Quemadmo*|*dùm in Currentibus Maris,*
35  *& quorundam Amnium, aliquandò euenit, vt Gurges in summitate*

---

31 *Amnes,*] / *c-t* has *Amne,* with copies corrected in pen and, in some cases, overwriting the *c-t* comma

10 effringentur] Effringentur     20 pyriu*m*] ~,     volunt),] ~,)     33 *Maris*;] ~:

another in the manner of bellows. But means currently in use relate only to simple compression.

4. Winds in the microcosm, and in ' animals, correspond very well to winds in the macrocosm. For they are generated from a humour and vary according to the humour, as winds and rain do; and they are dispersed and sweated off by a stronger heat. Thus from these we can transfer for sure this observation to winds: namely that they are generated from matter which provides a clinging vapour which cannot be easily resolved, matter such as peas, pulses, and fruit, and this works in the same way in the greater winds.

5. In the distillation of vitriol, and other fossils which are windier, it is necessary to use very capacious and large vessels; because otherwise they would break. '

6. Wind produced by nitre mixed in with gunpowder, when it bursts out and expands flame, not only mimics but surpasses winds in the world at large, except for those in thunderstorms.

7. Now the forces of this wind are put under pressure in man-made machines, as in cannon, mines, and powder mills when they blow up. But whether a great quantity of gunpowder let off in the open air would stir up the air and produce a wind lasting for many hours has not yet been tested by experiment.

8. A windy and expansive spirit lies hidden in quicksilver, such that (as some believe) it mimics gunpowder, and a little of ' it mixed with gunpowder, makes the latter stronger. The chemists also say that gold in certain preparations blows up dangerously and almost like thunder; but I have no certain knowledge of this.

## Major Observation

As in a mirror, the motion of winds is seen in the motions of water in many respects.

Great winds are the floods of the air such as can be seen in floods of ' waters, and both of them spring from increase of quantity. In the same way as water either falls from above or springs from the earth, so it is that some winds are cast down from above while others arise from below. Just as there are contrary motions in great rivers, the tide from the sea in one direction and the river's flow in the other, and yet there is only one motion because the tide prevails, so when contrary winds blow the greater reduces the lesser to order. In the same ' way that in currents in the sea and some great

*Aquæ, in contrarium vergat, Gurgiti in profundò; ita & in Aere,* flantibus simul contrarijs ventis, *alter alterum superuolat.* Quemadmodum sunt Cataractæ Pluuiarum, *in spatio angusto; similitèr* 4 *&* Turbines ventorum. *Quemadmodum Aquæ, vtcunque progredi-* [Q3ᵛ] *antur, tamen si perturbatæ fuerint, interìm vndulant,* modò ascen- dentes, *& cumulatæ, modò descendentes, & sulcatæ; similiter faciunt & venti, nisi quod absit* Motus Grauitatis. *Sunt & aliæ similitudines, quæ ex ijs, quæ inquisita sunt, notari possunt.*

## Canones mobiles de Ventis.

10 **Connexio.** Canones *aut* Particulares *sunt, aut* Generales; *vtrique* [Q4ʳ] Mobiles *apud nos. Nil enim adhûc* pronunciamus. At Particulares *ex singulis fere* Articulis, *possunt decerpi, aut expromi;* Generales, *eosque paucos ipsi iam excerpemus, & subiungemus.*

    **1.** Ventus *non est aliud quippiam ab Aere moto, sed ipse* Aer 15 Motus; *aut per* Impulsionem simplicem, *aut per* Immistionem Vaporum.

    **2.** Venti *per* Impulsionem *Aeris simplicem, fiunt quatuor modis,* [Q4ᵛ] *aut per* Motum Aeris *naturalem; aut per* Expansionem Aeris in *vijs solis; aut per* Receptionem Aeris *ex frigore subitaneo; aut per* 20 Compressionem Aeris *per Corpora externa.*

    *Possit esse & quintus Modus, per* Agitationem, *&* Concussionem Aeris ab Astris: *sed sileant paulispèr huiusmodi res, aut audiantur parcâ fide.*

[Q5ʳ]     **3.** Ventorum, *qui fiunt* per Immistionem vaporum, *præcipua* 25 *causa est* Super-oneratio Aeris, *per Aerem nouitèr factum ex vaporibus; vnde moles Aeris excrescit, & noua spatia quærit.*

    **4.** Quantum *non magnum Aeris super-additi, magnum ciet tumorem in Aere circumquaquè; ita vt Aer ille nouus, ex resolutione* 29 *Vaporum, plus conferat ad Motum, quam ad Materiam;* Corpus [Q5ᵛ] *autem magnum* venti *consistit ex Aere priore, neque Aer nouus Aerem veterem ante se agit, ac si Corpora separata essent; sed vtraque commista ampliorem locum desiderant.*

---

1 *profundò;*] ~:    8 *similitudines,*] ~, / BNF copy with semicolon; Yale copy with comma

rivers it sometimes happens that the eddies on the surface move in the opposite direction to those below, so it is in the air when contrary winds blow together, one of them flies above the other. In the same way that there are cataracts of water in a confined space, so likewise there are whirlwinds. In the same way that waters, if they have been disturbed, move forward as well as undulate, | alternately rising in ridges and falling in furrows, so do the winds except that they lack motion of gravity. There are also other resemblances which can be identified from those already investigated.

## Provisional Rules concerning Winds

*Connection.* The rules are either particular or general, and for us both are provisional. For as yet | I do not lay down the law about anything. But particular rules can be harvested or elicited from almost every article; the general ones, just a few of them, I shall pick out and subjoin below.

1. Wind is nothing other than air in motion, air moved either by simple impulsion or by mixing of vapours.

2. Winds brought about by simple impulsion of the air happen in four ways: by means of the air's | natural motion, or the expansion of the air by the Sun's approaches, or the retention of air by sudden cold, or by air's compression by external bodies.

A fifth way may exist, i.e. when the air is stirred and disturbed by the stars; but let's say nothing about these things for now, or listen to them with scant conviction.

3. The main cause of winds | arising from mixture of vapours is the overburdening of the air with air newly made from vapours, and from that the mass of the air increases and needs more room.

4. When a small amount of extra air is added it causes a great swelling in the air all around, such that the new air resulting from resolution of vapours does more to increase motion than matter. For the great body | of the wind consists of the old air, and the new air does not drive the old before it as if it were a separate body, but the two of them mixed together demand more room.

**5.** *Quando aliud concurrit Principium Motûs, præter ipsam Super-onerationem Aeris, Accessorium quippiam est illud, & Principale fortificat, & auget; vndè fit, vt* venti *magni, & impetuosi, rarò* [Q6ʳ] *oriantur ex Super-oneratione Aeris simplici.* |

5  **6.** *Quatuor sunt Accessoria ad Super-onerationem Aeris, Expiratio è subterraneis, Deiectio ex mediâ regione Aeris (quam* vocant); *Dissipatio ex Nube factâ; & Mobilitas, atque Acrimonia Exhalationis ipsius.*

**7.** *Motus* venti *quasi sempèr lateralis est; verum is qui fit per*
10  *Super-onerationem simplicem, vsque à principio; is qui fit per*
[Q6ᵛ] *Expirationem è terrâ, aut Repercussionem* | *ab alto, non multò post; nisi Eruptio, aut Præcipitium, aut Reuerberatio, fuerint admodùm violenta.*

**8.** *Aer nonnullam compressionem tolerat, antequam Super-*
15  *onerationem percipiat, & Aerem contiguum impellat; ex quo fit, vt* omnes venti *sint paulò densiores, quam Aer quietus.*

**9.** *Sedantur* venti *quinque modis; aut Coeuntibus vaporibus; aut*
[Q7ʳ] *Incorporatis; aut Sublimatis; aut Trans* | *uectis; aut Destitutis.*

**10.** *Coeunt vapores, atque adeò ipse Aer in pluuiam, quatuor*
20  *modis: aut per Copiam aggrauantem; aut per Frigora condensantia; aut per* Ventos *contrarios compellentes; aut per Obices repercutientes.*

**11.** *Tam Vapores, quam Exhalationes, Materia* ventorum *sunt. Etenim ex Exhalationibus nunquam* Pluuia, *ex vaporibus sæpissimè*
[Q7ᵛ] Venti. *At illud inter* | *est, quod facti* venti *ex Vaporibus, faciliùs se*
25  *incorporant Aeri puro, & citiùs sedantur, nec sint tam obstinati, quam illi ex Halitibus.*

**12.** *Modus, & diuersæ conditiones* Caloris, *non minùs possunt in* Generatione ventorum, *quam Copia aut Conditiones* Materiæ.

29  **13.** *Solis calor, in* Generatione ventorum, *ita proportionatus esse*
[Q8ʳ] *debet, vt eos excitet, sed non tantâ copiâ, vt coeant in* pluuiam; nec |
*tantâ paucitate, vt prorsùs discutiantur, & dissipentur.*

7 *vocant*);] ~:)    9 *est*;] ~:    10 *principio*;] ~:    11 *post*;] ~:

5. When some other principle of motion comes into play in addition to the overburdening of the air, it acts as an accessory which strengthens and increases the main principle; whence it happens that great and turbulent winds seldom arise from simple overburdening of the air. |

6. There are four accessories to overburdening of the air: expiration from subterranean regions; forcing down from what they call the middle region of the air; dispersal caused by cloud production; and the mobility and acrimony of the exhalation itself.

7. Wind motion is almost always lateral and, indeed, winds made by plain overburdening move laterally from the start; wind arising from expiration from the earth or from repercussion | from on high becomes so not much later—unless the eruption, precipitation, or reverberation are extremely violent.

8. Air puts up with some compression before it perceives overburdening, and pushes against the air next to it. It follows from this that all winds are a little denser than air when it is still.

9. Winds are calmed in five ways: either by vapours coming together, or by their incorporation, sublimation, transportation, | or scantiness.

10. The coming together of vapours, and indeed of the air itself into rain, takes place in four ways: either by increase of abundance, or by condensation by cold, or by compulsion of contrary winds, or by bouncing off obstacles.

11. Vapours as much as exhalations are the matter of winds. For rain never comes from exhalations but very often from vapours. But this is the case: | that winds made from vapours more easily incorporate themselves with pure air, and are calmed more swiftly, and they are not so obstinate as those generated from exhalations.

12. The mode and different conditions of heat are no less effective in the generation of wind than are the abundance or condition of matter.

13. The heats of the Sun in the generation of winds should be

**14.** Venti *spirant ex parte Fomitum suorum; cumque Fomites variè disponantur, diuersi* venti, *vt plurimùm, simul spirant; sed fortior debiliorem, aut obruit, aut flectit in Currentem suum.*

**15.** *Vbique generantur* venti, *ab ipsa terræ superficie, vsque ad frigidam Regionem Aeris; sed frequentiores in proximo, fortiores in sublimi.*

[Q8ᵛ] **16.** Regiones, *quæ habent* | ventos Asseclas *ex tepidis, sunt calidiores, quam pro ratione* Climatis *sui; quæ ex gelidis, frigidiores.*

Charta humana; siue optatiua cum proximis, circa ventos.

**Optatiua. 1.** Vela Nauium, *ita componere & disponere, vt minore flatu, maiorem conficiant viam. Res insignitèr vtilis, ad compendia* [R1ʳ] *itinerum per Mare, & parcendum impensis.* |

*Proxim.* Proximum *non occurrit adhùc inuentum, præcisè in practicâ. Sed Consule de eo* Obseruationes Maiores, *super* Art. 26.

**Optatiu. 2.** Molendina *ad* ventum, *& vela ipsorum ita fabricari, vt minore flatu, plùs molant. Res vtilis ad lucrum.*

*Proxim. Consule de hoc,* Experimenta *nostra, in* Responso *ad* Artic. 27, *vbi videtur res quasi peracta.*

**Optatiu. 3.** Ventos *orituros, & occasuros, & tempora ipsorum prænoscere. Res vtilis ad Nauigationes, & Agriculturam; maximè* [R1ᵛ] *autem* |*ad electiones temporum, ad prœlia Naualia.*

*Proxim. Hùc multa pertinent eorum, quæ in Inquisitione, præsertìm in* Responso *ad* Articulum 32 *notata sunt. At Obseruatio in posterum diligentior (si quibus ea cordi erit) patescente iam causâ* Ventorum, *longè exactiora* prognostica præstabit.

**Optatiu. 4.** Iudicium, *& Prognostica facere per* ventos, *de alijs rebus: veluti primò, si sint* Continentes, *aut* Insulæ *in* Mari, *in* [R2ʳ] *aliquo loco, vel potiùs mare liberum? Res vtilis ad* Nauigationes | *nouas, & incognitas.*

---

1 *suorum;*] ~:     5 *Aeris;*] ~:     23 32] ~.

so proportioned that they raise them, but not of such abundance that they gather them into rain, nor | of such scantiness that they dispel and disperse them.

14. Winds blow from the direction where their nurseries lie; and when the nurseries lie in different places, different winds generally blow together, with the stronger overcoming the weaker or bending it into its own course.

15. Winds are generated everywhere, from the surface of the earth right up to the air's frigid region; but more frequent winds arise near at hand, and stronger ones up above.

16. Regions which have | warm prevailing winds are hotter, and those with cooler winds are colder to a degree greater than their climates would lead one to expect.

### A Table of Human Goods, or Desiderata with their Closest Approximations, concerning the Winds

*Desideratum* 1. That ships' sails be so set up and arranged that with less wind they travel further. This is a matter of great utility for shortening voyages and saving money. |

*Approximation.* No approximation has yet been discovered precisely to satisfy this. But on this consult the major observations on article 26.

*Desideratum* 2. That windmills and their sails be so constructed that they grind more with less wind; a matter useful for profit.

*Approximation.* On this consult my response to article 27, where this business is practically put into effect.

*Desideratum* 3. That ways be found to predict the risings, fallings, and times of winds; this would be useful for navigation and agriculture, and most of all | for choosing times for sea battles.

*Approximation.* Many things relevant to this have been noted in the inquiry, especially in the response to article 32. But more careful observation by posterity (if it can be bothered) will discover much more accurate prognostics now that the cause of winds has been made clear.

*Desideratum* 4. That judgement and prognostication be made by the winds in other matters, as for instance whether there be continents or islands in the sea anywhere or open water. This is a matter useful for new | navigations into unknown regions.

***Proxim.*** Proximum *est, Obseruatio circa* Ventos statos: *id quo vsus videtur* Columbus.

***Optatiu.* 5.** *Itidem de vbertate, aut Caritate fructuum, &*
*segetum, annis singulis.* Res *vtilis ad lucrum, & venditiones*
5 *anticipantes, & coemptiones, vt proditum est de* Thalete, *circa*
Monopolium Oliuarum.

***Proxim.*** *Hùc pertinent nonnulla, in* Inquisitione *posita, de* Ventis, *aut*
*malignis, aut decussiuis, & temporibus, quandò nocent, ad* Artic. 29.

[R2ᵛ]     ***Optatiu.* 6.** *Itidem, de Morbis, &* | *Pestilentijs, annis singulis.* Res
10 *vtilis ad existimatio*n*em* Medicorum, *si illa prædicere possint; etiam*
*ad Causas, & Curas morborum; & nonnulla alia Ciuilia.*

***Proxim.*** *Hùc pertinent etiam nonnulla in* Inquisitione *posita, ad*
Art. 30.

***Monitum.*** *De* Prædictionibus *ex* Ventis, *circa segetes, fructus, & mor-*
15 *bos; Consule* Historias Agriculturæ, *&* Medicinæ.

***Optatiu.* 7.** *Ventos excitare, & sedare.*

***Proxim.*** *De his, habentur quædam superstitiosa, &* Magica: *quæ non*
[R3ʳ] *videntur digna, quæ in* Historiam Naturalem *seriam, & seueram re* | *cipian-*
*tur. Neque occurrit nobis, aliquid* Proximum, *in hoc genere. Designatio ea*
20 *esse poterit, vt Natura Aeris penitùs introspiciatur, & inquiratur, si possit*
*inueniri aliquid, quod in quantitate non magnâ, in Aerem immissum,*
*possit excitare, & multiplicare Motum, ad Dilatationem, aut Contrac-*
*tionem, in Corpore Aeris; Ex hoc etenim (si fieri possit) sequentur Excita-*
*tiones, & Sedationes* Ventorum; *quale est illud Experimentum* Plinij, *de*
25 Aceto *iniecto in occursum* Turbinis, *si verum foret. Altera designatio possit*
*esse, per Emissionem* Ventorum *ex subterraneis, si congregentur alicubi, in*
[R3ʳ] *magnâ copiâ, quale est illud receptum de* Puteo *in* Dalmatiâ; *verum &* |
*loca huiusmodi Carcerum nosse difficile.*

***Optatiu.* 8.** *Complura Ludicra, & mira, per motum* Ventorum
30 *efficere.*

***Proxim.*** *De his cogitationem suscipere, nobis non est otium.* Proximum
*est illud vulgatum* Duellorum *ad* Ventum. *Proculdubiò, multa eiusmodi*
*iucunda reperiri possunt; & ad* Motus, *& ad* Sonos.

---

14–15 *morbos,*] / BNF copy with a semicolon; Yale copy with a comma     23 *Aeris,*] ~:
26 *subterraneis*] subterrraneis     32 Ventum.] ~:

*Approximation.* An approximation is the observation about recurrent winds which Columbus seems to have used.

*Desideratum* 5. Likewise that a way be found to forecast abundance or dearth of corn and fruit every year. This is a useful matter for profit, and speculative buying and selling, as is related of Thales, who cornered the olive market.

*Approximation.* Relevant to this are some issues raised in article 29 concerning either malign or destructive winds, and when they may do damage.

*Desideratum* 6. Likewise forecasting every year diseases and | plagues. This is a matter useful to the standing of physicians, if these things could be predicted, as also for the causes and cures of diseases, and for other matters relating to civil life.

*Approximation.* Relevant to this are some issues raised in connection with article 30.

*Advice.* For predictions from winds in the matter of crops, fruits, and illnesses, consult the histories of agriculture and medicine.

*Desideratum* 7. That winds be prompted and calmed down.

*Approximation.* In this connection there exist certain superstitious and magical practices which seem to be unworthy of inclusion in a serious and strict natural | history. And no approximation occurs to me in this kind. But it could take us in the right direction if the nature of the air were looked into and investigated to see if anything can be found, a small amount of which introduced into the air can stimulate and multiply the motions of dilatation and contraction in the body of the air. From this (if it could be done) the raising and subduing of the winds would follow, an example of which (if it be true) is Pliny's experiment of casting vinegar in the face of a whirlwind. Another step might be to let out winds from places beneath the ground if there are such where the winds gather in great abundance, as for example is reported of the Dalmatian wind. However, it is | hard to find these places of imprisonment.

*Desideratum* 8. That many amusements and wonders can be performed by the motions of the winds.

*Approximation.* I have no leisure to think about these things. The approximation is the common game of battledore and shuttlecock. Doubtless many other such delights can be found out, both in relation to motion and to sound.

[R4ʳ]            | ADITVS AD TITVLOS IN
proximos quinque Menses destinatos.

Historia Densi & Rari.
Aditus.

5 [Here follows (R4ʳ–R8ᵛ) the preface to the *Historia densi*. The preface, as
it appears in the version published with the history itself, is reproduced
with facing-page translation in *OFB*, XIII, pp. 36–9. There are no substantive differences between that version and the one presented in
*Historia naturalis*. As for accidental differences, the latter version has
10 fewer initial capitals than the former; it has more diacriticals, slightly
heavier punctuation, and prefers initial *vs* to *us*.]

[R8ᵛ]            |

Historia Grauis & Leuis.
Aditus.

*Motum* Grauitatis *&* Leuitatis, *Veteres Motûs Naturalis nomine*
15 *insigniuerunt. Scilicet nullum conspiciebant Efficiens externum;*
[S1ʳ] *nullam etiam Resistentiam appareni* tem. *Quinimò citatior videbatur*
*Motus iste in progressu suo. Huic Contemplationi, vel Sermoni*
*potiùs, Phantasiam illam* Mathematicam *de Hæsione Grauium ad*
*Centrum Terræ (etiam si perforata foret ipsa Terra), nec non Com-*
20 *mentum illud* Scholasticum, *de Motu Corporum ad loca sua, veluti*
salem *asperserunt. His positis, perfunctos se credentes, nil amplius*
[S1ᵛ] *quærebant, nisi quod de* Centro Graui tatis, *in diuersis figuris, & de*
*ijs, quæ per Aquam vehuntur, paulò diligentiùs quispiam ex illis*
*quæsiuit. Neque ex Recentioribus quisquam operæ pretium circa hoc*
25 *fecit, addendo solummodò pauca* Mechanica, *eáque per* Demonstra-
tiones *suas detorta. Verùm missis verbulis, certissimum est, Corpus*
*non nisi à Corpore pati; nec vllum fieri Motum Localem, qui non*
[S2ʳ] *sollicitetur, aut à partibus Corporis ipsius, quod moue* tur, *aut à*
30 *Corporibus adiacentibus, vel in contiguo, vel in proximo, vel saltem*

---

3 Historia Densi] / for *tns* to this preface see *OFB*, XIII, pp. 36–8      19 *Terra*),] ~,)
24 *operæ pretium*] *operæpretium*

<sup>|</sup> PREFACES TO THE TITLES
planned for the next five months

History of Dense and Rare
Preface

[See note on facing page.]

|

History of Heavy and Light
Preface

The ancients identified motion of heavy and light with the name *natural motion*, since they could discern neither an external efficient nor apparent resistance. <sup>|</sup> Moreover this motion seemed to get quicker as it progressed. They made this thought, which was really all talk, tastier by adding that mathematical fancy that heavy bodies would stick to the centre of the Earth (even if the Earth had a hole all the way through it), as well as the scholastic fiction of the motion of bodies to their natural places. Having laid down these principles they assumed they had done their job and made no further inquiry, except for the man <sup>|</sup> who studied a little more closely the centre of gravity of bodies with different shapes, and bodies floating on water. Nor has one of the recent writers done very much on this topic, having added only a few mechanical exercises and those mangled by his demonstrations. But leaving aside mere words, it is quite certain that a body is not affected except by another body, and that no local motion occurs which is not prompted either by the parts of the moving body itself; by <sup>|</sup> adjacent bodies, be they contiguous or close at hand; or at least by ones within their orb of virtue. Therefore not ignorantly did Gilbert raise the matter of magnetic powers; even though he

*intra Orbem Actiuitatis suæ. Itaque vires* Magneticas *non inscitè introduxit* Gilbertus, *sed & ipse factus* Magnes; *nimiò scilicet plura, quam oportet, ad illas trahens, &* Nauem *ædificans ex* Scalmo.

[S2$^v$]            <sup></sup>| Historia Sympathiæ & Antipathiæ rerum.

5            Aditus.

Lis, *& *Amicitia *in Naturâ, stimuli sunt Motuum, & Claues Operum. Hinc Corporum Vnio & Fuga, hinc Partium Mistio & Separatio, hinc altæ atque intimæ Impressiones Virtutum, & quod vocant, Coniungere actiua cum passiuis, denique Magnalia Naturæ.*

[S3$^r$] *Sed impu*|*ra est admodùm hæc pars Philosophiæ, de* Sympathiâ *&*
11 Antipathiâ *Rerum, quam etiam* Naturalem Magiam *appellant, atque (quod semper ferè fit) vbi diligentia defuit, spes superfuit. Operatio autem eius in hominibus, prorsùs similis est soporiferis nonnullis* Medicamentis, *quæ somnum conciliant, atque insuper læta, &*
15 *placentia somnia immittunt. Primò enim Intellectum humanum in*
[S3$^v$] *soporem conjicit, decan*|*tando Proprietates specificas, & Virtutes occultas, & cælitùs demissas; vnde homines ad veras causas eruendas non ampliùs excitantur, & euigilant, sed in huiusmodi otijs acquiescunt; deinde innumera* Commenta, *Somniorum instar, insinuat, &*
20 *spargit. Sperant etiam Homines vani,* Naturam *ex fronte, & personâ cognoscere, & per Similitudines extrinsecas, Proprietates internas*
[S4$^r$] *detegere. Practica quoque Inquisiti*|*oni simillima. Præcepta enim* Magiæ Naturalis *talia sunt, ac si confiderent homines terram subigere, &* Panem suum comedere, *absque sudore vultus; & per*
25 *otiosas, & faciles Corporum Applicationes, rerum potentes fieri; semper autem in ore habent, & tanquam sponsores appellant* Magnetem, *& consensum* Auri *cum* Argento viuo; *& pauca huius*
[S4$^v$] *generis, ad fidem aliarum Rerum, quæ neutiquàm simili*|*contractu obligantur. Verùm optima quæque laboribus, tùm Inquirendi, tùm*
30 *Operandi, proposuit Deus. Nos in iure Naturæ enucleando, & rerum fœderibus interpretandis, paulò diligentiores erimus; nec Miraculis fauentes, nec tamen Inquisitionem instituentes humilem, aut angustam.*

18–19 *acquiescunt,*] ~:    22 *simillima*] *similima* / *nld* thus in *SEH* (II, p. 81)    25 *fieri;*] ~: /
BNF copy with colon; Pierpont Morgan with semicolon

turned himself into a loadstone, i.e. he drew towards them more phenomena than he should have done, and so built a ship from a shell.

| History of the Sympathy and Antipathy of things
Preface

Strife and friendship in nature are the spurs of motion and keys of works. From these spring the union and flight of bodies, the mixture and separation of parts, the high and intimate impressions of virtues, and what they call the conjoining of actives with passives, and in short the *magnalia naturæ*. But | this branch of philosophy (also known as natural magic) which concerns itself with the sympathy and antipathy of things is pretty corrupt; and, as ever, where diligence has been short, hope has been extravagant. Hope works on men very like some soporific drugs, which not only bring sleep but send sweet and happy dreams. In the first place it lulls the human intellect to sleep with | enchanting talk of specific properties, and occult virtues sent from the heavens, as a result of which men are no longer encouraged or alive to the business of unearthing real causes but give in to idleness of this kind; and then it insinuates and spreads countless fictions like so many dreams. Men also foolishly expect to get to know nature from her face and outward form, and to detect her inner properties from superficial resemblances. Their practice too is very like their methods | of inquiry. For the teachings of Natural Magic are such as if men expected that they could till the soil and eat their bread without the sweat of their brow, and by idle and facile applications of bodies to achieve great things. And they are never silent about the magnet, the consent of gold and quicksilver, and a few other things of the kind, and appeal to them as guarantees of their faith in other effects which do not at all | spring from any such similar cause. But God has laid it down that whatever is best shall be achieved by hard labour both in operation and investigation. And I shall be a little more diligent in examining nature's laws, and interpreting the true alliances of things, nor shall I indulge in marvels, nor yet set up a shallow or narrow inquiry.

[S5$^r$]  | Historia Sulphuris, Mercurij, & Salis.
Aditus.

Principiorum Trias *istud à* Chimistis *introductum est, atque quoad*
*Speculatiua, est ex ijs, quæ illi afferunt, Inuentum optimum. Sub-*
5 *tiliores ex ijs, quíque philosophantur maximè,* Elementa *volunt esse*
Terram; Aquam; Aerem; Æthera; *Illa autem non* Materiam *rerum*
[S5$^v$] *esse po|nunt, sed* Matrices, *in quibus specifica Semina rerum*
*generant, pro naturâ* Matricis. *Pro* Materiâ *autem* primâ (*quam*
*spoliatam, & adiaphoram ponunt* Scholastici) *substituunt illa tria*
10 Sulphurem, Mercurium, *&* Salem; *ex quibus omnia Corpora sint*
*coagmentata, & mista. Nos vocabula ipsorum accipimus, Dogmata*
*parûm sana sunt. Illud tamen non malè cum illorum Opinione*
[S6$^r$] *conuenit, quod duo ex illis,* Sulphu|rem *scilicet, &* Mercurium
(*sensu nostro accepta) censemus esse Naturas admodùm primordiales,*
15 *& penitissimos* Materiæ *Schematismos; & inter* Formas *primæ*
classis *ferè præcipuas. Variare autem possumus Vocabula* Sulphuris
*&* Mercurij, *vt ea aliter nominemus;* Oleosum, Aqueum; Pingue,
Crudum; Inflammabile, Non inflammabile; *& huiusmodi.*
[S6$^v$] *Videntur enim esse hæ duæ Rerum Tribus* | *magnæ prorsus, & quæ*
20 *Vniuersum occupant, & penetrant. Siquidem in Subterraneis, sunt*
Sulphur, *&* Mercurius, *vt appellantur; in Vegetabili & Animali*
*genere, sunt* Oleum, *&* Aqua; *in Pneumaticis inferioribus, sunt* Aer,
*&* Flamma; *in Cœlestibus,* Corpus stellæ, *&* Æther purum; *verùm*
*de vltimâ hac Dualitate nil adhùc pronunciamus, licet probabilis*
[S7$^r$] *videatur esse Symbolizatio. Quod verò ad* Salem *atti|net, alia res est.*
26 *Si enim* Salem *intelligunt pro parte Corporis fixâ, quæ neque abit in*
*Flammam, neque in Fumum, pertinet hoc ad* Inquisitionem Fluidi,
*&* Determinati, *de quibus nunc non est sermo; sin* Salem *accipi*
*volunt, secundum literam, absque parabolâ, non est* Sal *aliquid*
30 *tertium à* Sulphure, *&* Mercurio, *sed mistum ex vtrisque per spiri-*
[S7$^v$] *tum acrem deuinctis. Etenim* Sal *omnis habet partes inflammabiles,* |

| | | |
|---|---|---|
| 7 Matrices,] ~; | 11 *mista.*] ~: | 18 Crudum;] ~: inflammabile;] ~: |
| 20 *penetrant.*] ~: | Siquidem] Si-quidem | 21 *appellantur;*] ~: 22 Aqua;] ~: |
| 23 Flamma;] ~: | 25 *atti|net,*] ~; | 27 *Fumum,*] ~; 28 Determinati,] ~; |
| 29 *parabolâ,*] ~; | | |

<sup>|</sup> History of Sulphur, Mercury, and Salt
Preface

This triad of principles was brought in by the chemists, and as a speculative doctrine it is among the best that they have put forward. The more subtle of them, philosophizing at their best, hold that the elements are earth, water, air, and ether. But they do not count these as the matter of things, <sup>|</sup> but as wombs in which the specific seeds of things are generated according to the nature of the womb. Instead of prime matter (which the scholastics take to be despoiled and without qualities) they substitute the triad of sulphur, mercury, and salt, from which all bodies are compounded and mixed. I adopt their terms but their dogmas are barely sensible. Nevertheless it is not entirely inconsistent with their opinion that two of these, i.e. <sup>|</sup> sulphur and mercury (in the meaning I ascribe to them), I take to be the most primordial natures, the most fundamental schematisms of matter, and among forms of the first class practically the chief. We can vary the terms of sulphur and mercury and give them different names: the oily, the watery; the fat, the crude; inflammable and non-inflammable; and the like. For it seems that these two unequivocally great tribes <sup>|</sup> of things occupy and penetrate through the universe; in that in the subterranean regions we have sulphur and mercury as they call them; in vegetable and animal beings we have oil and water; among the lower pneumatic bodies, air and flame; in the heavens, stellar body and pure ether. But on this last duality I do not as yet pronounce definitively, though this correspondence seems quite probable. As for salt, <sup>|</sup> that is another matter. For if by salt they mean the rigid part of a body, the part that does not go up in flame or fume, this belongs to the investigation into fluid and determinate, a matter which does not concern me now. But if they want to take salt in its literal and not in its figurative sense, salt is not a third principle distinct from sulphur and mercury, but a mixture of those two held together by an energetic spirit. For all salt has inflammable parts, <sup>|</sup> and other parts which not only do not conceive flame but shrink and fly

*habet alias, flammam non solùm, non concipientes, sed eam*
*exhorrentes, & strenuè fugientes.* Nihilominùs cum Inquisitio *de*
Sale, *sit quiddam Affine* Inquisitioni *de duobus reliquis, atque insu-*
*per sit eximij vsûs, vtpote vinculum vtriusque Naturæ,* Sulphureæ
5  & Mercurialis, & *Vitæ ipsius rudimentum; illum etiam in hanc*
Historiam, & Inquisitionem *recipere visum est.* At illud interìm
[S8$^r$]  *monemus, de* Pneumaticis *illis,* Aere, $^|$ Flamma; Stellis, Æthere; *nos*
*illa* (*prout certe merentur*) Inquisitionibus *proprijs reseruare, & de*
Sulphure, & Mercurio *tangibili* (*nimirùm vel Minerali, vel*
10  *Vegetabili, & Animali*) *hîc tantùm* Historiam *instituere.*

[S8$^v$]  $^|$ Historia Vitæ & Mortis.
Aditus.

[In the copy-text (S8$^v$–T7$^r$) the preface to the *Historia vitæ* now follows.
This preface, in the version published with that history, appears below,
15  and the textual notes printed therewith record the differences between
the *Historia vitæ* version and the one that stood here in the *Historia*
*naturalis.*]

*FINIS.*

5 Mercurialis] Salis / emended thus in *SEH* (II, p. 83) following the 1638 edition (Gibson nos.
196–7), i.e. the one presented in *Operum . . . tomus,* 4I4$^r$     7 Flamma] Aqua / emended
thus in *SEH* (II, p. 83) following the 1638 edition (see *tns* to S7$^v$ above). Water is not of course a
pneumatic substance

7 Stellis,] ~;

from it with all speed. Nevertheless since the inquiry into salt has a certain kinship with the other two things, and is in addition of great utility—given that salt combines the sulphureous and mercurial natures, and is a rudiment of life itself—this too I have decided to bring into this history and inquiry. But in the meantime I must let it be known that I reserve those pneumatic substances—air, [1] flame, the stars, and ether—for their own proper inquiries (as they certainly deserve) and that here I only set up a history of tangible sulphur and mercury as they manifest themselves in mineral, vegetable, or animal bodies.

[1] History of Life and Death
Preface

[See note on facing page.]

*THE END.*

| FRANCISCI
BARONIS DE VERVLAMIO,
VICE-COMITIS SANCTI ALBANI,

Historia *Vitæ* & *Mortis.*

SIVE, TITVLVS SECVNDVS
in Historiâ Naturali & Experimentali
ad condendam Philosophiam:

Quæ est
*INSTAVRATIONIS MAGNÆ* PARS TERTIA.

[1] FRANCIS
LORD VERULAM,
VISCOUNT ST ALBANS,

The History *of Life and Death*

OR THE SECOND TITLE
in the Natural and Experimental History for the
building up of Philosophy

Which is
THE THIRD PART OF *THE GREAT INSTAURATION*

[A1$^v$: blank]

[A2$^r$]

## ᵎ VIVENTIBVS ET POSTERIS
Salutem.

Cvm *Historiam Vitæ* & *Mortis*, inter *Sex Designationes Menstruas*,
5 vltimo loco posuerimus; omninò hoc præuertere visum est, &
[A2$^v$] secundam ede⌐re, propter eximiam rei vtilitatem; in quâ, vel
minima temporis iactura, pro pretiosâ haberi debet. Speramus
enim, & cupimus futurum, vt id plurimorum bono fiat; atque
vt *Medici Nobiliores* animos nonnihil erigant, neque toti sint in
10 *Curarum sordibus*; neque solùm propter *Necessitatem honorentur*,
sed fiant demum Omnipotentiæ, & Clementiæ diuinæ admini-
[A3$^r$] stri, in *vitâ Hominum* ᵎ prorogandâ, & instaurandâ; præsertìm
cum hoc agatur, per vias tutas, & commodas, & ciuiles, licet
intentatas. Etsi enim nos *Christiani*, ad *Terram Promissionis* per-
15 petuò aspiremus, & anhelemus; tamen interìm itinerantibus
nobis, in hâc *Mundi Eremo*, etiam *Calceos* istos, & tegmina
(Corporis scilicet nostri fragilis) quam minimum atteri, erit signum *Fauoris Diuini*.

6 in] In

|

## ' TO PRESENT AND FUTURE GENERATIONS
### Greetings

Although I placed the *History of Life and Death* last on the list of the six monthly histories that I planned, I have decided to bring it forward and publish it in second place ' on account of the exceptional utility of the matter, a matter in which the slightest loss of time should be counted precious. For I hope and wish that it works for the good of many, and that the more outstanding physicians lift up their minds somewhat, and not immerse themselves in mercenary cures, nor acquire honour only out of necessity, but become servants of God's omnipotence and mercy in prolonging and ' renewing the life of man, especially as this is achieved by ways that are safe, convenient, and civil, though untried. For although we Christians ever aspire and thirst after the Promised Land, yet in the meantime it will be a mark of Divine Favour if, in our pilgrimage in this world's wastes, these our shoes and clothes (our frail bodies, that is) be as little worn out as possible.

[A3ᵛ: blank]

[A4ʳ]

Historia Vitæ & Mortis.
Aditus.

*De* Vita breui, *&* Arte longâ, *vetus est Cantilena, & querela.*
*Videtur igitur esse tanquam ex congruo, vt nos qui pro viribus*
5 *incumbimus, ad* Artes perficiendas, *etiam de* Vitâ Hominum pro-
ducendâ, *cogitationem suscipiamus, fauente &* Veritatis, *&* Vitæ
[A4ᵛ] Authore. *Etsi enim* Vita Morta¹lium, *non aliud sit, quàm Cumulus,*
*& accessio* Peccatorum, *&* Ærumnarum, *quíque ad* Æternitatem
*aspirant, ijs leue sit lucrum* vitæ; *tamen non despicienda est, etiam*
10 *nobis* Christianis, *Operum Charitatis Continuatio. Quinetiam*
Discipulus Amatus *cæteris superstes fuit; & complures ex* Patribus,
*præsertìm* Monachis *sanctis, &* Eremitis, *longæui fuerint; vt isti*
[A5ʳ] Benedictioni (*toties in* Lege veteri *repetitæ*), ¹ *minus detractum*
*videatur post æuum* Seruatoris, *quam reliquis* Benedictionibus
15 *terrenis. Verùm vt hoc pro maximo* Bono *habeatur, procliue est. De*
*Modis assequendi, ardua* Inquisitio; *eóque magis, quod sit &*
Opinionibus *falsis, &* Præconijs *vanis deprauata. Nam & quæ à*
*turbâ* Medicorum, *de* Humore Radicali, *&* Calore Naturali *dici*
[A5ᵛ] *solent, sunt seductoriæ: & laudes immodicæ* Medici¹narum
20 Chymicarum, *primò inflant* Hominum *spes, deindè destituunt.*
*Atque de* Morte, *quæ sequitur ex* Suffocatione, Putrefactione, *&*
*varijs* Morbis, *non instituitur præsens* Inquisitio; *pertinet enim ad*
Historiam Medicinalem; *sed de eâ tantum* Morte, *quæ fit per*
24 Resolutionem, *ac* Atrophiam *senilem. Attamen de vltimo passu*
[A6ʳ] Mortis, *atque de ipsâ* Extinctione vitæ, *quæ tot modis, &* ¹ *exteriùs,*
*& interiùs, fieri potest* (*qui tamen habent quasi* Atriolum *commune,*
*antequàm ad* Articulum Mortis *ventum sit*) *inquirere, affine*
*quiddam præsenti* Inquisitioni *esse censemus, sed illud postremo loco*
*ponemus.*
30 *Quod reparari potest sensìm, atque primo* Integro *non destructo,*
*id potentiâ æternum est, tanquam* Ignis vestalis. *Cum igitur viderent*

---

8 *Ærumnarum*] / cf. *HNE*, T1ʳ: *Aerumnarum*  13 *Benedictioni*] ~, repetitæ),] ~,)
22 *varijs*] / cf. *HNE*, T2ʳ: *variis*  23 Medicinalem;] ~:  26 *potest*] ~,
27 *inquirere,*] ~;  28 *censemus,*] ~:

| The History of Life and Death
Preface

Ancient is the refrain and complaint that life is short and art long. So it seems right that I, who devotes his utmost strength to perfecting the arts, should also, by the grace of the Author of Truth and Life, apply my mind to the prolongation of human life. For though this mortal life | is nothing other than an accumulation of sin upon sin, and affliction upon affliction, and though they who long for eternity set little store by this life, even so keeping works of charity going should not be held in contempt by us Christians. Besides, the beloved disciple outlived the others, and many of the Fathers, especially the holy monks and hermits, were long-lived; so that this blessing (repeated so often in the Old Law) | seems to have been removed after our Saviour's time less than other earthly blessings. Now it is easy to take this as the greatest good, but an inquiry to come up with means of achieving it is hard, the more so because it has been corrupted by false opinions and groundless reports. For what the medical rabble generally says about the radical moisture and natural heat is deceitful, while the extravagant praise heaped | on chemical medicines only raises men's hopes to dash them.

Now I have not put the present inquiry in hand to concern itself with death resulting from suffocation, putrefaction, and the various diseases, for that belongs to the history of medicine; here I am concerned only with death caused by the disintegration and atrophy of old age. Nevertheless I judge that to investigate the last step of death and the very extinction of life, which can come about | by so many internal and external factors (but which still lead as it were to a common anteroom before reaching the point of death), is also relevant to the present inquiry, but I shall leave that until the end.

Anything which can be repaired gradually, without destroying the original whole, is, like the vestal flame, potentially eternal.

[A6ᵛ] Medici, *& Philosophi, ali prorsùs Ani¹malia, eorúmque Corpora reparari, & refici; neque tamen id diù fieri, sed paulò post senescere ea, & ad Interitum properè deduci;* Mortem *quæsiuerunt in aliquo, quod propriè reparari non possit;* existimantes Humorem *aliquem*
5 Radicalem, *&* Primigenium *non reparari in solidum, sed fieri iam vsque ab Infantiâ, Appositionem quandam degenerem, non*
[A7ʳ] *Reparationem iustam; quæ sensìm cum Ætate deprauetur, & de¹mùm* Prauum *deducat ad* Nullum. *Hæc cogitârunt imperitè satis, & leuitèr. Omnia enim in Animali, sub Adolescentiâ, & Iuuentute,*
10 *reparantur integrè; quinetiam ad tempus, Quantitate augentur, Qualitate meliorantur; vt Materia Reparationis, quasi æterna esse posset, si Modus Reparationis non intercideret. Sed reuerâ hoc fit. Vergente ætate, inæqualis admodùm fit Reparatio; aliæ partes*
[A7ᵛ] *reparantur satis fæticitèr, aliæ ¹ ægrè, & in peius; vt al eo tempore,*
15 Corpora humana *subire incipiant tormentum illud* Mezentij, Vt viua in amplexu mortuorum immoriantur, *atque facilè reparabilia, propter ægrè reparabilia copulata, deficiant. Nam etiam post Declinationem, & decursum ætatis,* Spiritus, Sanguis, Caro,
19 Adeps, *facilè reparantur; at quæ sicciores, aut porosiores sunt partes,*
[A8ʳ] Membranæ, *&* Tunicæ *omnes,* Nerui, Arteriæ, Venæ, ¹ Ossa, Cartilagines, *etiam* Viscera *pleraque, denique* Organica *ferè omnia, difficiliùs reparantur, & cum iacturâ. Illæ autem ipsæ partes, cum ad illas alteras Reparabiles partes, actu reparandas, omninò officium suum præstare debeant; actiuitate suâ, ac viribus imminutæ, func-*
25 *tiones suas ampliùs exequi non possunt. Ex quo fit, vt paulò pòst, omnia ruere incipiant, & ipsæ illæ partes, quæ in naturâ suâ sunt*
[A8ᵛ] *valdè reparabiles, ta¹men deficientibus Organis Reparationis, nec ipsæ similiter ampliùs commodè reparentur, sed minuantur, & tandem deficiant. Causa autem* Periodi *ea est; quod* Spiritus, *instar Flammæ*
30 *lenis, perpetuò prædatorius, & cum hoc conspirans Aer externus, qui etiam corpora sugit, & arefacit, tandèm officinam Corporis, & Machinas, & Organa perdat, & inhabilia reddat ad Munus*

---

7 *Ætate*] / cf. *HNE*, T3ʳ: *Aetate*    9 *leuitèr*] / cf. *HNE*, T3ᵛ: *leuiter*    14 *peius,*] ~:
16 mortuorum] / cf. *HNE*, T4ʳ: mortuorun    21 *pleraque*] / cf. *HNE*, T4ᵛ: *pleráque*
31 *tandèm*] / cf. *HNE*, T5ʳ: *tandem*

When therefore the physicians and philosophers saw that |
animals were fed and that their bodies were repaired and
refreshed, and that this happened only for a while, and that not
long after they began to grow old, and were promptly dragged to
their destruction, these same physicians sought the cause of death
in something that could not be properly repaired, supposing that
some radical and primigenial moisture could not be thoroughly
repaired, but even from childhood took on a kind of defective
apposition instead of due repair, and with time it grew worse and
| eventually reduced a bad condition to nothing at all. Their
thinking was ignorant and light-minded enough. For in animals
all things are completely repaired while they are growing up and
still youthful; indeed for a time they increase in size and improve
in quality, so that the matter of repair could be practically ever-
lasting, if the means of repair did not break down. But this is
what really happens: in our declining years repair becomes
extremely patchy, some parts being repaired well enough, others |
with difficulty and less well, so that from then on human bodies
begin to suffer the torment devised by Mezentius, *that the living
perish in the embrace of the dead,* and that the parts easily reparable
give out because they are coupled with ones difficult to repair. For
even after the passage of time and declining years, the spirit,
blood, flesh, and fat repair easily, but the drier and more porous
parts, the membranes, and all tunicles, nerves, arteries, veins, |
bones, cartilages, and most of the innards too, as well as nearly all
the organic structures, are repaired with difficulty and some cost.
Now these very parts, when they ought to carry out the work of
repairing those reparable parts, can no longer discharge their
proper functions because their activity and powers have been
impaired. The upshot of this is that not long after all parts begin
to collapse, and those very parts which are intrinsically most
reparable are | nevertheless, once the organs of repair have given
out, no longer likewise able suitably to repair themselves, but
grow weaker, and at last themselves give out. Now the cause of
this conclusion is this: that the spirit, like a gentle flame, forever
predatory, with the external air colluding with it—air which also
sucks and dries bodies out—, at last destroys the workshop of the

[B1ʳ] *Reparationis. Hæ sunt veræ viæ* Mortis naturalis | *benè & diligentèr animo voluendæ. Etenim qui Naturæ vias non nouerit, quomodo is illi occurrere possit, eamque vertere?*

*Itaque duplex debet esse* Inquisitio, *altera de* Consumptione, *aut*
5 Deprædatione *corporis humani; altera de eiusdem* Reparatione, *aut* Refectione: *eo intuitu, vt altera, quantum fieri possit, inhibeatur, altera confortetur. Atque prior istarum pertinet præcipuè ad Spiritus,*
[B1ᵛ] *& Aerem exter|num, per quos fit Deprædatio; Secunda ad vniuersum processum Alimentationis, per quem fit Restitutio. Atque quoad*
10 *primam* Inquisitionis *partem, quæ est de* Consumptione, *omninò illa cum Corporibus Inanimatis, magnâ ex parte, communis est. Etenim quæ Spiritus innatus (qui omnibus Tangibilibus, siue viuis, siue mortuis inest) & Aer ambiens, operatur super Inanimata, eadem*
[B2ʳ] *& tentat super Animata; licet super-addi|tus Spiritus vitalis, illas*
15 *operationes partìm infringat, & compescat, partìm potentèr admodùm intendat, & augeat. Nam manifestissimum est Inanimata complura, absque Reparatione, ad tempus benè longum durare posse; At Animata absque Alimento, & Reparatione, subitò concidunt, &*
19 *extinguuntur, vt & Ignis. Itaque* Inquisitio *duplex esse debet; primò*
[B2ᵛ] *contemplando Corpus humanum, tanquam Inanimatum, & | Inalimentatum; deinde tanquam Animatum, & Alimentatum. Verùm hæc præfati, ad* Topica Inquisitionis *iam pergamus.*

---

2 *non*] / cf. *HNE*, T5ᵛ: *non*      3 *eamque*] / cf. *HNE*, T5ᵛ: *eamque*      5 *eiusdem*] / cf. *HNE*, T5ᵛ: *eiusdem*      15 *potentèr*] cf. *HNE*, T6ᵛ: *potenter*      17 *posse;*] ~:

body and its machinery and instruments, and renders them incapable of doing the job of repair. These are the true ways of natural death, | and ones which we must consider diligently and well. For how can someone who does not know nature's ways counteract and reverse them?

Therefore the inquiry is twofold: on the one hand into the consumption and ravaging of the human body, and on the other into its repair or refreshment; with the object of curbing the former as far as possible, and of strengthening the latter. The first of these relates chiefly to the spirit and external | air which cause the ravaging; the second to the whole process of aliment which brings about renewal. As far as the first part of the inquiry goes, which is concerned with consumption, it has a great deal in common with what happens in inanimate bodies. For what the innate spirit (present alike in all tangible bodies living and non-living), together with the ambient air, does to inanimate things, it also tries to do to animate ones, though here the superadded | vital spirit partly tempers and blocks its operations, and partly intensifies and increases them no end. For it is perfectly obvious that many inanimate bodies can last for a very long time without repair; but animate ones without aliment and repair rapidly break down and die out just like fire. Therefore the inquiry should be twofold: first considering the human body as something inanimate and | unnourished; and secondly as animate and nourished. But having said that, I now move on to the Topics of Inquiry.

[B3ʳ] | Topica Particularia;
Siue,
*Articuli Inquisitionis* de *Vitâ* & *Morte.*

1. De Naturâ *Durabilis,* & *minùs Durabilis,* in Corporibus *Inanimatis,*
5 atque simul in *Vegetabilibus,* inquisitionem habeto; non copiosam, aut
legitimam, sed strictìm, & per Capita, & tanquam in transitu.

[B3ᵛ] 2. De *Desiccatione, Arefactione,* & | *Consumptione* Corporum
*Inanimatorum,* & *Vegetabilium;* & de *Modis,* & *Processu* per quos fiunt;
atque insupèr de *Desiccationis, Arefactionis,* & *Consumptionis Prohibi-*
10 *tione,* & *Retardatione,* Corporumque in suo statu *Conseruatione;* atque
rursùs de Corporum, postquàm semel arefieri cœperint, *Inteneratione,*
& *Emollitione,* & *Reuirescentiâ,* diligentiùs inquirito.

*Neque tamen de his ipsis, perfecta, aut accurata facienda est Inquisi-
tio, cùm ex proprio Titulo* Durabilis *hæc depromi debeant, cúmque non*
15 *sint in Inquisitione præsenti principalia, sed lumen tantummodò*
[B4ʳ] *præbeant ad Prolongationem,* & Instauratio|nem *vitæ, in* Animalibus.
*In quibus ipsis (vt iam dictum est) eadem ferè vsu veniunt, sed suo
modo. Ab Inquisitione autem circa* Inanimata, & Vegetabilia, *transeat
Inquisitio ad* Animalia *præter* Hominem.

20 3. De *Animalium Longæuitate,* & *Breuitate vitæ,* cum circumstantijs
debitis, quæ ad huiusmodi Æuitates videantur facere, inquirito.

4. Quoniam verò duplex est Duratio Corporum, altera in Identitate
simplici, altera per Reparationem; quarum prima in Inanimatis tantùm
obtinet, secunda in Vegetabilibus, & Animalibus, & perficitur per *Ali-*
[B4ᵛ] *mentationem;* ideò de *Alimentatione,* | eiusque vijs, & processu inquirito:
26 neque id ipsum exactè (pertinet enim ad *Titulos Assimilationis,* &
*Alimentationis*) sed vt reliqua in transitu.

*Ab Inquisitione circa* Animalia, *atque* Alimentata, *transeat illa ad*
Hominem: *cùm verò iam deuentum sit, ad Subiectum Inquisitionis*
30 *principale, debet esse in omnibus Inquisitio magis exacta,* & *numeris suis*
*absoluta.*

---

11 arefieri] arifieri / the unusual *c-t* form also occurs in some copies on Z2ʳ; see *tns* p. 318
below        17 *ipsis*] ~,

## $^|$ Particular Topics
or
## Articles of Inquiry on Life & Death

1. Inquire into the nature of durable and less durable in inanimate bodies, and likewise in vegetables, but undertake the inquiry not in a full or legitimate way but summarily, by headings, and as if in passing.

2. Inquire more diligently into the desiccation, arefaction, and $^|$ consumption of inanimate and vegetable bodies, and into the ways and process by which they happen; and in addition into the prevention and retarding of the same, and into keeping bodies in their proper state; and more diligently again into the softening, mellowing, and revival of bodies once they have started drying out.

But we should not conduct a perfect or detailed inquiry even into these things, since these ought to be fetched out of the proper title of Durable since they are not the main issues in the present inquiry but only illuminate the prolongation and instaurat$^|$ion of life in animals. And in these things (as I have said) the same effects occur but in their own way. Now from the inquiry into inanimate and vegetable bodies, we pass to animals other than man.

3. Inquire into the length and shortness of life in animals with appropriate factors which seem to affect their life expectancy.

4. Because the lasting of bodies is twofold, i.e. in simple identity or by repair, with the first belonging to inanimate bodies alone, and the second to vegetables and animals and accomplished by alimentation, $^|$ look into alimentation and its ways and process, yet not too closely (for it belongs to the title of assimilation and alimentation) but, like the above, by the way.

From the inquiry concerning animals and bodies sustained by alimentation, we pass on to man; and since we arrive here at the main subject of investigation, the inquiry should be in every way more precise and perfect in every detail.

5. De *Longæuitate,* & *Breuitate vitæ* in *Hominibus,* secundum *Ætates Mundi, Regiones,* & *Climata,* & *Loca Natiuitatis,* & *Habitationis,* inquirito.

[B5ʳ]    6. De *Longæuitate,* & *Breuitate vitæ* in *Hominibus,* secundum |
5  *Propagines,* & *Stirpes* suas (tanquam esset hæreditaria); atque etiam secundum *Complexiones, Constitutiones,* & *Habitus* Corporis, *Staturas,* nec-non *modos,* & *spatia grandescendi,* atque secundum *Membrorum Facturas,* & *Compages* inquirito.

7. De *Longæuitate,* & *Breuitate vitæ* in *Hominibus,* secundum
10  *Tempora Natiuitatis,* ita inquirito, vt *Astrologica* & *Schemata cœli,* in præsentiâ omittas; recipito tantùm Obseruationes (si quæ sint) plebeias, & manifestas, de *partubus septimo, octauo, nono,* & *decimo Mense;* etiam *Noctù, Interdiù,* & quo *Mense Anni?*

[B5ᵛ]    8. De *Longæuitate,* & *Breuitate vitæ* in *Hominibus,* secundum |
15  *Victum, Diætas, Regimen vitæ, Exercitia,* & similia, inquirito; Nam quatenùs ad *Aerem,* in quo viuunt, & morantur *Homines,* de eo, in *Articulo* superiore de *Locis Habitationis,* inquiri debere intelligimus.

9. De *Longæuitate,* & *Breuitate vitæ* in *Hominibus,* secundum *Studia,* & *Genera Vitæ,* & *Affectus Animæ,* & varia *Accidentia,* inquirito.

20    10. De *Medicinis,* quæ putantur vitam prolongare, seorsùm inquirito.

11. De *Signis* & *Prognosticis* vitæ longæ, & breuis, non illis quæ *Mortem* denotant in propinquo (id enim ad *Historiam Medicinalem* [B6ʳ] pertinet), sed de ijs quæ etiam in | *Sanitate* apparent, & obseruantur,
25  inquirito, siue sint *Physiognomica,* siue alia.

*Hactenùs instituta est Inquisitio de* Longæuitate, *&* Breuitate vitæ, *tanquam* Inartificialis, *& in confuso; huic adijcere visum est* Inquisitionem Artificialem, *atque innuentem ad Praxim, per* Intentiones. *Eæ genere sunt tres. Distributiones autem magis Particulares*
30  Intentionum *earum proponemus, cùm ad ipsam* Inquisitionem *ventum erit. Tres illæ* Intentiones *generales sunt:* Prohibitio Consumptionis; Perfectio Reparationis; Renouatio Veterationis.

[B6ᵛ]    12. De ijs quæ *Corpus* in *Homine,* | ab *Arefactione,* & *Consumptione* conseruant, & eximunt, aut saltem *Inclinationem* ad eas, remorantur, &
35  differunt, inquirito.

5 hæreditaria);] ~;)    23 propinquo] ~,    24 pertinet),] ~)ₐ    34 eximunt,] ~;

5. Inquire into the length and shortness of life in men according to the ages in which they lived, and their regions, climes, birthplaces, and habitations.

6. Inquire into the length and shortness of life in men according to ' their family and stock (as if life expectancy were hereditary); and also according to the complexions, constitution, and disposition of the body, their stature, means, and intervals of growth; and according to the fashion and framework of their members.

7. Inquire into the length and shortness of life in men according to the times of their nativities, but for the time being leave astrological figures, and only adopt ordinary and obvious observations (if such there be) concerning the month the births came to term (e.g. in the seventh, eighth, ninth, or tenth month), whether by night or day, and in what month of the year.

8. Inquire into the length and shortness of life in men according ' to their food, diet, mode of life, exercise, and the like. As for the air in which they live and breathe, I think that this should be inquired under the article above about their places of habitation.

9. Inquire into the length and shortness of life in men according to their studies, the lives they lead, the affections of their souls, and other accidents.

10. Inquire separately into the medicines which supposedly prolong life.

11. Inquire into the signs and prognostics of long or short life; not into those which indicate that death is nigh (for this belongs to the history of medicine) but into those which are ' open to observation even in good health, be those signs physiognomical or whatever else.

Hitherto the inquiry into length and shortness of life has been conducted in an unskilful and muddled way, and I mean to supplement this by systematic inquiry directed towards practice by means of Intentions. Of these there are three kinds, and when I come to inquire into them I shall set out their particular distributions. The three general intentions are the prohibition of consumption, the accomplishing of repair, and the renovation of what has grown old.

12. Inquire into those things which ' keep and relieve the body of man from drying out and consumption, or at least curb and defer those tendencies.

13. De ijs quæ pertinent ad vniuersum *Processum Alimentationis* (vnde fit *Reparatio* in Corpore *Hominis*) vt sit proba, & minimâ cum iacturâ, inquirito.

14. De ijs quæ *purgant inueterata*, & *reponunt Noua*, quæque etiam 5 ea, quæ iam arefacta & indurata sunt, rursùs intenerant, & humectant, inquirito.

*Quoniam verò difficile est, vias ad* Mortem *nosse, nisi ipsius* Mortis Sedem, *&* Domicilium (*vel* Antrum *potiùs*) *perscrutatus sis, &* [B7ʳ] *inueneris; de hoc facienda est* Inqui|sitio, *neque tamen de omni genere* 10 Mortis, *sed tantùm de ijs* Mortibus, *quæ inferuntur per* Priuationem, *&* Indigentiam, *non per* Violentiam; *Illæ enim sunt tantùm, quæ ad* Atrophiam Senilem *spectant.*

15. De *Articulo Mortis*, & de *Atriolis Mortis*, quæ ad illum ducant, ab omni parte (si modò id fiat per *Indigentiam*, & non per *violentiam*), 15 inquirito.

*Postremò, quoniam expedit nosse* Characterem, *& formam* Senectutis, *quod fiet optimè, si differentias omnes, in* Statu Corporis, *&* Functionibus, *inter* Iuuentutem, *&* Senectutem, *diligentèr collegeris;* [B7ᵛ] *vt ex ijs perspicere possis, quid sit illud tan*|*dem, quod in tot effectus* 20 *frondescat; etiam hanc* Inquisitionem *ne omittito.*

16. De differentijs *Statûs Corporis*, & *Facultatum* in Iuuentute, atque in *Senectute;* & si quid sit eiusmodi, quod in Senectute maneat, neque minuatur, diligentèr inquirito.

## Natura Durabilis.
25        *Historia.*

*Ad Artic.* 1. 1. *Metalla* in tantum æuum durant, vt tempus *Durationis* [B8ʳ] ipsorum, Hominum obseruationem fugiat. | Etiam quandò soluuntur per ætatem, in Rubiginem soluuntur, non per Perspirationem; *Aurum* autem per neutrum.

30    2. *Argentum viuum*, licet humidum sit, & fluidum, atque per Ignem facilè fiat volatile, tamen (quod nouimus) absque Igne, per ætatem solam, nec consumitur, nec contrahit Rubiginem.

3. *Lapides*, præsertìm duriores, & complura alia ex *Fossilibus*, longi sunt æui; idque licet exponantur in Aerem, multò magis dùm

---

1 *Alimentationis*] ~,        8 Domicilium] ~,        14 parte] ~,    *violentiam*),] ~,)
31 volatile,] ~;

13. Inquire into those things which relate to the whole process of alimentation (whence comes repair of the human body), so that it may be suitable and with the least possible wasting.

14. Inquire into those things which purge old stuff and put back new; and also those which make soft and moist those parts which are now dried out and hardened.

But since it is difficult to know the ways to death, unless you first examine and search out the seat and residence (or rather vault) of death, an inquiry into this should be made, | yet not into every kind of death but only of those deaths which come by privation and want, and not by violence. For only the former relate to the atrophy of old age.

15. Inquire into the moment of death and into the antechambers which lead to it on every hand (so long as it is caused by want and not violence).

Lastly, since it helps to know the character and form of old age, the inquiry into it should not be left out; which will be best accomplished if you make a careful collection of all differences between the condition and functions of the body in youth and old age; so that from these differences you will be able identify from what it is in the end | that so many effects ramify.

16. Inquire carefully into the differences in bodily condition and faculties in youth and old age, and if there is anything which lasts into old age without loss.

## The Nature of Durable
### *History*

*To Article* 1. 1. Metals last for so long that their lasting outstrips human observation. | Even when time undoes them they decay into rust, and not through perspiration. But gold does neither.

2. Quicksilver, though it be moist and fluid, and easily rendered volatile by fire, still does not (as far as we know) get consumed or take on rust by age alone and without fire.

3. Stones, especially the harder ones, and many other fossils last a long time, even when exposed to the air, and much more while they lie

conduntur sub Terrâ; Attamen Nitrum quoddam colligunt *Lapides*, quod illis est instar *Rubiginis*. *Gemmæ* autem, & *Crystalla, Metalla* ipsa

[B8ᵛ]  æuo supe⸍rant; Attamen clarore suo nonnihil, a longâ ætate, mulctantur.

  **4.** Obseruatum est, *Lapides* ex parte *Boreæ,* citiùs temporis edacitate

5  consumi, quam *Austro* expositos, idque & in *Pyramidibus,* & in *Templis,* & alijs Ædificijs manifestum esse; *Ferrum* contrà, ad *Austrum* expositum, citiùs *Rubiginem* contrahere, ad *Septentrionem* tardiùs, vt in *Bacillis* illis ferreis, aut *Cratibus,* quæ ad *Fenestras* apponuntur, liquet. Nec mirum, cum in omni *Putrefactione* (qualis est *Rubigo*) *Humiditas*

10  acceleret *Dissolutionem;* in *Arefactione* simplici, *Siccitas.*

[C1ʳ]  **5.** In *Vegetabilibus* (loquimur de auulsis nec vegetantibus) *Stem⸍mata Arborum* duriorum, siue trunci, atque ligna, & *Materies* ex ipsis, per sæcula nonnulla durant. Partes autem *Stemmatis* variè se habent; Sunt enim quædam *Arbores* fistulosæ, vt *Sambucus,* in quibus *Pulpa* in medio

15  mollior sit, exterius durius; at in *Arboribus* solidis, qualis est *Quercus,* interius (quod *Cor Arboris* vocant) durat magis.

  **6.** *Folia Plantarum,* & *Flores,* etiam *Caules,* exiguæ sunt *Durationis,* sed soluuntur in puluerem, seseque incinerant, nisi putrefiant. *Radices* autem sunt magis durabiles.

20  **7.** *Ossa Animalium* diù durant, vt videre est in *Ossuarijs,* scilicet

[C1ᵛ]  *Repositorijs Ossium* Defunctorum. ⸍ *Cornua* etiam valdè durant; nec-non *Dentes,* sicut in *Ebore,* & *Dentibus Equi Marini.*

  **8.** *Pelles* etiam, & *Corium* valdè durant, vt cernere est in *Pergamenis* antiquorum Librorum: quinetiam *Papyrus,* complura sæcula tolerat,

25  licet *Pergamenæ* duratione cedat.

  **9.** *Ignem passa* diù durant, vt *Vitrum, Lateres;* etiam *Carnes,* & *Fructus Ignem passi,* diutiùs durant, quam crudi; neque ob id tantùm, quod huiusmodi *Coctio* arceat *Putredinem;* sed etiam quod emisso *Humore*

29  *aqueo, Humor oleosus* diutiùs se sustineat.

[C2ʳ]  **10.** *Aqua* omnium *Liquorum* citissimè sorbetur ab *Aere,* *Oleum* ⸍ contrà tardiùs euaporat; vt cernere est, non solùm in *Liquoribus* ipsis,

---

4–10 Obseruatum. . . Siccitas.] / this is one of two passages (for the other see *tns* to C2ʳ⁻ᵛ) which exist in Bacon's holograph (see Introduction, pp. lxxvi–lxxvii above) which reads, 'Obseruatu*m* est lapides ex parte Boreæ citius <<putres>> temporis edacitate consumj quam Austro expositos, idque et in Pyramidibus, et in Templis et alijs ædificijs manifestum esse. <<n>> Ferru*m* <<autem>> contra ad austru*m* expositum citius rubiginem contrahere, ad septentrionem tardius, vt in paruis illis bacillis ferreis <<quæ>> aut Cratibus quæ ad fenestras apponunt⸍ liquet. Nec miru*m* cum in omnj putrefactione (qualis est rubigo) humiditas acceleret dissolutionem, in Arefactione simplicj siccitas.'

| | | | |
|---|---|---|---|
| 1 Terrâ;] ~: | 2 Rubiginis.] ~: | 6 esse;] ~: | 10 *Dissolutionem,*] ~, |
| 16 interius] ~, | vocant)] ~,) | 18 putrefiant.] ~: | 21 Defunctorum.] ~: |
| 27 crudi;] ~: | | | |

underground. Nevertheless stones collect a kind of nitre, which to them is like rust. Now gems and crystals outlast metals, | though length of time takes some of the shine off them.

4. It has been observed that stones facing north are eaten away by time more quickly than those facing south, as is apparent in pyramids, temples, and other buildings. Iron on the other hand becomes rusty more quickly when facing south than north, as may be seen in iron bars and gratings of windows. And that is no wonder seeing that in all putrefaction (of which rust is a kind) moisture speeds up dissolution, while dryness is hastened by simple arefaction.

5. In vegetables (and I speak of ones harvested not growing) be they | the stems or trunks of hardwoods, and the building timbers made from them, last for some ages. But the parts of trunks are not all the same, for there are some, like elder, which are full of tubes in which the pith in the middle is softer but towards the outside harder. But in solid trees, such as oak, the inside (which people call the tree's heart) lasts longer.

6. The leaves, flowers, and even the stalks of plants do not last, but if they do not putrefy they crumble to dust and ashes. However, the roots last longer.

7. Animal bone lasts a long time, as we see in ossuaries where the bones of the dead are kept. | Horns too last for ages, as do teeth, as we see in ivory and hippopotamus teeth.

8. Skins and leather last for ages, as we see in the vellum of old books. Paper too survives for long ages though not as long as vellum.

9. Things fired last for ages, like glass, bricks. Meat and fruits last longer cooked than raw, and not just because cooking curbs putrefaction but because once the watery moisture has been discharged the oily can keep going longer.

10. Of all liquors water is absorbed into the air most quickly. Oil, | on the other hand, evaporates more slowly, as we see not just in the

verum etiam in Mistis. Etenim Papyrus *Aquâ* madefacta, atque inde nonnihil Diaphaneitatis nacta, paulò post albescit, & Diaphaneitatem suam deponit, exhalante scilicet *Vapore Aquæ*; At contrà, Papyrus *Oleo* tincta, diu Diaphaneitatem seruat, minimè exhalante *Oleo*; vnde qui
5 Chirographa adulterant, *Papyrum Oleatam* Autographo imponunt, atque hâc industriâ lineas trahere tentant.

11. *Gummi* omnia, valdè diù durant; Etiam *Cera*, & *Mel.*

12. At *Æqualitas*, & *Inæqualitas* eorum, quæ Corporibus accidunt,
[C2ᵛ] non minùs quam res ipsæ, ad | *Durationem*, aut Dissolutionem valent.
10 Nam *Ligna, Lapides*, alia, vel in Aquâ, vel in Aere perpetuo manentia, plus durant, quàm si quandóque alluantur, quandóque afflentur. Atque *Lapides* eruti, & in Ædificijs positi, diutiùs durant, si eodem situ, & ad easdem Cœli plagas ponantur, quibus iacebant in Mineris: id quod *plantis* etiam è loco motis, & aliò transplantatis, accidit.

15                          Obseruationes maiores.

1. *Loco* Assumpti *ponatur, quod certissimum est: Inesse omni*
[C3ʳ] *Tangibili* | Spiritum, *siue* Corpus pneumaticum, *Partibus tangibilibus obtectum, & inclusum*; Atque ex illo Spiritu *initium capi omnis* Dissolutionis, *&* Consumptionis; *Itaque earundem*
20 *Antidotum est,* Detentio Spiritus.

2. *Spiritus detinetur duplici modo: Aut per* Compressionem *arctam, tanquam in Carcere; aut per* Detentionem *tanquam spontaneam. Atque ea* Mansio *etiam duplici ratione inuitaˈtur, videlicet,*
[C3ᵛ] *si* Spiritus *ipse non sit mobilis admodùm, aut acer; atque si insupèr*
25 *ab* Aere *ambiente minùs sollicitetur ad exeundum. Itaque duo sunt* Durabilia: Durum, *&* Oleosum; Durum *constringit* Spiritum; Oleosum *partìm demulcet* Spiritum, *partìm huiusmodi est, vt ab* Aere *minùs sollicitetur;* Aer *enim* Aquæ *consubstantialis,* Flamma *autem* Oleo. *Atque de* Naturâ Durabilis, *& minùs* Durabilis *in*
30 Inanimatis, *hæc inquisita sint.*

8 At *Æqualitas. . . accidit.*] / holograph version: 'At <<in>> Aequalitas et inæqualitas eoru*m* quæ corporibus accedunt, non minus quam res ipsæ ad durationem aut dissolutionem valent. Nam ligna, lapides, alia vel in aqua vel in aere perpetuo manentia plus durant quàm sj quandoque alluantʳ quandoque afflentur. Atque lapides erutæ et ædificijs positæ diutius durant sj 'eodem situ et' easdem coelj plagas <<loc>> ponantʳ, quibus jacebant in mineris, id quod plantis etiam 'remotis et' alio transplantis accidit.'

1 Mistis.] ~:     4 *Oleo;*] ~:     16 *est.*] ~;     21 *modo.*] ~:     22–3 *spontaneam.*] ~:
26 Durabilia:] ~;     28 *sollicitetur,*] ~:

liquors themselves but also in their mixtures. For paper moistened with water takes on a certain transparency which not long after disappears and turns white as the water evaporates. By contrast paper stained with oil keeps its transparency for ages since the oil is given off slowly, whence forgers lay oiled paper on an original and in that way try to trace its handwriting.

11. All gums last for a very long time; so too do wax and honey.

12. But the stability or instability of circumstance contributes no less than the body itself¹ to its lasting or decay. For wood, stones, and other things last longer if they always stay in water or air, than if they were moistened and dried alternately. Stones dug up and placed in buildings last longer if they stay in the same position and orientation that they had in the quarry. This also happens with plants when they are moved or otherwise transplanted.

## Major Observations

1. Let us take our stand on this most certain proposition, that in every tangible thing ¹ there exists a spirit or pneumatic body hidden and enclosed in the tangible parts, and that this spirit is the source of all dissolution and consumption. Thus the antidote to these ills is to detain the spirit.

2. Spirit is detained in two ways: either by close confinement as if in a prison, or by a kind of voluntary detention. And two conditions likewise induce them to stay, ¹ namely if the spirit itself is not too mobile or sharp, and if, moreover, it is not encouraged to leave by the air outside. Thus bodies that last are of two kinds: hard and oily. The hard holds the spirit down; the oily partly calms the spirit, and partly works in such a way that it is encouraged less by the air. For air and water are consubstantial, as oil and flame are. So much then for the nature of durable and less durable in inanimate bodies.

[C4'] | *Historia.*

**13.** *Herbæ* quæ habentur ex Frigidioribus, annuæ sunt, & quotannis moriuntur, tam Radice, quam Caule: vt *Lactuca, Portulaca,* etiam *Triticum,* & *Frumenti* omne genus. Sunt tamen etiam ex frigidis, quæ per
5 tres, aut quatuor Annos durant, vt *Viola, Fragaria, Pimpinella, Primula veris, Acetosa;* At *Borago,* & *Buglossa,* cùm videantur viuæ tam similes, Morte differunt; *Borago* enim annua, *Buglossa* Anno superstes.

**14.** At *Herbæ* calidæ plurimæ ætatem, & annos ferunt; *Hyssopus,*
[C4'] *Thymus, Satureia, Maiorana altera,* | *Melissa, Absynthium, Chamædrys,*
10 *Saluia,* &c. At *Feniculum* Caule moritur, Radice repullulat: *Ocymum* verò, & *Maiorana* (quam vocant), *suauis,* non tam Ætatis, quam Hyemis sunt impatientes; satæ enim in loco valdè munito, & tepido, superstites sunt. Certè notum est *Schema* (qualibus in Hortis vtuntur ad ornamentum) ex *Hyssopo,* quotannis bis tonsum, vsque ad quadraginta
15 Annos durasse.

**15.** *Frutices* & *Arbores* humiliores, ad sexagesimum Annum, aliæ etiam duplò magis, viuunt. *Vitis* sexagenaria esse potest, & ferax est etiam in Senectute. *Rosmarinus* fœlicitèr collocatus, etiam Sexagesimum
[C5'] Annum complet. | At *Acanthus,* & *Hedera* vltra centesimum durant.
20 Sed *Rubi* Ætas, non percipitur, quia flectendo caput in terram nouas nanciscitur Radices, vt veterem à nouâ distinguere haud facile sit.

**16.** Ex *Arboribus* grandioribus annosissimæ sunt *Quercus, Ilex, Ornus, Vlmus, Fagus, Castanea, Platanus, Ficus Ruminalis, Lotos, Oleaster, Olea, Palma, Morus:* ex his nonnullæ vsque ad octingentesimum
25 Annum; etiam earum minùs viuaces, vsque ad ducentesimum perueniunt.

**17.** At *Arbores Odoratæ,* & *Resinosæ,* materiâ suâ siue Ligno, etiam illis, quas diximus, magis durabiles; Ætate paulò minùs viuaces;
[C5'] *Cupressus, Abies, Pinus,* | *Buxus, Iuniperus;* At *Cedrus* Corporis magnitu-
30 dine adiutus, etiam superiores ferè æquat.

**18.** *Fraxinus* prouentu alacris & velox, Ætatem ad centesimum annum, aut nonnihil vltra producit; quod etiam quandoque facit *Ferula,* & *Acer,* & *Sorbus;* At *Populus,* & *Tilia,* & *Salix,* & (quam appellant) *Sycomôrus,* & *Iuglans,* non adeò viuaces sunt.

35 **19.** *Malus, Pyrus, Prunus, Malus Punica, Malus Medica,* & *Cytria, Mespilus, Cornus, Cerasus,* ad Quinquagesimum, aut Sexagesimum

---

4 genus.] ~:     11 vocant),] ~,)     13 sunt.] ~:     *Schema*] ~,     14 ornamentum)] ~;)     19 complet.] ~:     33 *Sorbus;*] ~:

## ¹ *History*

**13.** Herbs reputed to be of the colder sort are annual and die every year both in root and stalk, like lettuce, purslane, and also wheat and grain of all kinds. Yet there are also cold plants which last for three or four years, like the violet, strawberry, burnet, primrose, and sorrel. But borage and bugloss, though they look so alike in life, differ in death, for the former is annual, while the latter lives longer.

**14.** But most hot plants better put up with age and time, like hyssop, thyme, savory, winter marjoram, ¹ balm, wormwood, germander, sage, etc. But fennel dies on the stalk and springs again from the root; but basil and what they call sweet marjoram are not so much intolerant of age as winter; for planted in a warm and very sheltered place they keep going. Certainly it is known that in knots (of the kind used to beautify gardens) hyssop pruned twice a year lasts for up to forty years.

**15.** Bushes and shrubs last for sixty years, some even twice as long. The vine can live into its sixties and even bear fruit in old age. Rosemary in a good spot will live to a similar age. ¹ But acanthus and ivy last for over a hundred years. The age of bramble cannot be made out because by dipping its head to the ground it puts out new roots so that it is difficult to tell old from new.

**16.** Of the larger trees the longest lived are the oak, holly, ash, elm, beech, chestnut, plane, fig, lotus, oleaster, olive, palm, and mulberry. Of these some live for up to eight hundred years, and the less vigorous up to two hundred.

**17.** But fragrant and resinous trees are, as timber and wood, even longer lasting than those just mentioned but not quite as long-lived, e.g. cypress, fir, pine, ¹ box, and juniper. But cedar, helped by its sheer size, lasts almost as long as these.

**18.** The ash, an eager, quick grower, lives for a hundred or more years, as do the birch, maple, and service-tree. But the poplar, lime, willow, sycamore (as they call it), and walnut are not so long-lived.

**19.** The apple, pear, plum, pomegranate, citron, lemon, medlar,

Annum peruenire possunt; præsertìm si à Musco nonnullas ipsarum vestiente, aliquandò purgentur.

[C6<sup>i</sup>] 20. Generalitèr Magnitudo Cor|poris in *Arboribus*, cum Diuturnitate vitæ (cæteris paribus) nonnihil habet commune; & similitèr durities
5 Materiæ: quin & *Arbores* Glandiferæ, & Nuciferæ, Fructiferis & Bacciferis, sunt plerunque viuaciores: atque etiam Præcocibus, vel Fructu, vel Foliis, serotinæ & tardiùs frondescentes, atque tardiùs etiam Folia deponentes, Ætate diuturniores sunt: quin & syluestres Cultis; & in eâdem specie, quæ acidum Fructum ferunt, illis quæ dulcem.

[C6<sup>v</sup>] | Obseruatio maior.

11 *3. Benè admodùm notauit* Aristoteles, *discrimen inter* Plantas, *& Animalia, quoad Alimentationem, & Renouationem; quòd scilicet Corpus Animalium suis Claustris circumseptum manet; atque insupèr postquam ad iustam Magnitudinem peruenerit, Alimento*
15 *continuatur, & conseruatur, sed nihil nouum excrescit, præter Capillos, & Vngues, quæ pro* | *Excrementis habentur; adeò vt necesse sit Succos Animalium citiùs veterascere. At in* Arboribus, *quæ nouos subinde Ramos, noua Vimina, nouas Frondes, nouos Fructus immittunt, euenit vt & ipsæ, quas diximus, partes nouæ sint, nec Ætatem*
20 *passæ; cùm verò, quicquid viride sit, & adolescens, fortiùs & alacriùs Alimentum ad se trahat, quàm quod incœperit desiccari; euenit vnà & simul, vt* Truncus *ipse, per quem* | *huiusmodi Alimentum transit ad* Ramos, *vberiore, & lætiore Alimento in transitu irrigetur, perfundatur, & recreetur. Id quod etiam insignitèr patet, ex hoc (licet*
25 *illud non annotauerit* Aristoteles, *qui nec ea ipsa, quæ iam diximus, tam perspicuè explicauit) quod in* Sepibus, Syluis cæduis, *Arboribus tonsis,* Amputatio Ramorum, *aut* Surculorum, Caulem *ipsum, aut* Truncum *confortat, illumque efficit longè diuturniorem.*

[C8<sup>r</sup>] | Desiccatio; Desiccationis prohibitio;
30 & Desiccati Inteneratio.

### *Historia.*

*Ad Artic.* 2. 1. *Ignis,* & *Calor* intensus alia desiccat, alia colliquat;
  *Limus vt hic durescit, & hæc vt* Cera *liquescit*
  *Vno eodemque Igne.*————

4 vitæ] ~,    17 *veterascere.*] ~:

cornel-cherry, and cherry can live to be fifty or sixty years old, especially if they are sometimes cleared of the moss that clothes some of them.

20.  In general sheer size ǀ in trees and likewise hardness of timber has (other things being equal) something in common with their longevity, but trees with acorns or nuts are commonly longer-lived than ones with fruit or berries; and trees which have late leaves and fruit live longer than those whose leaves and fruit come early. Wild trees live longer than cultivated ones, as do trees which bear sour fruit as against trees of the same kind which bear sweet.

ǀ Major Observation

3.  Aristotle did very well to note the difference between plants and animals as far as alimentation and renovation are concerned, namely that the animal body stays enclosed within its own bounds, and that once it has reached its right size food keeps it going and conserves it, but it puts out nothing new apart from hair and nails, which are regarded as ǀ excrements; so that of necessity the juices of animals age more quickly. But in trees, which now and then put out new branches, new shoots, new foliage, and new fruits, it happens that the parts that I have just mentioned are new and unstaled by time. But as everything fresh and youthful draws food to itself more strongly and eagerly than that which has started to dry out, it happens that along with this the very trunk by which ǀ this kind of aliment passes to the branches, is watered, suffused, and restored by that more abundant and plentiful aliment as it passes. This is evident to a remarkable extent by the fact (though Aristotle did not notice this, or clearly explain the things just mentioned) that in hedges, coppices, and pollarded trees, the cutting back of the branches or shoots strengthens the stalk or trunk itself, and makes it last longer.

ǀ Desiccation, the Prevention of Desiccation, and the
Softening of Things Desiccated

*History*

*To Article* 2. 1.  Fire and intense heat dry some things but melt others:

<blockquote>

*As by one and the same fire this clay grows hard,*
*That wax grows soft————*
</blockquote>

163

Desiccat *Terram*, & *Lapides*, & *Lignum*, & *Pannos*, & *Pelles*, & quæcunque non fluunt; Colliquat *Metalla*, & *Ceram*, & *Gummi*, & [C8ᵛ] *Butyrum*, & *Seuum*, & huiusmodi. ¹

2. Attamen in illis ipsis quæ colliquat *Ignis*, si vehementior fuerit,
5 ea in fine desiccat; nam & *Metalla*, ex *Igne* fortiore, emisso *Volatili*, minuuntur pondere (præter *Aurum*) & deueniunt magis fragilia; atque *Oleosa* illa & *pinguia*, ab *Igne* fortiore deueniunt frixa, & tosta, & magis sicca, & crustata.

3. *Aer*, præcipuè apertus, manifestò desiccat, nunquàm colliquat;
10 veluti cum viæ, & Superficies Terræ, Imbribus madefactæ desiccantur; Lintea lota, quæ ad *Aerem* exponuntur, siccantur; Herbæ, & Folia, & Flores, in vmbrâ siccantur. At multò magis, hoc facit *Aer*, si aut Solis [D1ʳ] radijs illustretur (modò non ¹ inducat Putredinem) aut moueatur; vt flantibus Ventis, & in Areis perflatilibus.

15 4. *Ætas* maximè, sed tamen lentissimè, desiccat; vt fit in omnibus Corporibus, quæ Vetustate (modò non intercipiantur à Putredine) arefiunt. *Ætas* autem, nihil est per se (cum sit Mensura tantùm Temporis) sed Effectus producitur, à *Spiritu* Corporum in-nato, qui Corporis Humorem exugit, & vnà cum ipso euolat; & ab *Aere* circumfuso,
20 qui multiplicat se super *Spiritus* innatos, & *Succos* Corporis, eosque deprædatur.

5. *Frigus* omnium maximè propriè exiccat; siquidem *Desiccatio* non [D1ʳ] fit nisi per *Contractionem*, ¹ quod est opus proprium *Frigoris*. Quoniam vero nos *Homines Calidum* potentissimum habemus in *Igne*, *Frigidum*
25 autem infirmum admodu*m*; nihil aliud scilicet quam *Hyemis*, aut fortassè *Glaciei*, aut *Niuis*, aut *Nitri*; ideò *Desiccationes Frigoris* sunt imbecillæ, & facilè dissolubiles; videmus tamen desiccari Faciem Terræ ex *Gelu*, atque ex *Ventis Martijs* plus, quam ex Sole; cùm idem *Ventus*, qui *Humorem* lambit, etiam *Frigus* incutiat.

30 6. *Fumus* Foci desiccat, vt in *Laridis*, & *Linguis Boûm*, quæ in Caminis suspenduntur; quinetiam Suffitus ex Olybano, aut Ligno [D2ʳ] Aloes, & similibus, desiccat Cerebrum, & Catarrhis medetur. ¹

7. *Sal*, morâ paulò longiore, desiccat, non tantùm in extimis, sed etiam in profundo; vt fit in Carnibus, aut Piscibus salitis, quæ per
35 diuturnam *Salitionem*, manifestò etiam intrinsecùs indurantur.

8. *Gummi* calidiora applicata ad Cutem eam desiccant, & corrugant; quod faciunt etiam *Aquæ* nonnullæ constringentes.

---

2 fluunt;] ~:      13 illustretur] ~,      16 Vetustate] ~,      17 arefiunt.] ~:
23 *Contractionem*,] ~;      27 dissolubiles;] ~:      30 *Laridis*,] ~;

It desiccates earth, stones, wood, cloth, skins, and whatever does not melt. It melts metals, wax, gums, butter, tallow, and the like. |

2. Yet in the very things that fire melts, desiccation takes place in the end if the fire be fiercer. For metals (gold excepted) by giving off their volatile part in stronger fire lose weight and become more brittle; and oily and fatty bodies become arid, charred, and more dry and flaky.

3. Air, especially the open air, plainly dries things out but never liquefies them: as when roads and the surface of the ground after rain, linen washed and hung out to dry, and herbs, leaves, and flowers left in the shade all dry out. But air does this much more when the Sun shines (provided that | it does not cause putrefaction), or when it is moved by the blowing of the wind, or sharp draughts.

4. Age dries things out most of all, but very slowly, as happens in all bodies dried out by lapse of time (provided they are not cut off by putrefaction first). Now age is nothing in itself (it is after all only a measure of time) but the effect is caused by the innate spirit of bodies which soaks up bodily moisture and escapes with it, and by the ambient air which multiplies itself on the innate spirits and juices of the body and preys on them.

5. Of all things cold has the greatest capacity to dry things, for desiccation only takes place with contraction, | which is the particular work of cold. But because we humans have a very powerful source of heat in fire, and extremely weak sources of cold (with nothing besides winter, or perhaps ice, snow, or nitre), cold's desiccations are feeble and easily dispelled. Nevertheless we see that the surface of the earth is dried out more by ice and March winds than by the Sun, for that same wind which absorbs moisture also chills to the marrow.

6. Smoke from the fire dries things out, as in bacon, and tongue hung in chimneys; and so fumigants of olibanum, lignum aloes, and the like dry the brain and cure catarrhs. |

7. Salt, though it takes longer, also dries things, and not just outside but deep inside too, as happens with salted meat or fish, which are manifestly hardened inside by long salting.

8. The hotter gums applied to the skin dry and wrinkle it, as do some toning waters.

9. Spiritus Vini fortis, in tantum desiccat instar *Ignis*, vt & Albumen Oui immissum candefaciat, & Panem torreat.

10. *Pulueres* desiccant instar Spongiarum, sugendo Humidum, vt fit in *Puluere*, Atramento iniecto, post Scriptionem. Etiam *Læuor*, & *Vnio* [D2ᵛ] Corporis (qui non | permittit Vaporem *Humidi* ingredi per poros) 6 per Accidens desiccat, quia ipsum *Aeri* exponit; vt fit in Gemmis, & Speculis, & Laminis Ensium; in quæ si spires, cernuntur illa primò vapore obducta, sed paulò post euanescit ille vapor, vt Nubecula. Atque de *Desiccatione* hæc inquisita sint.

10 11. *Granaria* in vsu sunt hodiè ad partes *Germaniæ Orientales*, in Cellis subterraneis, in quibus *Triticum*, & alia Grana conseruantur, substrato, & circumposito vndique stramine, ad nonnullam altitudinem, quod Humiditatem Cauernæ arceat, & sorbeat; Quâ industriâ seruantur Grana [D3ʳ] | etiam ad vicesimum, aut tricesimum Annum; neque seruantur tantum 15 à Putredine, sed (quod ad præsentem *Inquisitionem* pertinet) in tali Viriditate, vt Panibus conficiendis optimè sufficiant; idemque fuisse in vsu, in *Cappadociâ*, & *Thraciâ*, & nonnullis locis *Hispaniæ*, perhibetur.

12. *Granaria*, in Fastigijs Ædium, cum Fenestris ad *Orientem* & *Septentrionem* commodè collocantur; quinetiam constituunt quidam 20 duo Solaria, superiùs, & inferiùs; Superius autem foraminatum est, vt Granum per Foramen (tanquam Arena in Clepsidrâ), continuè descen- [D3ᵛ] dat, & subinde palis, post aliquot dies, | reponatur; vt Granum sit in continuo Motu. Notandum autem est, etiam huiusmodi res, non tantùm Putredinem cohibere, verùm etiam Viriditatem conseruare, & 25 Desiccationem retardare; Cuius causa est ea, quam etiam superiùs notauimus; quòd Euolatio Humoris Aquei, quæ Motu, & Vento acceleratur, Humorem Oleosum in suo esse conseruat; qui aliàs in consortio Humoris Aquei fuisset vnà euolaturus. Etiam in quibusdam *Montibus* vbi Aer est purus, *Cadauera* ad plures dies manent, non 30 multùm deflorescentia.

[D4ʳ] 13. *Fructus*, veluti *Granata*, *Cytria*, *Mala*, *Pyra*, & huiusmodi; etiam | & *Flores*, vt *Rosa*, *Lilium*, in Vasis Fictilibus benè obturatis, diutiùs seruantur; neque tamen non officit *Aer* ambiens ab extimis, qui etiam per Vas, inæqualitates suas defert, & insinuat; vt in Calore, & Frigore 35 manifestum est. Itaque si & Vasa diligentèr obturentur, atque obturata

---

4 Atramento] Attramento / *nld* thus in *SEH* (II, p. 116)     Scriptionem.] ~:
8 Nubecula.] ~:     Atque] lc     10 hodiè] ~,     13 sorbeat;] ~:     15 pertinet)] ~,)
21 Foramen] ~,     Clepsidrâ),] ~,) / *nld* in *SEH* (II, p. 116) as clepsydra     35 est.] ~:
Itaque] lc

9. Strong spirits of wine dry like fire, so that egg whites put in them congeal and bread is toasted.

10. Powders dry things in the manner of sponges by sucking up moisture, as happens in powder cast on ink after writing. Also the smoothness and compactness of the body (which | stops the moisture's vapour getting into the pores) dries it *per accidens* when exposed to the air, as happens in gems, glasses, and sword blades which, if you breathe on them, you first see a vapour covering them, which vanishes soon after like a little cloud. So much then for desiccation.

11. Granaries in underground cellars are used to this day in parts of eastern Germany to store wheat and other grain. In these straw of some depth is laid at the bottom and round the sides which checks and absorbs the moisture of the vault, and in this way they keep the grain | for twenty or thirty years not just from putrefaction but (and this is relevant to the present inquiry) fresh enough to make first-rate bread. And people say that this practice was current in Cappadocia, Thrace, and some regions of Spain.

12. Granaries in house tops with windows to east and north are well positioned. People also set them up with two floors, an upper and a lower, with the upper perforated so that the grain falls continually through the holes (like sand in a glass) only to be shovelled up after a few days | and returned whence they came, so that grain never stops moving. Now it should be noted that a setup of this kind not only restrains putrefaction, but also maintains freshness and curbs desiccation. As we noted above, the reason for this is that the escape of the watery moisture, hastened by motion and wind, keeps the oily moisture, which would otherwise escape with the watery, in its proper state. Again, on some mountains where the air is pure, corpses last for many days without going off.

13. Fruits such as pomegranates, lemons, apples, pears, and the like, and | flowers such as the rose and lily keep for a long time when sealed in earthenware pots; yet the ambient air still interferes by carrying and insinuating its inequalities through the sides of the pot, as is obvious in heat and cold, so that it is best if the pots be well sealed to bury them below the ground as well. No less effective is to put them if not under

sub Terram insupèr condantur, optimum erit; neque minùs vtile est, si non sub Terrâ, sed sub Aquis condantur, modo sint vmbrosæ, vt Putei, & Cisternæ in Domibus; sed quæ sub Aquis conduntur, meliùs reponuntur in Vasis Vitreis, quam in Fictilibus.

[D4ᵛ]    **14.** Generalitèr quæ sub *Terrâ*, & in *Cellis* subterraneis, aut in pro-| fundo Aquarum reponuntur, virorem suum diutiùs tuentur, quam quæ supra Terram.

   **15.** Tradunt in *Conseruatorijs Niuium* (siue sint in montibus, in Foueis naturalibus, siue per Artem in Puteis ad hoc factis) obseruatum fuisse, quod aliquandò *Malum*, aut *Castanea*, aut *Nux*, aut simile quippiam inciderit, quæ post plures Menses liquefactâ niue, aut etiam intra Niuem ipsam, inuenta sunt recentia, & pulchra, ac si pridiè essent decerpta.

   **16.** *Vuæ* apud Rusticos seruantur in *Racemis* coopertis intra *Farinam*; quod licet gustui eas reddat minùs gratas, tamen Humorem & Viridi-
[D5ʳ] tatem conseruat; | etiam omnes *Fructus* duriores, non tantùm in *Farinâ*, sed in *scobe Lignorum*, etiam inter Aceruos *Granorum* integrorum, diù seruantur.

   **17.** Inualuit Opinio, *Corpora* intra *Liquores* suæ speciei, tanquam *Menstrua* sua, conseruari recentia; vt *Vuas* in *Vino*, *Oliuas* in *Oleo*, &c.

   **18.** Seruantur *Mala Granata*, & *Cotonea* tincta paulispèr in *Aquam Marinam*, aut *salsam*, & paulò post extracta, & in *Aere* aperto (modò fuerit in vmbrâ) siccata.

   **19.** In *Vino*, *Oleo*, aut *Amurcâ* suspensa diù seruantur; multò magis in *Melle*, & *Spiritu vini*; atque etiam omnium maximè (vt quidam tradunt) in *Argento Viuo*.

[D5ᵛ]    **20.** *Incrustatio* etiam *Fructuum*, *Cerâ*, | *Pice*, *Gypso*, *Pastâ*, aut alijs oblinimentis, aut capsulis, diutiùs eos virides conseruat.

   **21.** Manifestum est *Muscas* & *Araneas*, & *Formicas*, & huiusmodi, casu in *Electro*, aut etiam *Arborum Gummis* immersas, & sepultas, nunquam posteà marcescere; licet sint Corpora mollia, & tenera.

   **22.** *Vuæ* seruantur pensiles; & sic de alijs *Fructibus*; duplex est enim eius rei commoditas; vna, quod absque vllâ contusione, aut compressione fiat, qualis contrà fit cum super dura collocantur; altera, quod *Aer* vndequáque ipsas æqualitèr ambit.

   **23.** Notatum est, tam *Putrefactionem*, quam *Desiccationem* in *Vegeta-*
[D6ʳ] *bilibus*, non similitèr, ex omni | parte, incipere; sed maximè ex eâ parte, per quam solebant, cum essent viua, attrahere *Alimentum*; itaque iubent

---

5–6 pro|fundo] / see Introduction, p. lxxv above    22 extracta,] ~;

ground but under water, provided that the water be out of the light as in wells, and domestic cisterns; but if you put them under water it is better to use glass rather than earthenware pots.

14. In general, things put underground and in cellars, or in the depths ˈ of the waters, stay fresh longer than things above ground.

15. They say that in conservatories of snow (be they in pits on mountains, or in artificial underground vaults made for the purpose) that once an apple, chestnut, nut or the like falls in, they have been found many months later when the snow has melted, or even in the snow itself, to be as fresh and fine as if they had been picked yesterday.

16. Country folk keep grapes in bunches covered with flour, which though it does nothing to improve their taste, it maintains their moisture and freshness. ˈ Likewise all harder fruits keep for a long time not only in flour but also in sawdust, and even heaps of unmilled grain.

17. The opinion has grown up that bodies are kept fresh in their own liquors, as if in their own menstrua, as grapes in wine, olives in olive oil, etc.

18. Pomegranates and quinces keep when briefly washed in sea or salt water, and soon taken out again and dried in the open air (provided that it is in the shade).

19. Bodies hung in wine, oil, or lees keep for a long time, but much longer in honey, and spirit of wine, but some say that they keep longest of all in quicksilver.

20. Covering fruit with a layer or ˈ coating of wax, resin, plaster, or paste keeps them fresh for a long time.

21. It is obvious that flies, spiders, ants, and so on, sunk and entombed in amber or even tree gums never afterwards shrivel up, though they be soft and tender bodies.

22. Grapes and other fruits are kept by hanging them up. This has two advantages: the one that they suffer none of the bruising or pressure they receive when laid on hard bodies; the other that they are aired uniformly, and on all sides.

23. It should be noted that in vegetable bodies neither putrefaction nor desiccation starts ˈ in every part alike; but mainly from that part through which they were accustomed to draw their aliment when they

aliqui *Pediculos Malorum,* aut *Fructuum, Cerâ,* aut *Pice* liquefactâ, obducere.

24. *Fila Candelarum* aut Lampadum maiora citiùs absumunt Seuum, aut Oleum, quam minora; Etiam *Flamma* ex *Gossipio* citiùs quam ex Scirpo, aut Stramine, aut Vimine ligneo; atque in *Baculis Cereorum,* citiùs ex *Iunipero,* aut *Abiete,* quam ex *Fraxino;* Etiam omnis *Flamma* mota, & vento agitata, citiùs absumit, quam tranquilla; itaque intra Cornu minùs citò, quam in aperto. Tradunt quoque *Lychna* in [D6ᵛ] *Sepulchris,* admodùm diù durare.

25. *Alimenti* etiam *Natura,* & præparatio non minùs facit ad diuturnitatem *Lychnorum,* quàm *Natura Flammæ.* Nam *Cera, Seuo* diuturnior est; & *Seuum* paulò madidum, *Seuo* sicciore; & *Cera* dura, *Cerâ* molliore.

26. *Arbores,* si quotannis circa *Radices* earum Terram moueris, breuiùs durant; si per Lustra aut Decennia, diutiùs; Etiam *Germina,* & *Surculos* decerpere, facit ad Longæuitatem; Item *Stercoratio* aut *Substratio cretæ* & similium, aut multa Irrigatio, *Feracitati* confert, Ætatem minuit. Atque [D7ʳ] de *Prohibitione Desiccationis,* et *Consumptionis,* hæc inquisita sunt.

*Inteneratio Desiccati* (quæ res est præcipua) *Experimenta* præbet pauca; ideoque nonnulla quæ in *Animalibus* fiunt, atque etiam in *Homine,* coniungemus.

27. *Vimina Salicis,* quibus ad ligandas Arbores vtuntur, in Aquâ infusa, fiunt magis flexibilia; Similiter *virgarum Ferulæ* extremitates in Vrceis cum Aquâ imponuntur, ne siccescant; quin et *Globuli* lusorij, licet per Siccitatem rimas collegerint, positi in Aquâ rursùs implentur, et consolidantur.

28. *Ocreæ* ex *Corio* vetustate duræ, et obstinatæ, per illinitionem Seui ad Ignem, molliuntur; etiam Igni simplici admotæ, nonnihil. [D7ᵛ] *Vesicæ,* et *Membranæ,* postquam fuerint induratæ, ab Aquâ calefactâ admixto Seuo, aut aliquo Pingui, intenerantur; meliùs autem, si etiam paululùm conﬁcentur.

29. *Arbores* veteres admodùm, quæ diu steterunt immotæ, fodiendo, & aperiendo terram, circa *Radices* ipsarum, manifestò tanquam iuuenescunt, nouis & teneris Frondibus emissis.

30. *Boues Aratores* veteres, & laboribus penitùs exhausti, in læta Pascua inducti, *Carnibus* vestiuntur nouis, & teneris, & iuuenilibus, vt etiam ad Gustum *Carnem Iuuencorum* referant.

were alive; so some people suggest covering the stalks of apples or other fruit with molten wax or resin.

**24.** The larger the wicks of candles or lamps the quicker they soak up tallow or oil; flame from cotton quicker than from rush, straw, or switch; and in wax tapers made of juniper or fir more quickly than ones from ash; and all flame soaks it up more quickly when stirred and fanned by wind than when still, and so it is consumed less quickly in a lantern than in the open. They say that lamps in tombs last for ages. <sup>|</sup>

**25.** The nature and preparation of the aliment contributes no less to the lasting of lamps than the nature of the flame. For wax lasts longer than tallow, slightly moistened tallow longer than dry, and hard wax longer than soft.

**26.** If you stir the earth about their roots every year, trees live less long, but longer if you do it every five or ten years. Cutting back of buds and shoots contributes to long life; yet manuring or putting down chalk and the like, or plenty of watering, makes them more fruitful but less long-lived. So much then for the prevention of desiccation and consumption. <sup>|</sup>

The business of making desiccated parts soft (which is the main point) furnishes us with few experiments; accordingly, I shall combine them with some that happen in animals and even in man.

**27.** Willow switches, which people use to tie trees, are more flexible when soaked in water. Likewise, the ends of birch rods are put in pots of water to stop them drying out. Bowls, though split by drying, fill out and become solid again when put in water.

**28.** Leather boots hard and stiff with age are softened when smeared with tallow in front of the fire; and even when just put by the fire they become somewhat softer. Bladders and skins, <sup>|</sup> once they have become hard, are softened when treated with hot water mixed with tallow or other fatty substance; and even more if they are rubbed in a little.

**29.** Very ancient trees, which have long stood unchanged, manifestly become young again and put out new and tender foliage when the earth about their roots is dug over and opened up.

**30.** Old plough oxen, thoroughly worn out with work, put to fresh pasture, are furnished with new flesh, tender and youthful, so that their meat even tastes like young beef.

**31.** *Diæta* stricta consumens, & emacians ex *Guaiaco*, *Pane bis cocto*,
[D8ʳ] & similibus (quali ad cuˈrandum *Morbum Gallicum*, & inueteratos
*Catarrhos*, & *Leucophlegmatiam* vtimur) *Homines* ad summam
*Macilentiam* deducit, consumptis *Succis* corporis; qui postquam
5 cœperint instaurari, & refici, manifestò cernuntur, magis Iuueniles &
Virides; quinetiam existimamus *Morbos emaciantes*, posteà benè curatos,
compluribus Vitam prolongasse.

## Obseruationes maiores.

9 **1.** *Miris modis Homines*, *more* Noctuarum, *in Tenebris* Notionum
[D8ᵛ] *suˈarum acutè vident*, *ad Experientiam*, *tanquam lucem diurnam*,
*nictant*, & *cœcutiunt*. Loquuntur *de* Elementari Qualitate
Siccitatis; & *de Desiccantibus*; & *de naturalibus Periodis Corporum*,
*per quas corrumpuntur*, & *consumuntur*; *sed interìm*, *nec de Initijs*,
*nec de Medijs*, *nec de Extremis* Desiccationis, & Consumptionis,
15 *aliquid*, *quod valeat*, *obseruant*.
[E1ʳ] **2.** Desiccatio & Conˈsumptio, *in Processu suo*, *tribus Actionibus*
*perficitur*, *atque originem ducunt Actiones illæ*, *à* Spiritu *innato*
*Corporum*, *vt dictum est*.
**3.** *Prima Actio est*, Attenuatio *Humidi in* Spiritum; *secunda*
20 *est*, Exitus, *aut Euolatio* Spiritûs; *tertia est*, Contractio *partium*
*Corporis crassiorum*, *statìm post* Spiritum *emissum. Atque hoc*
*vltimum est illa* Desiccatio, & Induratio *de quâ præcipuè agimus.*
[E1ᵛ] *priora duo consumunt tantùm.* |
**4.** *De* Attenuatione, *res manifesta est*; Spiritus *enim*, *qui in omni*
25 *Corpore Tangibili includitur*, *sui non obliuiscitur*, *sed quicquid*
*nanciscitur in corpore* (*in quo obsidetur*) *quod digerere possit*, &
*conficere*, & *in se vertere*, *illud planè alterat*, & *subigit*, & *ex eo se*
*multiplicat*, & *nouum Spiritum generat*. Hoc ex probatione eâ,
29 *instar omnium*, *euincitur*, *quòd quæ plurimùm siccantur*, Pondere
[E2ʳ] *minuuntur*, & *deueniunt Caua*, | Porosa, & *ab intùs Sonantia.*
*certissimum autem est*, Spiritum *rei præ-inexistentem*, *ad Pondus*
*nihil conferre*, *sed illud leuare potius*; *ergo necesse est*, *vt* Spiritus
*præ-inexistens*, *Humidum*, & *succum Corporis*, *quæ anteà pondera-*
*uerant*, *in se verterit*; *quo facto pondus minuitur. Atque hæc est prima*

---

20 Spiritûs;] ~:　　　34 *minuitur.*] ~;　　*Atque*] lc

31. A rigid consuming and wasting diet of guaiac, biscuit, and the like (of the kind given to treat ˈ the French pox, chronic catarrhs, and the tendency to dropsy) reduces people to extreme thinness by consuming the juices of the body, which when they begin to be renewed and refreshed, appear manifestly younger and fresher, on which account I think that wasting diseases properly cured have prolonged the lives of many.

## Major Observations

1. Marvellous it is that men, in the manner of owls, see keenly in the darkness of their own notions, ˈ but blink blindly in the daylight of experience. They speak of the elementary quality of dryness, of desiccants, of the natural periods of bodies by which they are corrupted and consumed; but in the meantime of the beginnings, intermediate, and final phases of desiccation and consumption they observe nothing that matters.

2. Desiccation and conˈsumption, when it is in process, is accomplished by three actions which take their origins (as I have said) from the innate spirit of bodies.

3. The first action is the attenuation of moisture into spirit; the second is the exit or escape of the spirit; the third is the contraction of the grosser parts as soon as the spirit has gone out. This last action is that desiccation and hardening which is our main business, for the first two only consume. ˈ

4. On the question of attenuation, the issue is clear: for the spirit enclosed in every tangible body does not forget itself, but whatever it finds in the surrounding body that it can digest, work up, and convert into itself, it alters, tames, and multiplies itself on it, and generates new spirit. This is demonstrated by one proof which can stand for all: that bodies dried thoroughly lose weight, become hollow, ˈ porous, and resonant inside. Now it is perfectly certain that the pre-existent spirit of the thing contributes nothing to its weight, but rather makes it lighter, and therefore it necessarily follows that the pre-existent spirit, by converting the moisture and juice of the body which had weight before, makes

*Actio, scilicet* Attenuationis *Humoris, & Conuersionis eius in* Spiritum.

[E2ᵛ] **5.** *Secunda Actio, quæ est* Exitus, *siue* Euolatio spi|ritûs, *res etiam manifestissima est. Etenim illa* Euolatio, *cùm fit confertìm,*
5 *etiam sensui patet; in vaporibus Aspectui, in Odoribus Olfactui; verum si sensìm fiat* Euolatio, *vt fit per Ætatem, tum demùm peragitur sine sensu; sed eadem res est. Quinetiam, vbi Corporis Compages, aut ita arcta est, aut ita tenax, vt* Spiritus *Poros, &*
9 *Meatus non inueniat, per quos exeat, tum verò etiam partes ipsas*
[E3ʳ] *crassiores Corporis, in nixu suo* | *exeundi, ante se agit, easque vltra Corporis superficiem, extrudit; vt fit in* Rubigine Metallorum, *& in* Carie omnium Pinguium. *Atque hæc est secunda Actio, scilicet* Exitûs, *&* Euolationis Spiritûs.

**6.** *Tertia Actio paulò magis obscura, sed æquè certa est; Ea est*
15 *Contractio Partium crassiorum post* Spiritum *emissum. Atque primò videre est Corpora post* Spiritum *emissum manifestò arctari, &*
[E3ᵛ] *minorem locum* | *complere, vt fit in* Nucleis Nucium, *qui siccati non implent testam; & in Trabibus, & Palis Ligni, quæ primò contiguæ sunt ad inuicem, ex* Desiccatione *autem hiant; atque ex Globulis*
20 *lusorijs, & similibus, qui per siccitatem rimosi euadunt; cùm Partes se contrahant, & contractæ necessariò spatia inter se relinquant. Secundò patet ex Rugis Corporum siccatorum; Nixus enim se con-*
[E4ʳ] *trahendi, tantum valet, vt Partes contrahen|do interìm adducat, & subleuet; Quæ enim in extremitatibus contrahuntur, in Medijs*
25 *subleuantur, atque hæc cernere est in Papyris, & Membranis vetustis; atque in Cute Animalium; atque in Extimis Casei mollioris, quæ omnia vetustate corrugantur. Atque tertiò se ostendit ampliùs hæc* Contractio, *in illis, quæ à Calore non tantùm corrugantur, verùm*
29 *etiam complicantur, & in se vertuntur, & quasi rotulantur; vt*
[E4ᵛ] *cernere est in* | Membranis, *& Papyris, & Folijs ad ignem admotis. Etenìm* Contractio *per Ætatem, cùm tardior sit, Rugas ferè parit; at* Contractio *per Ignem, quæ festina est, etiam Complicationes. At in plurimis, vbi non datur* Corrugatio *aut* Complicatio, *fit simplex*

---

3 *est.*] ~:  12 Pinguium.] ~:  *Actio,*] ~;  15 *emissum.*] ~;  *Atque*] lc
21 *relinquant.*] ~:  27 *corrugantur.*] ~:  33 Corrugatio] ~,  Complicatio,] ~;

the body lighter. And that is the first action, namely that of the attenuation of moisture and its conversion into spirit.

5. The second action, which is the going out or escape of the spirit, | is also perfectly plain. For this escape happens all at once and is even evident to sense in visible vapours and detectable smells. But if escape happens gradually, as is the case with age, then it goes on imperceptibly, but is still the same thing. But if the body's structure is so close and tight that the spirit cannot find any pores or passages to get out, then in its struggle to escape it drives | the grosser parts of the body before it, and extrudes them above the body's surface, as you see in the rusting of metals and the rotting of all fat substances. And this is the second action, namely the going forth and escape of the spirit.

6. The third action is a little more obscure but just as certain, i.e. the contraction of the grosser parts once the spirit has left; and the first thing to be seen after the spirit has left is that bodies become manifestly shrunken and fill less | space, as happens in the kernels of nuts which when dried do not fill the shell, in beams and wooden planks which fit snug at first but after desiccation split apart, and in bowls, and the like, cracks appear when they dry out, since the parts contract and on contraction necessarily leave gaps between them. In the second place, it is evident from the wrinkles of dried bodies; for the effort to contract has the effect of making the parts | meanwhile draw together and rise up; for what gets contracted at the edges gets raised up in the middle, as can be seen in paper, old vellum, and in animal skin, and in the rind of softer cheese, all of which buckle with age. In the third place this contraction shows itself more abundantly in things which are not only wrinkled by heat but also get tangled up, turn in, and as it were, roll up on themselves; as can be seen in | vellum, paper, and leaves when they are exposed to fire. For contraction by age, as it is slower, generates wrinkles as a rule; but contraction by fire, which is faster, also produces crumpling. But in most things, where wrinkling and crumpling do not occur, simple contraction, tightening, hardening, and desiccation take place, as I

Contractio, *&* Angustiatio, *&* Induratio, *&* Desiccatio, *vt primò*
*positum est; Quòd si eoúsque inualescat* Euolatio Spiritûs, *&*
[E5ʳ] Absumptio Humidi, *vt* | *non relinquatur satis Corporis, ad se*
*vniendum, & Contrahendum, tum verò cessat* Contractio *ex necessi-*
5 *tate, & Corpus redditur putre, & nihil aliud quam Puluisculus*
*cohærens, qui leui tactu dissipatur, & abit in Aerem; vt fit in Cor-*
*poribus cunctis valdè absumptis; & Papyro, & Linteo, ad vltimum*
*combustis; & Cadaueribus imbalsamatis post plura sæcula.* Atque
[E5ᵛ] *hæc est tertia illa Actio; scilicet* Contractionis *Partium* | *crassiorum,*
10 *post* Spiritum *emissum.*

7. *Notandum est Ignem, & Calorem, per Accidens tantùm*
*desiccare; proprium enim eorum Opus est, vt* Spiritum *& Humida*
*attenuent, & dilatent; sequitur autem ex Accidente, vt partes reliquæ*
*se contrahant; siue ob Fugam Vacui tantùm, siue ob alium Motum*
15 *simul; de quo nunc non est Sermo.*
[E6ʳ] 8. *Certum est etiam* Putrefactionem, *non minùs* | *quam* Arefac-
tionem, *à* Spiritu *innato originem ducere, sed longè aliâ viâ*
*incedere; nam in* Putrefactione, Spiritus *non emittitur simplicitèr,*
*sed ex parte detentus, mira comminiscitur; atque etiam partes*
20 *crassiores, non tam localitèr contrahuntur, quam coeunt singulæ ad*
Homogeniam.

[E6ᵛ] | Longæuitas, & Breuitas Vitæ in
Animalibus.

*Historia.*

25 **Ad Artic. 3. Connexio.** De Diuturnitate, *&* Breuitate Vitæ *in*
Animalibus, *tenuis est* Informatio *quæ haberi potest; Obseruatio*
*negligens; Traditio fabulosa. In* Cicuribus *Vita degener corrumpit;*
[E7ʳ] *in* Syluestribus | *Iniuria* Cœli *intercipit.*

*Neque quæ Concomitantia videri possint, huic* Informationi
30 *multùm auxiliantur* (Moles Corporis; Tempus Gestationis in
vtero; Numerus Fœtûs; Tempus Grandescendi; *alia*), *proptereà*
*quod complicata sunt ista, atque aliàs concurrunt, aliàs*
*disiunguntur.*

---

5 *aliud*] ~,        27 *fabulosa.*] ~:        31 *alia*),] ~:)

said at the start. But if the escape of spirit and the absorption of moisture gets to the point where not enough of the body ꞌ is left to pull itself together and contract, then contraction necessarily ceases, and the body is reduced to putrefaction, and nothing other than a pinch of dust hanging together which can be dispersed by the merest touch, and off it goes into the air, as happens in bodies thoroughly consumed, in paper and linen burnt to a crisp, and in embalmed corpses after long ages. And this then is the third action, namely, contraction of the grosser parts ꞌ following emission of spirit.

7. It should be noted that fire and heat only desiccate *per accidens*; for their proper work is to attenuate and dilate the spirit and moist substances. Now it follows by accident that the remaining parts contract themselves, either to avoid a vacuum alone, or from another motion at the same time—of which I say nothing now.

8. It is certain that putrefaction, no less ꞌ than arefaction takes its origin from the innate spirit, but that it proceeds by a quite different route. For in putrefaction the spirit is not just sent out but, in part kept in, works wonders; and the grosser parts too do not just contract locally, but come together like to like.

ꞌ Length and Shortness of Life in
Animals

*History*

**To Article 3. Connection.** Regarding length and shortness of life the available information is scanty, observation careless, and tradition full of fables. In domesticated animals a degenerate life spoils them; in wild ꞌ animals a hard life cuts them off.

Nor do factors which can be regarded as accompaniments (bodily mass, time in the womb, number of offspring, how long it takes to grow up, and so on) do much to eke out this information; because these factors are not easy to separate, and in some cases they coincide, but in others they diverge.

1. *Hominis* Æuum cæterorum *Animalium* omnium superat (quantum *Narratione* aliquâ certâ constare potest) præter admodùm [E7$^r$] paucorum. Atque Concomitantia in eo, satis æqualitèr se habent; *Statura* & *Magnitudo* grandis; *Gestatio* in *Vtero* nouimestris; *Fœtus* vt
5 plurimùm vnicus; *Pubes* ad Annum Decimum quartum; *Grandescentia* ad Vigesimum.

2. *Elephas*, fide haud dubiâ, curriculum Humanæ Vitæ ordinarium transcendit: *Gestatio* autem in vtero Decennalis, fabulosa; Biennalis, aut saltem supra annuam, certa: At *Moles* ingens, & *Tempus Grandescendi*
10 vsque ad Annum tricesimum, *Dentes* Robore firmissimo. Neque etiam Obseruationem *Hominum* fugit, quod *Sanguis Elephanti* omnium [E8$^r$] sit frigidissimus. *Ætas* autem Ducentesimum Annum nonnunquam compleuit.

3. *Leones* viuaces habiti sunt, quod complures ex ijs reperti sint
15 edentuli; Signo nonnihil fallaci; cùm illud fieri possit ex *Grauitate Anhelitûs*.

4. *Vrsus* magnus Dormitor est; *Animal* pigrum, & iners, neque tamen Viuacitatis notatum: Illud autem signum breuis Æui; quod *Gestatio* eius in *vtero* sit festina admodum, vix ad Quadraginta dies.
20 5. *Vulpi* multa se benè habere videntur ad *Longæuitatem*; optimè tecta est, Carniuora, & degit in Antris; neque tamen *Viuacitatis* notata. [E8$^v$] Certè est Generis *Canini*, quod Genus breuioris est vitæ.

6. *Camelus* longæuus est; *Animal* macilentum, & neruosum; ita vt Quinquaginta Annos ordinariò, centum quandóque compleat.
25 7. *Equi* vita mediocris, vix Quadragesimum Annum attingit; Ordinarium autem curriculum, Viginti Annorum est: Sed hanc *Breuitatem vitæ* fortassè *Homini* debet; Desunt enim iam nobis *Equi Solis*, qui in Pascuis liberi, & læti degebant. Attamen crescit *Equus* vsque ad Sextum Annum, & generat in Senectute. Gestat etiam in vtero *Equa*
30 diutiùs, quam *Fœmina*, & in *Gemellis* rarior est. *Asinus* similis ferè æui [F1$^r$] vt *Equus*; *Mulus* vtroque viuacior.

8. *Ceruorum* vita celebratur vulgò ob Longitudinem; neque tamen *Narratione* aliquâ certâ. Nescio quid de *Ceruo Torquato*, coopertâ Torque ipsâ pinguedine carnis, circùm ferunt. Eò minùs credibilis est
35 Longæuitas in *Ceruo*, quod Quinto Anno perficitur; atque non multò

10 firmissimo.] ~:    12 frigidissimus.] ~:    21 notata.] ~:    25 Annum] / BNF copy with semicolon; Huntington 601102 with comma    33 certâ.] ~:    34 circùm ferunt] / *nld* in *SEH* (II, p. 123) as circumferunt

1. As far as any certain account be established, man's lifespan surpasses (with very few exceptions) that of all other animals. And the attendant circumstances ¹ of this pretty much bear this out: large size and stature, nine-month gestation in the womb, offspring generally single; puberty at fourteen, growth up to twenty.

2. The elephant, on good authority, lives longer than the ordinary course of human life. That its gestation in the womb lasts ten years is a fable; that it lasts two years, or at least more than one, is certain. But it has great bulk, grows right up to its thirtieth year, has extremely strong teeth, and it has not escaped men's notice that its blood is the coldest of all. It sometimes lives ¹ for two hundred years.

3. Lions are regarded as long-lived because many are toothless; but this is a deceptive sign, since it could be caused by their rank breath.

4. The bear is a great sleeper; a lazy and idle animal but yet not noted for longevity. This too is a sign of short life: that its gestation in the womb is very fast and hardly lasts forty days.

5. Foxes seem very well cut out for long life; they are very well clad, carnivorous, and live in earths; and yet are not noted for longevity. They belong to the canine race, which is short-lived. ¹

6. The camel is long-lived, a lean, stringy animal such that it usually lives for fifty and sometimes a hundred years.

7. The horse is of middling longevity, and lasts barely forty years, and usually just twenty. Its short life is perhaps due to man, for now we have no horses of the sun which live at large in open pastures. And yet they go on growing for six years, and breed in old age. Moreover a mare's pregnancy is longer than a woman's, and she has twins less often. The ass lives almost as long as the horse. The mule lives longer than either of them. ¹

8. Stags are commonly famed for long life, yet no certain evidence supports this. I know nothing about the story of a stag whose collar became buried in fat; but the stags' longevity is less credible because they are at their best at five years old, and not much later their antlers

post, Cornua (quæ annuatìm decidunt, & renouantur) succedunt magis coniuncta fronte, & minùs ramosa.

9. *Canis* breuis est æui; non extenditur Ætas vltra Annum Vicesimum; neque sæpè attingit ad Decimum quartum: *Animal* ex calidissimis, atque inæqualitèr Viuens; cùm, vt plurimùm, aut vehementiùs [F1$^v$] moue¦at, aut dormiat. Etiàm Multiparum est, & Nouem Septimanas gestat in vtero.

10. *Bos* quoque, pro Magnitudine & Robore, admodùm breuis est æui, quasi Sex-decim Annorum; Maresque Fœminis nonnihil viuaciores. Attamen vnicum plerunque edit Partum, & gestat in vtero circa sex Menses. *Animal* pigrum, & carnosum, & facilè pinguescens, & *Herbis* solis pastum.

11. At Decennalis ætas in *Ouibus* etiam rara est; licet sit *Animal* mediocris Magnitudinis, & optimè tectum; atque quod mirum, cum minimum in illis reperiatur Bilis, Capillitium habent omnium [F2$^r$] crispissimum; ne¦que enim *Pilus* alicuius Animalis, tam tortus est, quam *Lana*. *Arietes* ante Tertium Annum non generant, atque habiles sunt ad generandum, vsque ad Octauum; *Fœmellæ* pariunt quamdiù viuunt. Morbosum *Ouis Animal*, nec *Ætatis* suæ curriculum ferè implet.

12. *Caper* etiam similis est Æui cum *Oue*, nec dispar multùm in cæteris; licet sit *Animal* magis agile, & carne paulò firmiore, eóque debuerit esse viuacius; attamen salacius est multò, eóque breuioris Æui.

13. *Sues* ad Quindecim Annos quandóque viuunt, etiam ad Viginti; [F2$^v$] cúmque sint Carne, inter *Animalia* omnia, humidissimâ, ta¦men nihil videtur hoc proficere ad Longitudinem vitæ. De *Apro* aut *Sue Siluestri*, nil certi habetur.

14. *Felis* Ætas est inter Sextum Annum, & Decimum; Agile *Animal*, & Spiritu acri, cuius Semen (vt refert *Ælianus*) *Fœmellam* adurit; vnde increbuit opinio, Quod *Felis* concipit in Dolore, & parit cum Facilitate; vorax est in *Cibis*, quos potiùs deglutit, quam mandit.

15. *Lepores*, & *Cuniculi* vix ad Septem Annos perueniunt; *Animalia* Generatiua, etiam Superfœtantia; In hoc disparia, quod *Cuniculus* sub [F3$^r$] terrâ viuit, *Lepus* in aperto; quodque *Leporis* Carnes atriores sint.

16. *Aues* mole Corporis, *Quadrupedibus* longè sunt minores. Pusilla

---

10 viuaciores.] ~:          25 vitæ.] ~:          29 opinio,] ~;    Facilitate;] ~:
34 [blank line] ] / this is editorial for, if the next item had not begun on a new page, a blank line would probably have been introduced as at, for instance, the transition between items on birds and fish (see below p. 186, after l. 3)          35 minores.] ~:

(which they shed and regrow annually) come up closer to the forehead and with fewer branches.

9. The dog does not last long; never beyond twenty, and seldom fourteen years. It is one of the hottest animals, living very irregularly, being mostly in furious motion | or fast asleep. Yet it gives birth to many pups at once which spend seven weeks in the womb.

10. The ox too, given its size and strength, lives a very short life of sixteen years or so, with the male living rather longer than the female. Yet they generally give birth to only one calf at a time, after a six-month pregnancy. It is a sluggish, fleshy animal, and easily fattened up, eating only grasses.

11. In sheep even ten years of life is rare, though it is an animal of average size and very well clad, and, strangely enough, though we find that they have little bile, their coat is the curliest of all, for no | other animal's hair is so tightly twisted as wool. Before their third year rams do not breed, but are able to do to their eighth; ewes give birth all their lives. The sheep is a sickly creature, and never lives out its full span.

12. The goat lives about as long as the sheep, and differs little from it in other ways, though it is more agile and with firmer flesh, which should make it more long-lived, but it is more highly sexed and that shortens its life.

13. Pigs sometimes live for fifteen years or indeed twenty as their flesh is the most moist of all animals; yet | this seems not to extend their lives. Nothing certain is known about the boar or wild pig.

14. The cat lives for six to ten years. It is an agile animal with a sharp spirit, and (as Ælian has it) with seed which burns the female, whence has grown the notion that the cat conceives in pain and gives birth with ease. It is voracious and bolts its food rather than chews it.

15. Hares and rabbits scarcely reach the age of seven; they breed a lot and have lots of young; but they differ in that the rabbit lives underground, the hare in the open, and hare's flesh is darker. |

16. The bodily mass of birds is much less than that of quadrupeds.

enim res est *Aquila* aut *Cignus,* præ *Boue,* aut *Equo;* itèm *Struthio* præ *Elephanto.*

17. *Aues* optimè tectæ sunt: Pluma enim tepore, & incubitu presso ad Corpus, & Lanam, & Capillitia excedit.

5  18. *Aues,* cùm plures pariant, eos simul in Aluo non gestant, sed Oua excludunt per vices: vnde liberaliùs sufficit Alimentum Fœtui.

19. *Aues* parùm, aut nihil Alimenta mandunt, vt integrum sæpe reperiatur in gulis ipsarum. Attamen frangunt Fructuum Nuces, & Nucleum
[F3ᵛ] excerpunt. Existimantur autem esse Con|coctionis fortis, & calidæ.

10  20. Motus *Auium,* dùm volant, mixtus est, inter Motum Artuum, & Gestationem; Saluberrimum Exercitationis genus.

21. De *Auium* Generatione *Aristoteles* benè notauit (sed malè ad alia *Animalia* traduxit) minùs scilicet conferre Semen *Maris* ad Generationem; sed Actiuitatem potiùs indere, quam Materiam; vnde etiam
15  Oua fœcunda, & sterilia, in plurimis non dignoscuntur.

22. *Aues* quasi omnes, ad Magnitudine*m* suam iustam perueniunt, primo Anno, aut paulò post; verum est, quoad plumas in nonnullis,
[F4ʳ] quoad Rostrum in alijs, Annos numerari; ad Mag|nitudinem autem Corporis, minimè.

20  23. *Aquila* pro *Longæuâ* habetur; Anni non numerantur. Etiam in Signum trahitur *Longæuitatis,* quòd Rostra renouet, vndè iuuenescat; ex quo illud, *Aquilæ Senectus.* Attamen res fortassè ita se habet; vt Instauratio *Aquilæ* non mutet Rostrum, sed contrà, Mutatio Rostri instauret *Aquilam*; Postquam enim Rostrum, aduncitate suâ nimiùm
25  increuerit, pascit *Aquila* cum difficultate.

24. *Vultures* etiam *Longæui* perhibentur, adeò vt vitam ferè ad centesimum Annum producant; *Milui* quoque, atque adeò omnes *Volucres*
[F4ᵛ] *Carniuoræ,* & *Ra|paces,* diuturnioris sunt æui. De *Accipitre* autem, quia vitam degit degenerem, & seruilem, ex vsu humano, minùs certum fieri
30  possit iudicium, circa periodum eius Vitæ naturalem. Attamen ex Domesticis, deprehensus est *Accipiter,* aliquandò ad Annos Triginta vixisse; ex Syluestribus, ad Quadraginta.

25. *Coruus* traditur esse similitèr *longæuus,* aliquandò Centenarius; Carniuora *Auis,* neque admodùm frequens in Volatu; sed magis
35  Sedentaria, & Carnibus admodùm atris. At *Cornix,* cætera (præterquam Magnitudine, & Voce) similis, paulò minùs diù viuit, sed tamen
[F5ʳ] habetur ex Viuacibus. |

For an eagle or swan is a tiny thing compared to an ox or horse; as is an ostrich compared to an elephant.

17.  Birds are very well clad; for feathers for warmth and close tailoring to the body are better than wool or hair.

18.  Though they have many chicks, birds do not carry them in their bellies at once, but lay their eggs at intervals; whence the offspring get more generous nourishment.

19.  Birds chew their food little if at all, so that we often find their food whole in their crops. Yet they break the shells of fruit, and extract the nuts. They are also | thought to have hot and strong digestions.

20.  Birds fly with a mixed motion combining motion of the limbs and gliding along; and very healthy exercise it is too.

21.  Aristotle well observed concerning the generation of birds (though he wrongly transferred it to other animals) that in breeding the male's semen contributed activity rather than matter, whence in many cases it is not possible to tell fertile from sterile eggs.

22.  Nearly all birds reach their full size in the first year or just after. It is true that years pass when it comes to the growth of feathers in some birds, and to the beak in others, but | not to body size.

23.  The eagle is regarded as long-lived, but its years have not been reckoned up. It is also taken as a sign of longevity that it renews its beak, which rejuvenates it; whence the expression *the eagle's old age*. But the matter stands thus: that perhaps the renewal of the eagle does not change the beak, but change of beak renews the eagle; for once the beak becomes very hooked the eagle eats with difficulty.

24.  Vultures are also thought to be long-lived, such that they live for practically a hundred years. Kites, and all raptors and carnivores, | live long. As for the hawk, which lives a degenerate and servile life at the hands of man, its natural lifespan is uncertain. Yet domesticated hawks sometimes live to thirty, and wild ones to forty years.

25.  The raven is likewise thought to be long-lived, sometimes to a hundred. It is a carnivore which seldom flies but leads a sedentary life, and has very dark flesh. But the crow, like the raven in everything but size and cry, lives a little less long, but is still taken for a long-lived bird. |

**26.** *Cygnus,* pro certo, admodùm *longæuus* inuenitur, & Centesimu*m* Annum, haud rarò superat; *Auis* optimè plumata, *Icthyophaga,* & perpetuò in Gestatione, idque in *Aquis Currentibus.*

**27.** *Anser* quoque ex *longæuis,* licet *Herbâ,* & id genus Pabulo nutriatur; Maximè autem *Syluestris,* adeò vt in *Prouerbium* apud *Germanos* sit: *Magis senex quam Anser Niualis.*

**28.** *Ciconiæ longæuæ* admodum esse deberent, si verum esset, quod antiquitùs notatum fuit; Eas *Thebas* nunquam accessisse, quia vrbs illa sæpiùs capta esset. Id si cauissent, aut plusquam vnius sæculi memoriam [F5ᵛ] habebant, aut Parentes pullos suos *Historiam* | edocebant. Verùm omnia *Fabellis* plena.

**29.** Nam de *Phœnice* tantum accreuit *Fabulæ,* vt obruatur, si qua in eâ re fuit veritas. Illud autem, quod Admirationi erat, eum, magno aliarum *Auium* Comitatu, volantem semper visum, minùs mirum; Cùm hoc etiam in *Vlulâ* interdiù volante, aut *Psittaco* è Caueâ emisso, vbique cernere detur.

**30.** *Psittacus,* pro certo, vsque ad Sexaginta annos, cognitus est viuere apud nos, quotquot supra habuisset, cum hùc esset transuectus. *Auis* Cibi quasi omnigeni, atque etiam mandens Cibos, atque mutans [F6ʳ] subinde Rostrum; Aspera, & | ferocula, Carnibus atris.

**31.** *Pauo* ad Viginti annos viuit; Oculos autem *Argûs* non recipit ante Trimatum; Tardigrada *Auis,* Carnibus verò candidis.

**32.** *Gallus Gallinaceus,* Salax, Pugnax, & breuis Æui; Alacris admodùm *Ales,* & Carnibus etiam albis.

**33.** *Gallus Indicus,* aut *Turcicus* (quem vocant) *Gallinacei* Æuum parùm superat; Iracundus *Ales,* & Carnibus valdè albis.

**34.** *Palumbes* sunt ex viuacioribus, vt Quinquagesimum Annum aliquandò compleant; Aërius *Ales,* & in alto, & Nidificans, & Sedens. [F6ᵛ] *Columbæ* verò, ac *Turtures,* Vitâ breues, vsque ad Annum Octauum. |

**35.** At *Phasiani,* & *Perdices* etiam Decimum sextum Annum implere possunt. *Aues* numerosi Fœtûs, Carnibus autem paulò obscurioribus, quam *Pullorum* genus.

**36.** Fertur de *Merulâ,* quod sit, ex *Auibus* minoribus, maximè *Longæua;* Procax certè *Auis,* & Vocalis.

**37.** *Passer* notatur esse Æui breuissimi; Id quod ad Salacitatem refertur in *Maribus.* At *Carduelis* corpore haud Maior, deprehensus est viuere ad Annos Viginti.

6 sit:] ~;     7 esset.] ~:     10 edocebant.] ~:    Verùm] lc     11 plena] plæna
22 Trimatum;] ~:     23 Æui;] ~:     26 superat;] ~:     28 compleant;] ~:
36 *Maribus.*] ~:

26. We find that the swan is extremely long-lived and often lives more than a hundred years. It is a very well feathered bird, fish-eating, and perpetually gliding along on running waters.

27. The goose is also among the long-lived, although it lives on grass and that sort of food. The wild goose lives even longer, so that among the Germans they have a proverb, *older than a snow goose.*

28. Storks should be very long-lived, if what was said in antiquity be true: that they never went to Thebes as it was captured so often. If that were admitted, they could either recall more than one age or the parents taught the goslings ǀ history. Everything is full of fables.

29. And what they say about the Phoenix has grown so fat with fables that any truth in it (if there ever was any) has been buried. But what was an object of admiration, that when flying it was always accompanied by a great troop of other birds, is not so wonderful, for this everywhere happens to owls flying in daylight or to escaped parrots.

30. It is certain that the parrot is known to live up to sixty years here with us, in addition to the ones before its import here. It is omnivorous and chews its food, and from time to time changes its beak; it is coarse and ǀ ill-tempered, and has dark flesh.

31. The peacock lives for twenty years, but does not acquire its Argus eyes before its third year; it is slow-paced with white flesh.

32. The domestic chicken is lustful, aggressive, and short-lived; extremely active, it also has white flesh.

33. The Indian or (as they call it) Turkey chicken lives a bit longer; a hot-tempered bird, with very white flesh.

34. Wood-pigeons are among the longer-lived, sometimes living to fifty; a bird of the air, it roosts and nests aloft. But doves and turtle-doves are short-lived, i.e. up to eight years. ǀ

35. But pheasants and partridges live twice as long. They have large broods, and have flesh rather darker than has the chicken race.

36. The blackbird is regarded as the longest-lived of the small birds; a bold bird and tuneful.

37. The sparrow is seen to be very short-lived, which is ascribed to the males' lustfulness. But the linnet, with a body not much bigger, has been known to live up to twenty years.

**38.** De *Struthionibus* nihil certi habemus, qui Domi nutriuntur, adeò infœlices fuerunt, vt non deprehensi sint diù viuere; De *Aue Ibi* constat [F7ʳ] tantùm quod | sit *Longæua*, Anni non numerantur.

**39.** *Piscium* vita magis incerta est, quam *Terrestrium*, quùm sub *Aquis*
5 Degentes minùs obseruentur. Non respirant ex ipsis Plurimi; vnde *Spiritus* Vitalis magis conclusus est; Itaque licet refrigerium excipiant per Branchias, haud tamen ita continua fit Refrigeratio, quàm per Anhelitum.

**40.** In *Aquis* cùm degant, à *Desiccatione* illâ, & *Deprædatione*, quæ fit
10 per Aerem ambientem, immunes sunt; Neque tamen dubium est, quin *Aqua* ambiens, atque intra poros Corporis penetrans, & recepta, plus [F7ᵛ] no|ceat ad Vitam quàm *Aer*.

**41.** *Sanguinis* perhibentur esse minùs tepidi; Suntque Nonnulli ipsorum voracissimi, etiam Speciei propriæ. *Caro* autem ipsorum mol-
15 lior est, quàm *Terrestrium*, & minùs tenax. Attamen pinguescunt maiorem in modum, vt ex *Balænis* infinita extrahatur Quantitas Olei.

**42.** *Delphini* traduntur viuere Annos circa Triginta; capto Experimento in aliquibus à Caudâ præcisâ. Grandescunt autem ad Annos Decem.

20 **43.** Mirum est, quod referunt de *Piscibus*, quod Ætate, post annos no*n*nullos, plurimùm attenuantur Corpore, manente Caudâ, & Capite, [F8ʳ] in Magnitudine priore. |

**44.** Deprehensæ sunt aliquando, in *Piscinis Cæsarianis*, *Murænæ* vixisse ad Annum Sexagesimum. Certè redditæ sunt longo vsu tam
25 familiares, vt *Crassus Orator* Vnam ex illis defleuerit.

**45.** *Lucius*, ex *Piscibus* Aquæ dulcis, longissimè viuere reperitur; ad Annum quandóque Quadragesimum; *Piscis* vorax, & Carnibus siccioribus, & firmioribus.

**46.** At *Carpio*, *Abramus*, *Tinca*, *Anguilla*, & huiusmodi, non
30 putantur viuere vltra Annos Decem.

**47.** *Salmones* citò grandescunt, breui viuunt; Quod etiam faciunt *Trutæ*; At *Perca* tardè crescit, & viuit diutiùs.

[F8ᵛ] **48.** Vasta illa moles *Balænarum*, | & *Orcarum*, quamdiù Spiritu regatur, nil certi habemus; Neque etiam de *Phocis*, aut *Porcis Marinis*, &
35 alijs *Piscibus* innumeris.

**49.** *Crocodili* perhibentur esse admodùm Viuaces, atque Grandescendi Periodum itidèm habere insignem; adeò vt hos solos ex *Animalibus* perpetuò, dum viuunt, grandescere Opinio sit. *Animal* est

---

1 habemus,] ~;     5 obseruentur.] ~:     14 propriæ.] ~:     18 præcisâ.] ~;

**38.** Of ostriches we know nothing for sure, and we find that those brought up under domestication have been unfortunate enough not to live long. As for the ibis we know only that | it is long-lived but not exactly how long-lived.

**39.** The age of fish is less certain than that of land animals since, living under water, they have been less closely observed. Most of them do not breathe, so their vital spirit is more confined; therefore though they get some refrigeration from their gills, yet that happens with more interruption than by breathing.

**40.** In the waters where they live they are better armed against the ambient air's depredations and desiccating powers. Yet there is no doubt but that the ambient water, getting into the pores of the body and staying there, is more | harmful to life than air.

**41.** They are not regarded as warm-blooded, and some of them are very voracious, even of their own species. Their flesh is softer and less tight than that of land animals. And yet they grow very fat, so that vast amounts of oil are extracted from whales.

**42.** They say that dolphins live for about thirty years, an experiment having been performed of clipping the tails of some of them. They go on growing for ten years.

**43.** It is remarkable what they say of fish, namely that after some years many get thinner in body, while their heads and tails keep their original size. |

**44.** People found that in Cæsar's fishponds lampreys sometimes lived for sixty years. Certainly by long habit they became so friendly that Crassus the orator shed tears for one of them.

**45.** The pike is found to live longest of all freshwater fish, sometimes up to forty years old. It is a voracious fish with dry and firm flesh.

**46.** But carp, bream, tench, eel, and the like are not thought to live beyond ten years.

**47.** Salmon grow large quickly but are short-lived; the same goes for trout; but perch grow slowly and live long.

**48.** I know nothing for certain | about how long the vast bulk of whales and orcae is governed by the spirit, nor likewise of seals, sea-pigs, and countless other kinds of fish.

**49.** Crocodiles are held to be extremely long-lived, and take remarkably long to grow up, so that opinion has it that they are unique among animals in growing throughout their lives. The crocodile lays eggs, is

Ouiparum, vorax, & sæuum, & optimè tectum contra Aquas. At de reliquo
*Testaceo* genere, nihil certi, quod ad Vitam ipsorum attinet, reperimus.

| Obseruationes maiores.

*Normam aliqua*m Longæuitatis, *& Breuitatis vitæ in* Animalibus,
5 *inuenire difficile est, propter Obseruationu*m *negligentiam, &
Causarum Complicationem. Pauca notabimus.*
   1. *Inueniuntur Plures ex* Auibus longæuæ, *quàm ex* Quadrupedibus (*sicut* Aquila, Vultur, Miluus, Pelicanus, Coruus, Cornix,
[Gɪ<sup>v</sup>] Cygnus, Anser, Ciconia, Grus, Ibis, Psitta|cus, Palumbis, *&c.*) *licet*
10 *intra Annum perficiantur, & minoris sint Molis. Tegumentum certè*
*ipsarum* Auium, *contra Intemperies Cœli, optimum est: cúmque in*
Aere libero *plerunque degant, similes sunt* Habitatoribus Montium
*puriorum, qui* longæui *sunt. Etiam Motus ipsarum, qui (vt alibi*
14 *dictum est) mixtus est ex* Gestatione, *atque* Motu Artuum, *minûs*
[G2<sup>r</sup>] *fatigat, aut concutit, & magis Salubris est. Neque in* vte|ro Matrum,
*Compressionem, aut Penuriam Alimenti patiuntur Initia* Volatilium;
*quia Oua per vices excluduntur. Maximè verò omnium, illud in causâ*
*esse arbitramur, Quod fiant* Aues, *magis ex Substantiâ* Matris,
*quam* Patris; *vnde* Spiritum *nanciscuntur minùs acrem, & incensum.*
20    2. *Poni possit,* Animalia, *quæ crea*ntur *magis ex Substa*ntia
Matris, *quam* Patris, *esse* Longæuiora, *quemadmodùm* Aues, *vt*
[G2<sup>v</sup>] *dictum est.* | *Etiam, quæ longiore tempore gestantur in Aluo, plus*
*habere ex Substantiâ* Matris, *minus è Semine* Patris; *ac proinde*
*diuturnioris Æui esse: adeò vt existimemus etiam inter* Homines
25 (*quod in aliquibus notauimus) eos qui similiores sunt* Matribus,
*diutiùs viuere; Nec-non* Liberos Senum, *qui ex* Vxoribus Adole-
scentulis *progignuntur, modò fuerint* Patres *sani, & non morbidi.*
[G3<sup>r</sup>]    3. Initia Rerum, & iniu|riæ, & auxilio maximè subjiciuntur.
*Itaque minorem* Compressionem, *& liberaliorem* Alimentationem
30 Fœtûs in Vtero, *ad* Longæuitatem *multùm conferre par est. Id fit,*
*aut cùm exeunt* Fœtus *per vices, vt in* Auibus; *aut cùm pariuntur*
Vnici, *vt in* Animalibus vniparis.
   4. *At Tempus longius* Gestationis in vtero, *tripliciter facit ad*

---

6 *Complicationem.*] ~:          7–8 Quadrupedibus] ~;          15 *est.*] ~:
17 *excluduntur.*] ~:          18 *Substantiâ*] Sustantiâ          24 Homines] ~,
28 *subjiciuntur.*] ~:          30 *est.*] ~:

voracious, ferocious, and very well covered against the water. But of the rest of the testaceous race we find nothing reliable relating to their lives.

## Major Observations

Any rule concerning the length and shortness of life in animals is difficult to discover on account of neglect of observations and entangled causes. But I shall note a few things.

1. We find that more birds (for instance, the eagle, vulture, kite, pelican, raven, crow, swan, goose, stork, crane, ibis, parrot, wood-pigeon, etc.), are longer-lived than some of the quadrupeds, although they mature within a year, and are smaller in mass. The plumage of birds certainly protects them very well against changes in the weather; and as they usually live in the open air, they are like the inhabitants of the undefiled mountains who are long-lived. Their motion (as I said elsewhere), which combines gliding along and movements of the limbs, tires and impairs them less, and is healthier. Nor do they suffer compression or lack of nourishment in their mothers' wombs in the beginning because the eggs are laid individually. But most of all I judge that we should take this into account: that birds are made more of the mother's than of the father's substance, whence they acquire a spirit less sharp and inflamed.

2. It can be laid down that animals with more of the mother's than the father's substance are longer-lived, in the same way (as I said) birds are. Again, those with longer gestation have more of the mother's substance than of the father's seed, and are therefore longer-lived, such that I believe that even among men (which in some cases I have noted) those who are more like their mothers live longer; as is also the case with children of old men begotten on adolescent wives, so long as the fathers are fit and not ill.

3. The beginnings of things are most susceptible both of injury and assistance. Therefore less compression and more liberal nourishment of the foetus in the womb contributes more to longevity. And this happens either when the offspring come at intervals (as in birds), or when they are born singly, as happens in animals which give birth to one at a time.

4. But longer gestation in the womb aids longevity in three

[G3ᵛ] Longitudinem vitæ. *Primò, quod plus habet* Fœtus *ex Sub|stantiâ* Matris, *vt dictum est; Deindè, quod prodit confirmatior; Postremò, quod* Aeris *vim prædatoriam tardiùs experitur. Quinetiam denotat Periodos ipsius Naturæ, per maiores fieri Circulos. Atque licet, &*

5 Boues, *&* Oues, *qui in* Vtero *manent circiter Sex Menses, breuioris sint Æui, tamen id ex alijs causis ortum habet.*

**5.** Comestores Graminis *&* Herbæ *simplicis, breuis sunt Æui; lon-*
[G4ʳ] *gioris autem* Animalia carniuo|ra, *aut etiam* Seminum, *&* Fructuum Comestores, *sicut* Aues: *Nam etiam* Cerui, *qui longæui sunt, quasi*

10 *dimidium Pabuli* (*vt vulgò loquuntur*) supra caput petunt; Anser *autem, præter* Gramen, *etiam aliquid inuenit ex* Aquis, *quod iuuet.*

**6.** Integumentum Corporis *ad* Longæuitatem *multùm conferre arbitramur.* Aeris *enim Inæqualitates* (*quæ miris modis corpus labe-*
[G4ᵛ] *factant, & subruunt*), *propulsat, & longiùs| arcet; Id quod in* Auibus

15 *præcipuè viget. At quod* Oues, *licet benè tectæ sint, parùm viuant; id* Morbis (*qui illud* Animal *obsident*) *atque simplici Esui* Graminis, *imputandum est.*

**7.** Spirituum Sedes *principalis proculdubiò est in* Capite; *atque licet ad* Animales Spiritus *tantùm, hoc vulgò referatur, tamen illud*

20 *ipsum ad omnia pertinet. Neque illud dubium, quod* Spiritus
[G5ʳ] *maximè* Corpus *lambunt, & consumunt; adeò | vt aut maior Copia ipsorum, aut maior Incensio, & Acrimonia, plurimùm Vitam abbreuiet. Itaque existimamus magnam causam* Longæuitatis *in* Auibus *esse, quod pro Mole* Corporis, Capita *habeant tam minuta;*

25 *adeò vt etiam* Homines, *qui valdè magnum habent* Cranium, *minùs diù viuere existimemus.*

**8.** Gestationem (*vt priùs notauimus*) *omne aliud genus Motús,*
[G5ᵛ] *ad* Longitudinem vitæ *superare arbitra|mur; Gestantur autem* Aues Aquatiles, *vt* Cygnus; *atque* Aues *omnes in* Volatu, *sed cum Artuum*

30 *Motu subindè contentiore; &* Pisces, *de quorum vitæ Longitudine, parùm certi sumus.*

**9.** *Quæ longiore tempore perficiuntur* (*non loquendo de* Grandescentiâ *solâ, sed de alijs* Gradibus *ad* Maturitatem; *sicut* Homo, *primò emittit* Dentes, *deindè* Pubem, *deinde* Barbam,

---

2 *est,*] ~:    *confirmatior,*] ~:         13 *Inæqualitates*] ~,            14 *subruunt*),] ~)ₐ
15 *viget.*] ~;          16 *Morbis*] ~,            20 *pertinet.*] ~:            23 *abbreuiet.*] ~:
34 *Homo,*] ~ₐ

ways: in the first place, as I said, the foetus has more of the ᛁ mother's substance; secondly, it is born stronger; thirdly, it suffers the predatory influence of the air later. Besides, it signifies that the periods of nature itself run on longer cycles. And although both oxen and sheep, who stay in the womb for around six months, are relatively short-lived, yet this springs from other causes.

5. Consumers of grass and plain herbaceous matter have short lives; carnivores or even consumers of seeds and ᛁ fruit (like birds) live longer. For stags which are long-lived look (as they commonly say) for half their food above their heads. The goose besides grass also finds something in the waters, which helps.

6. In my view, the body's covering is very important for longevity. For it repels and curbs the inequalities of the air (which in wonderful ways enfeeble and undermine the body). ᛁ This works very well in birds. But sheep, although they are well wrapped up, do not live long, and that must be attributed to the diseases which afflict that creature, and to a diet of plain grass.

7. The principal seat of the spirits is doubtless in the head, and although this is commonly meant of the animal spirits alone, it nevertheless applies to all. And it is undoubtedly true that the spirits lap up and consume the body most, such ᛁ that either its abundance or its greater inflammation and acrimony greatly shortens life. I think therefore that a major cause of long life in birds is that relative to their body mass they have small heads, such that even in men those with very big heads live (I believe) less long.

8. I judge that gliding along (as I observed before) more than any other motion contributes to longevity. ᛁ Aquatic birds like the swan glide along on the water, and all birds in flight, but with strenuous motion of the limbs at intervals; and fish do the same though I do not know how long they live.

9. Animals that take longer to reach maturity live longer (and I am not speaking of getting bigger alone but also of the other stages of maturation; as in man's case, the teeth come first, then puberty, then the beard, etc.). ᛁ For this indicates that the stages are accomplished through longer cycles.

[G6ʳ] *&c.*) *longæuiora sunt; In* ⏐*dicat enim Periodos confici per maiores Circulos.*

10. Animalia mitiora *longæua non sunt, vt* Ouis, Columba; Bilis enim complurium *Functionum in Corpore, veluti* Cos *est, &* Stimulus.

5    11. Animalia, *quorum Carnes sunt paulò atriores, longioris sunt vitæ, quam quæ Carnibus sunt candidis; Indicat enim Succum Corporis, magis firmum, & minùs dissipabilem.*

[G6ᵛ]   12. *In omni* Corruptibili, ⏐ Quantitas *ipsa multùm facit ad* Conseruationem Integri. *Etenim* Ignis *magnus longiore tempore*
10 *extinguitur;* Aquæ *Portio parua citiùs euaporat;* Truncus *non tam citò arescit, quam* Vimen; *Itaque generalitèr* (*in* Speciebus *dico, non in* Indiuiduis) *quæ Mole grandiora sunt* Animalia, *Pusillis sunt* longæuiora; *nisi aliqua alia Causa potens Rem impediat.*

[G7ʳ]                    ⏐ ALIMENTATIO
15                              &
                   Via Alimentandi.

                        *Historia.*

*Ad Artic.* 4. 1. *Alimentum* erga *Alimentatum* debet esse Naturæ inferioris, & simplicioris substantiæ: *Plantæ* ex Terrâ & Aquâ nutriuntur;
20 *Animalia* ex Plantis; *Homines* ex Animalibus. Sunt & *Animalia Carniuora,* atque *Homo* ipse Plantas sumit in partem *Alimenti; Homo* verò, & *Carniuora Animalia,* ex Plantis solis ægrè nutriuntur; possunt
[G7ᵛ] fortassè ex Fructibus, & Seminibus Igne ⏐ coctis, multo vsu nutriri, sed Folijs Plantarum, aut Herbarum, minimè; vt *Ordo Foliatanorum*
25 Experimento comprobauit.

2. At nimia *Proximitas,* aut Consubstantialitas *Alimenti* erga *Alimentatum,* non succedit. Etenim *Animalia,* quæ Herbis vescuntur, Carnes non tangunt. Etiam ex *Carniuoris Animalibus,* pauca Carnes propriæ Speciei sapiunt. *Homines* vero, qui *Anthropophagi* fuerunt, ordinariò
30 tamen Humanis Carnibus non vescebantur; sed aut ex Vltione in Inimicos, aut prauis Consuetudinibus in illud desiderium lapsi sunt. At
[G8ʳ] *Aruum,* Grano ex ipso proueniente, fœlicitèr non se⏐ritur; neque in Insitione, Surculus aut Virgultum, in proprium Truncum, immitti solet.

3. Quò *Alimentum* meliùs est præparatum, & paulò propiùs accedit

9 Integri.] ~:        20 Animalibus.] ~:        27 succedit.] ~:        28 tangunt.] ~:
29 sapiunt.] ~:        31 sunt.] ~:

10. The gentler animals, like the sheep or dove, are not long-lived. For bile is as it were the whetstone and goad of many bodily functions.

11. Animals with darker flesh live longer than ones with whiter. For that indicates that the juice of the body is firmer and less easy to disperse.

12. In every corruptible thing ˈ quantity itself does a lot for the conservation of the whole. For a big fire takes longer to put out; a small amount of water evaporates more quickly; the trunk does not dry out as quickly as the twig. Thus in general (I speak of species not individuals) animals of great mass live longer than ones with smaller, unless some other more powerful cause gets in the way.

ˈ ALIMENTATION
and
the Way of Providing Aliment

*History*

*To Article* 4. **1.** As compared with the body nourished, the nourishment should be of a lower nature and a simpler substance. Plants feed on earth and water, animals on plants, and man on animals. There are also carnivorous animals, and even man himself lives partly on plants. But man and the carnivorous animals can hardly feed on plants alone. They can perhaps by long habit be fed ˈ by fruits and grains which have been cooked, but not at all by the leaves of plants and herbs, as the experiment of the Feüillan order has proved.

**2.** Now too much closeness or consubstantiality of nourishment to the body nourished does not turn out well. For herbivores do not touch meat; and among carnivores very few dine on their own species. And anthropophagus men do not eat human flesh as a matter of course, but do it for revenge on enemies or because they have slipped into that corrupt practice. Arable land is not successfully sown with the seed that grew there in the first place, ˈ and in grafting, a cutting or slip is not generally grown on its own stock.

**3.** The better the nourishment is prepared, and the nearer it approaches to the substance of the thing nourished, the more productive do plants become, and the more do animals fatten. For no slip or cutting stuck in the ground gets fed as well as if it were grafted on a

ad Substantiam *Alimentati*, eò & *Plantæ* feraciores sunt, & *Animalia*
Habitu sunt pinguiora; Neque enim Virgultum aut Surculus, in Terram
immissus, tam benè pascitur, quam si idem immittatur in Truncum,
cum naturâ suâ, benè consentientem, vbi inuenit *Alimentum* digestum,
5 & præparatum; Neque etiam (vt tradunt) *Semen Cepæ*, aut Similium, in
Terram immissum, tam magnam producit Plantam, quam si Semen in
[G8ᵛ] aliam *Cepam* indatur, Insitione qua'dam in Radicem, & Subterranea:
Quinetiam nuper inuentum est, Virgulta *Arborum Syluestrium*, veluti
*Vlmi, Quercûs, Fraxini*, & Similium, in Truncos insita, longè maiora
10 proferre Folia, quam quæ sine Insitione proueniunt. Etiam *Homines*
Carnibus crudis, non tam benè pascuntur, quam Ignem passis.

  **4.** *Animalia* per Os nutriuntur; *Plantæ* per Radices; *Fœtus Animal-*
*ium* in *Vtero* per Vmbilicum; *Aues* ad parum temporis, ex Vitellis Ouorum suorum; quorum nonnulla pars, etiam postquam exclusæ sunt, in
15 gulis earum inuenitur.

[H1ʳ]   **5.** Omne *Alimentum* mouet maximè à Centro ad Circumferen'tiam,
siue ab Intrà ad Extrà; Attamen notandum est, *Arbores* & *Plantas* potiùs
per Cortices, & Extima, quam per Medullas, & Intima, nutriri; Etenim
si circum circa decorticatæ fuerint, licet ad Spacium paruum, non
20 viuunt ampliùs. Atque *Sanguis* in Venis *Animalium*, non minùs Carnes
sub illis sitas nutrit, quam supra illas.

  **6.** In omni *Alimentatione* duplex est Actio, *Extrusio*, & *Attractio*;
Quarum prima à Functione interiore, altera ab exteriore procedit.

24   **7.** *Vegetabilia* assimilant Alimenta sua simplicitèr, absque Excretione:
[H1ᵛ] Etenim *Gummi*, & *Lachrymæ* potius Exuberantiæ, quam ' Excretiones
sunt; *Tuberes* autem Morbi potiùs. At *Animalium* Substantia magis sui
similis est perceptiua; Itaque cum Fastidio coniuncta est, & Inutilia
reijcit, Vtilia assimilat.

  **8.** Mirum est de *Pediculis Fructuum*; quòd omne *Alimentum*, quod
30 tantos quandóque producit Fructus, per tam angusta *Collula* transire
cogitur; Fructus enim nunquam Trunco inhæret, absque Pediculo aliquo.

  **9.** Notandum *Semina Animalium* Nutritionem non excipere, nisi
recentia; At *Semina Plantarum* manent Alimentabilia ad longum
34 Tempus. Attamen *Virgulta* non germinant, nisi indantur recentia;
[H2ʳ] neque *Radices* ipsæ lon'giùs vegetant, nisi sint Terrâ coopertæ.

  **10.** In *Animalibus* Gradus sunt Nutrimenti pro Ætate; *Fœtui* in *vtero*
sufficit Succus Maternus; *A Natiuitate* Lac; posteà Cibi, & Potus; Atque
sub *Senectute* crassiores ferè Cibi, & sapidiores placent.

10 proueniunt.] ~:          20 ampliùs.] ~:          26 sunt;] ~:     potiùs.] ~:
34 Tempus.] ~:

stock which agrees well with its nature, where it finds nourishment digested and prepared. Neither too (as they report) will an onion seed or the like produce a plant as big when planted in the ground, as when it is set by a kind | of grafting on the root of another onion beneath the ground. And furthermore, it has been discovered lately that slips of wild trees, like elm, oak, ash, and so on, set on stocks produce much larger leaves than if they had come up without grafting. Men too are better fed on cooked rather than raw meats.

4. Animals take food by the mouth; plants by their roots; animal young in the womb by the umbilical cord; birds for a short time from their egg yolks, a bit of which is found in their crops even after they are hatched.

5. In the main all nourishment moves from the centre to the circumference, | or from the inside to the outside. Yet we should note that trees and plants are nourished rather through their bark and outer layers than via their pith and innards; for if even a narrow band of bark be stripped off all round, they do not live much longer. As for blood in the veins of animals it nourishes the flesh below them no less than that above.

6. In all nourishing there is a twofold action: extrusion and attraction, of which the former stems from an interior function, the latter from an exterior.

7. Vegetable bodies take their nourishment straight, and without excretion. For gums and tears are rather superabundances than | excretions; and excrescences are really diseases. But the substance of animals is more sensitive to what is similar to it, and so is associated with a degree of fastidiousness, and rejects the useless and assimilates the useful.

8. It is a wonder concerning fruit stalks that all nourishment, which sometimes produces such large fruit, is obliged to pass through so narrow a constriction; for fruit never hangs from the stem without a stalk.

9. We should note that animal seed only sustains increase if it is fresh, but plant seed remains capable of sustenance for a long time. Yet slips do not germinate unless planted fresh, and the roots themselves do not long | keep their vigour unless they are buried in the earth.

10. In animals nutriment varies with age: for young in the womb the mother's juices are enough; at birth, milk; afterwards food and drink; and in old age heavier, tastier foods will do.

***Mandatum.*** *Præcipuè omnium ad* Inquisitionem *præsentem facit, diligentèr, & attentè indagare, vtrùm non possit fieri* Nutritio *ab extrà; aut saltem non per Os? Certè* Balnea *ex* Lacte *exhibentur in Marasmis, & Emaciationibus; neque desunt ex* Medicis, *qui existimant* Alimentationem
5   *nonnullam fieri posse per* Clysteria. *Omninò huic rei incumbendum; si*
[H2ᵛ] *enim* Nutritio *fieri possit, aut per* | *Extrà, aut aliàs quam per Stomachum, tum verò Debilitas Concoctionis, quæ ingruit in Senibus, illis auxilijs compensari possit, & tanquam in Integrum restitui.*

# LONGÆVITAS
10   & Breuitas vitæ in
Homine.

### *Historia.*

***Ad Artic.*** **5, 6, 7, 8, 9, & 11. 1.** Ante *Diluuium*, plura Centenaria
Annorum, vixisse *Homines*, refert sacra *Scriptura.* Nemo tamen *Patrum*
[H3ʳ] Millesimum Annum compleuit. Neque hæc Vitæ Diuturnitas, | *Gratiæ*,
16   aut *Lineæ Sanctæ*, attribui possit; cùm recenseantur, ante *Diluuium*,
*Patrum Generationes* vndecim; at *Filiorum Adami* per *Cain* tantùm
*Generationes* octo; vt *Progenies Cain* etiam Longæuior videri possit.
Ista vero *Longæuitas*, immediatè post *Diluuium*, dimidio corruit; sed in
20   *Post-natis;* Nam *Noah* qui antè natus erat, Maiorum Ætatem æquauit; &
*Sem* ad Sexcentesimum Annum peruenit. Deinde post tres *Generationes*
à *Diluuio*, *Vita Hominum*, ad quartam quasi partem Ætatis primitiuæ,
reducta est; videlicet, ad annos circiter Ducentos.
   **2.** *Abraham* Annos Centum Septuaginta Quinque vixit; Vir
[H3ᵛ] Magnanimus, & cui cuncta ce|debant prosperè. *Isaac* autem ad annum
26   Centesimum, & Octogesimum peruenit; Vir Castus, & Vitæ quietioris.
At *Iacob*, post multas Ærumnas, & numerosam Sobolem, ad annum
Centesimum Quadragesimum Septimum durauit; Vir Patiens, &
Lenis, & Astutus. *Ishmael* autem, Vir Militaris, Annos Centum Triginta
30   Septem vixit. At *Sarah* (cuius vnicæ ex Fœminis, Anni recensentur),
mortua est Anno Ætatis suæ, Centesimo Vicesimo Septimo; Mulier
Decora, & Magnanima, optima *Mater* & *Vxor*, neque tamen minùs
libertate, quam Obsequio erga Maritum, clara. *Ioseph* etiam, Vir

2 *indagare*,] ~;          5 Clysteria.] ~;      *Omninò*] lc          13 9,] ~.
14 *Scriptura*.] ~:      22 *Diluuio*] / BNF copy with semicolon; Folger copy 2 with comma
30 *Sarah*] ~,      recensentur),] ~,)

*Direction.* The most important thing of all for the present inquiry is to examine carefully and attentively whether nutrition can be achieved from outside, or at least not by mouth. Certainly milk baths are provided for consumptions and wasting complaints. And there is no lack of physicians who think that some degree of alimentation can be administered by clysters. We must pay particular attention to this: for if nutriment can be absorbed either from ǀ outside, or by ways other than the stomach, then indeed the weakness of concoction that threatens old men can be made good by these aids, and as it were made whole again.

LENGTH

and Shortness of life in

Man

*History*

*To Articles* 5, 6, 7, 8, 9, *&* 11. 1. According to Holy Scripture men lived for many hundreds of years before the Flood. Nevertheless none of the patriarchs lived to a thousand. Now this length of life ǀ cannot be ascribed to Grace or the holy line, for before the Flood we count eleven generations of patriarchs, but of the sons of Adam by Cain only eight; so that the descendents of Cain would be longer-lived. But straight after the Flood this longevity was cut by half but only in those born after that event, for Noah who was born before it lived as long as his ancestors, and Shem lived 600 years. And then within three generations of the Flood, men's lives had dropped to a quarter of their original length, i.e. to about 200 years.

2. Abraham lived for 175 years, a man of mighty spirit for whom everything ǀ turned out well. Isaac lived to be 180, an upright man of quiet life. But Jacob, after many sorrows and a large family, lived to be 147; a patient, gentle, and shrewd man. As for Ishmael, a military man, he lived for 137 years. And Sarah (the only woman whose age is mentioned) died in her 127th year; she was a handsome woman with a great heart, and the best of wives and mothers, and no less famous for her plain speaking than for her deference to her husband. Joseph

[H4ʳ] Prudens, & Politicus, in Aᴵdolescentiâ afflictus, posteà in magnâ fœlicitate Ætatem transigens, ad Annos Centum, & Decem vixit. *Leui* autem Frater eius, natu maior, Centesimum Tricesimum Septimum Annum compleuit; Vir contumeliæ impatiens, & Vindicatiuus. Eandemque ferè
5 ætatem attigit *Filius Leui*; itemque *Nepos* eius, Pater *Aaronis*, & *Mosis*.
  **3.** *Moses* Centum Viginti Annos vixit; Vir Animosus, & tamen *Mitissimus*, linguâ autem impeditus. Ipse verò *Moses*, in *Psalmo* suo, Vitam *Hominis* pronunciauit Annorum tantùm Septuaginta, & si quis
9 Robustior fuerit, Octaginta esse; quæ certè Mensura Vitæ, vsque ad
[H4ᵛ] hodierᴵnum diem, maximâ ex parte, durat. *Aaron* autem Tribus Annis Senior, eodem, cum Fratre, Anno mortuus est; Vir Linguâ promptior, Moribus facilior, & minùs constans. At *Phineas Aaronis Nepos* (ex Gratiâ fortassè extraordinariâ), ad Trecentesimum Annum vixisse colligitur, si modò *Bellum Israelitarum* contra *Tribum Beniamin* (in quâ Expeditione
15 *Phineas* consultus est) eâdem serie Temporum gestum sit, quâ res in Historiâ narratur; Vir erat omnium maximè *Zelotes*. *Ioshua* autem, Vir Militaris, & Dux egregius, & perpetuò florens, ad Annum Centesimum, & Decimum vixit. Cui *Caleb* fuit *Contemporaneus*, &
[H5ʳ] videᴵtur fuisse Æquæuus. *Ehud* autem *Iudex*, etiam Centenarius ad
20 minimum fuisse videtur, cum post deuictos *Moabitas*, Octoginta Annos, sub eius Regimine *Terra Sancta* quieuisset; Vir Acer, & Intrepidus, quíque pro Populo se quodammodò deuouisset.
  **4.** *Iob* post Instaurationem Fœlicitatis suæ, Annos Centum, & Quadraginta vixit, cùm ante Afflictiones suas, eorum Annorum fuisset,
25 qui Filios habuerit Ætatis Virilis; Vir Politicus, & Eloquens, & Euergetes, & *Exemplum Patientiæ*. *Eli Sacerdos* vixit Annos Nonaginta octo; Vir Corpore obesus, Animo placidus, & indulgens in Suos.
[H5ᵛ] *Elizæus* autem *Propheta*, videtur morᴵtuus esse Centenario maior; cùm reperiatur vixisse post *Assumptionem Eliæ* Annos Sexaginta; Tempore
30 vero *Assumptionis*, talis fuerit, vt Pueri eum tanquam *Vetulum Caluum* subsannauerint; Vir Vehemens, & Seuerus, & austeræ vitæ, & Contemptor Diuitiarum. *Isaias* etiam *Propheta* videtur esse Centenarius; nam Prophetiæ Munus exercuisse septuaginta annos reperitur; Annis tùm quo cæpisset prophetizare, tùm quo mortuus esset, incertis.
35 Vir admirabilis Eloquentiæ, & *Propheta Euangelizans*, Promissis Dei Testamenti Noui (tanquam Vter Musto) plenus.

---

12 *Nepos*] ~,  13 extraordinariâ),] ~,)  18 *Contemporaneus*] *Contemperaneus* / cf. *HVM*, I8ᵛ, p. 210 below  19 Æquæuus.] ~:  31 subsannauerint;] ~.
34 tùm] ~, / second occurence  36 Musto)] ~,)

likewise, a prudent and politic man, afflicted ¹ in his youth, but very fortunate later in life, lived to be 110. Levi, his elder brother, lived to be 137, a man intolerant of insults, and vengeful. Levi's son, and grandson too, who was the father of Moses and Aaron, lived to nearly the same age.

3. Moses lived for 120 years; a bold man and yet very mild, and hesitant of speech. That same Moses in his Psalm says that man's life lasts for only threescore years and ten, or if he be more robust, fourscore; which has been for the most part ¹ the measure of life right down to our own time. Aaron, who was three years older, died in the same year as his brother; he was a better speaker, more flexible, and less steadfast. But Phineas, Aaron's grandson, is reckoned to have lived (perhaps by extraordinary grace) for 300 years, if it be that the Israelites' war against the tribe of Benjamin (on which expedition Phineas was consulted) happened in the temporal order given in the history. Phineas was a great zealot. Now Joshua, the military man and great, ever successful commander, lived to 110. Caleb his contemporary seems to ¹ have lived to much the same age. Ehud the judge seems to have lived to at least a hundred; for after the defeat of the Moabites the Holy Land was at peace for eighty years under his rule; he was a powerful and fearless man, who, so to speak, devoted himself to his people.

4. After the restoration of his good fortune Job lived for 140 years; and before his afflictions he was old enough to have sons who had grown to manhood. He was a politic man, eloquent, beneficent, and an example of patience. Eli the priest lived to 98, a fat man, with an equable temper, and indulgent to his kin. Elisha the prophet seems to ¹ have died after his 100th year, since it appears that he lasted sixty years after the assumption of Elijah, which was around the time that the boys ridiculed his baldness. He was a harsh, unbending man of austere life, and scornful of wealth. Isaiah the prophet seems to have been a centenarian, for he appears to have practised the gift of prophecy for seventy years; but when he began to do so, and when he died are uncertain. He was a man of marvellous eloquence, and the evangelizing prophet full (like a skin of new wine) with God's promises of the New Testament.

[H6$^r$]   **5.** *Tobias Senior,* Annos Centum ⏐ Quinquaginta Octo; *Iunior* Centum Viginti Septem vixerunt; Viri Misericordes, & Eleemosynarij. Videntur etiam Tempore *Captiuitatis,* Complures ex *Iudæis,* qui è *Babylone* reuersi sunt, longæui fuisse; cum vtriusque *Templi* (interiecto
5   Annorum Septuaginta spatio) dicantur meminisse, & Disparitatem ipsorum deplorasse. Posteà defluxis Sæculis compluribus, Tempore *Seruatoris, Simeon* inuenitur Nonagenarius; Vir Religiosus, & Spei, & Expectationis plenus. Et eodem tempore *Anna Prophetissa,* vltra Centenariam vixisse manifestò deprehenditur; cum *Septem Annis Nupta* fuis-
9   set, *Vidua* autem per *Annos* ⏐ *Octoginta Quatuor,* quibus addendi sunt
[H6$^v$]   Anni Virginitatis, & qui Prophetiam eius de *Seruatore* insecuti sunt; Mulier Sancta, & Vitam degens, in Orationibus, & Ieiuniis.

**6.** *Longæuitates Hominum,* qui apud *Ethnicos Auctores* inueniuntur, parùm certæ memoriæ sunt; tum propter Fabulas, in quas huiusmodi
15   Narrationes procliues admodùm sunt; tum propter Fallaciam in Calculationibus Annorum. Certè de *Ægyptijs,* nil magni refertur, in his quæ exta*n*t, quoad *Longæuitatem;* cùm Reges ipsorum, qui longissimè
[H7$^r$]   regnârunt, Quinquagesimum, aut Quinquagesimum quintum an⏐num, non excesserint; quod pro nihilo est; cùm etiam Temporibus modernis,
20   huiusmodi Spatia nonnunquam compleantur. At *Arcadum Regibus,* Vitæ longissimæ fabulosè tribuuntur: *Regio* certè illa Montana, & Pastoralis, & Victûs incorrupti. Attamen cum sub *Pane,* tanquam *Deo Tutelari,* fuerit, videntur etiam omnia, quæ ad Eam pertinent, fuisse tanquam *Panica,* & Vana, & ad Fabulas idonea.

25   **7.** *Numa Romanorum Rex* Octogenarius fuit; Vir Pacificus, & Speculatiuus, & Religioni addictus. *M. Valerius Coruinus* centum annos impleuit, interiectis, inter primum & Sextum *Consulatum,* Annis
[H7$^v$]   Quadraginta sex; Vir ⏐ Bello, & Animis fortissimus, Ingenio ciuilis, & Popularis, & Fortunâ perpetuo florens.

30   **8.** *Solon Atheniensis, Legislator,* & Vnus ex Septem, supra Annos Octoginta vixit; Vir Magnanimus, sed Popularis, & amans Patriæ; item eruditus, & non alienus à Voluptatibus, & Vitâ teneriore. *Epimenides Cretensis* Centum Quinquaginta Septem Annos vixisse traditur; mixta Res cum Portento, quia Quinquaginta septem ipsorum, sub Antro eum
35   delituisse ferunt. At dimidio Sæculi post, *Xenophanes Colophonius,* Annos Centum & duos, aut etiam diutiùs vixit; vtpote qui Viginti

---

2 vixerunt;] ~.          4 *Templi*] ~,          5 spatio)] ~,)          7 Nonagenarius;] ~:
15 sunt;] ~.          22 incorrupti.] ~;          31 Patriæ;] ~:

5. Tobias the Elder lived | for 158 years; the Younger for 127; both men of mercy and charity. At the time too of the Captivity many of the Jews who returned from Babylon seem to have been long-lived, since after seventy years had elapsed they were said to have remembered both temples and lamented the difference between the two. After the passing of several ages, in the time of our Saviour, we find that Simeon lived to be 90, a man of religion, filled with hope and expectation. At the same time Anna the prophetess clearly lived to more than a hundred, since she lived a married woman for seven years, and a widow | for eighty-four, on top of which must be added her years of virginity, and those which followed her prophecy of the Saviour. She was a holy woman, spending her life in prayer and fasting.

6. The ages given for men in the heathen authors are not very reliable, because of the fables to which such narratives are too much given, and of errors in reckoning of the years. And for sure there is not much remarkable concerning longevity in the surviving records about the Egyptians; since of their kings the longest reigns did not exceed fifty | or fifty-five years, which amounts to very little since in modern times some reigns have been as long. The kings of Arcadia are said to have had very long lives but that is just a fable. Yet it is certainly a mountainous and pastoral region and with an uncorrupted way of life. But since it was subject to Pan, as to its tutelary god, everything to do with it seems to have been panic, vain, and fit for fables.

7. Numa king of the Romans was an octogenarian; a peaceful and speculative man, devoted to religion. M. Valerius Corvinus lived for a hundred years, since forty-six years intervened between his first and sixth consulships. He was | a very warlike and bold man, by nature courteous and popular, and ever fortunate.

8. Solon the Athenian lawgiver, and one of the seven wise men, lived for more than eighty years; a man of noble heart, but popular and a great patriot, and at the same time learned and not immune to pleasure-seeking and a softer life. Epimenides the Cretan is said to have lived for 157 years; but fact and fiction may merge here, since for fifty-seven of them he is supposed to have been hidden in a cave. But half a generation after this Xenophanes of Colophon lived for 102 years or

[H8<sup>r</sup>] quinque annos natus, Patriam reliquit, Septua|ginta septem totos Annos
est peregrinatus, ac posteà redijt; sed quamdiù à reditu vixerit, non
constat; Vir non magis Itineribus, quam Mente oberrans; vtpote cuius
Nomen propter Opiniones, à *Xenophane,* in *Xenomanem* traductum est;
5 Vasti proculdubiò Conceptûs, & nihil spirans nisi Infinitum.

    **9.** *Anacreon Poeta* maior Octogenario fuit; Homo lasciuus, &
Voluptuarius, & Bibax. *Pindarus Thebanus* Octogesimum Annum
compleuit, Poeta sublimis, cum quâdam nouitate Ingenii, & multus in
9 Cultu Deorum. *Sophocles Atheniensis* similem Ætatem compleuit; Poeta
[H8<sup>v</sup>] grandiloquus, totus in scri|bendo, & Familiæ negligens.

    **10.** *Artaxerxes Persarum Rex,* Annos Nonaginta quatuor vixit; Vir
hebetioris Ingenij, neq*ue* Curarum magnaru*m* patiens, amans Gloriæ,
sed Otij magis. Eodem tempore *Agesilaus Rex Spartanus* Octoginta
quatuor annos impleuit; Vir moderatus, vt inter *Reges Philosophus;*
15 sed nihilominùs Ambitiosus, & Bellator, & tam Militiâ, quam Rebus
gerendis strenuus.

    **11.** *Gorgias Leontinus* Annos Centum & Octo vixit; Vir Rhetor, &
prudentiæ suæ ostentator, & qui Adolescentes Mercede acceptâ vt
19 instetueret, multùm Peregrinator fuit, & paulò ante Mortem, *nihil se*
[I1<sup>r</sup>] *habere quod Se|nectutem incusaret,* dixit. *Protagoras Abderites* Nonaginta
Annos vixit. Iste similitèr Rhetor fuit, sed non tam Cyclopædiâ vsus,
quàm Ciuiles Res, & instructionem, ad Remp. tractandam, docere pro-
fessus; attamèn Circum-cursator Ciuitatum, æquè ac *Gorgias.* At *Isocra-*
*tes Atheniensis* Nonagesimum Octauum Annum compleuit; Rhetor
25 itèm, sed Vir valdè modestus, & Lucem Forensem fugiens, atq*ue* Domi
tantùm Scholam aperiens. *Democritus Abderites,* ad Annos Centum &
Nouem Ætatem produxit; Magnus *Philosophus,* & si quis alius ex
[I1<sup>v</sup>] *Græcis,* verè *Physicus;* Regionum complurium, & multò magis <sup>|</sup> Naturæ
ipsius Perambulator; Sedulus quoque Experimentator, & (quod *Aristo-*
30 *teles* ei objicit) *Similitudinum* potius *Sector,* quam *Disputationum Leges*
*seruans. Diogenes Synopeus,* ad Nonaginta Annos vixit; Vir erga alios
Liber, in se Imperiosus; Victu sordido, & Patientiâ gaudens. *Zeno*
*Cittieus,* Centenarius, duobus tantùm demptis Annis, fuit; Vir Animo
excelso, & Opinionum Contemptor, magni itidem Acuminis, neque
35 tamen molesti, sed quod animos magis caperet, quam constringeret;
quale etiam posteà fuit in *Seneca. Plato Atheniensis* Annum Octogesi-
[I2<sup>r</sup>] mum primu*m* impleuit; Vir Magnanimus, sed <sup>|</sup> tamen Quietis aman-
tior, Contemplatione sublimis, & Imaginatiuus, Moribus Vrbanus, &

---

    3 constat;] ~:    21 vixit.] ~:

more, for he left his country at the age of 25, roamed | the world for seventy-seven years, and then came home; but how long he lived after that is not recorded. He was a man no less wandering in mind than on his journeys, so that on account of his opinions he changed his name from Xenophanes to Xenomanes; he was a man no doubt of high conceptions, breathing nothing save the infinite.

9. Anacreon the poet lived to be more than eighty; a naughty, pleasure-loving man who liked a drink. Pindar the Theban completed his eightieth year; a sublime poet with a certain originality of wit, and much given to the cult of the gods. Sophocles the Athenian reached a similar age; a poet of high speech, a passion for | writing, and neglectful of his family.

10. Artaxerxes king of Persia lived 94 years; a man of weak intellect, impatient of heavy responsibilities, loving glory but preferring idleness. In the same age Agesilaus king of Sparta lived for 84 years; he was a moderate man, and a philosopher among kings, but still ambitious, warlike, and vigorous in military and civil matters.

11. Gorgias of Leontini lived for 108 years; a rhetorician, a parader of his prudence, who taught youths for money, and who shortly before his death said | that *he could find no fault in old age*. Protagoras of Abdera lived for 90 years; he too was a rhetorician but professed to teach not so much the whole range of knowledge, as civil business and matters to do with government; and yet like Gorgias he loved doing the rounds of other cities. But Isocrates the Athenian reached the age of 98. He too was a rhetorician but a very modest man who shunned the glare of public places, and only taught at home. Democritus of Abdera lived for 109 years; he was a great philosopher and, if ever Greek was, a physicist in the true sense; he visited many countries but much more | did he visit nature itself; an indefatigable experimenter, and (as Aristotle objected against him) rather a follower of similitudes than an upholder of the laws of disputation. Diogenes of Sinope lived for 90 years; generous to others but strict with himself, he rejoiced in a vile diet and putting up with hardship. Zeno of Citium lived for two years short of a hundred; a man of sublime soul, a despiser of opinions, very sharp-minded, yet not hurtful but of the sort which engaged rather than inhibited other minds; as was later the case with Seneca. Plato of Athens lived to be 81; a great-hearted man but | yet a lover of tranquillity, sublime in thought and imaginative, in his manners urbane and graceful, and yet more calm

elegans; attamen magis Placidus, quam Hilaris; & Maiestatem quandam præ se ferens. *Theophrastus Eresus* Annum Octogesimum quintum
compleuit; Vir dulcis Eloquio, dulcis etiam Rerum varietate; quíque
ex Philosophiâ, Suauia tantùm decerpserit, Molesta & Amara non
5 attigerit. *Carneades Cyrenæus* multis posteà Annis, ad Octogesimum
quintum Ætatis Annum similitèr peruenit; Vir Eloquentiæ profluentis,
quíque gratâ, & amœnâ Cognitionis varietate, & se ipsum, & alios
[I2ᵛ] delectaret. At *Ciceronis* Tempore *Orbilius*, non ᴵ *Philosophus*, aut *Rhetor*,
sed *Grammaticus*, ad Centesimum ferè annum vixit; primò Miles,
10 deinde Ludimagister; Vir naturâ acerbus, & Linguâ & Calamo, &
versus Discipulos etiam Plagosus.

12. *Q. Fabius Maximus* Sexaginta tribus Annis *Augur* fuit; vnde constat eum Octogenario maiorem occubuisse; licet verum sit, in *Auguratu,*
*Nobilitatem* magis spectari solitam, quam *Ætatem*; Vir prudens &
15 Cunctator, & in omnibus Vitæ partibus Moderatus, & cum Comitate
seuerus. *Masinissa Rex Numidarum* Nonagesimum Annum superauit, &
[I3ʳ] filium genuit, post Octogesimum quintum; Vir acer, & Fortunæ ᴵ
fidens, & Iuuentute multas rerum vicissitudines expertus, decursu
Ætatis constanter fœlix. At *M. Porcius Cato*, vltra Annum Nonagesi-
20 mum vixit; Vir ferei propè Corporis, & Animi; Linguæ acerbæ, &
Simultates amans; Idem Agriculturæ deditus, Sibique & Familiæ suæ
Medicus.

13. *Terentia Ciceronis Vxor*, ad Annum Centesimum tertium vixit;
Mulier multis ærumnis conflictata, primò Exilio Mariti, deinde Dis-
25 sidio, & rursus Calamitate eius extremâ; etiam Podagrâ sæpiùs vexata.
*Luceia* Annum Centenarium, haud parùm superauit; cum dicatur,
[I3ᵛ] Centum Annis totis, in Scenâ Mimam ᴵ agens pronunciasse; Puellæ
fortassè primò partes suscipiens, postremò Anûs decrepitæ. At *Galeria*
*Copiola*, Mima etiam & Saltria, pro Tyrocinio suo producta est in
30 Scenam, quoto Anno Ætatis incertum est; verùm post Annos Nonaginta Nouem ab ea productione, rursùs reducta est in Scenam, non iam
pro *Mimâ*, sed pro *Miraculo*, in *Dedicatione Theatri* à *Pompeio Magno*;
Neque hic finis, cum etiam in *Ludis Votiuis* pro *Salute Diui Augusti*,
iterùm monstrata sit in Scenâ.

35 14. Fuit & alia *Mima*, Ætate paulò inferior, Dignitate sublimior, quæ
[I4ʳ] ad Nonagesimum Annum Ætatem ferè produxit; *Liuia Iulia* ᴵ *Augusta*,
*Cæsaris Augusti* vxor, *Tiberij* Mater. Etenim si *Fabula* fuit *Vita Augusti*

2 *Eresus*] *Etesius* / Ἐρέσιος in Diogenes Laertius' Greek; see *cmt*, p. 432 below
14 *Ætatem*,] ~:

than light-hearted, conducting himself with a certain dignity. Theophrastus of Eresus lived for 85 years; a man of engaging words and of engaging variety of matter too; and who only chose the sweets of philosophy, and did not touch the harsh or bitter things. Many years later Carneades of Cyrene also reached his 85th year; a man of abundant eloquence, who delighted both himself and others with the pleasing and attractive variety of his knowledge. But Orbilius in Cicero's time, who ǀ was not a philosopher nor a rhetorician but a grammarian, lived for almost a hundred years; at first a soldier, then a schoolmaster; a man by nature rough of tongue and pen, and much given to thrashing his pupils.

12. Q. Fabius Maximus was augur for sixty-three years and therefore he must have been over 80 when he died; though the truth is that for that office rank was generally more highly regarded than age. He was a wise man and master of delay, moderate in all aspects of his life, and courteous as well as severe. Masinissa king of the Numidians lived for more than ninety years and begot a son when he was over eighty-five; a bold man who trusted ǀ to fortune, who experienced many ups and downs in his youth but was uniformly lucky in age. M. Porcius Cato lived to more than ninety; in mind and body a man of iron, sharp-tongued, and a lover of quarrels. He was also a great farmer, and doctor to himself and his family.

13. Terentia, Cicero's wife, lived for 103 years; a woman burdened with sorrows, first by her husband's exile, then by their quarrel, and lastly by his ultimate disaster; she was also a martyr to gout. Luceia lived well past a hundred since she is said to have acted for a whole century on the stage, ǀ perhaps at first playing young girls' roles and at the end those of a decrepit old woman. But we do not know at what age Galeria Copiola, an actress and dancer, was introduced to the stage, yet ninety-nine years after her debut she returned to the stage again, not now as a performer but as a prodigy, when the theatre was dedicated by Pompey the Great. And that was not the end of it, for she appeared on stage again at the votive games in honour of the divine Augustus.

14. There was yet another actress slightly younger but of higher rank, who almost reached ninety. She was Livia Julia ǀ Augusta, wife of Cæsar Augustus, and mother of Tiberius. For if the life of Augustus was a play

(Id quod ipse voluit, cùm decumbens Amicis præcepisset, vt postquam expirarit, sibi *Plaudite* exhiberent) certè & *Liuia* optima *Mima* fuit; quæ cum Marito Obsequio, cum Filio, Potestate quâdam, & Prædominantiâ tam benè congrueret. Mulier Comis, & tamen Matronalis, Negotiosa,

5 & Potestatis tenax. At *Iunia C. Cassij* Vxor, *M. Bruti* Soror, etiam Nonagenaria fuit; cum post *Aciem Philippensem* Sexaginta quatuor Annos vixisset; Mulier Magnanima, Opibus fœlix, Calamitate Mariti &

[I4ᵛ] Proximorum, & longâ Viduitate mœsta, sed tamen Honorata. |

**15.** Memorabilis est *Annus Domini* Septuagesimus Sextus, Tempore

10 *Imperatoris Vespasiani,* quo reperiuntur *Longæuitatis,* tanquam *Fasti;* Eo enim Anno peractus est *Census* (*Census* autem de Ætatibus Auctoritatem, & Informationem habet fidissimam); atque in eâ parte *Italiæ,* quæ iacet inter *Apenninum* & *Padum,* inuenti sunt Homines, qui Annum Centesimum æquarunt, & superarunt, Centum & Viginti

15 quatuor; videlicet Annorum Centum, Homines Quinquaginta quatuor; Annorum Centum & decem, Homines Quinquaginta septem; Annorum Centum, & viginti quinque, Homines duo; Annorum

[I5ʳ] centum & triginta, Ho|mines quatuor; Annorum Centum & triginta quinque aut triginta septem, Homines itèm quatuor; Annorum centum

20 & quadraginta, Homines tres. Præter hos, Speciatìm *Parma* edidit Quinque, quorum Tres centum viginti Annos, Duo centum triginta compleuerunt; *Brixillum* Vnum, Annorum centum viginti quinque; *Placentia* Vnum Annorum centum triginta vnius; *Fauentia* Vnam Mulierem, Annorum centum triginta duorum; Oppidum quoddam

25 (tunc dictum *Velleiacium*) in Collibus circa *Placentiam,* decem dedit, quorum Sex Annum Ætatis centesimum decimum, Quatuor centesi-

[I5ᵛ] mum | vicesimum compleuerunt; *Ariminum* denique Vnum, Centum & quinquaginta Annorum, nomine *M. Aponium.*

**Monitum.** *Ne res in longum procederet, visum est, tam in illis, quos iam*

30 *recensuimus, quam in his, quos mox recensebimus, Nullum adducere Octogenario Minorem. Apposuimus autem singulis* Characterem, *siue* Elogium, *verum & perbreue; at eiusmodi, quod iudicio nostro, nonnullam habeat ad* Longæuitatem (*quæ Moribus, & Fortunâ non parùm regitur*) *relationem; Sed duplici modo; Aut quod tales Longæui esse plerúnque*

35 *soleant, aut quod tales, licet minùs aptè dispositi, tamen Longæui esse aliquandò possint.*

---

22 *Brixillum*] *Bruxella* / this emendation noted but not incorporated in *SEH* (II, p. 139)

7 vixisset;] ~.     11 *Census* ] ~,     12 fidissimam);] ~)ₐ     25 dedit,] ~;

(as he himself acknowledged when, while dying, he told his friends to give him a round of applause when he expired), so surely Livia was a brilliant actress who could so well reconcile obedience to her husband with power and ascendancy over her son. She was a courteous woman, and yet matron-like, involved in affairs of state and tenacious of power. As for Junia, wife of C. Cassius, and sister of M. Brutus, she too lived to 90, since she lived for sixty-four years after the battle of Philippi. A great-hearted woman, and wealthy, she was unhappy because of the disaster that overtook her husband and close kin, and her long widowhood; yet she was held in high esteem. |

15. Memorable is the year of our Lord 76 in the emperor Vespasian's day, which gave us a kind of calendar of long life. For in that year a census (which provides the most authoritative data about people's ages) was carried out in the part of Italy which lies between the Apennines and the Po, where they found one hundred and twenty-four who had attained or passed their centenary; namely fifty-four men of 100; fifty-seven of 110; two men of 125; four of 130; | four men of 135 or 137; three men of 140. Besides these Parma in particular yielded five men, of whom three were 120, and two 130; Brixillum, one man of 125; Placentia, one of 131; Faventia, one woman of 132. A certain town (then called Velleiacium), in the hills around Placentia, yielded ten of whom six had completed their 110th year, and four their | 120th; and finally Ariminum yielded one, named M. Aponius, who had reached 150.

*Advice.* To avoid dragging things out, I have thought it best both in the matters already recounted, and in those soon to be, to adduce no age less than 80; and I have adjoined to each a true and very concise character sketch or biography such, in my judgement, as bears on longevity (which is governed to no small degree by habit and fortune); and this in two ways; for they are either generally long-lived, or are not so predisposed, but sometimes could be.

[16ʳ]    16. Inter *Imperatores Romanos*, & ⏐ *Græcos*, itèm *Francos*, & *Germanos*, vsque ad nostram Ætatem, qui numerum prope Ducentorum Principum complêrunt, Quatuor tantùm inuenti sunt Octogenarij; quibus addere liceat *Imperatores* duos primos, *Augustum*, & *Tiberium*,
5   quorum hic Septuagesimum Octauum, ille Septuagesimum Sextum Annum impleuit, & ad Octogesimum fortè peruenire Vterque potuisset, si placuisset *Liuiæ*, & *Caio*. *Augustus* (vt dictum est) Annos vixit Septuaginta Sex; Vir moderatus Ingenio, Idem ad res perficiendas
9   Vehemens, cætera Placidus & Serenus, Cibo & Potu Sobrius, Venere
[16ᵛ]  intemperantior, per omnia Fœlix; Quíque Anno ⏐ Ætatis tricesimo, grauem & periculosum passus est Morbum, adeò vt Salus eius pro desperatâ esset: quem *Antonius Musa Medicus*, cùm cæteri *Medici* Calida Medicamenta, tanquam Morbo conuenientia, adhibuissent, contrariâ ratione Frigidis curauit; Quod fortassè ei, ad Diuturnitatem Vitæ pro-
15   fuit. *Tiberius* duos amplius Annos vixit; *Vir lentis maxillis* (vt Augustus aiebat) Sermone scilicet tardus, sed validus; Sanguinarius, Bibax, quique Libidinem etiam in Diætam transtulit; Attamen Valetudinis suæ Curator probus, vt qui solitus esset dicere, Stultum esse, qui post
[17ʳ]  Triginta Annorum Vitam, Medicum consuleʹret, aut aduocaret. *Gordia-*
20   *nus Senior*, Octoginta Annos vixit, & tamen violentâ Morte perijt, postquam vix degustasset Imperium; Vir Magnanimus, & Splendidus, Eruditus, & Poeta, & constanti vitæ tenore (ante ipsum obitum) Fœlix. *Valerianus Imperator*, Septuaginta sex Annos vixit, antequam à *Sapore Rege Persarum*, captus esset; post Captiuitatem autem Septem Annos
25   vixit, inter contumelias; etiam violentâ Morte præreptus; Vir mediocris Animi, nec strenuus; Existimatione tamen paulò eminentior, & euectus, Experimento minor. *Anastasius* cognomine *Dicorus*, Octoginta Octo
[17ᵛ]  Annos vixit; Homo animi sedati, sed ⏐ Humilior, & Superstitiosus, & Timidus. *Anicius Iustinianus*, Annos Octoginta tres vixit; Vir Gloriæ
30   appetens, Personâ propriâ socors, Ducum suorum virtute fœlix, & celebris; Vxorius, neque suus, sed aliorum ductu circumactus. *Helena Britanna Constantini Magni Mater*, Octogenaria fuit; Mulier ciuilibus Rebus minùs se immiscens, nec Mariti, nec Filij Imperio, sed tota Religioni dedita; Magnanima, & sempèr florens. *Theodora Imperatrix*
35   (quæ *Zoes* soror erat, *Monomachi* vxoris, ipsa autem post obitum eius sola regnauit) Annos supra Octoginta vixit; Mulier negotiosa, &
[18ʳ]  Imperio delectata, Fœlix adʹmodùm, & ex Fœlicitate Credula.

---

15 *maxillis*] ~,      22 tenore] ~,

16. Among the Roman, Greek, ' French, and German emperors right down to our time, who number about two hundred princes, we find that only four reached the age of 80, to whom may be added the first two, Augustus and Tiberius, of whom the latter reached 78, the former 76, and who could, had Livia and Caius Caligula wished it, perhaps have reached 80. Augustus (as I said) lived for seventy-six years; a man of moderate disposition, pressing in pursuit of his ends but in other things placid and serene, temperate in food and drink but not in his sexual appetites, and in all things fortunate. In his ' thirtieth year he suffered a severe and dangerous illness such that his case seemed desperate until the doctor Antonius Musa, after all the other physicians had applied the hot medicines appropriate to the disease, cured him on the principle of contraries with cold ones—which perhaps contributed to his longevity. Tiberius lived two years longer than Augustus, who said Tiberius was a slow-jawed man, i.e. slow but formidable in speech; he was a bloodthirsty drunkard who made lechery part of his diet; and yet he took good care of his health, being accustomed to say that only a fool would call out or consult a physician ' once he had passed thirty. Gordianus the Elder lived to 80 and yet died a violent death before he had tasted the fruits of empire. He was a large-hearted and brilliant man, learned and a poet, and ever fortunate up to the time of his death. The emperor Valerian was 76 before the Persian king Sapor got his hands on him, after which he lasted another seven amid abuse, and he died a violent death. He was a man of average mind and not given to exertion, yet more highly esteemed than was justified in the event. Anastasius, surnamed Dicorus, lived for 88 years; a man of steady temper but ' insignificant, superstitious, and timid. Anicius Justinianus lived to 83; a man eager for glory, sluggish in himself, but fortunate and famous through the ability of his commanders; uxorious, and not his own man, he acted under the tutelage of others. Helena of Britain, mother of Constantine the Great, was an octogenarian, a woman who meddled in civil business not at all during the reigns of her husband and son, but one wholly devoted to religion; she was a great soul and always fortunate. The empress Theodora (the sister of Zoe, and wife of Monomachus, after whose death she reigned alone) lived for more than eighty years. She was a woman engaged in affairs of state, a lover of empire, and extremely fortunate, ' and on that account credulous.

17. Iam à *Sæcularibus*, ad *Principes Viros in Ecclesiâ* Narratione*m* conuertemus. *S. Ioannes Apostolus Seruatoris*, & *Discipulus Amatus*, Nonaginta tres Annos vixit; Verè *Aquilæ Emblemate* notatus, nihil spirans nisi Diuinum, & tanquam *Seraph* inter *Apostolos*, propter Feruo-
5 rem Charitatis. *S. Lucas Euangelista* Octoginta quatuor Annos compleuit; Vir eloquens, & Peregrinator, *S. Pauli* Comes indiuiduus, & Medicus. *Symeon Cleophæ, Frater Domini* dictus, *Episcopus Hierosolymi-*
[18ᵛ] *tanus*, Annos Centum & viginti vixit, licet *Martyrio* præreptus fuerit; |
Vir Animosus, & Constans, & bonorum Operum Plenus. *Polycarpus*
10 *Apostolorum Discipulus, Smyrnensis Episcopus*, videtur ad Centum annos & ampliùs, ætatem produxisse; licet *Martyrio* interceptus; Vir excelsi Animi, & *Heroicæ* patientiæ, & Laboribus indefessus. *Dionysius Areo-pagita, Paulo Apostolo* Contemporaneus, ad Nonaginta Annos vixisse videtur; *Volucris cœli* appellatus, ob Theologiam sublimem, neque
15 minùs Factis, quam Meditationibus insignis. *Aquila* & *Priscilla Pauli Apostoli* primò Hospites, deinde Coadiutores, Coniugio fœlici & cele-bri, ad Centum ad minimum Annos vixerunt; cum sub *Xisto primo*
[Kıʳ] superstites fueⁱrint; Nobile Par, & in omnem Charitatem effusum; qui-bus inter maximas consolationes (quales proculdubiò primos illos
20 Ecclesiæ Fundatores sequebantur), etiam illud Coniugalis Consortij tanquam magnus cumulus accesserat. *S. Paulus Eremita*, Annos Centum & tredecim vixit; vixit autem in Speluncâ, Victu tam simplici, & duro, vt eo Vitam tolerare supra Humanas Vires videri possit; in Meditationi-bus & soliloquijs tantummodò *Æuum* transigens; qui tamen non
25 illiteratus, aut Idiota, sed eruditus fuit. *S. Antonius Cœnobitarum primus Institutor*, aut (vt alij volunt) *Restitutor*, ad Centesimum quintum
[Kıᵛ] Annum | peruenit; Vir Deuotus, & Contemplatiuus, & tamen Ciuilibus Rebus vtilis; Vitæ genere austero, & aspero; attamen in gloriosâ quâdam Solitudine degens, nec sine Imperio; cùm & *Monachos* suos sub se
30 habuisset, atque insupèr à compluribus, & *Christianis*, & *Philosophis*, veluti Viuum aliquod Simulachrum, non sine Adoratione quâdam, visitatus esset. *S. Athanasius* mortuus est Octogenario Maior; Vir Inuincibilis Constantiæ, Famæ sempèr imperans, nec Fortunæ suc-cumbens; Idem erga Potentiores liber, erga Populum gratiosus, &
35 acceptus; exercitatus Contentionibus, in ijsque & Animosus, & Solers.
[K2ʳ] *S. Hieroⁱnymus*, plurimorum consensu, Annum Nonagesimum super-auit; Vir Calamo potens, & virilis Eloquentiæ; variè eruditus & Linguis,

---

17. I shall now direct my account away from temporal rulers to princes of the Church. St John, apostle and dear disciple of our Saviour, lived for 93 years; rightly represented by the symbol of the eagle, breathing nothing but the divine, and as a seraph among the apostles by reason of the ardour of his charity. St Luke the Evangelist lived for 84 years, an eloquent man, a traveller, inseparable companion of St Paul, and a physician. Simeon son of Cleophas, called our Lord's brother, and Bishop of Jerusalem, lived to 120, though cut off by martyrdom. | He was a bold man, steadfast and full of good works. Polycarp, disciple of the apostles, and Bishop of Smyrna, seems to have lived for a hundred years or more, though he died a martyr. He was a man of noble mind, heroic endurance, and tireless in his labours. Dionysius the Areopagite, contemporary of Paul the Apostle, seems to have lived for 90 years; called the Hawk of Heaven for his sublime theology, he was no less famous for his acts than for his contemplations. Aquila and Priscilla, first the hosts of St Paul, and then his co-workers, lived for at least a hundred years in their famous wedded bliss, since they lived into the time of Sixtus I. | They were a noble pair, abundant in all charity; to whom among other very great consolations (of the kind doubtless given to the first founders of the Church) there also accrued the great addition of their married companionship. St Paul the Hermit lived for 113 years; he lived in a cave and on such simple and hard fare, that it would appear to be beyond human capacity to put up with it; he spent his time just meditating and talking to himself, yet he was not an illiterate or half-wit but a man of learning. St Antony, the founder or (as some have it) the restorer of coenobitic institutions, lived to 105; | a devout and contemplative man, and yet able in civil affairs; austere and harsh in his way of life, yet living in a kind of splendid isolation not without authority, for he kept his monks under him, and he was visited, not without a certain reverence, by many Christians and philosophers as if he were a living image. St Athanasius was over eighty when he died; a man of unconquerable steadfastness, always commanding fame and never giving in to fortune. He told the truth to power, and was affable and welcome to the people; skilled in disputes, and bold and resourceful in them. St Jerome, | as many agree, lived beyond ninety; a man with a mighty pen and masculine eloquence; learned

& Scientiis; Peregrinator item, atque Vitæ versus senium austerioris; sed in Vitâ priuatâ Spiritus gerens altos, & latè fulgens ex Obscuro.

18. At *Papæ Romani* numerantur Ducenti Quadraginta Vnus; Ex tanto Numero Quinque solummodò Octogenarii, aut supra reperiun-
5 tur; Primitiuis autem compluribus, iusta Ætas *Martyrij Prærogatiuâ* anticipata est. *Ioannes vicesimus tertius, Papa Romanus,* Nonagesimum Ætatis Annum compleuit, Vir Ingenii inquieti, & Nouis Rebus
[K2ᵛ] stu¦dens, & Multa transferens, nonnulla in Melius, haud pauca in Aliud; Magnus aute*m* Opu*m*, & Thesauri Accumulator. *Gregorius* dictus *duo-*
10 *decimus,* creatus *Papa* in *Schismate,* & quasi *Interrex,* Nonagenarius obiit; De eo, propter breuitatem *Papatûs,* nihil inuenimus, quod annotemus. *Paulus Tertius,* ad Octoginta & vnu*m,* Annos vixit; Vir sedati Animi, & profundi Consilii, idem Doctus, & Astrologus, & Valetudinem impensè regens; more autem veteris *Sacerdotis Eli,* Indul-
15 gens in suos. *Paulus Quartus* Octoginta tres Annos vixit; Vir Naturâ asper & seuerus, altos gerens Spiritus, & Imperiosus, Ingenio com-
[K3ʳ] motior, Sermone ¦ eloquens, et expeditus. *Gregorius Decimus tertius,* similem Ætatem Octoginta trium Annorum impleuit; Vir planè Bonus, Animo & Corpore sanus, Politicus, Temperatus, Euergetes &
20 Eleemosynarius.

19. Quæ sequentur, Ordine promiscua, Fidei magis dubiæ, Obserua-tione magis ieiuna, erunt. *Rex Arganthonius,* qui regnauit *Gadibus* in *Hispaniâ,* Centum & Triginta, aut (vt alij volunt) Quadraginta Annos
24 vixit; ex quibus Octoginta regnauit. De Moribus eius, & Vitæ genere, &
[K3ᵛ] Tempore quo vixit, siletur. *Cyniras Cypriorum Rex,* ¦ in *Insulâ* illâ tunc habitâ Beatâ, & *Voluptuariâ,* Centum Quinquaginta, aut Sexaginta Annos vixisse perhibetur. *Reges* duo *Latini* in *Italiâ,* Pater & Filius, alter Octingentos, alter Sexcentos Annos vixisse traduntur; verùm hoc nar-ratur à *Philologis* quibusdam, quibus & ipsis (cætera satis credulis) Fides rei suspecta est, imò damnata. *Arcadum Reges* nonnullos Trecentos
30 Annos vixisse alii tradunt; Regio certè ad Vitam longam satis idonea; Res fortassè Fabulis aucta. Narrant *Dandonem* quendam in *Illyrico,* absque Incommodis Senectutis, Quingentos Annos vixisse. Apud *Epios*
[K4ʳ] *Ætoliæ* videlicet Par¦tem, narrant Vniuersam Gentem admodùm long-
35 æuam fuisse; vt multi ex his Ducenûm Annorum inuenti sint; Inter eos Præcipuum quendam, nomine *Litorium,* virum *Giganteæ Staturæ,* qui Trecentos Annos cumulauerat. In *Tmoli Montis Fastigio* (*Tempsi* anti-quitùs vocato) Homines complures, Centum quinquaginta Annos vixisse

24 regnauit.] ~:

in languages and the sciences, and a traveller who lived more austerely in old age; private in life but lofty in spirit, his light blazed from his obscurity.

18. There have been two hundred and forty-one Popes of Rome. Of such a great number only five have been octogenarians or more. Many of the early ones were cut short by the privilege of martyrdom. John XIII, Pope of Rome, lived to 90; a man of restless mind, who, favouring novelty, changed many things, | some for the better, and not a few into something other; he was also a great accumulator of wealth and treasure. Gregory, called the twelfth, made pope during the Schism, and a kind of interrex, died in his nineties; but because of the shortness of his papacy I find nothing to say about him. Paul III lived to be 83; a man of calm mind and fine judgement; a learned man and astrologer, and very correct in matters of health; and in the manner of Eli the priest, indulgent to his kin. Paul IV lived to 81; a man harsh and severe by temperament, an imperious spirit harbouring high ambitions, a restless mind and eloquent and prompt of | speech. Gregory XIII likewise reached 83; a really good man, of sound body and mind, politic, temperate, beneficent, and charitable.

19. The following instances are in no particular order, more doubtful in reliability, and more deficient in detail. King Arganthonius, who ruled over Cadiz in Spain, lived for 130 or (as others have it) 140 years, of which he spent eighty on the throne. Of his habits, mode of life, and his times history is silent. Cinyras king of the Cypriotes | is said to have lived to 150 or 160 on that island, which was then thought blessed and delightful. Two Latin kings in Italy, father and son, are reported to have lived the one 800 and the other 600 years; but this was stated by certain philologists who (credulous enough in other matters) have themselves doubted the truth of the story, and in fact condemned it. Others say that some of the kings of Arcadia lived for 300 years; the area is certainly cut out for longevity but maybe its reputation has been inflated by fables. They say that a certain Dando in Illyria lived for 500 years with none of the ailments of age. They say that among the Epii, i.e. a part of | Ætolia, the whole population is very long-lived, and that many of them are found to live for 200 years. One of them stands out in particular, a man of gigantic stature called Litorius, who chalked up 300 years. On the top of Mount Tmolus (anciently called Tempsis) many are said to have lived to 150. They report that the sect of the Essenes among the

traditur. *Sectam Essæorum* apud *Iudæos*, vltra Centum Annos communitèr vixisse tradunt. *Secta* autem illa simplici admodum Diætâ
vtebatur, ad Regulam *Pythagoræ*. *Apollonius Tyaneus* Centum Annos
[K4$^v$] excessit, aspectu (vt in tantâ Ætate) pulcher, Vir $^|$ certè Mirificus, apud
5  *Ethnicos Diuinus* habitus, apud *Christianos Magus*; Victu *Pythagoricus*,
magnus peregrinator, magnâ etiam Gloriâ florens, & tanquàm pro
Numine cultus; Attamen sub finem Ætatis, Accusationes & Contumelias passus, vnde nihilominùs incolumis quoquo modo euasit. Attame*n*
ne *Longæuitas* sua *Diætæ Pythagoricæ* solùm tribuatur, sed etiam è
10  Genere suo aliquid traxisse videatur; *Auus* eius etiam Centum Triginta
Annos vixit. *Q. Metellum* vltra Centum Annos vixisse certa res est;
Atque post *Consularia Imperia* fœlicitèr administrata, *Pontificem Maxi-*
[K5$^r$] *mum*, iam Senem creatum esse, & sacra per Viginti duos $^|$ Annos
tractasse; neque Ore in Votis nuncupandis hæsitante, neque in Sacri-
15  ficiis faciendis tremulâ Manu gerentem. *Appium Cœcum*, Annosis-
simum fuisse constat; Annos non numerant; quorum partem maiorem,
postquam Luminibus orbatus esset, transegit; Neque proptereà mol-
litus, Familiam numerosam, Clientelas quam plurimas, quinetiam
Remp. fortissimè rexit; Extremâ vero Ætate Lecticâ in Senatum delatus,
20  Pacem cum Pyrrho vehementissimè dissuasit; cuius Principium Orationis admodùm memorabile, & inuincibile quoddam Robur, et impe-
[K5$^v$] tum Animi spirans. *Magnâ*, inquit, *impatientiâ* (*patres con$^|$scripti*),
*Cœcitatem meam, per plures iam Annos tuli; at nunc etiam me Surdum*
*quoque optauerim, cùm vos tam deformia Consilia agitare audiam. M.*
25  *Perpenna* vixit Annos Nonaginta Octo; omnibus, quos *Consul* senten-
tiam in *Senatu* rogauerat (hoc est, omnibus *Senatoribus* sui Anni), superstes fuit: etiam omnibus, quos paulò post, *Censor* in Senatum legerat,
Septem tantùm exceptis. *Hiero Rex Siciliæ* Temporibus *Belli Punici*
*secundi*, ad Centesimum ferè annum vixit; Vir & Regimine, & Moribus
30  moderatus, Numinum Cultor, & Amicitiæ Conseruator religiosus,
Beneficus, & constantèr Fortunatus. *Statilia* ex nobili Familiâ, *Claudij*
[K6$^r$] tempo$^|$re, vixit Annos Nonaginta Nouem. *Clodia Ofilij Filia* Centum &
quindecim. *Xenophilus*, antiquus Philosophus è *Sectâ Pythagoræ*, Centum & sex Annos vixit, sanâ & viuidâ Senectute, & magnâ apud
35  Vulgum Doctrinæ famâ. *Insulani Corcyrei*, habebantur olim Viuaces,
sed hodiè communi aliorum Sorte viuunt. *Hippocrates Côus* Medicus
insignis, Centum & quatuor Annos vixit; Artemque suam tam Longâ

2 tradunt.] ~:        22 *impatientiâ*] ~,    *con$^|$scripti*),] ~,)      26 Anni),] ~,)
30 moderatus,] ~;    religiosus,] ~;

214

Jews generally lived to more than a hundred. Now that sect maintained a very simple diet on the Pythagorean model. Apollonius of Tyana lived more than a hundred years, a man very good-looking for his age; he evoked ¹ admiration, and was regarded by the heathens as a god, and by the Christians as a magus; by diet a Pythagorean, he was a great traveller who also flourished in great fame, and was venerated as a deity. Yet towards the end he suffered accusation and insult, but managed to escape unharmed. Yet lest his long life be ascribed to his Pythagorean diet alone, he can be shown to have derived some of it from his family; his father lived to 130. That Q. Metellus lived to more than a hundred is a matter of certainty; having discharged consulships successfully, he was made Pontifex Maximus in old age and discharged that sacred function for twenty-two ¹ years; and neither did his voice shake when taking the vows, nor his hands tremble when he offered sacrifices. Appius the Blind lived a very long time but the number of his years is unknown, though most of them came after his light was spent—not that he was weakened by that, for he governed a large family, many clients, and even the republic itself with great firmness. Indeed in extreme old age, borne on a litter to senate, he argued with great passion against peace with Pyrrhus; the start of his speech being very memorable for its invincible strength and intellectual force: *Conscript fathers, with great impatience* ¹ *and for many years have I borne my blindness but now I could even wish myself deaf when I hear you propose such shameful policies.* M. Perpenna lived to 98, outlasting all those in the senate whose support he sought for the consulship (i.e. all the senators of that year; he also survived all those (save seven) whom, shortly after, he had as censor selected for the senate. Hiero king of Sicily at the time of the second Punic war lived to almost a hundred; a man moderate in regimen and habits, votary of the gods, and one who cultivated his friends assiduously; beneficent and unfailingly fortunate. Statilia, of noble birth, lived ¹ in Claudius' time to 91. Clodia, Ofilius' daughter, reached 115. Xenophilus, an ancient philosopher of the Pythagorean sect, lived to 106; fit and vigorous in old age, he had a great reputation among the common people for learning. The islanders of Corcyrea were once thought to be long-lived but nowadays their lifespan seems nothing unusual. Hippocrates of Cos, the famous physician, lived for 104; and by his length of life justified and

Vitâ comprobauit, & honestauit; Vir cum Prudentiâ quâdam doctus; in
Experientiâ, & Obseruatione multus; non Verba, aut Methodos
[K6ᵛ] captans, sed Neruos tantùm Scientiæ separans, & proponens. *De¹monax*
*Philosophus* (non solùm Professione, sed Moribus) Tempore *Adriani*,
5 ad Centenarium ferè Annum vixit; Vir magni Animi, atque Animi
victor, idque verè sine affectatione, & in maximo humanarum rerum
Contemptu, Ciuilis, & Vrbanus. Is cum Amici de Sepulturâ ipsius,
verba inijcerent, *Desinite*, inquit, *de Sepulturâ curare; Cadauer enim*
*Fœtor sepeliet;* atque illi; *Placet ergo Auibus, aut Canibus exponi?* Ille
10 rursus, *Cum,* inquit, *viuus Hominibus prodesse pro viribus contenderim,*
*quæ inuidia est, si mortuus etiam Animalibus aliquid præbeam? Populus*
*Indiæ, Pandoræ* appellati, admodùm Longæui; etiam vsque ad Annum
[K7ʳ] Ducentesimum. ¹Addunt rem magis miram; Scilicet cum Pueri ferè
candido Capillo fuerint, Senectute ante Canitiem eos nigrescere solitos:
15 Id tamen vbique vulgare est, vt Pueris Capillitio candidiore, virili Ætate,
Pili mutentur in obscurius. Etiàm *Seres Indorum* Populus, cum Vino suo
ex Palmis, Longæui habiti sunt, vsque ad Annum Centesimum Tricesimum. *Euphranor Grammaticus* consenuit in Scholâ, & docebat Literas,
vltra Annum Centesimum. *Ouidius Senior, Poetæ Pater,* Nonaginta
20 Annos vixit; Diuersus à Moribus Filii, vtpote qui Musas contempsit, &
Poeticem Filio dissuasit. *Asinius Pollio Augusti* Familiaris, Centum
[K7ᵛ] annos ¹superauit; Vir ingentis Luxûs, Eloquens, Literarum cultor, attamen Vehemens, Superbus, Crudelis, & tanquam sibi Natus. Inualuit
Opinio, de *Senecâ,* quod admodùm Annosus fuerit, vsque ad Annum
25 Centesimum decimum quartum; Quod verum esse non potest, cùm
tantùm absit, vt Senex decrepitus ad *Neronis* tyrocinium admotus sit, vt
contra Rebus gerendis strenuè suffecerit; quinetiam paulò antè, medio
tempore *Claudij,* exulârit, ob Adulteria aliquarum Principum Fœmi-
29 narum; quod in talem Ætatem non competit. *Ioannes de Temporibus,* ex
[K8ʳ] omnibus posterioribus Sæculis, Traditi¹one quâdam, & Opinione vulgari, vsque ad Miraculum, vel potius vsque ad Fabulam, Longæuus
perhibetur, Annorum supra Trecentos; Natione fuit *Francus,* militauit
autem sub *Carolo Magno. Gartius Aretinus, Petrarchæ* Proauus, ad Centum & quatuor Annos peruenit, prosperâ semper vsus Valetudine,
35 atque in Extremis, Vires labantes sentiens potiùs quàm Morbum; quæ
vera est Resolutio per Senium. Ex *Venetis* reperiuntur haud pauci
Longæui, etiam Gradu eminentiori; *Franciscus Donatus Dux; Thomas*
*Contarenus Procurator S. Marci; Franciscus Molinus* itèm *Procurator S.*

---

4 Philosophus] ~,     8 *Desinite*,] ~ₐ     13 Ducentesimum.] ~;     33 *Aretinus*,] ~ₐ

graced his art; a learned man with a certain prudence; rich in experience and observation, and no slave to words or methods but one who isolated and displayed the very nerves of knowledge. Demonax, | a philosopher (in his habits as well as his profession), in Hadrian's time lived for almost a hundred years; a man who mastered a high soul, and that without affectation; utterly scornful of human affairs but civil and urbane. When his friends dropped hints about his burial, he replied, *Stop worrying about that, for the stench will bury the corpse.* Then they said, *Do you want it to be left out for the carrion birds and dogs, then?* He answered them thus: *Since I strove while alive to benefit men, what's wrong with doing a good turn to the animals when dead?* The people of India called Pandoræ are very long-lived, even surviving up to 200 years. | They also add something even more strange, namely that they have hair practically white when young, but in old age it turns black before greying. Yet it is common that whiter hair in boyhood grows darker in maturity. Likewise the Seres, an Indian people, with their palm-wine, are held to be long-lived, for as much as 130 years. Euphranor the Grammarian kept his school and taught letters beyond his hundredth year. Ovid the elder, father of the poet, lived for 90 years; his outlook was different from his son's, for he had no time for the Muses, and tried to put his son off poetry. Asinius Pollio, the friend of Augustus, lived over one hundred | years; an extravagant man, eloquent, a devotee of letters, yet fiery, proud, cruel, and only out for himself. The idea has arisen that Seneca was very old, and up to 114 years old, but this cannot be true, for far from being a broken-down old man when he became Nero's tutor, he was very active in public affairs. Moreover, not long before, in the middle of Claudius' time, he was exiled for adultery with some high-ranking women, behaviour not compatible with such advanced years. Joannes de Temporibus, of all men of latter times, is by tradition and | common opinion said to have been long-lived to a miraculous, or rather fabulous, degree. Above three hundred years old, he was a Frenchman who fought under Charlemagne. Gartius of Aretium, Petrarch's great-grandfather, lived to 104, always enjoying good health, but in the end he experienced declining powers rather than illness, which is the true resolution of old age. Among the Venetians we find quite a few long-lived individuals who were also men of rank. Francesco Donato the Doge; Tommaso Contarini and Francesco Molino, both

[K8ᵛ] *Marci*; Alij. At maximè memorabile est, illud de *Cornaro Veneto*, qui Corpore, sub initio, valetudinario, cœpit primùm metiri Cibum & Potum, ad certum Pondus, in curam Sanitatis. Ea Cura transijt vsu in Diætam, & ex Diætâ in magnam Longæuitatem, vsque ad Annum
5 Centesimum, & vltra, integris Sensibus, & constanti Valetudine.

*Guilielmus Postellus*, nostrâ ætate, *Gallus*, ad Centesimum & propè Vicesimu*m* Annu*m* vixit; etia*m* summitatibus Barb*æ* in labro superiore nonnihil nigrescentibus, neque prorsùs canis; Vir Capite motus, & non
9 integræ omnino Phantasiæ; magnus Peregrinator, & Mathematicus, &
[L1ʳ] Hæreticâ Prauitate nonnihil aspersus.

20. Apud nos in *Angliâ*, arbitror non existere Villulam, paulò populosiorem, in quâ non reperiatur aliquis Vir, aut Mulier ex Octogenariis; Etiam ante paucos Annos in Agro *Herefordiensi* inter *Ludos Florales*, instituta erat *Chorea*, & *Saltatio* ex Viris octo, quorum Ætas simul
15 computata, Octingentos Annos complebat; cùm quod Alteris eorum ad Centenarium deesset, Alteris aliquibus superesset.

21. In *Hospitali Bethleem*, ad *Suburbia Londini*, quod in Sustentationem, & Custodiam Phreneticorum, institutum est, inueniuntur de
[L1ᵛ] tempore in tempus, multi ex Mente captis, fuisse longæui.

20 22. Ætates, de quibus fabulantur, *Nympharum*, & *Dæmonum Aëreorum*, qui Corpore mortales essent, sed admodùm longæui (id quod & Antiquâ, & inter quosdam Recenti Superstitione, & Credulitate, receptum est) pro Fabulis, & Somniis habemus; præsertìm cum sit Res, nec cum Philosophiâ, nec cum Religione benè consentiens. Atque de
25 *Historia Longæuitatis* in *Homine*, per Indiuidua, aut Indiuiduis proxima, hæc inquisita sint. Iam ad Obseruationes per Capita transibimus.

23. *Decursus Sæculorum*, & *Successio Propaginis*, nihil videntur
[L2ʳ] omninò demere, de Diuturnitate Vi'tæ; quippe Curriculum Humanæ Ætatis videmus, vsque à Tempore *Mosis* ad nostra, circa Octogesimum
30 Annum stetisse; neque sensìm & paulatìm (vt quis crederet) declinasse. Sunt certè Tempora, in singulis Regionibus, quibus Homines diutiùs, aut breuiùs degunt. Diutiùs plerúnque, cum Tempora fuerint Barbara, & simplicioris Victûs, & Exercitationi Corporis magis dedita; Breuiùs, cum magis Ciuilia, & plus Luxuriæ & Otii: verùm ista transeunt per
35 Vices, Propago ipsa nihil facit. Neque dubium est, quin idem fiat in Animalibus cæteris, siquidem nec *Boues*, nec *Equi*, aut *Oues*, & Similia,
[L2ᵛ] Æuo, vltimis his Sæculis, minuuntur. Itaque *Præcipitatio Ætatis* facta est per *Diluuium*; & fieri fortassè potest, per Similes Maiores Casus (vt

---

3 Sanitatis.] ~:     21 longæui] ~;     37 minuuntur.] ~;     38 Casus] ~,

proctors of St Mark's; and so on. But the most memorable │ case is that of Cornaro of Venice who, after an unhealthy youth, began for health's sake to restrict his intake of food and drink to a certain weight. This regimen turned into a regular diet, from which grew great longevity, right up to a hundred and beyond, with unimpaired senses and unflagging health. In our time, Guillaume Postel, a Frenchman, lived close on 120 years, with his moustache remaining black with no greying. He was a man of shifting mind and unsound fancy, a great traveller, mathematician, and somewhat touched with wrong-headed heresy. │

20. With us in England I judge that there is no village of any size where you cannot find some man or woman over eighty. And not so many years ago in Herefordshire they managed to organize a dance during the May games for eight men whose combined age was 800, with some of them falling short of a hundred by as much as some others exceeded it.

21. In Bedlam Hospital on the outskirts of London, which was set up for the care and custody of the insane, we find from time to time many long-lived mad people. │

22. Stories about the ages of nymphs and demons of the air who are mortal in body but extremely long-lived (an idea received by superstition and credulity of the ancients and some of the moderns) I take for fables and dreams, especially as they agree neither with philosophy nor religion. So much then for the history of human longevity considered in particular or near-particular cases. Now I shall pass on to observations under general headings.

23. The passing of ages, and the succession of generations, seem not to have lessened life expectancy, │ for we see that the span of human life from the time of Moses down to our age has stayed at around eighty, without (as some have thought) imperceptibly and gradually declining. Certainly there are times in particular countries when men's lives are longer or shorter: longer generally in barbarous times with their simpler diets and greater dedication to exercise; shorter with more civilization, good living, and idleness. But these things come and go, and over the generations make no difference. Nor can there be any doubt but that the same is true for other animals, since neither oxen, nor horse, nor sheep, and so on, have lived shorter lives │ in these last ages. Thus was the great drop in age caused by the Flood, and perhaps could be caused again by similar great disasters (as they call them), like less

loquuntur) veluti Inundationes particulares, Combustiones per longas Siccitates, Terræ motus, & similia. Quinetiam videtur, similis esse ratio in *Magnitudine Corporum*, siue *Staturâ*; quæ nec ipsa per Successionem propaginis defluit; licet *Virgilius* (communem opinionem secutus) diuinasset Posteros futuros Præsentibus minores; vnde ait de *Campis Emathijs*, & *Emonensibus subarandis*:

> *Grandia*que *effossis mirabitur Ossa sepulchris.*

[L3ⁱ] Etenim cùm constet fuisse, | quondam Homines *Staturis Gigantæis* (quales & in *Siciliâ*, & alibi, in vetustis Sepulchris, & Cauernis, pro certo reperti sunt), tamen iam per tria ferè Millenaria Annorum, ad quæ producitur Memoria satis certa, in ijsdem Locis, nil tale continuatur; licet etiam hæc res, per Mores, & Consuetudines ciuiles, vices quasdam patiatur, quemadmodùm & illa altera. Atque hæc magis notanda, quia insêdit Animis Hominum penitùs Opinio, quod sit perpetuus Defluxus per Ætatem, tum quoad *Diuturnitatem Vitæ*, tum quoad *Magnitudinem*
[L3ᵛ] & *Robur Corporis*, Omniáque labi & ruere in Deterius. |

**24.** *Regionibus Frigidioribus* & *Hyperboreis*, diutiùs Homines viuunt plerunque, quàm *Calidioribus*; Quod necesse est fieri, cum & Cutis sit magis astricta, & Succi Corporis minùs dissipabiles; & Spiritus ipsi minùs acres ad consumendum, & magis fabriles ad reparandum, & Aer (vtpote modicè calefactus, à Radijs Solis) minùs Prædatorius. At sub *Lineâ Æquinoctiali*, vbi Sol transit, & duplex sit Hyems, & Æstas, sitque etiam maior Æqualitas inter Spatia Dierum, & Noctium (si cætera non impediant), etiam benè diù viuunt; vt in *Peruuiâ*, & *Taprobanâ*.

[L4ⁱ] **25.** *Insulani Mediterraneis*, vt pluri|mùm sunt Longæuiores; Neque enim tam diù viuunt in *Russiâ*, quam in *Orcadibus*; neque tam diù in *Africâ* eiusdem Paralleli, quam in *Canarijs*, & *Terceris*; *Iaponenses* etiam *Chinensibus* (licet hi Longæuitatis appetentes sint vsque ad Insaniam) sunt viuaciores; nec mirum, cum Aura Maris, & in Regionibus Frigidioribus foueat, & in Calidioribus refrigeret.

**26.** *Loca Excelsa* potiùs edunt Longæuos, quam *Depressa*; præsertìm si non sint *Iuga Montium*, sed *Terræ Altæ*, quatenus ad Situm eorum generalem; qualis fuit *Arcadia* in *Græciâ*, & *Ætoliæ* Pars, vbi Longæui admodùm fuerunt. At de *Montibus* ipsis | eadem foret ratio, propter
[L4ᵛ]

5 minores;] ~:     8 *Gigantæis*] ~,     10 sunt),] ~,)     16 *Corporis*,] ~;
19 astricta,] ~;     20 reparandum,] ~;   Aer] ~,     21 Prædatorius.] ~;
23 Noctium] ~,     24 impediant),] ~)ₐ     35 fuerunt.] ~:

220

general floods, combustions wrought by long droughts, earthquakes, and so on. The same seems to be true of body height or stature, which has not been reduced as the generations have passed, though Virgil (following common opinion) forecast that later generations would be shorter than his own; whence he said of the ploughing of the Æmathian and Æmonian fields that the farmer:

> *will marvel at the giant bones in upturned graves.*

For it is true that <sup>|</sup> men of gigantic stature once lived (evidence of which has been found in old tombs and caves in Sicily and elsewhere), but now for almost three thousand years of which we have memory certain enough, nothing of the kind has gone on in those places; though in this matter, just as in the other case, certain alterations have happened by reason of the general practices of society. And this should be the more observed because the opinion has become embedded in men's minds that a perpetual decline has been going on both in life expectancy, and in bodily size and strength, and that everything is going to the dogs. <sup>|</sup>

24. In colder northern regions men generally live longer than in the warmer ones, and of necessity, for in the former the skin is tighter, the bodily juices less prone to dispersal, and spirits less keen to consume, and more handy at repair; while the air (but slightly heated by the Sun's rays) is less predatory. But below the equinoctial line, where the Sun passes and where there are two summers and winters, and the days and nights are of more equal length, people live (if nothing else stops them) a long time, as in Peru and Taprobane.

25. As compared with inland folk, islanders generally <sup>|</sup> live longer. For they do not live as long in Russia as in the Orkneys, nor so long in Africa as in the Canaries or Azores, which are on the same latitude; and the Japanese longer than the Chinese (though the latter are mad keen on longevity). And all this is no wonder, as the sea air keeps people snug in colder regions, and cools them down in hotter ones.

26. High places produce more long-lived people than low ones, especially if those places are not mountain tops but high plains, such as were Arcadia in Greece and part of Ætolia, where they were extremely long-lived. The same would <sup>|</sup> be true of the mountains themselves,

Aerem videlicet puriorem, & Limpidiorem, nisi hoc labefactaretur per Accidens; interuentu scilicet Vaporum ex Vallibus eò ascendentium, & ibi acquiescentium. Itaque in *Montibus Niualibus*, non reperitur aliqua insignis Vitæ Longitudo; non in *Alpibus*, non in *Pyrenæis*, non in *Apen-*

5 *nino*; sed *Medij Colles*, aut etiam *Valles* dant Homines Longæuiores; At in *Montium Iugis protensis* versus *Æthiopiam*, & *Abyssinos*, vbi propter Arenas subiectas, parum aut nihil incumbit in *Montes* Vaporis, diutissimè viuunt, etiam ad Hodiernum Diem, Annum non rarò Centesi-

[L5ᵛ] mum, & Quinquagesimum implentes.

10 **27.** *Paludes* & *Tractus* earum, præsertìm exporrecti in Plano, Natiuis propitii, Aduenis maligni, quoad Vitæ prorogationem, aut decurtationem; Quodque mirum videri possit, *Paludes* Aquâ Salsâ, per vices inundatæ, minùs salubres, quam quæ Aquâ Dulci.

**28.** *Regiones Particulares*, quæ notatæ sunt Longæuos produxisse,

15 sunt *Arcadia*, *Ætolia*, *India* cis *Gangem*, *Bresilia*, *Taprobana*, *Britannia*, *Hybernia*, cum *Insulis Orcadibus*, & *Hebridibus*; nam de *Æthiopiâ*, quod ab Aliquo ex Antiquis refertur, quod Longæui fuerint, Res vana est.

[L5ᵛ] **29.** Occulta est res, Salubritas, præsertim perfectior, *Aeris*; & potiùs Experimento, quam Discursu, & Coniecturâ, elicitur. Capi possit

20 *Experimentum* ex *Vellere Lanæ*, per expositionem in Aerem cum morâ aliquâ Dieru*m*, minùs aucto Pondere; Aliud ex *Frusto Carnis*, diutiùs manente non putrefacto; Aliud ex *Vitro Calendari* minori spatio reciprocante. De his, & Similibus ampliùs inquiratur.

**30.** *Aeris* non tantùm *Bonitas*, aut *Puritas*, verùm etiam *Æqualitas*

25 quoad Long*æ*uitatem spectatur. *Collium* & *Vallium Varietas* Aspectui & Sensui grata, Longæuitati suspecta; At *Planities* modicè sicca, nec tamen nimis sterilis, aut Arenosa, nec prorsùs sine Arboribus, & Vmbrâ,

[L6ʳ] Diu|turnitati Vitæ magis commoda.

**31.** *Inæqualitas Aeris* (vt iam dictum est), in *Loco Mansionis*, mala;

30 verùm *Mutatio Aeris* in *Peregrinatione*, postquàm quis assueuerit, bona; Vnde & magni *Peregrinatores* Longæui fuere; similitèr etiam Longæui, qui in *Tuguriolis suis*, eodem loco, perpetuò Vitam degerunt; *Aer* enim Assuetus, minùs consumit, at Mutatus magis alit, & reparat.

**32.** Vt *Series*, & *Numerus Successionum*, ad Diuturnitatem, aut Breui-

35 tatem vitæ, nihil est (vt iam diximus); ita *Conditio* immediata *Parentum*, tam ex parte *Patris*, quam *Matris*, proculdubiò multùm potest. Alij

---

8 Diem,] ~;          21 Pondere;] ~:          23 reciprocante.] ~:          25 spectatur.] ~:
29 *Aeris*] ~,          est),] ~)ₐ          31 fuere;] ~:          32 degerunt;] ~:
35 diximus);] ~;)

because of the purer, clearer air, were it not undermined by an accident, i.e. of the vapours rising up from the valleys and staying there. Thus in snowy mountains we find no special longevity: not in the Alps, Pyrenees, or Apennines; but in the foothills or valleys men live longer. However, on mountain tops reaching down towards Ethiopia and Abyssinia where, situated among the sands, little or no vapour settles on the mountains, they live a very long time, and even today quite often live to 150. $^|$

27. Marshes, and flat expanses of them, are good for those born there but bad for incomers as far as length and shortness of life is concerned. And, strangely enough, marshes regularly flooded by salt water are less healthy than freshwater ones.

28. The particular regions remarkable for longevity are Arcadia, Ætolia, India on this side of the Ganges, Brazil, Taprobane, Britain, Ireland, and the Orkneys and Hebrides. As for the remark of one of the ancients that the Ethiopians were long-lived, it is of no substance.

29. The wholesomeness of the air, especially when it is more perfect, is a tricky matter and $^|$ better sorted out by experiment than by talk or guesswork. An experiment can be tried with a hank of wool to see whether it loses weight when put out in the air for some days; or with a piece of meat lasting longer without going off; or with a calendar glass not rising and falling by much. Look into these and the like more closely.

30. The equality of the air, and not just its goodness or purity, is a factor in longevity. Variety of hill and vale, though pleasant to look at and to sense, is untrustworthy from the point of view of longevity. But flat land moderately dry, yet not entirely barren or sandy, nor quite without trees or shade, $^|$ is more conducive to long life.

31. As I just said, inequality of the air is bad in your place of habitation; but a change of air in travelling is, once you have got used to it, a good thing, and for that reason great travellers have been long-lived. Likewise too men who have spent their whole lives in their cottages in the same place are long-lived, for air to which you are accustomed consumes the body less, whereas a change nourishes and repairs it more.

32. As I have already said, the number and succession of generations have nothing to do with length or shortness of life, but the immediate condition of the parents, whether on the father's or the mother's side, is

[L6ᵛ] siquidem ge|nerantur ex *Senibus*, Alii ex *Adolescentulis*, Alii ex *Viris*
Ætate iustiore; Itèm Alii à *Patribus*, cum *Sani* fuerint, & benè dispositi,
Alii à *Morbidis*, & *Languidis*; Item Alii à *Repletis*, & *Ebrijs*, Alii post
*Somnum*, & *Horis Matutinis*; Item Alii post longam *Intermissionem*
5 *Veneris*, Alii post *Venerem repetitam*; Item Alii Flagrante Amore Patrum
(vt fit plerunque in *Spurijs*), Alii deferuescente, vt in *Coniugijs Diu-*
*turnis*. Eadem etiam ex parte *Matris* spectantur: Quibus addi debent,
*Conditio Matris*, dum gestat Vterum, quali sanitate, quali diætâ? Et
9 *Tempus Gestationis*, ad Decimum Mensem, aut celeriùs? Hæc ad Nor-
[L7ʳ] mam reducere, quatenùs ad Longæui|tatem difficile est; Atque eò dif-
ficilius, quod fortassè quæ optima quis putaret, in contrarium cedent;
Etenim *Alacritas* illa in *Generatione*, quæ Liberos Corpore robustos, et
agiles producit, ad Longæuitatem minùs vtilis erit, propter *Acrimoniam*,
et *Incensionem Spirituum*. Diximus anteà, *Plus habere ex Materno San-*
15 *guine*, conferre ad Longæuitatem; Etiam *Mediocria* simili ratione
optima esse putamus, *Amorem* potiùs *Coniugalem*, quam *Meretricium*;
*Horas Generationis matutinas*; *Statum Corporis* non nimis *Alacrem*, aut
*Turgidum*, et Similia. Illud etiam benè obseruari debet, quod *Habitus*
[L7ᵛ] *Parentum robustior*, ipsis magis est propitius, | quàm Fætui; Præcipuè in
20 Matre. Itaque satis imperitè *Plato* existimauit, Claudicare virtutem
Generationum, quod Mulieres similibus cum Viris Exercitiis tam
Animi, quam Corporis, non vtantur; Illud contrà se habet: Distantia
enim Virtutis inter Marem, et Fœminam, maximè vtilis est Fœtui;
Atque Fœminæ teneriores magis Præbitoriæ sunt, ad alendum fœtum,
25 quod etiam in Nutricibus tenet. Neque enim *Spartanæ Mulieres*, quæ
ante Annum Vicesimum secundum, aut (vt alii dicunt) Quintum,
nubere non solebant (ideóque *Andromanæ* vocabantur) generosiorem,
[L8ʳ] aut longæuiorem Sobolem edide|runt, quam *Romanæ*, aut *Athenienses*,
aut *Thebanæ*, apud quas, Anni Duodecim, aut Quatuordecim Nubiles
30 erant. Atque si in *Spartanis* aliquid fuerit Egregium, id magis Victûs
Parsimoniæ debebatur, quam Nuptiis Mulierum serotinis. Illud verò
Experientia docet, Esse quasdam *Stirpes*, ad Tempus longæuas, vt
Longæuitas sit, quemadmodùm Morbi, *Res Hæreditaria*, in aliquibus
periodis.

35 **33.** *Candidiores Genis, Cute, & Capillis*, minùs Viuaces; *Subnigri*, aut
Rufi, aut *Lentiginosi* magis. Etiam *Rubor* nimius in *Iuuentute*, Longæui-
[L8ᵛ] tatem minùs promittit, quam *Pallor*. *Cutis durior* | Longæuitatis signum,

doubtless very influential. For some ǀ are begotten by old men, and some by very young ones, and others by men in their prime. Some again are begotten by fathers who are healthy and well set up, others by the sick and enfeebled; others by men full of food and wine; others after sleep or at morn; some after a long break from sex, some after regular sex, some in their fathers' flagrant lust (as is often the case with bastards), others when long marriage has cooled it. The same things affect the mother's part, to which we should add her condition, health, and diet while she is pregnant, as well as the length of pregnancy, be it up to ten months or less. It is difficult to reduce this to a rule ǀ as far as longevity is concerned, and the more so because what one might think best, may perhaps turn out to be the opposite. For that enthusiasm for breeding which begets fit, bold, and energetic children may not be so good for long life because of the acrimony and inflammation of the spirits. As I said before, to have more of the mother's blood is good for longevity; and for the same reason I think that moderation is also best: married love rather than meretricious; making love in the morning; being in a bodily state not too lustful or inflamed; and so on. It should be well noted too that fitness in parents is better for them ǀ than their offspring, especially in the mother's case. Therefore Plato was ignorant enough when he thought that the capacity to conceive was inhibited because women did not pursue the same intellectual and bodily exercises as men. But the opposite is true: a big difference between the powers of man and woman is most beneficial to their offspring, and the more delicate the mother the better she is for supplying nourishment to the foetus, as is the case with nurses too. For Spartan women, who were not accustomed to marry before they were 22 or (as others say) 25 (and so were called Andromanæ), did not produce more or longer-lived ǀ children than the Romans, Athenians, or Thebans, who were marriageable at 12 or 14. And if there were anything special about the Spartans, it should be put down more to their austere way of life than to the women marrying late. But experience teaches that some stocks are long-lived for a time, so that longevity like illnesses runs for certain periods in families.

33. Lighter cheeks, skin, and hair betoken shorter life; dark or red or freckly longer. Too much redness in youth is not so propitious as pallor for a long life. A sign of longevity ǀ is a firmer skin rather than a softer,

potiùs quam *mollior*; neque tamen hoc intelligitur de *Cute Spissiori* (quam vocant *Anserinam*) quæ est tanquam Spongiosa; sed de Durâ simul, & Compactâ; Quin & *Frons* maioribus Rugis sulcatus, melius signum, quam Nitidus, & Explicatus.

5　34. *Pili* in *Capite* Asperiores, & magis setosi, ostendunt vitam longiorem, quam Molles, & delicati; Crispi verò eandem prænunciant, si sint simul Asperi; contrà si sint Molles, & splendentes. Item si sit Crispatio potiùs Densa, quam per largiores Cincinnos.

[Mɪ<sup>r</sup>]　35. Citiùs aut Seriùs *Caluescere*, Res est quasi indifferens; cum
10　Caluastri plurimi longæui fuerint; Etiam citò *Canescere* (vtcunque videatur *Canicies* præcursor ingruentis Senectutis) Res fallax est; cùm haud pauci præproperè *Canescentes*, diù posteà vixerint; Quinetiam præmatura *Canicies*, absque vllâ *Caluitie*, signum est Longæuitatis; Contrà, si concomitetur Caluities.

15　36. *Pilositas* Partium Superiorum signum Vitæ minùs longæ; atque Pectore hirsuti, & quasi iubati, minùs Viuaces; at Inferiorum *Pilositas*, vt *Femorum*, *Tibiarum*, signum longæ Vitæ.

37. *Proceritas Staturæ* (nisi fuerit enormis) Compage commodâ, &
[Mɪ<sup>v</sup>]　sine Gracilitate, præsertìm si concomitetur *Corporis Agilitas*, | signum
20　longæ Vitæ; At contrà, Homines breuioris Staturæ magis viuaces, si fuerint minùs Agiles, & Motu tardiores.

38. In *Corporis Analogiâ*: Qui Corpore aliquantò breuiores sunt, Tibiis longioribus, longæuiores sunt, quàm qui Corpore magis demisso, Tibiis autem breuioribus; Item, qui *Inferioribus Partibus* largiores sunt,
25　& *Superioribus* contractiores (Structurâ Corporis, quasi surgente in Acutum) longæuiores, quàm qui Humeros lati, deorsùm sunt tanquam attenuati.

39. *Macies* cum *Affectibus Sedatis*, tranquillis, & facilibus; *Pinguior*
[M2<sup>r</sup>]　autem Habitus cum *Cholerâ*, vehementiâ, & pertinaciâ, Diutur|nitatem
30　Vitæ significant. *Obesitas* autem in *Iuuentute*, breuiorem Vitam præmonstrat, in *Senectute*, Res est magis indifferens.

40. Diù & sensìm *Grandescere*, signum Vitæ longæ; si ad *Staturam magnam*, magnum signum; sin ad *minorem*, signum tamen; At contrà velocitèr *Grandescere* ad *Staturam magnam*, signum malum est; sin ad
35　*staturam breuem*, minùs malum.

41. *Carnes Firmiores*, & Corpus Musculosum, & Neruosum, & Nates minùs tumentes (quantum Sedendo tantùm sufficiant) & Venæ

but by this I do not mean a thick and spongy skin (like, as they say, goose-skinned) but at once hard and compact. Moreover, a furrowed brow is a better sign than a smooth and clear one.

34. Rougher, more bristly hair on the head suggests a longer life than soft and delicate. Curly hair if it is also rough presages the same, but the opposite if it is soft and shiny; and tight curls are better than loose ringlets.

35. Early or late balding is a matter of no account, as ¦ many bald men have lived long. Early greying (though they seem to be the forerunners of old age) is deceptive, for many people prematurely grey have lived long afterwards. Indeed, premature greying is without baldness a sign of longevity; with baldness, the opposite.

36. Hairiness of the upper parts of the body is a sign of a shorter life, so too is a chest hairy like a mane; but hairiness lower down, as on thighs and calves, is a sign of longevity.

37. Tallness (unless it is extreme), suitable compactness without thinness, especially if it is accompanied by bodily agility, ¦ is a sign of longevity. On the other hand, short men live longer if they are less agile and more sluggish.

38. In bodily proportion, those with short bodies and long legs live longer than those with long bodies and short legs. Likewise, those who are broader below and narrower above (the fabric of the body rising as it were to a point) live longer than those with broad shoulders but narrow below.

39. Leanness with the passions settled, calm, and easy to handle; and a plump habit together with a short, strong, and stubborn temper signify ¦ long life. Fatness in youth presages a short life, but in old age it signifies nothing.

40. Long and gradual growing up is a sign of longevity, and the taller the better the sign; but to a shorter height is still a good sign. On the other hand, quick growth to considerable height is a bad sign, but less bad to smaller stature.

41. Firmer flesh, a muscular and stringy body, and buttocks no

paulò eminentiores, Longæuitatem denotant; contraria Breuitatem Vitæ.

[M2ᵛ] 42. *Caput*, pro Analogiâ Corporis, | minutius; *Collum* mediocre, non oblongum, aut gracile, aut tumidum, aut tanquam Humeris impactum; 5 *Nares* patulæ, quâcunque forma Nasi; *Os* largius; *Auris* cartilaginea, non Carnosa; *Dentes* robusti, & contigui, non exiles, aut rari; Longæuitatem prænunciant; Et multò magis, si *Dentes* aliqui noui, Prouectiore Ætate proueniant.

43. *Pectus* latius, sed non eleuatum, quin potius adductius; 10 *Humerique* aliquantulùm gibbi, & (vt loquuntur) *Fornicati*; *Venter* planus, nec prominens; *Manus* largior, & *Palma* minùs Lineis exarata; *Pes* breuior, & rotundior; *Femora* minùs carnosa; *Suræ* non cadentes, sed [M3ʳ] se | altiùs sustentantes; Signa Longæuitatis.

44. *Oculi* paulò grandiores, atque *Iris* ipsorum cum quodam Virore; 15 *Sensus* omnes non nimis acuti; *Pulsus* Iuuentute tardior, sub Ætatem vergentem paulò incitatior; *Detentio Anhelitûs* facilior, & in plura Momenta; *Aluus* Iuuentute siccior, Vergente Ætate humidior; signa etiam Longæuitatis.

45. De *Temporibus Natiuitatis*, nihil obseruatum est, quoad Lon- 20 gæuitatem, memoratu dignum, præter *Astrologica*, quæ in *Topicis* rele- gauimus. *Partus Octimestris*, non solùm pro non Viuaci, verum etiam [M3ᵛ] pro non Vitali habe|tur. Etiam *Partus Hyemales* habentur pro Longæuioribus.

46. *Victus* siue *Diæta Pythagorica*, aut *Monastica*, secundum Regulas 25 strictiores, aut ad Amussim æqualis (qualis fuit illa *Cornari*) videtur potentèr facere ad Vitæ Longitudinem. At contrà ex ijs, qui liberè, & communi more viuunt, Longæuiores reperti sunt sæpenumerò *Edaces*, & *Epulones*, denique qui liberaliore Mensâ vsi sunt. *Media Diæta*, quæ habetur pro Temperatâ, laudatur, & ad *Sanitatem* confert, ad *vitam* 30 *Longæuam* parùm potest; Etenim, Diæta illa *Strictior*, *Spiritus* progignit [M4ʳ] paucos, & lentos, vndè minùs consumit; at illa | *Plenior*, Alimentum præbet copiosum, vnde magis reparat; *Media* neutrum præstat; Vbi enim Extrema nociua sunt, Medium optimum, verùm vbi Extrema iuuatiua, Medium nihili ferè est. *Diætæ* autem illi *strictiori*, conuenit 35 etiam *Vigilia*, ne Spiritus pauci, multo Somno opprimantur; *Exercitatio*

---

4 impactum;] ~:    5 Nasi;] ~:    largius;] ~:    6 rari;] ~,    9 adductius;] ~:
10 *Fornicati*;] ~:    11 prominens;] ~:    exarata;] ~:    12 rotundior;] ~,
13 sustentantes;] ~,    15 acuti;] ~:    17 humidior;] ~,    25 æqualis] ~,

bigger than you need for sitting, and rather prominent veins, indicate long life; the opposite a short one.

42. A head small in relation to the body; | an average neck not too long or thin, or too thick and set into the shoulders; broad nostrils, whatever the shape of the nose; a large mouth; gristly rather than fleshy ears; strong teeth set together, and not weak or gappy: all these presage long life, and the more so if new teeth come later in life.

43. A broad breast not uplifted but rather pulled in; shoulders rather rounded and (as they say) arched; a flat belly not at all protuberant; a large hand with few lines furrowing the palm; shorter, rounder feet; thighs not too fleshy; calves not sagging but | firmly braced up: all are signs of long life.

44. Eyes on the large side, with irises fresh; senses not too acute; a slow pulse in youth, speeding up with age; easy and long holding of breath; bowels drier in youth but moister in age; these are signs of longevity.

45. As far as longevity is concerned, nothing has been observed worth remembering about times of nativity, besides the astrological one which I put aside in the topics. An eighth-month birth is taken to be not only short-lived but unlikely to live. | Winter births are reckoned to be long-lived.

46. A Pythagorean or monastic diet or regimen, according to the stricter rules, or one more evenly precise (as Cornaro's was), seems to be extremely good for promoting long life. But on the other hand, of those who live freely and in the normal way, we often find that the gluttons, feasters, and lovers of a good spread are often long-lived. A moderate diet of the kind taken for temperate is praiseworthy and contributes to health, but does virtually nothing for longevity. For the more rigorous diet generates spirits scanty and sluggish, whence it consumes less; but the ampler | diet furnishes abundant nourishment, whence it supplies more repair; the moderate diet does neither. For where the extremes are harmful, the middle way is best; but where the extremes are good for you, the middle way is practically useless. For strict diets need vigilance, lest the scanty spirits be oppressed by too much sleep; slight exercise lest they be unloosed; and sexual abstinence lest they be exhausted. But a

item modica, ne exoluantur; *Veneris Abstinentia,* ne exhauriantur; At *Diætæ* Vberiori conuenit contrà *Somnus* largior, *Exercitatio* frequentior, *Vsus Veneris* tempestiuus. *Balnea,* & *Vnguenta* (qualia fuerunt in vsu)

4 Delicijs potiùs, quam Vitæ producendæ, accommodata fuerunt; Verùm

[M4ᵛ] de his omnibus, cum ad *Inquisitionem* | secundum *Intentiones* ventum erit, accuratiùs dicemus. Illud interim *Celsi Medici,* non solum Docti, verum etiam Prudentis, non contemnendum est; Qui *Varietatem* & *Alternationem Diætæ* iubet, sed cum inclinatione in *Partem Benigniorem:* Scilicet, vt quis *Vigilijs* quandóque se assuescat, aliàs *Somno* indulgeat,

10 sed *Somno* sæpiùs; Itidem, interdùm *ieiunet,* interdùm *epuletur,* sed *epuletur* sæpiùs; Interdùm *Animi Contentionibus* strenuè incumbat, interdùm *Remissionibus* vtatur, sed *Remissionibus* sæpiùs. Illud certè minimè dubium est, quin *Diæta* benè instituta, partes, ad prolongan-

[M5ʳ] dam Vitam, potiores teneat; neque conueni vnquam Aliquem valdè

15 Longæuum, qui interrogatus de *Victu* suo, non obseruasset, aliquid Peculiare; Alij alia. Equidem memini quendam *Senem* Centenario Maiorem, qui productus est Testis, de antiquâ quadam Præscriptione; Is cùm finito Testimonio, à *Iudice* familiaritèr interrogaretur, quid agens tam diù vixisset; respondit (præter expectatum, & cum Risu

20 Audientium) *Edendo, antequam esurirem, & Potando, antequàm sitirem.* Sed de his (vt dictum est) posteà.

**47.** *Vita Religiosa,* & in sacris, videtur ad Longæuitatem facere;

[M5ᵛ] Sunt in hoc Genere Vitæ, Otium; Admiratio & Contemplatio rerum diuinarum; Gaudia non Sensualia; Spes nobiles; Metus salubres;

25 Mœrores dulces; denique Renouationes continuæ per Obseruantias, Pœnitentias, & Expiationes; quæ omnia ad Diuturnitatem Vitæ, potentèr faciunt. Quibus si accedat, *Diæta* illa Austera, quæ Massam Corporis induret, Spiritus humiliet, nil mirum si sequatur Longæuitas insignis; qualis fuit *Pauli Eremitæ, Simeonis Stilitæ Anachoretæ Columnaris,* Et

30 complurium aliorum *Monachorum* ex Eremo, & *Anachoretarum.*

**48.** Huic proxima est *Vita in Literis, Philosophorum, Rhetorum,* &

[M6ʳ] *Grammaticorum;* Degitur hic quoque in Otio, & in ijs Cogitationibus, quæ cùm ad Negotia Vitæ nihil pertineant, non mordent, sed Varietate, & impertinentiâ, delectant. Viuunt etiam ad arbitrium suum, in quibus

35 maximè placeat, Horas & Tempus terentes; Atque in consortio plerúnque Adolescentium, quod paulò lætius est. In *Philosophijs* autem magna est Discrepantia, quoad Longæuitatem, inter *Sectas.* Etenim *Philosophiæ,* quæ nonnihil habent ex Superstitione, &

---

1 exhauriantur;] ~:    21 his] ~,    34 delectant.] ~;

more generous diet needs the opposite: plenty of sleep, lots of exercise, and sex at the right time. Baths and ointments (as have been in use) are better adapted to pleasure than prolonging life. But I shall say more about all these things ˡ when I come to the inquiry according to the intentions. In the meantime, we should not despise the advice of Celsus, a physician who was not just learned but practical, who advises variety and alternation of diet, but with a leaning towards the generous, i.e. that one should get used to wakefulness sometimes, at others to enjoy sleep but more often; sometimes starve, sometimes feast, but feast more often; sometimes weigh oneself down with worry, at other times to let oneself off, but the latter more often. But this much is certain: that a well-established diet does much to prolong life; and I have never ˡ met a very old man who, when asked about his diet, had not observed some practice peculiar to him, some one thing, others something else. For my part I remember an old man, a hundred and more years old who, when produced as a witness to some ancient prescription, and after giving his evidence, was casually asked by the judge how he had managed to live so long, replied unexpectedly and to the delight of the listeners, *by eating before I was hungry and drinking before I was dry*. But more of this (as I said) later.

47. The religious life spent in sacred duties seems conducive to longevity. In this kind of life there is ˡ leisure, wonder, and contemplation of divine things, joy not of the senses, noble hopes, salutary awe, sweet sadness, and lastly continual renewals by observances, penances, and expiations—all of which powerfully contribute to the prolongation of life. And if one adds to this a harsh diet to harden the mass of the body, and humble the spirits, it is no wonder that very long life is a consequence, as was the case with Paul the Hermit, Simeon Stylites the columnar anchorite, and many other hermit monks and anchorites.

48. Next to this comes the life of letters, of philosophers, rhetoricians, and grammarians. Here too days pass ˡ in leisure, and in thoughts which, since they have nothing to do with the business of life, do not distress but delight them with their variety and irrelevance. They also live as they please, and spend their time doing what they love, and often in the company of youth, which is pretty pleasant. But in philosophers there are great differences in longevity between the different sects. For the best philosophies for long life are those with a degree of superstition

Contemplationibus sublimibus, optimæ, vt *Pythagorica, Platonica*;
Etiam quæ Mundi Perambulationem, & Rerum Naturalium varietatem
[M6ʳ] complectebantur, & Cogitatioʲnes habebant discinctas, & altas, &
magnanimas (de *Infinito*, & de *Astris*, & de *Virtutibus Heroicis*, &
5 huiusmodi) ad Longæuitatem bonæ; quales fuerunt *Democriti, Philolai,
Xenophanis, Astrologorum & Stoicorum*; Etiam quæ nihil habebant Specu-
lationis Profundioris, sed ex Sensu Communi, & Opinionibus vulga-
tis, absque Inquisitione acriori, in omnem partem placidè disputabant,
similiter bonæ; quales fuerunt *Carneadis*, & *Academicorum*; itèm *Rheto-
10 rum*, & *Grammaticorum*. At contrà, *Philosophiæ* in Subtilitatum
Molestijs versantes, & *Pronunciatiuæ*, & singula ad Principiorum Truti-
[M7ʳ] nam examinantes, & torquenʲtes, denique Spinosiores; & Angustiores,
malæ; quales fuerunt plerúnque *Peripateticorum*, & *Scholasticorum*.

49. *Vita Rusticana* item ad Longæuitatem idonea; Frequens est sub
15 Dîo, & Aere libero; non Socors, sed in Motu; Dapibus plerúnq*ue*
Recentibus, & inemptis; sine Curis & Inuidiâ.

50. De *Vitâ Militari*, in Iuuentute, etiam bonam habemus Opi-
nionem; Certè complures *Bellatores* egregii longæui fuerunt: *Coruinus,
Camillus, Xenophon, Agesilaus*, & Alii tam Prisci, quam Moderni.
20 Prodest certè Longæuitati, si à Iuuentute, ad Ætatem Prouectam, omnia
[M7ᵛ] crescant in *Benignius*, vt Iuuentus laboriosa, ⁱ Dulcedinem quandam
Senectuti largiatur. Existimamus etiam *Affectus Militares*, ad Conten-
tionis studium, & spem Victoriæ, erectos, talem infundere Calorem
*Spiritibus*, qui Longæuitati prosit.

25 MEDICINÆ AD
Longæuitatem.

*Ad Artic.* 10. *Connexio*. Medicina *quæ habetur, intuetur ferè
tantùm* Conseruationem Sanitatis, *&* Curationem Morborum;
[M8ʳ] *de ijs autem,* ⁱ *quæ propriè spectant ad* Longæuitatem, *parua est
30 Mentio, & tanquam obitèr*. Proponemus *tamen ea* Medicamenta,
*quæ in hoc genere notantur*, Cordialia *scilicet quæ vocantur*. Etenim
*quæ sumpta in* Curationibus, Cor, *& (quod verius est)* Spiritus
*muniunt, & roborant, contra* Venena, *&* Morbos, *translata cum
Iudicio & Delectu in* Diætam, *etiam ad Vitam producendam,*

---

1 Contemplationibus] Comtemplationibus    *Platonica*;] ~:        4 magnanimas] ~,
6 *Stoicorum*;] ~:                10 *Grammaticorum*.] ~:          12 Spinosiores;] ~,
13 fuerunt] ~,      18 fuerunt:] ~;      19 Moderni.] ~:      22 largiatur.] ~:

or sublime contemplations in them, as for example the Pythagorean and Platonic. Good too for longevity are those which embraced the whole sweep of nature and the variety of natural things, and entertained high, ⏐ untrammelled, and great thoughts (of the infinite, the stars, the heroic virtues, and so on); such as were those of Democritus, Philolaus, Xenophanes, the astrologers, and the Stoics. And so too were those which entertained no deep speculations but which from common sense, vulgar opinion, and without any penetrating inquiry calmly discussed all sides of a question; such as were those of Carneades, and the Academics, as well as the rhetoricians and grammarians. By contrast, philosophies dealing with vexatious subtleties, given to laying down the law, and reducing or twisting single instances to demands of general principles, ⏐ and lastly the prickly and narrow ones, are bad; such as were in the main those of the Peripatetics and Scholastics.

**49.** A country life is also conducive to long life. It is often out in the open and the fresh air; not sluggish but energetic; with food generally fresh and home-made; and a life without care or envy.

**50.** I have a good opinion of military life in youth. Certainly many famous fighters have been long-lived, as were Corvinus, Camillus, Xenophon, Agesilaus, and others both ancient and modern. And certainly it is good for longevity if in the passage from youth to age everything grows more prosperous, so that youth spent in hardship ⏐ grants a certain sweetness to old age. I think too that military feelings, directed to the fight and hope of victory, infuse such a heat into the spirits as is conducive to long life.

## Medicines for Prolonging
## Life

*To Article* 10. *Connection.* Medicine as it is now is directed only towards the preservation of health and the treatment of illnesses; but of those ⏐ which properly refer to longevity, there is scant mention, and that only in passing. I shall nevertheless present those medicines which are notable in the latter category, namely those they call cordials. For those that are taken as treatments to fortify and strengthen the heart or (to put it more accurately) the spirits against poisons and disease could, when carried over with judgement and discretion into the diet, also help up to a point to

[M8ᵛ] *aliquâ ex parte, prodesse* | *posse, consentaneum est. Id faciemus non promiscuè ea cumulantes* (*vt moris est*) *sed excerpentes Optima.*

**1.** *Aurum* triplici Formâ exhibetur; Aut in *Auro* (quod appellant) *Potabili*; aut in *Vino Extinctionis Auri*; aut in *Auro in Substantiâ*; qualia sunt, *Aurum Foliatum*, & *Limatura Auri*. Quod ad *Aurum Potabile* attinet, cœpit dari in Morbis desperatis, aut grauioribus, pro egregio Cordiali, atque successu non contemnendo. Verùm existimamus *Spiritus Salis*, per quos fit Dissolutio, Virtutem illam, quæ reperitur, largiri [Nᵣ] potiùs, quàm ipsum *Aurum*; Quod tamen sedulò celatur. | Quòd si aperiri possit *Aurum* absque Aquis Corrosiuis, aut per Corrosiuas (modò absit Qualitas Venenata) benè posteà ablutas, Rem non inutilem fore arbitramur.

**2.** *Margaritæ* sumuntur aut in *Puluere læuigato*, aut in *Malagmate* quodam siue Dissolutione, per *Succum Limonum* impensè acerborum, & recentium; atque dantur aliquandò in *Confectionibus Aromaticis*, aliquandò in *Liquore*. *Margarita* proculdubiò affinitatem habet cum *Conchâ*, cui adhæret; & possit esse similis ferè Qualitatis, cum *Testis Cancrorum Fluuiatilium*.

[Nᵣ] **3.** Inter *Gemmas Crystallinas* habentur pro Cordialibus præci|puè duæ: *Smaragdus* & *Hyacinthus*; quæ dantur sub ijsdem Formis, quibus *Margaritæ*, excepto quod Dissolutiones earum (quod scimus) non sint in vsu. Verùm nobis magis suspectæ sunt *Gemmæ* illæ *Vitreæ*, ob Asperitatem.

*Monitum. De his, quæ memorauimus, quatenùs & quomodò Iuuamentum præbeant, posteà dicetur.*

**4.** *Lapis Bezoar* probatæ est Virtutis; quòd Spiritus recreet, & lenem Sudorem prouocet. *Cornu* autem *Monocerotis*, de Existimatione suâ decidit; ita tamen vt gradum seruet, cum *Cornu Cerui*, & *Osse* de *corde Cerui*, & *Ebore*, & similibus.

**5.** *Ambra Grisia* ex optimis est, ad Spiritus demulcendos, & con- [N₂ʳ] |fortandos. Sequuntur *Nomina* tantum *Simplicium*, cùm Virtutes ipsarum satis sint cognitæ.

| Calida. | Frigida. |
| --- | --- |
| *Crocus.* | *Nitrum.* |
| *Folium Indum.* | *Rosa.* |
| *Lignum Aloes.* | *Viola.* |

2 *cumulantes*] ~,     9 celatur.] ~;     10 Corrosiuas] ~,     20 duæ:] ~;

prolong | life. And this I shall do not (as the custom is) by piling them up indiscriminately but by picking out the best of the bunch.

1. Gold shows itself in three forms: in what they call potable gold; in wine in which gold has been quenched; or in its substance, as in gold leaf or filings. As for potable gold, it is given as a first-rate cordial in grave or desperate illnesses with no little success. But I think that rather than the gold the spirits of salt which bring the dissolution about confer the virtue we find in it—a fact which they are careful to keep in the dark. | But if gold could be opened without corrosive waters, or by corrosives (provided they lack any poisonous quality) well washed afterwards, I judge that the effort would be worthwhile.

2. Pearls are taken either as a fine powder or in a certain amalgam or solution with very sharp, fresh lemon juice. Now they are sometimes given in aromatic confections, sometimes in a liquor. The pearl is no doubt akin to the shell in which it sticks, and its quality may be almost the same as the shells of crayfish.

3. Among the crystalline gems two in particular are regarded | as cordials: emerald and jacinth, which are given in the same forms as pearl, except that their dissolutions (as far as I know) are not in use. But for me glassy gems should be regarded with suspicion on account of their harshness.

*Advice.* These things which I record here I shall return to later on the question of how far and in what ways they are helpful.

4. Bezoar stone has a tried and tested virtue for recreating the spirit and stimulating a light sweat. The unicorn's horn has fallen out of favour, yet not so far that it ranks beneath stag's horn, the bone of a stag's heart, ivory, and the like.

5. Ambergris is one of the best things for calming and comforting | the spirits. Here follows a list of simples, by bare name only as their virtues are well enough known.

| Hot | Cold |
|---|---|
| Saffron | Nitre |
| Folium Indum | Rose |
| Lignum Aloes | Violet |

| | |
|---|---|
| Cortex Citri. | Fragaria. |
| Melissa. | Fraga. |
| Ocymum. | Succus Limonum dulcium. |
| Gariophyllata. | Succus Arantiorum dulcium. |
| 5  Flores Arantiorum. | Succus Pomorum Fragrantium. |
| Rosmarinus. | Borago. |
| Menta. | Buglossa. |
| Betonica. | Pimpinella. |
| 9  Carduus Benedictus. | Santalum. |
| [N2ᵛ] | Camphora. |

***Monitum.*** *Cum de ijs iam Sermo sit, quæ in* Diætam *transferri possint;* Aquæ *illæ* Ardentiores, *atque* Olea Chymica (*quæ vt ait Quidam ex* Nugatoribus, *sunt sub* Planetâ Martis, *& habent Vim furiosam, & destructiuam*), *quinetiam* Aromata *ipsa Acria & Mordacia, reijcienda* 15 *sunt; & videndum, quomodò componi possint* Aquæ, *&* Liquores *ex Præcedentibus, non* Phlegmaticæ *illæ* Stillaticiæ; *neque rursùs Ardentes ex* Spiritu vini, *sed magis Temperatæ, & nihilominùs viuæ, & Vaporem benignum spirantes.*

19     6. Hæsitamus de *Frequenti Sanguinis Missione,* vtrùm ad Longæui-
[N3ʳ] tatem conferat; & potiùs in eâ sumus Opinione, quod hoc faciat, si in Habitum versa fuerit, & cætera sint accommodata; Etenim Succum Corporis Veterem emittit, & Nouum inducit.

    7. Arbitramur etiam *Morbos* quosdam *Emaciantes,* benè curatos, ad Longæuitatem prodesse; Succos enim Nouos præbent Veteribus con-
25 sumptis; atque (vt ait ille) *Conualescere est Iuuenescere.* Itaque inducendi sunt tanquam *Morbi* quidam *Artificiales,* id quod fit per *Diætas strictas,* & *Emaciantes,* de quibus posteà dicemus.

[N3ᵛ: blank]

[N4ʳ]                              Intentiones.

*Ad Artic.* 12. 13. *& 14. Connexio. Postquàm autem* Inquisitionem
30 *absoluerimus secundum subiecta, Videlicet* Corporum Inanimato-
rum, Vegetabilium, Animalium, Hominis; *propiùs accedemus, &* Inquisitionem *per* Intentiones *ordinabimus; Veras & proprias (vt omninò arbitramur) quæque sint tanquam Semitæ Vitæ Mortalis.*
[N4ᵛ] *Neque enim in hâc Parte, quicquam quod va⏐leat, hactenùs*

---

12 Chymica] ~,      14 *destructiuam*),] ~)ₐ      16 Stillaticiæ] / for alternative spelling
(stillatitia) see P1ʳ below      25 *Iuuenescere.*] ~;      32 *proprias*] ~,

236

| | |
|---|---|
| Citron peel | Strawberry plants |
| Balm | Strawberry |
| Basil | Sweet lemon juice |
| Gillyflowers | Sweet orange juice |
| Orange blossom | Fragrant apple juice |
| Rosemary | Borage |
| Mint | Bugloss |
| Betony | Burnet |
| Blessed Thistle | Sandal-wood |
| | Camphor |

*Advice.* Since the present discussion is concerned with remedies which may be carried over into diet, strong waters and chemical oils (which, as one of the triflers put it, are under the planet Mars, and have a furious and destructive force) and indeed all sharp and biting spices are to be rejected; and we must see how waters and liquids can be compounded from the aforementioned, not those phlegmatic distillates, nor again burning waters from spirit of wine, but ones more temperate, yet lively, and giving off a beneficial vapour.

6. I am not sure whether frequent blood-letting contributes to longevity; and I am rather of the view that it does so, if it has been made a habit of, other things being equal. For it rids the body of old juice and encourages new.

7. I judge too that some wasting diseases well cured promote long life, for the old juices being consumed, they supply new ones; and (as the man says) *To recover health is to recover youth.* We should bring on certain artificial (as it were) illnesses, as is done through strict and slimming diets—as I shall explain later.

## Intentions

*To Articles* 12, 13, *& 14. Connection.* Now that I have covered the inquiry by subjects, i.e. inanimate bodies, vegetable bodies, animal, and human, I will go deeper, and organize the inquiry by intentions, such as are (as I really believe) true and specific and, so to speak, the trackways of mortal life. For in this connection nothing of any worth has so far been inquired into; but men's

*inquisitum est; sed planè fuerunt Hominum Contemplationes, quasi simplices, & non proficientes.* Nam *cùm audiamus ex vnâ parte Homines, de confortando* Calore Naturali, *atque* Humore Radicali; *atque de* Cibis, *qui generant Sanguinem laudabilem, quíque sit nec*
5  *Torridus, nec Phlegmaticus; atque de* Refocillatione, *&* Recreatione
[N5$^r$] Spirituum, *verba facientes; existimamus sanè, Homines non malos* |
*esse, qui hæc loquuntur. Sed nihil horum, potentèr facit ad Finem. Cùm verò ex alterâ parte, audiamus Sermones inferri de* Medicinis *ex* Auro (*quia scilicet* Aurum *Corruptioni est minimè obnoxium*) &
10  *de* Gemmis, *ad recreandos Spiritus, propter Proprietates occultas, & Clarorem suum; Quódq*ue *si possint detineri, & excipi in Vasibus,* Balsama, *&* Quintæ Essentiæ Animalium, *Superba*m *faceret Spem*
[N5$^r$] Immortalitatis; *Quòdque,* | Carnes Serpentum, *&* Ceruorum, *consensu quodam valeant ad Renouationem Vitæ, quia Alter mutat*
15  Spolia, *alter* Cornua (*debuerant autem* Carnes Aquilarum *adiungere, quia* Aquila *mutat* Rostrum); *Quódque quidam, cùm* Vnguentum *sub terrâ defossum reperisset, eóque se à Capite ad Pedes vsque vnxisset* (*exceptis* Plantis Pedum) *ex huiusmodi Vnctione, tre-*
[N6$^r$] *centos Annos vixisset absque morbo* (*præter* | *Tumores* Plantarum
20  Pedum); *Atque de* Artefio, *qui cùm* Spiritum *suum labascere sensisset,* Spiritum Adolescentis *cuiusdam Robusti, ad se traxisset, eumque indè exanimasset; sed ipse complures Annos, ex Alieno illo* Spiritu *vixisset; Et de* Horis Fortunatis, *secundum* Schemata Cœli,
24  *in quibus* Medicinæ *ad Vitam producendam, colligi, & componi*
[N6$^r$] *debent; Atque de* Sigillis Planetarum, *per quæ* Virtutes *cœlitùs ad* | *prolongationem vitæ haurire, & deducere possimus; & huiusmodi* Fabulosis, *&* Superstitiosis; *prorsùs miramur Homines ita Mente captos, vt ijs huiusmodi Res imponi possint. Denique, subit Humani generis Miseratio, quod tam duro Fato, obsideatur, inter Res inutiles,*
30  *& ineptas. Nostræ autem* Intentiones, *& Rem ipsam premere, & procùl esse à Commentis vanis, & credulis, confidimus; & tales, vt*
[N7$^r$] Rebus, *quæ illis* In|tentionibus *satisfaciant, à Posteris quamplurima, Intentionibus autem ipsis, non multum addi posse existimemus.*

*Sunt tamen Pauca, sed magni prorsùs Momenti, quorum Homines*
35  *præmonitos esse volumus.*

---

9 Auro] ~,  15 Cornua] ~;  16 Rostrum);] ~;)  18 *vnxisset*] ~,
19 *morbo*] ~,  20 Pedum);] ~;)

reflections upon it have been shallow and unskilful. For when on the one hand I hear men talk of comforting the natural heat and radical moisture, of foods which beget excellent blood which is neither too torrid nor too phlegmatic, and of rekindling and recreating the spirits, I truly think that these are not bad men | who speak in this way; it is just that none of these things get them anywhere. When on the other hand I hear discussions about medicines made from gold (because (I ask you!) gold is not liable to corruption), and of gems to recreate the spirits with their occult properties and their brightness; and that if balsams and quintessences of animals could be captured and bottled, there would be high hopes of immortality; and that | snake and deer meat by a certain consent give you a new lease of life because the one sheds its skin and the other its antlers (so why don't they add eagle meat since eagles shed their beaks?); and that someone who found an ointment buried in the earth, and smeared himself with it from head to foot (his soles excepted) lived then for 300 years without any ailment (besides | bunions on his soles); and that Artefius, when he felt his spirit failing, took the spirit of a strong youth to himself, and so finished him off, but himself lived for many years because of the other's spirit; and of propitious hours indicated by the configurations of the heavens during which medicines for prolonging life should be collected and compounded; and of planetary seals with which we can draw and bring down heavenly virtues for prolonging life; | and of suchlike superstitious fables; when I hear all this I am altogether astonished that men are dim-witted enough to be imposed upon by such stuff. Lastly, let us bewail the cruel fate of the human race that it is besieged by such useless and foolish ideas. But I am confident that my own intentions get to the heart of the matter, and are remote from futile and credulous fictions; and I think that they are such that while our successors may be able to add a great deal to the things which satisfy | those intentions, they will not be able to add much to the intentions themselves.

Nevertheless, there are a few things, but ones of very great consequence, which I would like men to be warned of in advance.

Primò, nos in hac Sententiâ sumus, vt existimemus Officia Vitæ,
esse Vitâ Ipsâ potiora. Itaque, si quid sit eiusmodi, quod Intentioni-
[N7$^v$] bus nostris, magis exactè respondere possit, ita tamen $^|$ vt Officia, &
Munia Vitæ, omninò impediat; quicquid huius generis sit, reijcimus:
5 Leuem fortasse aliquam Mentionem huiusmodi Rerum facimus, sed
minimè illis insistimus. Neque enim de vitâ aliquâ, in Speluncis,
vbi Radij & Tempestates cœli non penetrent, instar Antri Epi-
menidis; Aut de perpetuis Balneis, ex Liquoribus præparatis; Aut
[N8$^r$] de Superpellicijs, & Ceratis ita applicandis, vt Corpus perpe$^|$tuo sit,
10 tanquam in Capsulâ; Aut de Pigmentis Spissis, more Barbarorum
nonnullorum; Aut de Ordinatione Victûs & Diætæ accuratâ, quæ
solùm hoc videatur agere, & nihil aliud curare, quàm vt quis viuat
(qualis fuit Herodici apud Antiquos, & Cornari Veneti nostro
Sæculo, sed maiore cum Moderatione); Aut de huiusmodi Portentis,
15 Fastidijs, & Incommodis, sermonem aliquem serium, & diligentem
[N8$^v$] instituimus; Sed $^|$ ea afferimus Remedia, & Præcepta, ex quibus
Officia vitæ non deserantur, aut nimias excipiant Moras, &
Molestias.

Secundò, ex alterâ parte, Hominibus denunciamus, vt nugari
20 desinant; nec existiment tantum Opus, quantum est Naturæ potentem Cursum remorari, & retrouertere, posse Haustu aliquo
Matutino, aut vsu alicuius pretiosæ Medicinæ, ad Exitum perduci;
[O1$^r$] sed vt pro certo habeant, necesse $^|$ esse vt huiusmodi Opus, sit planè
Res operosa, & quæ ex compluribus Remedijs, atque eorum inter se
25 Connexione idoneâ, constet; Neque enim Quisquam ita stupidus esse
debet, vt credat id quod nunquàm est factum fieri posse, nisi per
Modos etiam nunquàm tentatos.

Tertiò diserstè profitemur, nonnulla ex ijs, quæ proponemus,
29 Experimento nobis non esse probata (neque enim hoc patitur
[O1$^v$] nostrum Genus Vitæ) sed tantùm $^|$ summâ (vt arbitramur) Ratione,
ex Principijs nostris, & Præsuppositis (quorum alia inserimus, alia
Mente seruamus) esse deriuata; & tanquam ex Rupe, aut Minerâ
ipsius Naturæ, excisa & effossa. Neque tamen Curam omisimus,
eamque prouidentem & sedulam, quin (quandoquidem de Corpore

---

In the first place, I take the view that the duties of life are more important than life pure and simple. Therefore if there is something that more precisely suits my intentions but yet ¹ in any way at all gets in the way of those duties and obligations, I cast it aside. I may make passing reference to things of this kind, but I do not at all insist on them. For of life in caves, like the cave of Epimenides, where the rays and perturbations of the heavens never penetrate; or of constant baths prepared with liquors; or of shirts or insulation so applied that the body ¹ is always, so to speak, packaged up; or of thick paints in the manner of some barbarians; or of exact regulation of diet or regimen which aims only at lengthening life and disregards all else (such as that of Herodicus with the ancients, and, with greater moderation, Cornaro of Venice in our time); or of suchlike abnormal, repulsive, or inconvenient practices—of none of these do I start on any serious or considered discussion. Instead, ¹ I put forward remedies and precepts which do not involve abandoning the offices of life or delay or disable them too much.

Secondly, and on the other side, I put men on notice to give up their irresponsibility, and stop imagining that so great a work as slowing up or turning back the mighty course of nature can be accomplished by some morning potion, or the use of some precious medicine; but to know for certain that a work ¹ of this kind must of necessity be hard labour, and embrace many remedies appropriately related to each other. For no one can be so stupid as to believe that what has never been achieved can be accomplished except by means never yet tried.

Thirdly, I expressly declare that I have not tested some of the points I shall make by experiment (for my way of life does not allow for that), but they are only ¹ derived with (as I judge) very good reason, from my principles and presuppositions (some of which I have inserted and others kept back), and are cut and dug from the very rock and veins of nature itself. Nor have I spared

*Humano agatur, quod (vt ait* Scriptura) *est supra Vestimentum) ea*
[O2^r] *proponamus Remedia, quæ sint tuta* | *saltem, si forte non fuerint*
*fructuosa.*

*Quartò, illud Homines ritè, & animaduertere, & distinguere*
5 *volumus; Non eadem semper, quæ ad* Vitam Sanam, *ad* Vitam
Longam *conferre. Sunt enim nonnulla, quæ ad* Spirituum
Alacritatem, *&* Functionum Robur, *& Vigorem, prosunt, quæ*
*tamen, de* Summâ Vitæ *detrahunt. Sunt & alia, quæ ad* Prolonga-
[O2^v] tionem Vitæ *plurimùm iuuant, sed tamen* | *non sunt absque Peri-*
10 *culo* valetudinis; *nisi per Accommodata quædam huic Rei occurratur,*
*de quibus tamen (prout Res postulat) Cautiones, & Monita exhibere*
*non prætermittemus.*

*Postremò, visum est nobis varia Remedia, secundùm singulas*
Intentiones, *proponere; Delectum verò Remediorum, atque*
15 *Ordinem ipsorum, in medio relinquere. Etenim ex ipsis, quæ* Con-
[O3^r] stitutionibus Corporum *diuersis, quæ* Gene|ribus Vitæ *varijs, quæ*
Ætatibus *singulis, maximè conueniant, quæque* Alia *post* alia
*sumenda sint, & quomodo* Praxis *vniuersa harum Rerum sit*
*instruenda, & regenda, exactè perscribere, & nimis longum foret,*
20 *neque idoneum est quod publicetur.*

Intentiones *in* Topicis *proposuimus tres.* Prohibitionem Consumptionis; Perfectionem Reparationis; *&* Renouationem
[O3^v] Veterationis. *Verùm, cùm* | *quæ dicentur, nihil minus sint, quam*
Verba, Intentiones *illas tres, ad decem* Operationes *deducemus.*
25  1. *Prima est Operatio super* Spiritus, *vt reuirescant.*
  2. *Secunda Operatio est, super* Exclusionem Aeris.
  3. *Tertia Operatio est, super* Sanguinem, *&* Calorem
Sanguificantem.
[O4^r]  4. *Quarta Operatio est, super* Succos Corporis. |
30  5. *Quinta Operatio est, super* Viscera, *ad* Extrusionem
Alimenti.
  6. *Sexta Operatio est, super* Partes Exteriores, *ad* Attractionem
Alimenti.

---

II  *tamen*] ~,     21  *tres.*] ~.

any pains (as I deal with the human body which is, as Scripture says, above raiment) but have been provident and diligent to propose those remedies which are at least harmless, ⏐ even if by chance they are not fruitful.

Fourthly, I would like men rightly to understand and discern that the same things that contribute to a healthy life do not always make for a long one. For there are some things which help the alacrity of the spirits, and the strength and vigour of the bodily functions but yet diminish life expectancy. Again, there are others which help to prolong life, yet are health risks ⏐ which have to be guarded against by appropriate measures. But of these (as the occasion demands) I will not neglect to display proper precautions and warnings.

Lastly, I have decided to set out various remedies, according to the individual intentions, but to leave open the choice and ordering of them. For exactly to prescribe the things best adapted to different bodily constitutions, to different modes ⏐ of life and different ages, as well as which ones should be taken after which others, and how their whole practice should be arranged and regulated, would take too long to accomplish, and not worth making public.

In the topics I set forth three intentions: the prohibition of consumption, the accomplishing of repair, and the renovation of what has grown old. But since ⏐ what I shall say by no means amounts to mere words, I shall convert the three intentions into ten operations.

1. The first is the operation on the spirits to recover their strength.
2. The second is the operation on the exclusion of air.
3. The third is the operation on the blood, and on the heat that makes it.
4. The fourth is the operation on the juices of the body. ⏐
5. The fifth is the operation on the entrails to force out the food.
6. The sixth is the operation on the external parts to attract the food.

7. *Septima Operatio est, super* Alimentu*m ipsum, ad* Insinuationem *eiusdem.*

8. *Octaua Operatio est, super* Actum vltimum Assimilationis.

[O4$^v$]  9. *Nona Operatio est, super* Intenerationem Partium,
5  *postquàm cœperint desiccari.*

10. *Decima Operatio est, super* Expurgationem Succi veteris, *&* Substitutione*m* Succi noui.

*Harum* Operationum *Primæ Quatuor pertinent ad* Intentionem *primam*; *Quatuor Proximæ, ad* Intentionem *secundam*; *Duæ*
10  *vltimæ, ad* Intentionem *Tertiam.*

*Cum vero hæc Pars de* Intentionibus, *ad* Praxin *innuat; sub* His-
[O5$^r$] toriæ *No*mine, *non solum* Experimenta, *&* Obseruationes, *sed etiam* Consilia, Remedia, Causarum Explicationes, Assumpta, *& quæcunque hùc spectant, immiscebimus.*

15  Operatio super Spiritus, vt maneant
iuueniles, & reuirescant. I.

*Historia.*

1. *Spiritus* omnium, quæ in Corpore fiunt, *Fabri* sunt atque *Opifices.* Id
[O5$^v$] & Consen|su, & ex infinitis *Instantijs* patet.
20  2. Si quis posset efficere, vt in Corpore Senili, rursùs indantur *Spiritus,* quales sunt in Iuuene, *Rotam* hanc Magnam, *Rotas* reliquas Minores, circumagere, & *Naturæ Cursum* retrogradum fieri posse, consentaneum est.

3. In omni *Consumptione,* siue per Ignem, siue per Ætatem, quo plus
25  *Spiritus* Rei, siue Calor deprædatur Humorem, eò breuior est Duratio Rei. Id vbique occurrit, & patet.

4. *Spiritus* in tali Temperamento, & Gradu Actiuitatis ponendi sunt, vt Succos Corporis (vt ait ille) *non bibant, & sorbeant, sed pitissent.*

[O6$^r$]  5. Duo sunt Genera *Flammarum:* vna *Acris* & *Impotens,* quæ Tenu-
30  iora euolare facit, in Duriora parùm potest, vt *Flamma ex Stramine,* vel *Ramentis Ligni;* Altera *Fortis* & *Constans,* quæ etiam insurgit in Dura, & Obstinata, qualis est *Lignorum grandiorum,* & similium.

---

21 Iuuene,] ~;  29 *Flammarum:*] ~;  30 potest,] ~;  31 *Ligni*;] ~:
32 Obstinata,] ~;

7. The seventh is the operation on the food itself and its introduction.

8. The eighth is the operation on the last act of assimilation.

9. The ninth operation is on the softening | of the parts after they have begun to dry out.

10. The tenth is the operation on the elimination of old juice and its replacement with new.

Of these the first four operations relate to the first intention; the next four to the second intention, and the last two to the third.

But since this part on the intentions points towards practice, under the heading of | history I shall mix together not only experiments and observations but also counsels, remedies, explanations of causes, assumptions, and whatever else is relevant.

## I. The Operation on the Spirits to keep them young and help them recover their strength

### *History*

1. The spirits are the craftsmen and workers who do everything that happens in the body. This is affirmed by general consent | and countless instances.

2. If a man could arrange to put into an old body spirits of the kind characteristic of a young one, it is likely that this mighty wheel might put the other, lesser wheels into reverse, and turn back the course of nature.

3. In all consumption, be it by fire or age, the more a thing's spirit or heat preys upon its moisture the shorter it lasts. This appears and occurs everywhere.

4. The spirits should be adjusted to such a condition, and level of activity that (as the man says) *they do not drink and soak up the juices of the body but only sip them.*

5. There are two kinds of flames: | one is biting but weak, like the flame from straw or wood shavings, which makes finer substances fly but has little effect on harder ones; the other is strong and steady, which flares up in hard and tough bodies, such as larger pieces of wood, and the like.

6. *Flammæ Acriores,* & tamen minùs *Robustæ,* Corpora desiccant, & reddunt effœta, & exucta; at *Fortiores* Corpora intenerant, & liquant.

7. Etiam ex *Medicinis dissipantibus,* nonnullæ in Tumoribus, tenuia tantùm emittunt, ideóque indurant; nonnullæ potentèr discutiunt, ideóque emolliunt.

[O6ᵛ] 8. Etiam in *Purgantibus,* & Abstergentibus, quædam magis | Fluida raptìm asportant; quædam magis Contumacia, & Viscosa trahunt.

9. *Spiritus* tali *Calore* indui, & armari debent, vt potiùs ament *Dura,* & obstinata conuellere, & subruere; quam *Tenuia,* & Præparata emittere, & asportare: Eo enim modo, fit Corpus Viride, & Solidum.

10. Spiritus *ita subigendi, & componendi sunt, vt fiant* Substantiâ densi, *non* rari; Calore pertinaces, *non* acres; Copiâ, *Quanta sufficit* [O7ʳ] *ad Munia Vitæ, non* redundantes, *aut* turgidi; Motu sedati, *non* | subsultorij, *& inæquales.*

11. Super *Spiritus* plurimùm operari, & posse *Vapores,* ex Somno, & Ebrietate, & Passionibus Melancholicis, & Lætificantibus, & Recreatione *Spirituum* per Odores in Deliquijs, & Languoribus, patet.

12. *Spiritus* quatuor modis condensantur. Aut *Fugando;* Aut *Refrigerando;* Aut *Demulcendo;* Aut *Sedando:* Atque primùm de *Condensatione* per *Fugam* videndum.

13. Quicquid *fugat* vndequâque, cogit Corpus in Centrum suum, atque ideò *condensat.*

[O7ᵛ] 14. Ad *Condensationem Spirituum* | per *Fugam,* longè potentissimum, & efficacissimum est, *Opium;* & deindè *Opiata,* atque generalitèr *Soporifera.*

15. Efficacia *Opij* ad *Condensationem Spirituum* admodùm insignis est; cùm tria fortassè Grana eius, *Spiritus* paulò post, ita coagulent, vt non redeant, sed extinguantur, & reddantur immobiles.

16. *Opium,* & Similia non *fugant Spiritus* propter *Frigus* suum (habent enim partes manifestò calidas) sed è conuerso *refrigerant,* propter *Fugam Spirituum.*

17. *Fuga Spirituum* ex *Opio,* & *Opiatis,* optimè cernitur, in illis exteriùs Applicatis; quia subinde *Spiritus* statim se subducunt, nec [O8ʳ] ampliùs accedere volunt, sed | mortificatur Pars, & vergit ad *Gangrænam.*

---

11–14 Spiritus. . .] / this para. is set (mistakenly) in 16-line type in *c-t;* see pp. lxiii–lxiv above
29 suum] ~;      30 habent] Habent    sed] Sed

6. Sharper though less robust flames dry bodies out, and leave them weak and without juice; but stronger flames soften and melt them.

7. Again, among dissipating medicines some give off only the finer parts in swellings, and so harden them; but others disperse them by force, and so soften them.

8. Again, in purgatives and abstergents, some carry off the more fluid parts in a rush; others draw off the more stubborn and sticky ones.

9. Spirits should be invested and armed with such a heat that they are drawn rather to undermining and wrecking hard and unyielding bodies than to taking and carrying off fine and elaborated ones. For in that way the body becomes fresh and firm.

10. The spirits should be so worked on and modified that they become dense, not rare, in their substance; persistent, not biting, in their heat; their bulk should suffice for the functions of life, and not excessive, or swollen in their abundance; and steady, not twitchy or uneven in their motion.


11. Vapours can and do work very often on the spirits, as is clear from dreams, drunkenness, melancholy and joyful passions, and in the revival of the spirits by smells in cases of fainting and lassitude.

12. The spirits are condensed in four ways: by putting them to flight, cooling them, calming them, or sedating them. We will consider condensation by flight first.

13. Whatever puts to flight from all sides concentrates a body on its centre, and so condenses.

14. Opium is far and away the most powerful and effective means of condensing the spirits by flight, and next to it opiates, and soporifics in general.

15. The effectiveness of opium in condensing the spirits is quite remarkable, since maybe three grains of it in short order so curdle the spirits that they do not bounce back; instead they are rendered immobile.

16. Opium and the like do not put the spirits to flight by their coldness (for they have parts manifestly hot). On the contrary they cool the spirits by putting them to flight.

17. The flight of the spirits from opium and opiates can best be seen when they are applied externally; for the spirits at once take themselves off, and they do not want to come back again, but the part mortifies, and slips towards gangrene.

18. *Opiata* in magnis *Doloribus*, veluti *Calculi*, aut in *Abscissione Membrorum*, Dolores mitigant; maximè per *Fugam Spirituum*.

19. *Opiata* sortiuntur bonum Effectum, ex malâ Causâ: *Fuga* enim *Spirituum*, mala; *Condensatio* autem eorum à *Fugâ*, bona.

5    20. *Græci* multum posuerunt, & ad Sanitatem, & ad Prolongationem Vitæ, in *Opiatis*; *Arabes* verò adhùc magis; in tantùm vt *Medicinæ* suæ *Grandiores* (quas *Deorum manus* vocant) pro Basi suâ, & Ingrediente principali habeant *Opium*; reliquis admistis ad eius
[O8ᵛ] noxias Qualitates retundendas, & corrigendas; quales ˡ sunt *Theriaca*,
10 *Mithridatium*, & cætera.

21. Quicquid in *Curâ Morborum Pestilentialium*, & Malignorum, fœlicitèr exhibetur, vt *Spiritus* sistantur, & frænentur, ne turbent, & tumultuentur; id optimè transfertur ad Prolongationem Vitæ; cùm idem faciat ad vtrumque; *Condensatio* videlicet *Spirituum*. Id autem
15 præstant ante omnia *Opiata*.

22. *Turcæ Opium* experiuntur, etiam in bonâ Quantitate, innoxium, & confortatiuum; adeò vt etiam ante Prælia, ad Fortitudinem illud sumant; Nobis verò nisi in paruâ Quantitate, & cum bonis Correctiuis,
19 læthale est.

[Pr ˡ]    23. *Opium*, & *Opiata*, manifestò ˡ deprehenduntur excitare Venerem; quod testatur vim ipsorum ad roborandos *Spiritus*.

24. *Aqua Stillatitia* ex Syluestri *Papauere*, ad *Crapulam*, *Febres*, & varios Morbos fœlicitèr adhibetur; quæ proculdubiò est temperatum Genus *Opiati*. Neque de Varietate Vsûs eius miretur quispiam; Id enim
25 *Opiatis* familiare est; quia *Spiritus* roboratus & densatus, insurgit in quemcunque Morbum.

25. *Turcæ* habent etiam in vsu *Herbæ* Genus, quam vocant *Caphe*, quam desiccatam puluerizant, & in Aquâ calidâ propinant; quam
29 dicunt haud paruum præstare illis Vigorem, & in Animis, & in
[Pr ˡ] Ingenijs; Quæ ˡ tamen largiùs sumpta, Mentem mouet, & turbat; vnde manifestum est eam esse similis Naturæ cum *Opiatis*.

26. Celebratur in vniuerso *Oriente Radix* quædam vocata *Betel*, quam *Indi*, & reliqui in ore habere, & mandere consueuerunt; atque ex ea Mansione mirè recreantur, & ad Labores tolerandos, & ad Languores
35 discutiendos, & ad Coitum fortificandum; videtur autem esse ex *Narcoticis*, quia magnoperè denigrat Dentes.

27. Incœpit nostro sæculo in immensum crescere vsus *Tobacco*; atque afficit Homines occultâ quâdam delectatione, vt qui illi semel assueti

6 Opiatis,] ~:    7 Grandiores] ~,    24 Opiati.] ~:    30 Ingenijs;] ~:

18. Opiates bring relief in cases of great pain, as in the stone, or amputation of limbs, mainly by putting the spirits to flight.

19. Opiates get a good effect from a bad cause, for putting the spirits to flight is bad, but their condensation by flight is good.

20. The Greeks set a lot of store by opiates both for health and for prolonging life. The Arabs were even keener, to the degree that their grander medicines (which they call God's hands) have opium as their basic and main ingredient, with an admixture of other things to offset and correct its noxious qualities; theriac, | mithridate, etc. are of this type.

21. Whatever works well in the treatment of pestilential and malignant illnesses, to put a brake on the spirits lest they become turbulent and riotous, can very well be transferred to the prolongation of life. For the same thing helps in both cases, namely condensation of the spirits, which opiates do better than anything else.

22. The Turks find opium non-toxic and comforting even in large doses, so that they even take it before battle to get strength. But to us it is deadly other than in small doses and with good correctives.

23. We see that opium and opiates | manifestly stimulate sexual activity, and that shows their capacity for strengthening the spirits.

24. Distilled water from wild poppy, which is no doubt a moderated kind of opiate, works well in cases of over-indulgence, fevers, and various illnesses. No one should be amazed by the variety of its uses. For it is usual in opiates that once the spirits have been strengthened and condensed, they revolt against any illness.

25. The Turks also make use of a type of herb they call *caphe* which, when it has been dried, they pulverize, and drink in hot water. They say that it gives them no end of courage and mental energy. All the | same, taken in large doses it overstimulates and troubles the mind, whence it is obvious that it is similar in nature to opiates.

26. All over the East, they sing the praises of a certain root called betel, which the Indians and the rest have become accustomed to keep in their mouths to chew, and this perks them up a treat so that they can put up with hard labour, shake off lassitude, and get their strength up for sex. It seems to be a kind of narcotic because it makes the teeth very black.

27. In our time the use of tobacco began to increase enormously; and it stirs a kind of secret pleasure in men so that once they have got used to it,

[P2ᵛ] sint, difficilè ⌐ posteà abstineant; Et facit proculdubiò, ad Corpus alleuandum, & tollendas Lassitudines. Atque vulgò Virtus eius refertur, eò quod aperiat Meatus, & eliciat Humores; Attamen, rectiùs referri potest ad *Condensationem* Spirituum; cum sit *Hyoscyami* quoddam
5 genus, & Caput manifestò turbet, quemadmodùm *Opiata*.

**28.** Sunt aliquandò *Humores* generati in Corpore, qui et ipsi sunt tanquam *Opiati*; vt fit in aliquibus Melancholiis, quibus si quis corripiatur, admodùm fit longæuus.

9 **29.** *Opiata* (quæ etiam *Narcotica* vocantur, & *Stupefactiua*) *Simplicia*
[P2ᵛ] sunt, *Opium* ipsum, quod ⌐ est *Succus Papaueris*; *Papauer* vtrunque & in Herbâ, & in Semine; *Hyoscyamus*; *Mandragora*; *Cicuta*; *Tobacco*; *Solanum*.

**30.** *Opiata Composita* sunt, *Theriaca*, *Mithridatium*, *Triferæ*, *Ladanum Paracelsi*, *Diacodium*, *Diascordium*, *Philonium*, *Pillulæ* de
15 *Cyno-glossâ*.

**31.** Ex his, quæ dicta sunt, possent deduci quædam *Designationes*, siue Consilia ad prolongationem vitæ, secundùm hanc *Intentionem*, scilicet *Condensationis Spirituum* per *Opiata*.

**32.** Sit itaque quotannis à Iuuentute Adultâ, *Diæta* quædam *Opiata*.
20 Vsurpetur sub Fine *Maij*; quia *Spiritus* Æstate maximè soluuntur, &
[P3ʳ] attenuantur, & minor instat metus ab Humoribus ⌐ frigidis. Sit verò *Opiatum* aliquod *Magistrale*, debilius quam ea, quæ in vsu sunt, & quoad minorem Quantitate*m* Opii, et quoad parciore*m* Mixtura*m* impensè Calidoru*m*. Sumatur Manè inter Somnos; Victus sit simplicior,
25 & parcior, absque Vino, aut Aromatibus, aut Vaporosis. Sumatur autem *Medicina* alternis tantùm diebus, & continuetur *Diæta*, ad Quatuordecim Dies. Hæc *Designatio* Iudicio nostro *Intentioni* haud perperàm satisfacit.

**33.** Possit etiam esse Acceptio *Opiatorum*, non tantùm per *Os*, sed
30 etiam per *Fumos*; Sed talis esse debet, vt non moueat nimis Facultatem
[P3ᵛ] Expulsiuam, aut eliciat Humores; sed tantùm bre⌐ui morâ, operetur super *Spiritus* intra cerebru*m*; Itaq*ue Suffumigatio matutina*, per Os, & Nares excepta, cum *Tobacco*, admisto *ligno Aloes*, & Foliis Siccis *Rorismarini*, & parum *Myrrhæ*, vtilis foret.
35 **34.** In *Opiatis magnis*, qualia sunt *Theriaca*, *Mithridatium*, & cætera (præsertìm in Iuuentute) non malum foret, potiùs *Aquas* ipsorum

2 Lassitudines.] ~:    3 Humores;] ~:    18 *Opiata*] / BNF copy with stop; Folger
copy 1 with colon    20 Vsurpetur] lc    21 frigidis.] ~:    24 Calidoru*m*.] ~:
25 Vaporosis.] ~:    30 debet,] ~;

they find it hard ¦ to give up. No doubt it helps to comfort the body and dispel lassitude. And its virtue is commonly ascribed to the fact that it opens the passages and draws off the humours. But its action can more properly be ascribed to the condensation of the spirits, seeing that it is a kind of henbane, and clearly disorders the head in the same way as opiates do.

28. Sometimes humours are produced in the body which themselves act like opiates, as happens in some kinds of melancholy, which, if they light upon a man, he lives to a great age.

29. Simple opiates (which are also called narcotics and stupefactives) are opium itself, which ¦ is the juice of the poppy, the poppy as plant and seed, henbane, mandragora; hemlock, tobacco; nightshade.

30. Composite opiates are theriac, mithridate, triferæ, laudanum paracelsi, diacodium, diascordium, philonium, pills of hound's-tongue.

31. From what has been said in connection with this intention, viz. of condensing the spirits with opiates, certain specifications or guidance for prolonging life can be deduced,

32. Thus from youth to adulthood, let there be every year a certain opiate diet. Let it be taken at the end of May because in summer the spirits are at their most relaxed and attenuated, and there is less worry about cold ¦ humours. But let the opiate be some magistral one, weaker than the ones in use, either as far as a lesser quantity of opium goes or as far as a more moderate mixture of very hot substances is concerned. Let it be taken early in the day between sleeps. Let nutriment be simpler and more moderate, without wine, and without aromatic or vaporous substances. Let the medicine be taken only every other day, and the diet carried on for a fortnight. This specification seems in my judgement to match the intention pretty well.

33. Now the opiates can be taken not just by mouth but also by fumigation, but it should be in such a way as not to stimulate the expulsive faculty too much, or to draw down the humours, but only to work on ¦ the spirits within the brain for a short interval. Thus, swallowed or sniffed, a morning suffumigation, with tobacco mixed together with lignum aloes, dried rosemary leaves, and a pinch of myrrh, would be useful.

34. In major opiates such as theriac, mithridate, etc. (especially in youth) it would not be a bad idea to take distilled waters rather than the

*Stillatitias* sumere, quam *Corpora* ipsorum; Etenim *Vapor* in distillando surgit; *Calor Medicamenti* ferè subsidet: *Aquæ* autem *Stillatitiæ,* plerúnque in Virtutibus, quæ per Vapores fiunt, bonæ; in cæteris, enerues.

5 **35.** Sunt *Medicamenta,* quæ Gradum habent quendam debilem, &
[P4ʳ] occultum, & propterea tu'tum, ad Virtutem *Opiatam*; Ea immittunt Vaporem lentum & copiosum, sed non malignum, quemadmodùm *Opiata* faciunt. Itaque *Spiritus* non fugant, sed congregant tamen, & nonnihil inspissant.

10 **36.** *Medicamenta* in ordine ad *Opiata* sunt ante omnia *Crocus,* atque eius *Flores*; deinde *Folium Indum*; *Ambra-grisia*; *Coriandri Semen* præparatum; *Amomum,* & *Pseudamomum*; *Lignum Rhodium*; *Aqua Florum Arantiorum,* & multo magis Infusio *Florum* eorundem recentium in *Oleo Amygdalino*; *Nux Muscata* foraminata, & in *Aquâ Rosaceâ*
15 macerata.

[P4ᵛ] **37.** Vt *Opiata* parcè admodùm, & certis Temporibus (vt dictum ¦ est) ita hæc *Secundaria,* familiariter, & in Victu quotidiano, sumi possunt, & multùm conferent ad Prolongationem vitæ. Certè *Pharmacopæus* quidam *Calecutiæ,* ex Vsu *Ambræ,* ad Centum sexaginta Annos vixisse
20 perhibetur; Atque *Nobiles* in *Barbariâ,* ex eiusdem Vsu, longæui reperiuntur, cùm Plebs breuioris sit Æui; Et apud Maiores nostros, qui nobis fuerunt Viuaciores, *Crocus* magno in Vsu fuit, in *Placentis, Iusculis,* &c. Atque de primo Modo *Condensationis Spirituum,* per
24 *Opiata,* et *Subordinata,* hæc inquisita sint.

[P5ʳ] **38.** Iam verò de secundo Modo *Condensationis Spirituum* per *Frigus,* ¦ inquiremus; Proprium enim Opus Frigoris est *Densatio*; atque perficitur absque Malignitate aliquâ, aut Qualitate inimicâ: ideoque tutior est Operatio, quàm per *Opiata,* licet paulò minùs potens, si per vices tantùm, quemadmodum *Opiata,* vsurparetur. At rursùs, quia
30 familiariter, & in Victu quotidiano moderatè adhiberi potest, etiam longè potentior ad prolongationem vitæ est, quàm per *Opiata.*

**39.** *Refrigeratio Spirituum,* fit tribus modis: Aut per *Respirationem*; Aut per *Vapores*; Aut per *Alimenta.* Prima optima est, sed ferè extra
34 nostram potestate*m*; Secunda etia*m* potens, & tamen præstò est; Tertia
[P5ᵛ] debilis & per Circuitus. ¦

**40.** *Aer Lympidus* & purus, & nihil habens Fuliginis, antequam

---

8 faciunt.] ~:  17 possunt,] ~;  28 *Opiata,*] ~;  29 vsurparetur.] ~:
32 modis:] ~;  33 *Alimenta.*] ~:

bodies from which they spring. For in distilling the vapour rises up while the medicine's heat pretty well settles, and the distilled waters in the virtues borne by the vapours are mostly good, in others weak.

35. There are medicines which possess a certain weak, hidden, and therefore safe degree | of opiate virtue. They discharge a slow and abundant vapour, but not harmful as opiates do. Therefore the spirits do not fly but yet come together, and thicken somewhat.

36. Medicines for opiates put in order of priority are these: before all saffron and its flowers, then Indian leaves, ambergris, prepared coriander seed, amomum and pseudamomum, rhodium wood, orange-blossom water, and much more an infusion of fresh orange blossom, in almond oil, and nutmeg perforated and steeped in rose-water.

37. Though opiates (as has | been said) are to be used very seldom and only at certain times, their substitutes can be taken routinely in an ordinary diet, and they contribute much to the prolongation of life. In fact a certain apothecary of Calicut is reputed by taking amber to have lived for 160 years; and we find that the aristocrats of Barbary by doing the same live for a long time, whereas the common people have shorter lives. And our forebears, who lived longer than we do now, took a lot of saffron in cakes, broths, and so on. So much then for the first means of condensing the spirits, i.e. with opiates and their subordinates.

38. Let us now go on to examine the second means of condensing the spirits, | i.e. by cold. For densation is cold's very own work, and takes place without any harmful effect or hostile quality; therefore its operation is safer than by opiates though a little less powerful if taken at intervals like opiates. On the other hand, it can be used in moderation routinely and in an ordinary diet, and is much more effective for promoting longevity than opiates.

39. Cooling of the spirits happens in three ways: by breathing, or vapours, or by aliment. The first is best but mostly beyond our control; the second is also powerful and yet at our service; the third is weak and indirect. |

40. Air clear and pure, and without a trace of smoke, before it is

recipiatur in Pulmones, & minùs obnoxius Radijs solis, *Spiritus* optimè densat. Talis inuenitur aut in iugis Montium Siccis, aut in Campestribus perflatilibus, & tamen vmbrosis.

**41.** Quoad *Refrigerationem* & *Densationem Spirituum* per *Vapores,* 5 *Radicem* huius *Operationis,* ponimus in *Nitro,* veluti *Creaturâ* ad hoc *Propriâ,* & *Electâ;* his vsi & persuasi indicijs.

**42.** *Nitrum* est tanquam *Arôma* Frigidum; Idque indicat Sensus ipse. Mordet enim, & tentat Linguam, & Palatum Frigore, vt *Aromata* [P6$^v$] Calore; Atque inter ea quæ nouimus, vnicum est, $^|$ & solum, quod hoc 10 præstet.

**43.** *Frigida* ferè omnia (quæ sunt propriè Frigida, non per Accidens, vt *Opium*) habent *Spiritum* Exilem, et Paucum; Contrà *Spirituosa* sunt omnia ferè Calida. Solum inuenitur *Nitrum* in Naturâ Vegetabili, quod *Spiritu* abundet, et tamen sit *Frigidum*. Nam *Caphura* quæ est 15 Spirituosa, et tamen edit Actiones Frigidi, refrigerat per Accidens tantùm; nempe Tenuitate suâ, absque Acrimoniâ, iuuando perspirationem, in Inflammationibus.

**44.** In *Congelatione* & *Conglaciatione Liquoru*m, quæ nuper cœpit 19 esse in Vsu, per *Niue*m et *Glaciem* ad Exteriora Vasis appositas, illis [P6$^v$] immiscetur *Ni*$^|$*trum;* atque proculdubiò excitat, et roborat *Congelationem;* Verum est, etiam vsurpari ad hoc Salem Nigrum communem, qui potiùs Actiuitatem indit *Frigori Niuali,* quam per se infrigidat; Sed, vt accepi, in Regionibus calidioribus, vbi Nix non cadit, fit Conglaciatio à *Nitro* solo; Sed hoc mihi compertum non est.

25 **45.** *Puluis Pyrius,* qui præcipuè constat ex *Nitro,* perhibetur epotus conducere ad Fortitudinem, et vsurpari à Nautis sæpenumerò, et Militibus ante prælia, quemadmodùm à *Turcis Opium.*

**46.** Datur fœlicitèr *Nitrum* in *Causonibus,* et Febribus Pestilentiali- [P7$^r$] bus, ad leniendos, et frænandos Ardores earum perniciosos. $^|$

30 **47.** Manifestissimum est *Nitrum* in *Puluere Pyrio* magnoperè exhorrere Flammam; vnde fit admirabilis illa ventositas, et Exufflatio.

**48.** *Nitrum* deprehenditur esse veluti *Spiritus Terræ.* Etenim certissimum est, quamcunque Terram licet puram, neque Nitrosis admixtam, ita accumulatam, & tectam, vt immunis sit à Radijs Solis, neque emittat 35 aliquod Vegetabile, colligere etiam satis copiosè *Nitrum;* vnde liquet

---

19 appositas] apposita / silently emended thus in *SEH* (II, p. 166)   illis] / inserted as a correction in some copies (e.g. Folger copy 1) but absent from others (e.g. BNF), and from *SEH* (II, p. 166)

11 omnia] ~,   12 *Opium*)] ~,)   19 appositas,] ~;

breathed into the lungs, and not much exposed to the sun's rays, condenses the spirits best. We find such air either on dry mountain ridges, or on shady flatlands exposed to the wind.

41. With regard to the cooling and densation of spirits by vapours, I locate the root of this operation in nitre, as something especially created and chosen for this end. I am persuaded by and base my view on the following evidence.

42. Nitre is like a cold aromatic body, as the sense itself tells us. For its cold bites and attacks the tongue and palate, as aromatics do with heat. And of all substances known to us, it is unique $^|$ and set apart as the only one which does this.

43. Almost all cold bodies (which are cold by nature and not accident, like opium) have a meagre and scanty spirit; whereas on the other hand almost all bodies full of spirit are hot. In the vegetable nature we find that nitre alone seethes with spirit and yet is cold. For camphor, which is full of spirit, and yet carries out the actions of cold, only cools by accident insofar as in cases of inflammation it helps sweating by its thinness free of acrimony.

44. In the congealing and freezing of liquids which has recently come into use, when they apply snow and ice to the outside of the vessel, they mix in $^|$ nitre which no doubt stimulates and reinforces the congealing. It is true that common black salt is also used for this but it rather bestows activity on the cold of the snow than makes it cold per se. But I understand that in hotter places with no snowfall, congealing is done with nitre alone. But I have no experience of this.

45. Gunpowder, which is mainly made up of nitre, taken in a drink is said to inspire courage, and is often taken by soldiers and sailors before battle, just as the Turks take opium.

46. Nitre is prescribed successfully for burning and pestilential fevers to mitigate and curb their deadly heats.$^|$

47. It is perfectly clear that nitre in gunpowder recoils from flame very much, whence happens that astonishing windiness and blast.

48. We find that nitre is as the spirit of earth. For it is absolutely certain that any earth, though pure and with no admixture of nitrous matter, if it is so heaped up under cover that the Sun's rays cannot get at it, and no vegetable body be bred in it, it will accumulate a body of nitre

*Spiritum Nitri,* non tantum *Spiritui Animalium,* verum etiam *Spiritui Vegetabilium* esse inferiorem.

**49.** *Animalia* quæ potant ex *Aquâ Nitrosâ,* manifestò pinguescunt, [P7ᵛ] quod signum est Frigidi in *Nitro.* |

5 **50.** *Impinguatio Soli* maximè fit à *Nitrosis.* Omnis enim Stercoratio est *nitrosa;* Atque hoc signum est *spiritûs* in *Nitro.*

**51.** Ex his patet *Spiritus humanos* per *Spiritum Nitri* posse infrigidari, & densari; & fieri magis Crudos, & minùs Acres; Quemadmodùm igitur *Vina fortia,* & *Aromata,* & Similia, *Spiritus* incendunt, & Vitam 10 abbreuiant; ita & *Nitrum* è conuerso illos componit, & comprimit, & facit ad Longæuitatem.

**52.** *Vsus* autem *Nitri* potest esse in Cibo inter Salem, ad decimam partem salis; in Iusculis matutinis, ad Grana à tribus ad decem; etiam in [P8ʳ] Potu; sed qualitercunque vsurpatum cum mo|do, ad Longæuitatem 15 summè prodest.

**53.** Quemadmodùm *Opium* præcipuas Partes tenet, in *Condensatione Spirituum* per *Fugam,* atque habet simul sua *Subordinata,* minùs potentia, sed magis tuta, quæ & maiori Quantitate, & frequentiori Vsu sumi possunt, de quibus superiùs diximus, ita similitèr & *Nitrum,* quod 20 condensat spiritus per Frigus, & quandam (vt Moderni loquuntur) *Frescuram*; habet quoque & ipsum sua *subordinata.*

**54.** *Subordinata* ad *Nitrum* sunt, omnia quæ exhibent *Odorem* nonnihil *Terreum*; qualis est Odor *Terræ puræ,* & bonæ, recentèr effossæ [P8ᵛ] & versatæ. In his | præcipua sunt *Borago, Buglossa, Hippo-buglossa,* 25 *Pimpinella, Fragaria,* & *Fraga* ipsa, *Frambesia, Fructus Cucumeris Crudus, Poma Cruda Fragrantia, Folia* & *Gemmæ Vitis,* etiam *Viola.*

**55.** *Proxima* sunt, ea quæ habent quendam *Virorem Odoris,* sed paulò magis vergentem ad *Calidum,* neque omninò expertem virtutis illius *Refrigerij*; Qualia sunt *Melissa, Citrum viride, Arantium viride, Aqua* 30 *Rosacea stillaticia; Pyra* Assa Fragrantia; etiam *Rosa Pallida, Rubea,* & *Muscatella.*

**56.** Illud notandum est, *Subordinata* ad *Nitrum* plerúnque plus ad *Intentionem* conferre *Cruda,* quam *Ignem passa*; quia *Spiritus* ille [Q1ʳ] *Refrigerij,* ab *Igne* dissipatur; Itaque | benè sumuntur infusa in *Potu,* aut 35 *Cruda.*

**57.** Quemadmodùm *Condensatio Spiritûs* per *Subordinata* ad *Opium,* fit aliquatenùs per *Odores;* similitèr & illa, quæ fit per *subordinata* ad

---

4 *Nitro*] / some copies (e.g. Minnesota); other copies (e.g. Liverpool) *Nitra;* for same error see *tns* to X5ᵛ below    5 *Nitrosis.*] ~: Omnis] lc    17 *Fugam,*] ~;    19 diximus,] ~:
24 versatæ.] ~:

abundant enough, whence it is evident that the spirit of nitre is inferior not only to the spirit of animals but also to that of vegetables.

**49.** Animals that drink from nitrous water clearly fatten up, which is a sign of the cold in nitre. |

**50.** The fattening of the land is produced most of all by nitrous bodies: for all manure is nitrous; and that is a sign of the spirit in nitre.

**51.** From these things it is plain that human spirits can be cooled and condensed by the spirit of nitre, and become more raw and less biting; and just as strong wines, aromatic bodies, and the like inflame the spirits and shorten life, so conversely nitre composes and compresses them, and contributes to longevity.

**52.** Nitre can be taken in food with salt up to one part of nitre to ten of salt; in morning broths as well as drink in a ratio of three grains to ten; but however it is taken, in moderation it is | very good for longevity.

**53.** Now just as opium plays the main parts in condensing the spirits by flight and at the same time (as I said above) has its less powerful, safer subordinates which can be taken more often and in larger doses, so likewise nitre, which condenses the spirits by cold, and (as the moderns say) by a certain kind of cooling virtue, has its own subordinates too.

**54.** Subordinates for nitre are things which have a somewhat earthy smell, such as the smell of good and pure earth recently dug up and turned over. The main things in this class | are borage, bugloss, ox-tongue, burnet, strawberry plants and strawberries themselves, raspberries, raw cucumbers, raw fragrant apples, leaves and buds of the vine, and violets too.

**55.** Next to these are things which have a certain fresh, greeny smell, but tending a little more to heat, and yet not altogether lacking in that virtue of cooling. Of this sort are balm, green citron, green orange, distilled rose-water, fresh roasted pears, and pale, red, and musk roses.

**56.** We should note that subordinates to nitre do more for the intention when they are raw than when cooked because the spirit of cooling is dissipated by fire. Therefore | they are better taken in a drink, or raw.

**57.** Just as condensing of the spirit by subordinates to opium is effected to a degree by smells, so likewise is it effected by subordinates to

*Nitrum;* Itaque *Odor Terræ Recentis,* & puræ, *Spiritus* optimè compescit, siue Aratrum sequendo, siue Fodiendo, siue Herbas inutiles euellendo; etiam *Folia* in Syluis, & Sepibus, vergente Autumno, decidentia, bonum *Refrigerium* præstant *Spiritibus;* Et maximè omnium, *Fragaria*
5 *moriens.* Etiàm *Odor Violæ,* aut *Florum Parietariæ,* aut *Fabarum,* aut [Q1ᵛ] *Rubi suauis,* & *Madre-selue,* exceptus dum crescunt, similis est Naturæ. |

58. Quin & nouimus *Virum Nobilem* Longæuum, qui statim à Somno, *Glebam Terræ Recentis,* sub Nares apponi quotidiè fecit, vt eius *Odorem* exciperet.

10 59. Dubium non est, quin *Refrigeratio,* & *Attemperatio Sanguinis,* per *Frigida,* qualia sunt *Endiuia, Cichorea, Hepatica, Portulaca,* &c. per Consequens infrigidet quoque *Spiritus;* sed hoc fit per Circuitum; At *Vapores* operantur immediatè.

Atque de *Condensatione Spirituum* per *Frigus,* iam inquisitum est;
15 Tertiam diximus esse *Condensationem,* per id quod vocamus, *Demulcere* [Q2ʳ] *Spiritus;* Quartam, per *Sedationem Alacritatis,* & Motûs nimij ipsorum. |

60. *Demulcent Spiritus,* quæcunque illis sunt *Grata,* atque *Amica;* neque tamen prouocant eos nimiùm ad *Exterius;* sed contrà faciunt, vt *Spiritus* quasi seipsis contenti, se fruantur, & recipiant se in *Centrum*
20 suum.

61. De his, si repetas ea, quæ superiùs posita sunt, tanquam *subordinata,* & ad *Opium,* & ad *Nitrum,* nihil est opus aliâ Inquisitione.

62. Quod vero ad *Sedationem Impetûs Spirituum* attinet, de eâ mox dicemus, cum de *Motu* ipsorum inquiremus: Nunc igitur postquam de
25 *Densatione Spirituum* dixerimus (quæ pertinet ad *Substantiam* [Q2ᵛ] ipsorum), veniendum ad *Modum Caloris* in ipsis. |

63. *Calor Spirituum,* vt diximus, eius generis esse debet, vt sit *Robustus,* non *Acris;* & amet Obstinata subruere, potius quam Attenuata asportare.

30 64. Cauendum ab *Aromatibus, Vino,* & *Potu forti;* vt vsus ipsorum sit valdè temperatus, & abstinentiâ interpolatus; Etiam à *Satureiâ, Origano, Pulegio,* & omnibus, quæ ad Palatum *Acria* sunt, & Incensiua. Illa enim præstant *Spiritibus* Calorem non *Fabrilem,* sed *Prædatorium.*

65. *Robustum* præbent *Calorem* præcipuè *Enula, Allium, Carduus*
35 *Benedictus, Nasturtium adolescens, Chamedrys, Angelica, Zedoaria,* [Q3ʳ] *Verbena, Valeriana, Myrrha, Costum, Sambuci Flores, Myrrhis.* | Horum

---

2 euellendo;] ~:     12 *Spiritus,*] ~.     22 *Nitrum,*] ~;     25 dixerimus] ~,
26 ipsorum),] ~,)     31 interpolatus;] ~:

nitre. Thus the smell of earth fresh and pure when you are following the plough, or digging, or rooting out weeds, restrains the spirits very well. Leaves falling in woods and hedgerows in late autumn give the spirits good cooling; that is especially true of strawberry plants as they die back. The smell of violets, wallflower, sweet pea, and clary picked as they grow has the same effect. [|]

**58.** Besides I knew a long-lived aristocrat who, as soon as he woke, had a fresh lump of earth placed under his nose every day, to breathe in its smell.

**59.** There is no doubt but that cooling, and tempering of the blood by cold things, like endive, chicory, anemone, purslane, etc., also causes consequential cooling of the spirits; but this happens in a roundabout way, whereas vapours act directly.

So much then for condensing the spirits by cold. I said that the third way of condensing is by what I call soothing the spirits; the fourth by damping down their liveliness, their extravagant motions. [|]

**60.** Whatever is pleasing or friendly to the spirits soothes them, so long as they do not provoke them too much towards the outside but act in the opposite way, viz. to make the spirits enjoy their own company, and withdraw into their own proper centre.

**61.** There is no need for further inquiry concerning the latter, if you recall what I said above about subordinates both to opium and to nitre.

**62.** But as for damping down the spirits' impetuosity I shall speak of it soon when I come to look into their motion. Now, therefore, after I have spoken of the spirit's densation (which has to do with their substance), I must move on to the due measure of their heat. [|]

**63.** The spirits' heat, should, as I have said, be of the kind that is robust, not biting, and likes undermining unyielding bodies rather than carrying off tenuous ones.

**64.** Be careful with aromatic substances, wine, and strong drink, to see that they are used very moderately, and left off regularly, which also applies to savory, oregano, pennyroyal, and everything sharp and burning to the palate. For these give the spirits heat which is not workmanlike but rapacious.

**65.** Those that give a robust heat are especially horse-heel, garlic, blessed thistle, young watercress, germander, angelica, zedoary, vervain, valerian, myrrh, costum, elderflower, chevril. [|] Taking these with care

vsus cum Delectu, & Iudicio, aliàs in *Condimentis*, aliàs in *Medicamentis*, huic Operationi satisfaciet.

66. Benè etiam cedit, quod *Opiata* Magna huic quoque Operationi egregiè seruiunt; eò videlicet, quod exhibent *Calorem* talem per *Compositionem*, qualis in *Simplicibus* optatur, sed vix habetur. Etenim recipiendo Calida illa intensissima (qualia sunt *Euphorbium*, *Pyrethrum*, *Stauisagria*, *Dracuntium*, *Anacardi*, *Castoreum*, *Aristolochium*, *Opoponax*, *Ammoniacum*, *Galbanum*, & similia; quæ intùs per se sumi non possunt) ad retundendam vim *Narcoticam Opij*, constituunt demùm talem Complexionem Medicaᵐenti, qualem iam requirimus; Quod optimè perspicitur in hoc, quod *Theriaca*, & *Mithridatium*, & Reliqua, non sunt Acria, nec mordent Linguam; sed tantùm sunt paululùm Amara, & Odoris potentis, & produnt demùm Caliditatem suam in Stomacho, & Operationibus sequentibus.

67. Etiam ad *Calorem Robustum Spirituum*, facit *Venus* sæpè *Excitata*, rarò *Peracta*; atque nonnulli ex *Affectibus*, de quibus posteà dicetur. Atque de *Calore Spirituum*, Analogo ad Prolongationem vitæ, iam inquisitum est.

68. De *Copiâ* spirituum, vt non sint Exuberantes, & Ebullientes, sed potiùs Parci, & intra Modum (cùm *Flamma* parua ⎮ non tantum prædetur, quantum magna), breuis *Inquisitio* est.

69. Videtur ab Experientiâ comprobari, quod *Diæta tenuis*, & ferè *Pythagorica*, vel ex Regulis Seuerioribus vitæ *Monasticæ*, vel ex Institutis *Eremitarum*, quæ Necessitatem & Inopiam habebant pro Regulâ, vitam reddat Longæuam.

70. Hùc pertinent *Potus Aquæ*, *Stratum durum*, *Aer frigidus*, *victus tenuis* (scilicet ex *Oleribus*, *Fructibus*, atque *Carnibus* et *Piscibus*, Conditis et Salitis, potiùs quam Recentibus et Calidis) *Indusium Cilicij*, *Crebra Ieiunia*, *Crebræ Vigiliæ*, *Raræ Voluptates sensuales*, et huiusmodi. Omnia enim ista minuunt *Spiriᵗtus*, eósque redigunt ad *Quantitatem* eam, quæ tantummodò vitæ Munijs sufficiat; vnde minor fit Deprædatio.

71. Quod si *Diæta* fuerit huiusmodi Rigoribus, et Mortificationibus paulò benignior, sed tamen semper æqualis, et sibi constans, eandem opem præstat; Etenim etiam in Flammis videmus, Flammam nonnihil maiorem (modò fuerit constans et tranquilla) minùs absumere ex Fomite suo, quam Flamma minor agitata, et per vices intensior, et

and judgement, sometimes in seasoning, sometimes in medicines, is adequate for this operation.

**66.** It is fortunate too that the great opiates also help a great deal with this operation, namely that they produce such heat by composition, as one would wish for but hardly obtain from simples. For the introduction of things with intense heats (as euphorbium, pyrethrum, stachys-agra, dragonwort, pistachio, castor oil, birthwort, opoponax, ammoniac, galbanum, and the like, which cannot be taken internally by themselves) to blunt the narcotic power of opium, make a medicine of the complexion ⎮ needed here, as can best be seen in the fact that theriac, mithridate, and the rest, are neither sharp nor bite the tongue but are only slightly bitter and strong smelling, and only show their heat in the stomach, and the operations following.

**67.** Sexual desire often stimulated but seldom carried through contributes to the robust heat of the spirits, and so do some of the other feelings of which I shall speak later. So much then for the heat of the spirits as it bears on the prolongation of life.

**68.** Now for a brief inquiry concerning the abundance of the spirits, that they may not surge or boil up but rather be scanty and stay within limits (seeing that a small Flame ⎮ is not as predacious as a large one).

**69.** Experience seems to show that life is lengthened by an austere and almost Pythagorean diet of the kind prescribed by the harsher monastic rules, or customs of hermits, whose guiding light was need and poverty.

**70.** Belonging to this are water drinking, a hard bed, cold air, plain food (i.e. of vegetables, fruits, and meat and fish pickled and salted rather than fresh and hot), a hair shirt, frequent fastings, frequent vigils, infrequent sensual pleasures, and the like. For all these things diminish the spirits ⎮ and reduce them to a quantity sufficient for the offices of life, and this causes them to be less predacious.

**71.** But if the diet were to be a little more pleasurable than severities and mortifications of this kind, and yet always level and steady, the same thing would be achieved. For in flames we see that a rather larger one (provided it is steady and quiet) takes rather less from its kindling than does one which is agitated, and by turns fiercer and calmer. And this has

remissior: id quod planè demonstrauit Regimen, et Diæta *Cornari Veneti,* qui bibit et edit tot Annos ad iustum Pondus; vnde centesimum [Q5$^r$] Annum Vi|ribus et Sensibus validus superauit.

72. Etiam videndum est, ne Corpus, quod pleniùs nutritur, neque
5 per huiusmodi (quales diximus) Diætas emaciatur, *Veneris vsum tempestiuum* omittat; ne *Spiritus* nimis turgeant, & Corpus emolliant, & destruant. Itaque de *Copiâ Spiritûs* moderatâ, & quasi frugali, iam inquisitu*m* est.

73. Sequitur Inquisitio de *Frænatione Motûs Spiritûs; Motus* enim
10 manifestò eum attenuat & incendit; Illa *Frænatio* fit tribus modis: Per *Somnum;* per *Euitation*em Laboris vehementis, aut Exercitii nimii,
[Q5$^v$] denique omnis lassitudinis; & per *Cohibitionem Affe|ctuum molestorum.* Ac primò de *Somno.*

74. *Fabula* habet, *Epimenidem* in *Antro,* plures Annos dormiuisse,
15 neque Alimento eguisse, cùm *Spiritus* inter dormiendum minùs depascat.

75. Experientia docet *Animalia* quædam (qualia sunt *Sorices,* & *Vespertiliones*) in quibusdam Locis occlusis, per integram Hyemem dormire; adeò *Somnus* Deprædationem vitalem compescit; Quod
20 etiam facere putantur *Apes,* & *Fuci,* licet quandóque à Melle destituti; Itidem *Papiliones,* & *Muscæ.*

76. *Somnus* post Prandium, ascendentibus in Caput Vaporibus non
[Q6$^r$] ingratis (vtpotè primis | Roribus Ciborum) *Spiritibus* prodest, sed ad Alia Omnia, quæ ad Sanitatem pertinent, grauis est & noxius; Attamen
25 in extremâ Senectute eadem est Ratio Cibi, & *Somni;* quia frequens esse debet, & Refectio, & Dormitio, sed breuis & pusilla; quinetiam ad vltimam Metam Senectutis, mera Quies, & perpetuus quasi Decubitus prodest, præsertìm Temporibus Hyemalibus.

77. Verùm vt *Somnus moderatus* ad Prolongationem vitæ facit, ita
30 multò magis, si sit *Placidus,* & non turbidus.

78. *Somnum placidum* conciliant *Viola, Lactuca* (præsertìm cocta)
[Q6$^v$] *Syrupus* è *Rosis siccis, Crocus, Me|lissa, Poma* in introitu Lecti, *Offa panis ex vino Maluatico,* præsertìm infusâ priùs *Rosâ Muscatellâ.* Itaque vtile foret, conficere aliquam Pillulam, vel aliquem Haustum paruum, ex
35 huiusmodi Rebus, eoque vti familiaritèr. Etiam ea quæ Os Ventriculi benè claudunt, vt *Semen Coriandri* præparatum, *Cotonea,* & *Pyra*

17 quædam] ~,     20 destituti;] ~:     23 ingratis] ~,     26 pusilla;] ~:
33 *Muscatellâ.*] ~:

been clearly demonstrated by the regimen and diet of Cornaro the Venetian who for years measured out exact amounts of food and drink, and so lived for more than a century, | with his powers and senses still hearty.

72. We must also see to it that a body properly nourished, and not starved by the diets just mentioned does not give up timely sexual activity, lest the spirits swell up too much and soften and destroy the body. So much then for our inquiry concerning the moderate and as it were temperate supply of spirits.

73. Next comes the inquiry concerning curbing the spirit's motion. For motion manifestly attenuates and heats it. This curbing takes place in three ways: by sleep, by avoiding hard work, violent exercise, and all lassitude, and by checking distressing | feelings. Let us start with sleep.

74. According to the story, Epimenides slept for many years in a cave without eating, for the spirit is less predatory during sleep.

75. Experience teaches that certain animals (like shrews and bats) sleep the winter through in certain enclosed places, so much does sleep suppress consumption of the vital powers. People also think that this happens with bees and drones though sometimes stripped of their honey; ditto butterflies and flies.

76. Sleep after supper, when not unattractive vapours (as the first moistures | of food) ascend into the head, is good for the spirits but bad and injurious to all else to do with health. Yet in extreme old age the same principle applies to food and sleep, seeing that eating and sleeping should be done often but briefly. At the very end of life sheer rest and as it were endless repose help a lot, especially in wintry weather.

77. But moderate sleep contributes to long life, much the more so if it is peaceful and untroubled.

78. Peaceful sleep is encouraged by violets, lettuce (especially cooked), syrup of dried roses, saffron, balm, | apples at bedtime, toast dipped in Malmsey, especially when infused with musk-rose. It would therefore be useful to make some pill or some small drink of things of this kind, and take them regularly. Things too which close the mouth of the stomach well, as prepared coriander seed, quinces, and fragrant

Fragrantia Assata, *Somnum* inducunt placidum; Ante omnia, *Iuuenili*
*Ætate,* & maximè iis, qui habent *Ventriculum* satis *fortem,* prodest
*Haustus* bonus *Aquæ Puræ,* Crudæ, in introitu Lecti.

4    **Mandatum.** *De* Ecstasi voluntariâ, *siue* procuratâ, *atque de* Cogita-
[Q7$^r$] tionibus Defixis, *& Profundis* (*modò sint absque Molestiâ*) *nihil habeo* |
comperti; *Faciunt proculdubiò ad* Intentionem, *&* densant Spiritus,
etiam *potentiùs quam* Somnus: *cùm Sensus æque, aut magis sopiant, &*
*suspendant. De illis inquiratur vlteriùs. Atque de* Somno *hactenùs.*

    79. Quatenùs ad *Motum* & *Exercitia. Lassitudo* nocet, atque Motus
10 & Exercitatio, quæ est nimis *celeris* & *velox,* quales sunt *Cursus, Pila,*
*Gladiatoria,* et similia; Et rursùs cùm Impetus extenditur ad *vltimas*
*Vires,* et *Nixus,* quales sunt, *Saltus, Lucta,* et similia. Certum enim est,
*Spiritus* in Angustiis positos, vel per *Pernicitatem Motûs,* vel per *Vltimos*
[Q7$^v$] *Nixus,* fieri posteà magis *Acres,* & *Prædatorios.* Ex alterâ Parte, *Exerci*|*tia,*
15 quæ satis *fortem* cient Motum, sed non nimis *Celerem,* aut ad *vltimas*
*Vires* (quales sunt *Saltatio, Sagittatio, Equitatio, Lusus Globorum,* &
similia) nihil officiunt, sed prosunt potiùs.

    Veniendum iam ad *Affectus,* & *Passiones Animi,* & videndum, qui ex
ipsis, ad Longæuitatem sint Noxij, qui Vtiles.

20    80. *Gaudia Magna* attenuant, & diffundunt *Spiritus,* & vitam
abbreuiant; *Læticia Familiaris* roborat *Spiritus,* Euocando eos, nec
tamen Exoluendo.

    81. *Impressiones Gaudiorum sensuales,* malæ; *Ruminationes Gaudi-*
[Q8$^r$] *orum* in *Memoriâ,* aut *Prehensiones* eorum ex *Spe* vel *Phantasiâ,* bonæ. |
25    82. Magis confortat Spiritus *Gaudium pressum,* & parcè com-
municatum, quam *Gaudium effusum,* & publicatum.

    83. *Mœror* & *Tristitia,* si *Metu* vacet, et non nimiùm *angat,* vitam
potiùs prolongat: *Spiritus* enim contrahit, et est *Condensationis Genus.*

    84. *Metus Grauiores* vitam abbreuiant, licet enim et *Mœror,* et *Metus,*
30 *Spiritum* vterque angustiet; tamen in *Mœrore* est simplex *Contractio;* at
in *Metu,* propter *Curas* de Remedio, et *Spes* intermistas, fit *Æstus,* et
*Vexatio Spirituum.*

    85. *Ira compressa* est etiam *Vexationis* genus; et *Spiritum* Corporis
[Q8$^v$] *Succos* carpere facit; At sibi *permissa,* et *foràs prodiens,* iuuat; | tanquam
35 *Medicamenta* illa, quæ *Robustum* inducunt *Calorem.*

---

5 *Profundis*] ~,     9 *Exercitia*] ~;     11 similia;] ~:     21 abbreuiant;] ~:
24 *Spe*] ~,     29 abbreuiant,] ~:

pears roasted, encourage sound sleep. Above all, in youth, and especially in those who have strong stomachs, a good drink of pure, plain water at bedtime is beneficial.

*Direction.* Concerning voluntary or procured trances, and concentrated and deep thoughts (provided they are not harmful) I have no information. No doubt they contribute to the intention, and condense the spirit more powerfully than sleep does, seeing that they soothe and suspend the senses as much if not more. Look further into these. So much then for sleep.

79. Now as far as motion and exercise go, fatigue is bad for you, as is motion and exercise which is too fast and furious—such as running, ball games, fencing, and the like. So too are things which push your strength to the limit—like jumping, wrestling, and so on. For it is certain that the spirits, driven into a tight corner by nimble motion or extreme exertion, become sharper and more predacious afterwards. On the other hand, exercises which stir up moderately strong motion but are not too fast and do not tax one's strength to the limit (such as dancing, archery, riding, bowling, and so on) do no harm but rather do good.

I must move on now to the feelings and passions of the soul, and see which of them hinder longevity and which help.

80. Great joys attenuate and diffuse the spirits, and shorten life. Ordinary happiness strengthens the spirits by summoning them up without dispersing them.

81. The perceptible impressions of joy are bad, but joys recollected in tranquillity, or the consciousness of joys arising from hope or imagination, are good.

82. Joy suppressed and little spoken of comforts the spirits more than joy let loose and made common knowledge.

83. Grief and sadness, if devoid of fear and without too much anguish, tend to prolong life; for these contract the spirits, and are a kind of condensation.

84. Great dread shortens life; for although both grief and dread distress the spirit, grief nevertheless consists in simple contraction, whereas dread causes disquiet and vexation of the spirits by mixing worries about a remedy with hope.

85. Suppressed rage is also a species of vexation, and causes the spirit to prey on the body's juices. But anger left to itself and let out is helpful like those medicines which induce robust heat.

86. *Inuidia* pessima est, et carpit *Spiritus*, atque illi rursùs *Corpus*; eò magis, quod ferè Perpetua est, nec *agit* (vt dicitur) *Festos Dies*.

87. *Misericordia* ex *Malo alieno*, quod in nos ipsos cadere non posse videtur, bona; quæ verò *Similitudine quâdam* potest reflecti in
5 *Miserantem*, mala, quia excitat *Metum*.

88. *Pudor leuis* minimè officit, cum *Spiritus* paululùm contrahat, et subinde effundat; adeò vt *Verecundi* diù (vt plurimùm) viuant. At *pudor* ex *Ignominiâ Magnâ*, et diù Affligens, *Spiritus* contrahit, vsqu*e* ad
[Rɪᵛ] *Suffocatio*nem, et est perniciosus. |

10 89. *Amor*, si non fuerit *Infœlix*, & nimis *Saucians*, ex genere *Gaudij* est; & easdem subit *Leges*, quas de *Gaudio* posuimus.

90. *Spes* omnium *Affectuum* vtilissima est, & ad prolongationem vitæ plurimùm facit; si non nimium sæpè intercidat, sed Phantasiam *Boni Intuitu* pascat; Itaqu*e* qui *Finem* aliquem, tanqua*m* Metam vitæ
15 figunt, & proponunt, & perpetuò, & sensìm, in *Voto* suo proficiunt, Viuaces vt plurimùm sunt; Adeò vt, cùm ad *Culmen Spei* suæ venerint, nec habeant quod ampliùs sperent, ferè Animis concidant, nec diù superstites sint; vt *Spes*, videatur tanquam *Gaudiu*m *Foliatu*m, quod in
[Rɪᵛ] immensu*m* extenditur, sicut *Aurum*. |

20 91. *Admiratio*, & *Leuis Contemplatio*, ad vitam prolongandam maximè faciunt; Detinent enim *Spiritus*, in Rebus quæ placent, nec eos turbare, aut inquietè, & morosè agere sinunt: Vnde omnes *Contemplatores Rerum Naturalium*, qui tot & tanta habebant, quæ mirarentur (vt *Democritus, Plato, Parmenides, Apollonius*) longæui fuerunt; Etiam
25 *Rhetores*, qui Res degustabant tantùm, & potiùs *Orationis Lumen*, quam *Rerum Obscuritatem* sectabantur, fuerunt itidem Longæui (vt *Gorgias, Protagoras, Isocrates, Seneca*). Atque certè quemadmodùm *Senes* plerúnque *Garruli*, & Loquaces sunt; ita & *Loquaces* sæpissimè
[R2ʳ] *Senescunt*. Indicat enim, *Leuem* | *Contemplationem*, & quæ *Spiritum* non
30 magnoperè stringat, aut vexet. At *Inquisitio Subtilis*, & *Acuta*, & *Acris*, vitam abbreuiat; *Spiritum* enim lassat, & carpit.

Atque de *Motu Spirituum* per *Animi Affectus* hæc inquisita sint; subiungemus autem quasdam alias *Obseruationes* Generales circa *Spiritus*, præter Superiores, quæ non cadunt in Distributionem
35 præcedentem.

---

7 viuant.] ~:          15 proponunt,] ~;   proficiunt,] ~;          23 mirarentur] ~,
24 fuerunt;] ~:          26 Longæui ( ] ~;ʌ   vt] Vt          27 *Seneca*).] ~ ‸:
30 vexet.] ~:

86. Envy is the worst thing, and preys on the spirits which in turn prey on the body; and it is all the worse because it is incessant, and (as people say) it never takes a day off.

87. Pity for a misfortune which is unlikely to afflict us is good; but that which by a kind of partial identification rebounds on the person pitying is bad because it engenders dread.

88. Slight shame is harmless since it contracts the spirits a little and promptly dissipates them; consequently diffident people generally live long. But long-running shame arising from deep humiliation contracts the spirits to the point of suffocation, and it is lethal.<sup>|</sup>

89. Love, if it is not unlucky or too painful, is a kind of joy, and so subject to the laws I have laid down for joy.

90. Of all the emotions hope is the most beneficial, and contributes most to the prolongation of life, if it is not disappointed too often and feeds the fantasy with fair prospects. People who therefore shape and set up a purpose or goal in life and always work towards it bit by bit, are generally long-lived, but in such a way that when they reach the height of their hopes, and have nothing more to look forward to, their *joie de vivre* drains away and they do not last long; so the joy of hope seems to be like foil made from gold because it can be spread so thin.<sup>|</sup>

91. Admiration and idle reflection are wonderful for prolonging life, for they engage the spirits in business which pleases them, and stop them ending up turbulent, restless, or difficult; whence all those contemplators of natural things, who had (like Democritus, Plato, Parmenides, and Apollonius) so many and such marvellous things to entrance them, lived for a long time. The rhetoricians too, who enjoyed (like Gorgias, Protagoras, Isocrates, and Seneca) a mere taste of such matters, and chased after the brightness of words rather than the darkness of things, lived likewise for a long time. And just as old men are often loquacious to the point of garrulity, so the loquacious very often grow very old; for it indicates idle<sup>|</sup> reflection that does not greatly distress or vex the spirit. But subtle, acute, and sharp inquiry shortens life, for it wearies the spirit and preys on it.

So much then for the inquiry into the motions of the spirits and the soul's feelings; I shall now add some other general observations on the spirits, which did not fall into the distribution just examined.

**92.** Præcipuæ Curæ esse debet, vt *Spiritus* non exoluantur sæpiùs; *Solutionem* enim præcedit *Extenuatio*, neque *Spiritus* semel Extenuatus, ita facilè se recipit, & densatur. *Exolutio* autem fit per Nimios *Labores*, [R2ᵛ] Nimis vehementes | *Affectus Animi*; Nimios *Sudores*; Nimias *Euacu-* 5 *ationes*; *Balnea tepida*; & *Intemperatum*, aut *Intempestiuum vsum Veneris*; Etiam Nimias *Curas*, & *Sollicitudines*, & *Expectationes* anxias; Denique per *Morbos malignos*; & *Dolores*, & Cruciatus Corporis graues: quæ omnia, quantum fieri potest (vt etiam *Medici* vulgares monent) euitanda sunt.

10　**93.** *Spiritus* & *Consuetis* delectantur, & *Nouis*. Mirum autem in Modum facit ad conseruandum *Vigorem Spirituum*, vt nec *Consuetis* vtamur ad *Satietatem*, nec *Nouis*, ante *Appetitum viuidum*, & strenuum. Itaque & *Consuetudines* abrumpendæ sunt Iudicio quodam, & Curâ, [R3ʳ] antequam per|ueniant ad *Fastidium*; & *Appetitus* ad *Noua*, ad tempus 15 cohibendus, donec fiat *Fortior*, & *Alacrior*. Atque insupèr, *Vita*, quoad fieri potest, ita instituenda, vt Multas, & Varias habeat *Redintegrationes*, neque perpetuò in ijsdem versando *Spiritus* torpeant: Licet enim non malè dictum sit à *Senecâ*, *Stultus sempèr incipit viuere*, tamen illa *Stultitia*, vt & aliæ quamplurimæ, Longæuitati prodest.

20　**94.** Circa *Spiritus* obseruandum est (etsi contrarium fieri consueuerit) vt quandò percipiant Homines *Spiritus* suos esse in *Statu Bono*, & Placido, & Sano (id quod ex *Tranquillitate Animi*, et *Læticiâ* datur [R3ᵛ] perspici) eos fo|ueant, nec mutent; Sin in *Statu Inquieto*, et Maligno (id quod ex *Tristitiâ*, *Pigritiâ*, atque aliâ *Indispositione Animi* apparebit) 25 eos subinde obruant, et alterent. Continentur autem *Spiritus* in eodem *Statu*, per *Cohibitionem Affectuum*, *Temperamentum Diætæ*, *Abstinentiam à Venere*, *Moderationem à labore*, *Otium mediocre*; Alterant autem, & obruunt *Spiritus*, Contraria istis; scilicet, *Affectus vehementes*, *Epulæ profusæ*, *Venus immoderata*, *Labores ardui*, *studia intensa*, et *Negotia*. 30 Atqui consueuerunt Homines, cùm læti sunt, et sibi maximè placent, tum *Epulas*, *Venerem*, *Labores*, *Contentiones*, *Negotia*, maximè sequi, & [R4ʳ] affectare. Quòd si quis Longitu|dini Vitæ consulere velit, contrario modo (quod mirum dictu) se gerere debet; *Spiritus* enim *Bonos* fouere, et continuare, *malè Dispositos* exhaurire, et mutare, oportet.

35　**95.** Non ineptè ait *Ficinus*, *Senes* debere ad *Confortationem Spirituum* suorum, *Acta Pueritiæ* suæ, et *Adolescentiæ* sæpè recordari, et ruminare. Certè Recreatio est *Senibus* singulis, tanquam peculiaris, *Recordatio*

3 densatur.] ~:　　　8 potest] ~,　　　20 est] ~,　　　22 Sano] ~,
32 affectare.] ~,

92. Special care should be taken to ensure that the spirits are not often diffused, for attenuation precedes dissolution, and once the spirit has lost its density it does not easily get it back again. Diffusion is caused by labour too hard, feelings too ¹ strong, sweating too much, too much evacuation, warm baths, and sex too much and at the wrong times. Also bad are too much worry, anxiety, and terrible suspense; and lastly malignant diseases, and excruciating bodily pain—all of which should (as the common run of physicians counsel) be avoided as far as possible.

93. The spirits delight in both routine and novelty. Now conservation of the spirit's vitality is wonderfully enhanced if we neither follow routine until it palls nor take up novelty before we feel a strong and vigorous desire to do so. And so we need a certain judgement and discretion to break off routine before ¹ aversion sets in; and to check the appetite for novelty until it is stronger and more eager. Moreover, life should be organized as far as possible so that it has many and varied renewals, and that the spirits do not become sluggish from endless concentration on the same things. For although Seneca was not wrong to say that *A fool is always beginning to live*, it is nevertheless the case that this folly, like many other kinds, helps you to live long.

94. With regard to the spirit, it should be observed (though people generally do the opposite) that when men sense that their spirits are in a good, quiet, and healthy condition (which can be inferred from peace of mind, and cheerfulness) they should foster ¹ and not change them. But if their condition is restless and ill-disposed (which can be inferred from sadness, dullness, and other indispositions of mind), they should at once overwhelm and alter them. Now the spirits are kept in the same condition by restraining the feelings, eating temperately, abstaining from sex, working in moderation, and relaxing sensibly. The spirits are overwhelmed and altered in the opposite ways, i.e. by strong feelings, heavy banqueting, too much sex, hard work, intensive study, and business. But men have grown used, when they are most cheerful and pleased with themselves, to giving themselves to banqueting, sex, work, contention, and business. But if you wish for long ¹ life, you ought (though the advice seems odd) to do the opposite; for good spirits should be fostered and kept going, whereas bad ones should be depleted and changed.

95. Ficino was not wrong when he said that old men should comfort their spirits by often recalling and ruminating on the deeds of their childhood and youth. Now surely such recollection is a particular

talis. Itaque dulce est Hominibus *Societatem* habere eorum, qui olim
vnà educati fuerant, et *Loca* ipsa *Educationis* suæ inuisere. *Vespasianus*
autem huic Rei tantum tribuebat, vt cùm esset *Imperator,* nullo modo
[R4ᵛ] Animum inducere potuisset, vt | *Ædes Paternas,* licet humiles, mutaret;
5 ne aliquid deperiret *Consuetudini Oculorum,* et *Memoriæ Pueritiæ* suæ;
Quinetiam in *Scypho* quodam *Auiæ* suæ *Ligneo,* cum Labro argenteo,
Diebus Festis potabat.

    **96.** Illud ante omnia *Spiritibus* gratum est, vt fiat *Progressus* continuò
*in Benignius.* Itaque eo modo est instituenda Iuuentus, et Ætas virilis,
10 vt Senectuti *Noua Solatia* relinquantur; quorum præcipuum sit *Otium
Moderatum.* Itaque sibi ipsi Manus inferunt *Senes Honorati,* qui in
*Otium* non secedunt: Cuius Rei insigne reperitur Exemplum in
*Cassiodoro,* qui tantâ apud *Reges Italiæ Gothos* Auctoritate pollebat, vt
[R5ʳ] instar | Animæ esset erga eorum Negotia; postea autem ferè Octoge-
15 narius in *Monasterium* se recepit, vbi non ante Centesimum demùm
Annum vitam clausit. At huic rei duæ *Cautiones* adhibendæ sunt: vna vt
non expectet, donec *Corpus* omninò *Confectum* sit, & *Morbidum;*
Etenim in huiusmodi Corporibus omnis *Mutatio,* licet in *Benignius,*
Mortem accelerat; Altera, vt *Otio* planè *inerti* se minimè dedant,
20 sed habeant *Aliquid* quod Cogitationes, & Animum ipsorum placidè
detinere possit; In quo genere, præcipua Oblectamenta sunt *Literæ,*
deindè Studia *Ædificandi,* & *Plantandi.*

[R5ᵛ]     **97.** Postremò eadem *Actio, Con|tentio, Labor* libenter Susceptus,
& cum *bona Voluntate, Spiritus* recreat; Cum *Auersatione* autem, &
25 *Ingratijs, Spiritus* carpit, & sternit. Itaque ad Longæuitatem confert, si
quis *Arte* talem vitam instituat, quæ libera sit, & ad *Arbitrium suum*
traducatur; Aut tale *Obsequium* Animo suo conciliauerit, vt quicquid à
Fortunâ imponatur, eum potiùs *Ducat,* quam *Trahat.*

    **98.** Neque illud omittendum ad *Regimen Affectuum,* vt præcipua
30 Cura adhibeatur *Oris ventriculi;* Maximè ne sit relaxatum nimis; quia
plus dominatur illa *Pars* super *Affectus,* præsertim *quotidianos,* quam aut
[R6ʳ] *Cor,* aut *Cerebrum;* Exceptis tantummodo ijs, quæ | fiunt per potentes
*Vapores,* vt in *Ebrietate,* & *Melancholiâ.*

    **99.** De *Operatione* super *Spiritus,* vt *Iuueniles* maneant, et *reuirescant,*
35 hæc inquisita sunt: Quod eò diligentiùs præstitimus, quòd de his *Operationibus,* potiori ex parte, magnum est apud *Medicos,* & alios Auctores,

---

5 suæ;] ~:      8 est,] ~;      13 *Cassiodoro,*] ~;      16 clausit.] ~:    sunt:] ~;
17 *Morbidum;*] ~:     19 accelerat;] ~:     25 sternit.] ~:     29 *Affectuum,*] ~;

270

recreation of all old men; and so they love the company of people with whom they went to school, and visiting the actual places where they were educated. In fact Vespasian was so given to this that when he was emperor he could in no way bring himself | to alter his father's house, humble though it was, in case he lost forever those familiar sights, and the memories of his childhood. Indeed, on festal days he used to drink out of the silver-rimmed wooden cup which had been his grandmother's.

96. The thing above all that pleases the spirits is steady advance to a better condition. Youth and manhood should therefore be organized so as to leave new comforts for old age, the main one of which is moderate ease. Therefore old men in high office who do not give it up for ease, take up arms against themselves. We find an excellent example of this in Cassiodorus, who had so much influence with the Gothic kings of Italy that he | was the moving spirit of their affairs; yet afterwards, at almost eighty years of age, he retired to a monastery where he eventually died a centenarian. But in this matter two notes of caution should be introduced: the first is that they do not hang on until the body is completely worn out and ill, for in bodies of this kind all change, even for the better, hastens death; the second is that they do not give themselves over to complete sloth but have something peacefully to occupy their thoughts and minds, and the chief pleasures in this line are letters and next an enthusiasm for architecture and gardening.

97. Lastly the same action, | competition, and labour taken up with good will recreates the spirits; but with revulsion and against one's will, consumes the spirits and lays them low. Thus it is good for longevity if someone has the skill to organize his life in such a way that it is unconstrained and led as he wishes; or that he makes his mind obedient so that whatever fortune deals out may rather lead him than drag him along.

98. But we should not forget that in managing the emotions special care should be paid to the mouth of the stomach, most of all to stop it relaxing too much; for this part exerts more influence over the emotions (especially the everyday ones) than the heart or brain, excepting only those which | arise from powerful vapours, as in drunkenness and melancholy.

99. So much then for the operation on the spirits to keep them young and restore them; and I have considered this the more attentively because the physicians and other writers have, for the most part, passed them over in silence, and most of all because the operation on the

Silentium: Maximè autem, quia *Operatio* super *Spiritus,* eorumque
Recrudescentiam, ad Prolongationem Vitæ, est *Via* maxime *Procliuis,*
& *Compendiaria*; Propter duplex scilicet *Compendium*: Alterum, quod
*Spiritus* Compendiò operetur super *Corpus*; Alterum, quod *Vapores,* &
5 *Affectus* compendiò operentur super *Spiritus*; Adeò vt hæc *Finem* petant,
[R6ᵛ] qua|si in *Lineâ rectâ*; cætera magis per *Circuitum.*

## OPERATIO
### super Exclusionem Aeris. II.

#### *Historia.*

10 **1.** *Exclusio Aeris Ambientis,* ad Diuturnitatem Vitæ dupliciter innuit.
Primò, quod maximè omnium post *Spiritum Innatum, Aer Extrinsecus*
(vtcunque *Spiritum Humanu*m quasi animet, et ad *Sanitatem* plurimùm
[R7ʳ] conferat) *Succos* Corporis deprædatur, et *Desiccationem* | Corporis
accelerat; Itaque *Exclusio Aeris,* ad Longitudinem vitæ confert.

15 **2.** Alter *Effectus,* qui sequitur *Exclusionem Aeris,* subtilior multò
est, et profundior; Scilicet quod *Corpus Occlusum,* & non *perspirans,*
*Spiritum* inclusum detinet, & in *Duriora* Corporis vertit; vnde *Spiritus*
ea emollit, et intenerat.

**3.** Huius rei explicata est Ratio in *Desiccatione Inanimatorum*; Atque
20 est *Axiôma* quasi infallibile, Quod *Spiritus Emissus* Corpora desiccat;
*Detentus* colliquat, et intenerat: Atque illud insupèr simul *Assumendum,*
Quod *Calor* omnis *propriè* attenuat, & humectat, & per *Accidens*
[R7ᵛ] tan|tùm contrahit, et desiccat.

**4.** *Vita* in *Antris* & *Speluncis,* vbi *Aer* non recipit *Radios Solis,* possit
25 facere ad Longæuitatem; *Aer* enim per se ad *Prædationem* Corporis
non multùm potest, nisi *Calore* excitatus. Certè si quis Memoriam
Rerum recolat, ex pluribus *Reliquijs,* & Monumentis constare videtur,
fuisse Hominum *Magnitudines,* & *Staturas,* longè ijs quæ posteà
fuerunt, grandiores; vt in *Siciliâ,* & alijs nonnullis Locis. Istiusmodi
30 autem Homines, in *Speluncis,* plerunque Ætatem degebant: Atqui
*Diuturnitas Ætatis,* & *Amplitudo Membrorum,* habent nonnihil com-
[R8ʳ] mune. Etiam *Antrum Epimenidis* inter *Fabulas* | ambulat. Suspicor
etiam, *vitam Anachoretarum Columnariu*m, simile quippiam fuisse
*vitæ in Antris;* quippe vbi *Radij solis* parùm penetrabant, neque *Aer*
35 magnas *Mutationes,* aut *Inæqualitates* recipere poterat. Illud certum,

---

3 *Compendium*:] ~;          27 recolat,] ~;

spirits and on their renewal is the easiest and most direct route to the prolongation of life. It is direct for two reasons: the one that the spirit works directly on the body; the other that the vapours and emotions work directly on the spirit, so that these head for their destination as it | were in a straight line, whereas other things take a more roundabout route.

## II. The Operation to Exclude
## Air

### *History*

**1.** Exclusion of the ambient air holds out the promise of long life in two ways: first because after the innate spirit, the outside air (though it is as the animating principle of the human spirit, and does a great deal for health) preys more than anything else on the juices of the body, and hastens | its desiccation. Accordingly exclusion of the air contributes to length of life.

**2.** The other effect of the exclusion of the air is much more subtle and deep, viz., that a body closed up and not sweating restrains the enclosed spirit, and turns it on the harder parts of the body, whence the spirit makes them soft and tender.

**3.** The principle of this is unfolded in the drying out of inanimate bodies, and this is an infallible axiom: i.e. that spirit given off dries bodies up, and spirit held back liquefies and softens them. And this should be assumed in addition: that all heat attenuates and moistens by nature but only contracts | and dries up bodies by accident.

**4.** Life in dens and caves where the air does not admit the Sun's rays can contribute to long life. For air, unless stimulated by heat, can do little by itself to prey on the body. Certainly if one goes over the memorials of things, it appears from many remains and monuments (in Sicily and in some other places) that men's heights and sizes were once much greater than they were in later ages. And these men generally spent their lives in caves. Now length of life and length of limb have something in common. Indeed the cave of Epimenides figures | among the fables. And I also suspect that the life of anchorites on columns was rather like life in caves, inasmuch as the Sun's rays scarcely penetrated there, and the air did not admit of great changes or inequalities. It is

vtrunque *Simeonem Stylitam,* et *Danielem,* et *Sabam,* atque alios *Anachoretas Columnares,* admodùm Longæuos fuisse. Etiam *Anachoretæ Moderni,* intra *Muros* aut Colu*m*nas septi, et Clausi, Longæui sæpiùs reperiuntur.

5   **5.** Proxima *Vitæ in Antris,* est *Vita in Montibus.* Quemadmodùm enim in *Antra, Calores Solis* non penetrant; ita in *Fastigijs Montium,* [R8ᵛ] reflexione destituti, parùm possunt. Accipiendum autem ⏐ hoc est de *Montibus,* vbi *Aer Limpidus* est, & purus; Scilicet vbi propter *Ariditates Vallium, Nebulæ* & *Vapores* non ascendunt; Quod fit in *Montibus,* qui 10 *Barbariam* cingunt, Vbi etiam hodiè viuunt, sæpenumerò ad Annos Centum & quinquaginta, vt iam anteà notatum est.

  **6.** Atque huiusmodi *Aer, Antrorum,* aut *Montium,* ex suâ Naturâ propriâ, parùm aut nihil deprædatur. At *Aer,* qualis est noster, cùm sit propter *Calores Solis Prædatorius,* quantùm fieri potest, à Corpore est 15 excludendus.

  **7.** *Aer* verò prohibetur, et excluditur, duobus Modis: Primò, si [S1ʳ] *Claudantur* Meatus; secundò, si *Oppleantur.* ⏐

  **8.** Ad *Clausuram Meatuum* faciunt, ipsius *Aeris Frigiditas, Nuditas Cutis,* ex quâ illa induratur; *Lauatio* in *Frigidâ; Astringentia* Cuti 20 applicata; qualia sunt, *Mastyx, Myrrha, Myrtus.*

  **9.** Multò magis huic *Operationi* satisfiet per *Balnea,* sed rarò vsurpata (præsertìm Temporibus Æstiuis) quæ constent ex *Aquis Mineralibus Astringentibus,* quæ tutò exhiberi possunt; quales sunt *Chalybeatæ,* & *Vitriolatæ;* Hæ enim *Cutem* potentèr contrahunt.

25   **10.** Quod ad *Oppletionem* attinet, *Pigmenta,* & huiusmodi *Spissamenta Vnctuosa,* atque (quod commodissimè in vsu potest esse) [S1ᵛ] *Oleum,* & *Pinguia,* non minùs ⏐ *Corporis Substantiam* conseruant, quàm *Pigmenta* in *Oleo,* & *Vernix Ligna.*

  **11.** *Britones Antiqui,* Corpus *Glasto* pingebant, & fuerunt admodùm 30 longæui; Quemadmodum & *Picti,* qui indè etiam *Nomen* traxisse à nonnullis putantur.

  **12.** Hodiè se pingunt *Bresilienses,* & *Virginienses,* qui sunt (præsertìm illi priores) admodùm longæui; adeò vt quinque abhinc Annis, *Patres Galli* nonnullos conuenerint, qui *Ædificationem Fernamburgi,* Annis 35 abhinc centum & viginti, ipsi ad tunc *Virilis Ætatis* meminissent.

  **13.** *Ioannes* de *Temporibus,* qui dicitur ad Trecentesimum Annum,

---

10 cingunt,] ~;      13 deprædatur.] ~:      20 applicata;] ~,      21 vsurpata] ~,
25 attinet,] ~;

certain that both Simon Stylites, Daniel, Sabas, and the other column-dwelling anchorites were very long-lived. We find too that modern anchorites walled up or confined to columns are often long-lived.

5. The next best thing to life in caves is life in the mountains. For just as in caves the heats of the Sun do not penetrate, so in the heights of the mountains where reflection is lacking, those heats achieve little. Now we must understand that ¹ this is true of mountains where the air is clear and pure, i.e. where, on account of the dryness of the valleys, clouds and vapours do not ascend, as happens in the mountains girdling Barbary where even nowadays people often live for 150 years, as we noted earlier.

6. Now cave and mountain air of this kind is, by its very nature, not very or not at all predacious. But as far as possible air like ours should, since the Sun's heats make it predacious, be kept away from the body.

7. Air is kept at bay and shut out in two ways: first, if the passages are closed; and secondly, if they are filled up. ¹

8. Closing of the passages is helped by the coldness of the air itself; by nakedness which hardens up the skin; by cold baths; and by applying astringents, such as mastic oil, myrrh, and myrtle, to the skin.

9. But this operation may be better achieved by baths taken infrequently and especially in summertime, baths of those astringent mineral waters which can be safely applied; among these are waters suffused with iron and vitriol, for these firmly contract the skin.

10. As for filling up, paints and that sort of thick, greasy matter, and (because they can be most conveniently put to use) oil and fat, conserve ¹ the substance of the body no less than oil-based paints and varnish preserve wood.

11. The ancient Britons painted their bodies with woad, and were very long-lived. The Picts did the same; and some think that their very name was derived from that.

12. To this day the Brazilians and Virginians paint their bodies, and are (the former particularly) extremely long-lived, so much so that five years ago the French Fathers encountered some who remembered the building of Pernambuco in their maturity 120 years before.

13. Joannes de Temporibus, who is said to have lived to the age of

[S2$^r$] Ætatem produxisse, interroga'tus quomodò se conseruasset; respondisse
fertur, *Extrà, Oleo; intùs, Melle.*

14. *Hyberni,* præsertìm *Syluestres,* etiam adhùc sunt valdè viuaces;
Certè aiunt paucis abhinc Annis, *Comitissam Desmondiæ* vixisse ad
5 Annum Centesimum Quadragesimum, & ter per vices *Dentijsse.*
*Hybernis* autem Mos est, se nudos ante Focum *Butyro Salso,* & Veteri,
fricare, & quasi condire.

15. Iidem *Hyberni* in vsu habuerunt *Lintea,* & *Indusia Croceata;*
Quod licet ad *Arcendam Putrefactionem* introductum fuerat, tamen
10 (vtcunque) ad *vitæ Longitudinem* vtile fuisse existimamus. Nam *Crocus,*
[S2$^v$] ex omnibus quæ $^|$ nouimus, ad *Cutem,* & *Confortationem Carnis,* est Res
optima; cùm & notabilitèr *Astringat,* & habeat insupèr *Oleositatem,* &
*Calorem Subtilem* sine vllâ *Acrimoniâ.* Equidem memini quendam
*Anglum,* vt *Vectigalia* supprimeret, *Croci Saccum,* cùm transfretaret,
15 circa *Stomachum* portasse, vt lateret; Eúmque, cùm anteà ex Mari
grauissimè *ægrotare* solitus esset, optimè tunc *valuisse,* nec *Nauseam*
vllam sensisse.

16. *Hippocrates* iubet *vestes* ad *Cutem Hyeme* puras portare, *Æstate*
19 sordidas, & *Oleo* imbutas; Huius Ratio videtur, quod per *Æstatem,*
[S3$^r$] *Spiritus* exhalant maximè; Itaque *Pori Cutis* opplendi sunt. $^|$

17. Ante omnia igitur vsum *Olei,* vel *Oliuarum,* vel *Amygdalini
dulcis,* ad Cutem ab Extrà vngendam, ad Longæuitatem conducere
existimamus; Eáque *vnctio* debet fieri singulis Auroris, cùm exitur è
Lecto, cum *Oleo,* in quo admisceatur parum *Salis nigri,* & *Croci. Vnctio*
25 autem leuis debet esse, ex *Lanâ,* aut *Spongiâ* moliori, neque quæ stillet
super Corpus, sed Cutem tantùm intingat, & inficiat.

18. Certum est *Liquores* in *Maiori Quantitate,* etiam *Oleosos,* haurire
nonnihil ex Corpore; sed contrà, *Paruâ Quantitate* imbibi à Corpore:
29 Itaque *leuis Aspersio* facienda est, vt diximus; Aut planè *Indusium* ipsum
[S3$^v$] *Oleo* liniendum est. $^|$

19. Objici verò fortè possit, Istam *Vnctionem* ex *Oleo,* quam laudamus (licet apud Nos in Vsu nunquam fuerit, atque apud *Italos* in Desuetudinem abierit) olìm quidem apud *Græcos,* & *Romanos* familiarem
fuisse, & Diætæ partem; neque tamen ijs Sæculis Homines magis fuisse
35 longæuos. Sed respondetur rectissimè; *Oleum* in Vsu fuisse tantùm post
*Balnea,* nisi fortè inter *Athletas; Balnea* autem ex Calido, Operationi
nostræ tantò contraria sunt, quantò *Vnctiones* congruæ; cùm Alterum
*Meatus* aperiat, Alterum obstruat. Itaque *Balneum* absque *Vnctione*

---

10 existimamus.] ~:     31–2 laudamus] ~,

300, when asked | how he had kept himself going, replied, *With oil on the outside and honey within.*

14. The Irish, and particularly the wild ones, are even now very long-lived. In fact they say that within these last few years the Countess of Desmond lived to the age of 140, and cut new teeth three times. Now the Irish habit is to stand in front of the fire without a stitch on, and rub, and, so to speak, embalm themselves in rancid salt butter.

15. These same Irish usually wear linen and shirts dyed with saffron, a custom which, although it was brought in to hold back putrefaction, I believe to be useful for prolonging life. For saffron is the best thing of all | that I can think of for the skin, and for comforting the flesh, since it is a remarkable astringent, besides having an oiliness and subtle heat without any ferocity. Indeed, I remember a certain Englishman who, to get out of paying the duty carried a hidden bag of saffron round his stomach when he crossed the Channel; and he, who had on previous occasions suffered violent seasickness, now fared well and experienced no nausea at all.

16. Hippocrates prescribes the wearing of clean clothes next to the skin in winter, but dirty, oil-soaked ones in summer, on the principle that in summer the spirits exhale most and so the pores of the skin should be filled up. |

17. Thus I think that the practice of anointing the skin from outside with oil, be it of olive or almond, is better than anything else for promoting longevity. The anointing should be done every morning when you get out of bed, and with oil mixed with a little black salt and saffron. Now the anointing should be lightly applied with wool or a soft sponge, and should not settle in drops on the body but just colour and tint the skin.

18. It is certain that in large quantities liquors, even the oily ones, absorb something from the body; but in small quantities they are imbibed by the body. Thus, as I have said, the sprinkling should be light, or the shirt itself should be coated with oil. |

19. It could perhaps be objected that this anointing with oil which I commend (though here we have never practised it, and the Italians have given it up) was once common with the Greeks and Romans, and part of their diet, and yet in those times they were no more long-lived than now. But this can be very properly countered by observing that oil was used only after baths, except perhaps by athletes; and that hot baths are as opposite to our operation as anointings are in line with it, for the one opens the passages, while the other blocks them. Therefore a bath

sequenti, pessimum, *Vnctio* absque *Balneo* optima. Etiàm ad *Delicias*
[S4$^r$] potiùs adhibeatur ista *Vnctio*, atque (si in optimam Partem accipias)
ad *Sanitatem*; sed nullo Modo in Ordine ad *Vitam longæuam*;
Itaq*ue* simul adhibebantur *Vnguenta pretiosa*, quæ ad *Delicias* grata, ad
5 nostram *Intentionem* noxia sunt, ob *Calorem*; Vt benè dixisse videatur
*Virgilius*.

*Nec Casiâ liquidi corrumpitur vsus Oliui.*

**20.** *Inunctio* ex *Oleo*, & *Hyeme* confert ad *Sanitatem*, per Exclusionem Frigoris; & *Æstate*, ad detinendos *Spiritus*, & prohibendam
10 Exolutionem eorum, & arcendam vim *Aeris*, quæ tunc maximè est
Prædatoria.

**21.** Cùm *Inunctio* ex *Oleo*, *Operatio* sit ad vitam longam ferè
[S4$^v$] poten|tissima; visum est addere *Cautiones*, ne periclitetur Valetudo. Eæ
quatuor sunt, secundùm quatuor Incommoda, quæ exindè sequi
15 possint.

**22.** Primum *Incommodum* est, quod *reprimendo Sudores*, Morbos
inducere possit, ex Humoribus illis Excrementitijs. Huic *Remedium*
adhibendum est ex *Purgationibus*, & *Clysterijs*, vt Euacuationi debitè
consulatur. Certum enim est, *Euacuationem* per *Sudores*, Sanitati
20 plerunque conferre; Longitudini vitæ officere. *Purgatiua* autem
moderata in Humores agunt, non in *Spiritus*, quòd facit *Sudor*.

[S5$^r$] **23.** Secundum *Incommodum* est, quod *Corpus calefacere* possit, &
subindè *inflammare*; *Spiritus* enim *Occlusus*, nec *Perspirans* feruentior
est. Huic *Incommodo* occurritur, si *Diæta*, vt plurimum, vergat ad
25 *Frigidum*, & sumantur Propria quædam ad *Refrigerandum* per vices;
de quibus mox in *Operatione* super *Sanguinem* inquiremus.

**24.** Tertium est, quod *Caput grauare* possit, Omnis enim *Oppletio*
extrinsecùs, repercutit Vapores, & eos mittit versus *Caput*. Huic
*Incommodo* omninò occurritur, per *Cathartica*, præsertìm *Clysteria*; Et
30 claudendo *Os Ventriculi* fortitèr cum *Stipticis*; Et *pectendo*, & *fricando*
*caput*, etiam cum Lixiuijs idoneis, vt aliquid exhalet; Et non omittendo
[S5$^v$] *Ex|ercitationem bonam*, & qualem conuenit, vt etiam per *Cutem*
nonnihil perspiret.

---

24 si *Diæta*] sit *Diæta* / latter version in Bibliothèque de Fels copy only

1 pessimum,] / some copies (e.g. BNF) with comma; other copies (e.g. Folger copy 1) with
semicolon    13 Valetudo.] ~;    17 Excrementitijs.] ~:    19 consulatur.] ~;
24 est.] ~:    27 possit,] ~;    28 *Caput*.] ~:

without subsequent anointing is very bad; but anointing without the bath is very good. This anointing was also practised | rather as a pleasure and (if you take it on the best interpretation) for the sake of health, but not at all for the sake of longevity. Therefore they used at once precious ointments which give pleasure but are harmful to our intention because of their heat; so that Virgil appears to have spoken well:

*The use of pure olive oil is not corrupted by casia.*

20. Anointing with oil in winter contributes to health by keeping the cold out; and in summer by keeping in the spirits and preventing their dispersion, and restraining the power of the air, which is then at its most predatory.

21. Since anointing with oil is almost the most | powerful operation affecting longevity, some warnings should be added in case it endangers health. There are four of these corresponding to the four disadvantages which may follow from it.

22. The first disadvantage is that by suppressing sweats it is capable of hatching diseases from those waste humours. The cure for this is to use purges and clysters to bring about proper evacuation. For it is certain that evacuation by sweats is generally good for health but not for longevity. But moderate purgatives work on the humours, not on the spirits, which is what sweat does.

23. The second disadvantage is that it can heat up the body, and | forthwith inflame it. For spirit shut in and not sweating is fiercer. The cure for this is a diet which tends to coldness, and the taking of certain specifics to cool the body down, specifics which I shall soon come to when I look into the operation on the blood.

24. The third disadvantage is that it can burden the head. For all external blocking drives back the vapours and sends them towards the head. This disadvantage can be altogether countered by cathartics, especially clysters; by firmly closing the mouth of the stomach with styptics; by combing and massaging the head; and also employing suitable lyes to get the vapours to exhale, and by not forgetting | to exercise conveniently and well to work up a sweat from the skin.

**25.** Quartum *Incommodum* subtilius est Malum; videlicet, quod *Spiritus Detentus* per Clausuram Pororum, videatur posse seipsum nimis multiplicare; quia cùm parùm euolet, & continuò *Spiritus* nouus generetur, nimiùm increscit *Spiritus,* & sic Corpus etiam plùs prædari
5 possit; verùm hoc non prorsùs ita se habet; Nam *Spiritus* omnis conclusus Hebes fit (quandoquidem ventiletur Motu *Spiritus,* vt & *Flamma*); ideóque minùs Actiuus est, & minùs sui generans; Calore certè auctus
[S6ʳ] (vt et *Flamma*) sed Motu Piger. Sed | & huic *Incommodo* Remedium adhiberi possit, à Frigidis, *Oleo* quandóque admistis; qualia sunt, *Rosa,*
10 & *Myrtus*; nam Calidis omninò abstinendum, vt dictum est de *Casiâ.*

**26.** Neque inutilis est Applicatio, ad Corpus, *Vestium,* quæ & ipsæ in se habent aliquid *Vnctuosi,* siue *Oleosi,* non *Aquosi.* Illæ enim exhaurient Corpus minùs; quales sunt ex *Lanâ,* potius quam ex *Lino.* Certè manifestum est in *Spiritibus Odorum,* quod si ponas *Pulueres Odoratos*
15 inter *Lintea,* multo citiùs Virtutem perdunt, quàm inter *Lanea.* Itaque *Lintea* Tactu, & Mundicie iucunda; sed ad nostram *Operationem*
[S6ʳ] suspecta. |

**27.** *Hyberni Syluestres,* cùm incipiunt ægrotare, nihil priùs faciunt, quam vt *Lintea* è Stratis tollant, et in *Laneis* Pannis se conuoluant.

20 **28.** Referunt nonnulli, se magno Sanitatis suæ commodo, *Laneis Carmosinis* proximè ad Cutem, sub Indusijs suis, vsos fuisse, tam ad Bracas, quàm ad Corporalia.

**29.** Est & illud obseruandum, *Aerem* Corpori *Assuetum,* minùs illud deprædari, quam *Nouum,* & subinde mutatum. Itaque *Pauperes,* qui in
25 *Tugurijs* suis, intra proprios *Lares* perpetuò viuunt, nec Sedes mutant,
[S7ʳ] sunt plerunque Longæuiores; Veruntamèn quoàd alias *Operatio|nes,* *Mutationem Aeris* (præsertìm *Spiritibus* non omnino inertibus) vtilem esse iudicamus; Mediocritas autem adhibenda foret, quæ vtrinque satisfaciat; Illud fiet, si *Quatuor Temporibus* Anni, fiet per Stata tempora
30 *Mutatio Loci* ad sedes idoneas, neque sint Corpora, aut in *Peregrinatione* nimiâ, aut in *Statione.* Atque de *Operatione* per *Exclusionem Aeris,* & de euitandâ *Vi* eius *Prædatoriâ,* hæc dicta sint.

6 fit] ~,    *Flamma*);] ~;)    7 auctus] ~,    8 *Flamma*)] ~;)    Piger.] ~:
12 *Aquosi.*] ~:    13 *Lino.*] ~:    24 mutatum.] ~:    27 *Aeris*] ~,

**25.** The fourth disadvantage is a subtler ill, namely that the spirit held back by the closing of the pores may be able to multiply itself too much, because when little escapes, and new spirit is continually being generated, the spirit grows too much and so can prey on the body more. But this is not quite how the matter stands; for all spirit enclosed becomes inert (for like flame, the spirit is ventilated by motion) and so is less active, and able to multiply itself, and its heat (like flame) surely grows but its motion gets sluggish. But ¦ this disadvantage can also be cured by cold preparations, like rose and myrtle, at some time or other mixed with oil. For all hot things should be altogether avoided, as I said in connection with casia.

**26.** Nor is it pointless to apply to the body clothes which have something unctuous or oily but not watery to them. For these take less out of the body, and are made of wool rather than linen. For this is certainly obvious in the spirits of odours that if you put odoriferous powders in linen stuff they lose their virtue much quicker than in woollen. Therefore while its feel and cleanliness may be pleasant, linen is to be suspected for our operation. ¦

**27.** The wild Irish, when they start falling sick, do nothing more at first than take the linen off their beds and wrap themselves in the woollen blankets.

**28.** Some aver that their health was benefited greatly by wearing crimson woollies as drawers and vests next to the skin beneath their clothes.

**29.** This too should be observed, that air habituated to the body preys on it less than new air changed often. Therefore the poor who always keep to hearth and hovel, and never move, are generally long-lived. But as far as other operations go, ¦ I judge that a change of air is useful (especially with spirits not entirely inert). But moderation should be applied to serve both ends, and that may be done if at stated times in the four seasons of the year we move to other suitable lodgings, but not so that our bodies be either travelling or at rest too much. So much then for the operation to exclude the air, and avoid its predatory influence.

[S7$^r$]                    | OPERATIO
super Sanguinem, & Calorem
sanguificantem. III.

## *Historia.*

5  1. *Operationes* duæ Sequentes sunt *Operationibus* duabus Præcedentibus
tanquam *Antistrophæ*; atque ijs respondent, quemadmodùm *Passiua
Actiuis*; Præcedentes enim duæ, id agunt, vt *Spiritus* & *Aer* Actionibus
suis sint minùs deprædantes; Hæ verò, vt *Sanguis* & *Succus Corporis*, sint
[S8$^r$] minùs | deprædabiles. Quoniam verò *Sanguis* est *Irrigatio Succorum*, et
10 *Membrorum*, & *Præparatio* ad ea; *Operationem* super *Sanguinem* primo
loco collocamus. Circa hanc *Operationem* proponemus *Consilia*
Numero pauca, sed Vi valdè efficacia. Ea tria sunt.

2. Primò Dubium non est, quin si *Sanguis* sit aliquantò frigidior,
minùs futurus sit Dissipabilis: Quoniam verò, quæ per Os sumuntur
15 Frigida, cum reliquis Intentionibus haud paucis, malè conueniunt; ideò
optimum foret alia inuenire, quæ non sunt cum istiusmodi incommodis
complicata. Ea duo sunt.

[S8$^v$]     3. Alterum huiusmodi est: Ad|ducantur in Vsum, idque maximè in
*Iuuentute*, *Clysteria* nihil omninò *Purgantia*, aut *Abstergentia*; sed
20 solummodò *Refrigerantia*, et nonnihil *Aperientia*; Probata sunt quæ
fiunt ex Succis *Lactucæ*, *Portulacæ*, *Hepaticæ*, etiam *Sedi maioris*, & *Muci-
laginis Seminis Psillij*, cum *Decoctione* aliquâ temperatâ Aperiente,
admisto aliquanto *Caphuræ*: Verùm *vergente Ætate*, omittatur *Sedum
maius*, & *Portulaca*, & substituantur *Succi Boraginis*, & *Endiuiæ*, aut
25 similium; Atque retineantur *Clysteria* huiusmodi, quantùm fieri potest,
ad Horam scilicet, aut ampliùs.

[Tr$^r$]     4. Alterum est eiusmodi: In vsu sint, præsertim *Æstate*, *Balnea* | *Aquæ
dulcis*, & modicè admodùm tepidæ, prorsùs absque Emollientibus,
*Maluâ*, *Mercuriali*, *Lacte*, & similibus; Adhibeatur potius *Serum Lactis*
30 Recens in nonnullâ Quantitate, & *Rosa*.

5. Verùm, quod *Caput* rei est, & *Nouum*, illud præcipimus; Vt
ante *Balneationem*, inungatur Corpus cum *Oleo*, cum spissamentis, vt
*Qualitas Refrigerij* excipiatur, *Aqua* magis arceatur. Neque tamen *Meatus
Corporis* nimiùm occludantur; Etenim cum *Frigus Exterius* Corpus

---

18 est:] ~;        23 *Caphuræ*] *Caphoræ* / silently *nld* thus in *SEH* (II, p. 181)
27 eiusmodi:] ~;        32 spissamentis,] ~;        33 arceatur.] ~:

## | III. The Operation
## on the Blood, and on the Heat
## that makes it

### *History*

1. The two operations following stand as it were in inverse relation to the two previous ones, and correspond to them as passives do to actives. For the two previous ones work to make the actions of spirit and air less predatory, whereas the next two act to make the blood and sap of the body less | susceptible of predation. But because the blood irrigates the juices and members, and prepares them, I put the operation on the blood first. I lay down a few suggestions concerning this operation, suggestions small in number but extremely effective in their force. There are three of them.

2. First, there is no doubt but that if the blood be somewhat cooler, it will be less easy to dissipate. But because cold things taken by mouth barely agree with the rest of the intentions, it would be best to find other things which are less burdened with those disadvantages. Of these there are two.

3. The first | is this: that there be brought into use especially in youth clysters which are not in the least bit purgative or abstergent, but only cooling and somewhat opening. We approve of those originating from lettuce, purslane, anemone, as well as the greater houseleek and mould of fleawort seed with some mild opening decoction, mixed with a little camphor. But in old age leave off the houseleek and purslane and take up the juice of borage, and endive, or the like instead. As for the clysters, hold them in for as long as possible, i.e. for an hour or more.

4. The other one is this: the use, especially in summer, of baths of fresh water | very slightly warmed, with no emollients at all, such as mallow, dog's mercury, milk, and so on, but have recourse rather to a fair quantity of fresh whey and roses.

5. But the main thing I recommend is new: that before bathing the body be anointed with something to thicken it up, so that the cooling quality may be taken up, the water held back, with the passages of the body still not blocked up too much. For when the external cold shuts

fortitèr occludit, tantùm abest, vt promoueat *Infrigidationem*; vt etiàm eam prohibeat, & irritet *Calorem.*

[T1ᵛ] **6.** Similis est vsus *Vesicarum*, cum ¹ *Decoctionibus*, & *Succis Refrigerantibus*, applicatis circa inferiorem Regionem Corporis, videlicet sub
5 Costas, vsque ad Pubem; Nam & hoc est Genus *Balneationis*, vbi *Corpus Liquoris*, vt plurimùm, excluditur; *Refrigerium* tantùm excipitur.

**7.** Restat tertium *Consilium*, quod non ad *Sanguinis Qualitatem*, sed ad *Substantiam* eius pertinet; vt reddatur magis Firma, & minùs Dissipabilis, & in quam Calor *Spiritûs* minùs agere possit.

10 **8.** Atque de vsu *Limaturæ Auri*, aut *Auri Foliati*, aut *Pulueris Margaritarum*, *Gemmarum*, & *Coralli*, & similium, Hodiè nihil credimus, nisi
[T2ʳ] quatenùs præsenti *Operationi* satisfaciant. Certè, cùm ¹ *Arabes*, & *Græci*, & *Moderni*, ijs Rebus tantas Virtutes tribuerint, non omninò nihil videatur esse in istis, quæ tot Homines Experti obseruârunt. Itaque
15 missis Phantasticis circa illas Opinionibus, planè arbitramur, si vniuersæ *Substantiæ Sanguinis* aliquid insinuari possit per *Minima*, in quod *Spiritus*, & *Calor* parùm, aut nihil agere possint, omninò id non tantùm *Putrefactioni*, sed etiam *Arefactioni* obstiturum; & ad vitam prolongandam fore efficacissimum. In hoc tamen plures adhibendæ sunt
20 *Cautiones*: Primò, vt fiat admodùm *exacta Comminutio*; Secundò, vt
[T2ᵛ] huiusmodi *Dura* & *Solida*, sint omnis *Malignæ Qualitatis* ex'pertia, ne cùm in Venis dispergantur & lateant, aliquid Nocumenti inferant; Tertiò, vt nunquàm *sumantur cum Cibis*, nec ita excipiantur, vt diù hæreant, ne generent periculosas Obstructiones circa Mesenterium.
25 Quartò, vt *Rarus* sit eorum *vsus*, ne coeant, & cumulentur in Venis.

**9.** Itaque *Modus* excipiendi sit, *Stomacho ieiuno*, in *Vino Albo*, cui admistum sit parùm *Olei Amygdalini*, & fiat *Corporis Exercitatio* super *Haustum* eorum.

**10.** *Simplicia* autem, quæ Operationi huic satisfaciant, possint esse
30 loco omnium *Aurum*, *Margaritæ*, & *Corallus*; *Metalla* enim omnia,
[T3ʳ] præter *Aurum*, non sunt absque ¹ *Malignâ Qualitate*, in *Volatili* ipsorum; Neque etiam tam exquisitè comminuuntur, quam *Aurum Foliatum*; *Gemmæ* autem *Translucidæ*, & tanquam *Vitreæ*, minùs nobis placent (vt & anteà diximus) propter suspicionem *Corrosionis.*

35 **11.** At nostro iudicio, & tutior, & efficacior foret vsus *Lignorum*, in *Infusionibus*, & *Decoctionibus*; Satis enim in iis possit esse, ad Firmitudinem *Sanguinis*. Neque tamen simile Periculum est ab *Obstructione*;

9 Dissipabilis,] ~;    12 satisfaciant.] ~:    19 efficacissimum.] ~:
20 *Comminutio*;] ~:    22 inferant;] ~:    37 *Sanguinis.*] ~;

up the body tightly, it is so far from promoting cooling that it actually prevents it and stimulates heat.

6. Similar is the use of bladders when | they are applied with decoctions and cooling juices about the lower parts of the body, i.e. from below the ribs down to the privates. For this is a type of bathing but with the body of the liquid mostly kept away and the cooling quality alone selected.

7. The third suggestion remains, and that is concerned not with the quality of the blood but with its substance, i.e. to beef it up, and make it less susceptible of dissipation and the workings of the spirit's heat.

8. As for the use of gold filings or gold leaf, or powdered pearls, gems, and coral, and so on, we nowadays have no faith in them, except to the extent that they are up to the present operation. Certainly, as | the Arabs, Greeks, and Moderns have ascribed such virtues to these things, it would seem strange if there were absolutely nothing in what so many experienced men have observed. Thus, leaving aside the fantasies associated with those opinions, I distinctly believe that if something could be insinuated *per minima* into the whole substance of the blood, in which the spirit and heat could do little or nothing at all, it would block absolutely not just putrefaction but arefaction too, and be extremely effective in prolonging life. Nevertheless in this a few notes of caution should be sounded. In the first place, that a very fine breaking down of particles be achieved; secondly, that hard and solid bodies of this kind be stripped of their malign qualities | lest when distributed and hidden in the veins, they introduce something harmful; thirdly, that they never be taken with food, nor absorbed in such a way that they stay long and thereby risk producing blockages round the mesentery; and fourthly, that they are used infrequently, lest they assemble and build up in the veins.

9. The way to take them is therefore on an empty stomach, in white wine mixed with a little almond oil, and with bodily exercise after drinking them.

10. Now of simples suitable for this operation, these may do for all: gold, pearl, and coral. For all metals besides gold have a | malign quality in their volatile component. They also do not bear such fine division as gold leaf. But (as I said before) I have little regard for translucent and glassy gems as they may be corrosive.

11. But in my judgement, the use of wood in infusions and decoctions would be more efficacious and safer. For they have the wherewithal to beef up the blood, but do not carry the same risk of

præcipuè autem, quia possunt sumi in *Cibo* & *Potu*; vnde faciliùs Ingressum reperient in *Venas*, nec deponentur in *Fæcibus*.

[T3ᵛ]    **12.** *Ligna* ad hoc idonea sunt, *San|talum*; *Quercus*; & *Vitis*; *Ligna* enim *Calidiora*, aut aliquâ ex Parte *Resinosa*, reiicimus; Possint tamen
5   adiici Caules siccæ, & lignosæ *Roris marini*; cùm *Frutex* sit *Rosmarinus*, & Ætatem multarum *Arborum* æquet; Etiàm *Hederæ* caules Siccæ, & Lignosæ, sed eâ Quantitate, vt Saporem non reddant ingratum.

**13.** Sumantur verò *Ligna* aut in *Iusculis* decocta, aut infusa in *Mustum*, aut *Ceruisiam* antequàm sedeat. In *Iusculis* autem (vt fit in
10   *Guaiaco*, & similibus) sempèr infundantur diù, antequàm decoquantur, vt firmior Pars *Ligni*, & non tantùm ea, quæ leuitèr hæret, eliciatur.
[T4ʳ]   *Fraxinus* autem, licet ad Pocula adhibeatur, | nobis suspecta est. Atque de *Operatione* super *Sanguinem* hæc inquisita sint.

## OPERATIO
15   super Succos Corporis. IV.

### *Historia.*

**1.** Dvo sunt *Corporum* Genera (vt in *Inquisitione* de *Inanimatis* iam dictum est) quæ difficiliùs consumuntur: *Dura* & *Pinguia*; vt cernitur in
19   *Metallis*, & *Lapidibus*; atque in *Oleo*, & *Cerâ*.
[T4ᵛ]    **2.** Operandum itaque est, vt *Suc|cus Corporis* sit *Sub-durus*; atque etiam vt sit *Sub-pinguis*, aut *Sub-roscidus*.

**3.** Quatenùs ad *Duritiem*, ea efficitur tribus Modis. *Naturâ Alimenti firmâ*; *Frigore condensante Cutem & Carnes*; & *Exercitatione Succos fermentante*, & compingente, ne sint molles, & spumosi.
25   **4.** Quatenùs ad *Naturam Alimenti*, talis esse debet, vt sit minùs *Dissipabilis*; qualia sunt *Caro Bouina*, *Caro Suilla*, *Caro Ceruina*, etiam *Caro Caprearum*, *Hœdorum*, *Cygnorum*, & *Anserum*, & *Palumbium Syluestrium* (præsertìm si huiusmodi Carnes fuerint modicè Salitæ),
[T5ʳ]   *Pisces* itidem Saliti, & sicci; Etiàm *Caseus* subuetus, & huiusmodi. |
30   **5.** Quoad *Panem* autem, *Auenatus*, aut etiam paululùm *Pisatus*, aut *Secalicius*, aut *Hordeaceus*, solidior est, quam ex *Frumento*; Atque etiam in *Pane Frumentaceo*, solidior qui paulò plus habet ex *Furfure*, quam qui purioris est *Pollinis*.

**6.** *Orcades*, qui *Piscibus* vescuntur *Salitis*, atque generalitèr
35   *Icthyophagi*, Longæui sunt.

---

9 sedeat.] ~:        11 eliciatur.] ~:        15 Corporis.] ~ₐ        17 Genera] ~,
28 *Syluestrium*] ~,   Salitæ),] ~)ₐ

obstruction, especially because they can be taken with food and drink, whence they will gain easier access to the veins, and not be voided in excrements.

12. Woods suitable for this are ˈ sandalwood, oak, and vine; for I reject woods which are hotter or to some degree resinous. But I add the dry and woody stalks of rosemary, for rosemary is a shrub and it lives as long as many trees; the dry and woody stalks of ivy too, but in such an amount as does not make their taste disagreeable.

13. But these woods should be taken boiled in soups, or infused in new wine or beer before it has settled. For with soups (as is the case with guaiac and the like) let them always be infused for a good while before they are boiled, so that the firmer part of the wood and not just the less tenacious part can be drawn out. Now ash, ˈ though used for cups, I distrust. So much then for the operation on the blood.

## IV. The Operation
## on the Juices of the Body

### *History*

1. There are two kinds of bodies (as I have already said in the inquiry concerning inanimate bodies) which are consumed only with difficulty: hard and fat, as we see in metals and stones, and in oil and wax.

2. We must therefore operate to ˈ make the juice of the body relatively hard, as well as relatively fat or moist.

3. As for hardness, it is achieved in three ways: by the firm nature of the aliment; by cold condensing the skin and flesh; and by exercise fermenting and confining the juices to stop them being soft or frothy.

4. As for the nature of the aliment, it should be such as cannot be dissipated, like beef, pork, venison, as well as goat, kid, swan, goose, and wood-pigeon (especially if these meats have been slightly salted), fish dry and salted, as well as relatively old cheese, and the like. ˈ

5. Now as far as bread goes, oatmeal, grain laced with peas, rye, or barley makes a more solid bread than wheat; and in wheat bread, the more solid is made with a little more bran than that which contains more refined flour.

6. Orcadians, who live on salt fish, and fish eaters in general, are long-lived.

7. *Monachi*, & *Eremitæ*, qui parcè, & *Sicco Alimento* pascebantur, fuerunt vt plurimùm Longæui.

8. Etiam *Aqua pura* in Potu frequentèr vsurpata, reddit *Succos*
4 Corporis, minùs *Spumosos*; cui, si propter *Spiritûs Hebetudinem* (qui
[T5$^r$] proculdubio in *Aquâ* est $^|$ parùm *Penetratiuus*) admisceatur aliquid *Nitri*, vtile esse existimamus. Atque de *Firmitudine Alimenti* hactenùs.

9. Quatenus ad *Condensationem Cutis* & *Carnium* per *Frigus*, Viuaciores ferè sunt, qui sub *Dîo* viuunt, quam qui sub *Tecto*; Atque qui in *Regionibus Frigidis*, quam qui in *Calidis*.

10 10. *Vestes nimiæ*, siue in Lectis, siue Portatæ, corpus soluunt.

11. *Lauatio Corporis* in *Frigidâ*, bona ad Longitudinem vitæ; vsus *Balneorum tepidorum*, malus. De *Balneis* autem, ex *Aquis Astringentibus*
[T6$^r$] *Mineralibus*, superiùs dictum est. $^|$

12. Quatenùs ad *Exercitationem, Vita ociosa* manifestò reddit
15 Carnes Molles, & Dissipabiles: *Exercitatio* autem *Robusta* (modo absint nimii *Sudores*, aut *Lassitudines*) duras et compactas; Etiam *Exercitatio* intra *Aquas Frigidas*, qualis est *Natatio*, valdè bona; Atque generalitèr *Exercitatio* sub *Dîo*, melior quam sub *Tecto*.

13. De *Fricationibus* (quod est *Exercitationis* genus) tamen quia
20 Alimenta magis euocant, quam indurant, posteà suo loco inquiremus.

14. Iam verò, cum de *Duritie Succorum* dictum sit, veniendum ad
[T6$^v$] *Oleositatem* siue Roscidationem $^|$ ipsorum, quæ perfectior, et potentior est *Intentio*, quam *Induratio*; quia non habet *Incommodum*, neque Malum complicatum. Omnia enim, quæ ad *Duritiem Succorum* per-
25 tinent, eiusmodi sunt, vt cum *Alimenti Absumptionem* prohibeant, etiam eiusdem *Reparationem* impediant; vnde fit vt Diuturnitati vitæ, eadem & prosint, & obsint. At quæ ad *Roscidationem Succorum* pertinent, ex vtrâque parte iuuant; cùm reddant *Alimentum*, & minùs *Dissipabile*, & magis *Reparabile*.

30 15. Cùm verò dicimus, quòd *Succus Corporis* debeat fieri *Roscidus*, &
[T7$^r$] *Pinguis*, Notandum est, hoc nos non intelligere de *Pinguedi*$^|$ne, aut *Adipe* manifesto, sed de *Rore perfuso*, et (si placet) *Radicali*, in ipsâ Corporis substantiâ.

---

4 cui] Cui   *Hebetudinem*] ~,          7 *Frigus*,] ~;          12 malus.] ~:
14 *Exercitationem*,] ~;          15 *Robusta*] ~,          16 compactas;] ~:
19 *Fricationibus*] ~,          22 ipsorum,] ~;          24 complicatum.] ~:
27 obsint.] ~;     31 *Pinguis*,] ~;

7. Monks and hermits, who lived frugally and on dry food, were generally long-lived.

8. Pure water too, taken often, makes the juices of the body less frothy; but on account of the sluggishness of the spirit (which in water is doubtless | not very penetrating), I think that some nitre should be mixed in with it. And so much for the firmness of the aliment.

9. As for the condensing of the skin and flesh by cold, people live longer who live in the open rather than in houses, as do those who live in cold rather than hot regions.

10. Too much clothing, be it in bed or borne about, undoes the body.

11. Washing of the body in cold water is good for longevity; the use of warm baths, bad. Of baths drawn from astringent mineral waters I have spoken above. |

12. As for exercise, a life of ease manifestly makes the flesh soft, and liable to dissipation, whereas good exercise (without too much sweat or weariness) makes it hard and compact. Also exercise in cold water, like swimming, is extremely good; and in general exercise in the open is better than indoors.

13. Concerning massage (which is a kind of exercise), because it rather draws out aliment than hardens it, I shall inquire into it later in its proper place.

14. But now, since I have dealt with the hardness of the juices, I must come to their oiliness or moistening, | which is a more perfect and powerful intention than hardening because it has no inconvenience or ill effect bound up with it. For all the things associated with hardness of the juices are of such sort that while they stop squandering of aliment, they also get in the way of its repair, with the result that the same things both promote and hinder length of life. But things which relate to the moistening of the juices help in both ways, since they make the aliment at once less liable to dissipation and more capable of repair.

15. But when we say that the juice of the body should be made moist and fat, it should be noted that we do not mean fatness | or evident fatty tissue, but a dewiness spread through, and (if you like) radical, in the very substance of the body.

**16.** Neque rursùs existimet quispiam, *Oleum*, aut *Pinguia Ciborum*, aut *Medullas*, similia sibi generare, atque *Intentioni* nostræ satisfacere; Neque enim quæ perfecta semel sunt, retrò aguntur; sed talia debent esse *Alimenta*, quæ post *Digestionem* et *Maturationem*, tùm demùm 5 *Oleositatem* in Succis ingenerent.

**17.** Neque rursùs existimet quispiam, *Oleum* & *Pingue Coaceruatum*, & *Simplex*, difficilis esse Dissipationis; in *Mistione* autem non eandem [T7$^v$] retinere naturam; Etenim quemadmodum *Oleum* | per se multò seriùs Consumitur, quàm *Aqua*, ita etiam in *Papyro*, aut *Sudario*, diutiùs 10 hæret, et tardius desiccatur; vt priùs notauimus.

**18.** Ad *Irrorationem* Corporis meliùs faciunt *Cibi Assati*, aut *Furno Cocti*, quam *Elixi*: Atque omnis *Præparatio Ciborum*, cum *Aquâ*, incommoda est; Quinetiam et *Oleum*, copiosiùs elici videmus ex *Corporibus Siccis*, quam ex *Humidis*.

15 **19.** Generalitèr ad *Irrorationem* Corporis, prodest multus Vsus Dulcium, *Sacchari*, *Mellis*, *Amygdalarum dulcium*, *Pinearum*, *Pistaciorum*, *Dactilorum*, *Vuarum Passarum*, *Vuarum Corinthi*, *Ficuum*, et huius-[T8$^r$] modi; Contrà, omnia | *Acida*, et nimiùm *Salsa*, et nimiùm *Acria*, sunt Generationi *Succi Roscidi* opposita.

20 **20.** Neque *Manichæis*, eorúmque Diætæ fauere existimabimur, si *Semina* quæque, et *Nucleos*, et *Radices*, in Cibis, aut eorum Condimentis frequentia esse debere dicamus; quandoquidem omnis *Panis* (*Panis* autem Ciborum Firmamentum est) aut ex *Seminibus* est, aut ex *Radicibus*.

25 **21.** Ante omnia verò ad *Irrorationem Corporis* maximè facit *Natura Potûs*, qui Ciborum Vehiculum est. Itaque in vsu sint *Potus* illi, qui absque omni *Acrimoniâ*, aut *Acedine*, subtiles tamen sint, quales [T8$^v$] sunt *Vina* (vt ait *Anus* apud *Plautum*) *Vetustate* | *Edentula*, et *Ceruisia* eiusdem generis.

30 **22.** *Hydromel* (vt arbitramur) non foret malum, si fuerit Forte, et Vetus; Attamen quoniam omne *Mel* habit aliquid Acutum (vt patet ex acerrimâ illâ *Aquâ*, quam *Chymici* ex eo extrahunt, quæ etiam *Metalla* soluit), meliùs foret, si fieret similis Potio ex *Saccharo*, non infuso leuitèr, sed ita incorporato, quemadmodùm *Mel* solet esse in 35 *Hydromellite*, et quæ habeat Vetustatem Anni, aut sex Mensium; vnde *Aqua* Cruditatem deponat, et *Saccharum* Subtilitatem acquirat.

---

17–18 huiusmodi;] ~:      26 est.] ~:      28 *Vina*] ~,      31 Acutum] ~,
33 soluit),] ~)‸

16. Nor should anyone imagine that oil or the fat of food or marrow gives rise to something like itself, and fulfils my intention, for things once perfected only slip backwards; so aliment should be such as only after digestion and maturation produces oiliness in the juices.

17. Nor again should anyone imagine that oil and fat concentrated and plain is hard to dissipate, but does not keep the same nature in a mixture. For just as oil | by itself is consumed less rapidly than water, so also does it stick longer to paper or napkin, and dries out more slowly, as I noted above.

18. For moisturizing the body, roast and baked food does better than boiled. And all cooking of food in water is unhelpful; and furthermore we see that oil is drawn more abundantly from dry bodies than from moist ones.

19. For moisturizing the body the use in general of sweet things helps a lot; sugar, for instance, honey, sweet almonds, pine-nuts, pistachio nuts, dates, raisins, currants, figs, and so on. On the other hand, all | acid, very salty, and bitter things militate against the generation of moist juice.

20. Nor shall anyone think that I favour the Manichæans and their diet, if I speak up for frequent use of seeds, nuts, and roots in foods or their seasonings, since all bread (and bread is the mainstay of all meals) is made from seeds or roots.

21. But before everything else the nature of one's drink, which is the vehicle of food, does most for the moisturizing of the body. Thus let the drinks we use be quite without acrimony or acidity, and yet subtle, as are those wines which are (as the old woman in Plautus says) toothless | with age; and the same goes for beer.

22. Mead (in my judgement) would not be bad, if it were strong and old; yet since all honey has something acrid to it (as appears from that very vehement water which the chemists extract from it, and which even dissolves metals) it would be better if a similar drink could be made with sugar, not lightly infused but as firmly embodied as honey usually is in mead, and which has been aged for six months or a year, so that the water loses its crudity, and the honey gains subtlety.

**23.** Atque *Vetustas Vini,* aut *Potûs,* hoc habet: quod Subtilitatem
[Vɪʳ] generat in Partibus *Liquoris,* | Acrimoniam in *Spiritibus,* quorum
primum vtile, secundum noxium; Itaque ad hanc Complicationem
enodandam, mittatur in Dolium, priusquam resederit nonnihil Vinum
5 à Musto, *Caro suilla,* aut *Ceruina,* benè cocta, vt habeant *Spiritus vini,*
quod ruminent, & mandant, atque inde mordacitatem suam deponant.

**24.** Similitèr, si recipiat *Ceruisia,* non solùm Grana *Tritici, Hordei,*
*Auenarum, Pisarum,* &c. sed etiam Partem (puta tertiam) ex *Radicibus,*
aut *Pulpis* pinguibus (qualia sunt *Radices Potado, Medullæ Atriplicis,*
10 *Radices Bardanæ,* aut aliæ *Radices* Dulces, et Esculentæ) vtiliorem fore
[Vɪʳ] *Potum* | ad Longæuitatem existimamus, quam *Ceruisiam* ex *Granis*
tantum.

**25.** Etiam quæ in Partibus suis valdè tenuia sunt, & nihilominùs
nullâ prorsus sunt *Acrimoniâ,* aut *Mordacitate,* vtilia sunt, in *Condi-*
15 *mentis Ciborum;* Qualem Virtutem inesse deprehendimus in paucis
quibusdam ex *Floribus; Floribus* scilicet *Hederæ,* qui in Aceto infusi
etiam gustui placent; *Floribus Calendulæ,* qui in vsu sunt in *Brodijs;* &
*Floribus Betonicæ.* Atque de *Operatione* super *Succos Corporis* hæc
19 inquisita sunt.

[V2ʳ] | OPERATIO
super Viscera ad Extrusionem Alimenti. V.

*Historia.*

**1.** Qvæ *Viscera* illa *Principalia* (quæ *Concoctionis Fontes* sunt) *Stoma-*
*chum, Hepar, Cor, Cerebrum,* ad Functiones suas probè exercendas con-
25 fortant (vnde Alimenta in partes distribuuntur, Spiritus sparguntur,
[V2ʳ] atque inde Reparatio Corporis totius transigitur), à *Medicis,* | atque
eorum *Descriptis,* & *Consilijs,* petenda sunt.

**2.** De *Splene, Felle, Renibus, Mesenterio, Ilijs* & *Pulmonibus,* non
loquimur; Sunt enim *Membra* Ministrantia Principalibus; Atque cùm
30 de *Sanitate* tractatur, in Considerationem vel præcipuam, quandóque

---

10 Esculentæ] / some copies Esculentas; see Introduction, p. lxxiii above

1 habet:] ~;          7 Similitèr] / some copies (e.g BNF) with grave, others (e.g. Folger copy
2) without          9 pinguibus] ~,     *Atriplicis*] *Artiplicis* / a compositorial transposition?
This emendation suggested in *SEH* (II, p. 185 n. 3). See also *tns* to V8ᵛ, p. 300 below
10 *Radices*] / some copies (e.g. BNF) without comma; some copies (e.g. Folger copy 2) with
comma          21 Alimenti.] / some copies (e.g. BNF) with stop, others (e.g. Newberry)
without          24–5 confortant] ~,          26 transigitur),] ~)ᴀ

23. Now age in wine or drink has the power to generate subtlety in the parts of the liquor | but acrimony in its spirits, of which the first is useful, the second harmful. Thus to get out of this dilemma put well-cooked pork or venison in the barrel before the wine has somewhat settled out its must, so that the spirits of the wine have something to get their teeth into, and so lose their edge.

24. Likewise, if beer were made up not just of wheat, barley, oats, peas, etc., but (say) a third part of roots or fatty pulps (like potato roots, the marrow of orach, burdock roots, or other sweet and edible roots), the drink | would, I believe, do more for longevity than beer made from grain alone.

25. Moreover, whatever has extremely thin parts, but yet has pretty well no acrimony or sharpness, serves well for seasoning food; and we find that this virtue inheres in some few flowers, namely those of ivy, which, infused in vinegar, are even palatable; flowers of marigold, which are used for broths, and betony flowers. And so much for the operation on the juices of the body.

## | V. The Operation
## on the entrails to force out the food

### *History*

1. Of the things which comfort the principal internal organs (which are the basis of concoction), i.e. the stomach, liver, heart, and brain, in the proper discharge of their functions (whence aliment is distributed to the parts, the spirits are spread out, and repair of the whole body is thereby carried out) you must consult the physicians, | and their prescriptions and advice.

2. Of the spleen, gall-bladder, kidneys, mesentery, guts, and lungs, I say nothing, as they are members ministering to the principal ones. And though they sometimes come in for close examination in relation to health since each has its own particular ills which, if untreated, might

veniunt; quia patiuntur singula suos Morbos, qui nisi curentur, etiam in *Viscera Principalia* incurrunt. Quatenùs verò ad *Prolongationem Vitæ,* & Reparationem per *Alimenta,* & *Retardationem Atrophiæ Senilis,* si *Concoctiones,* & *Principalia* illa *Viscera* benè se habeant, cætera maximâ
5 ex parte ad Votum sequentur.

[V3ʳ] **3.** Atque ex *Medicorum* Libris, | qui de Quatuor *Membrorum Principalium Confortatione,* & *Commodis,* Sermones faciunt, decerpenda sunt ea vnicuique, quæ pro ratione *Statûs Corporis proprij,* in *Diætam* & *Regimen Vitæ* transferri poterint. Etenim *Sanitas Medicinis temporalibus*
10 plerunque indiget; at *Diuturnitas Vitæ* ex *Victûs ratione,* & constanti *Medicinarum* iuuantium serie, speranda est. Nos verò Pauca, eaque Selecta, & Optima proponemus.

**4.** *Stomachum* (qui (vt aiunt) est *Paterfamilias,* & cuius Robur ad reliquas *Concoctiones* est Fundamentale) ita munire decet, & con-
[V3ᵛ] firmare, vt sit absque *Intem|perie Calidus;* deinde *Astrictus,* non *Laxus;*
16 etiam *Mundus,* non *Humorum Fastidijs Oppressus;* Et nihilominùs (cùm ex seipso, potius quam ex Venis nutriatur) minimè prorsus *Inanis,* aut *Ieiunus;* postremò in *Appetitu* seruandus est, quia *Appetitus Digestionem* acuit.

20 **5.** Miramur quomodò illud, *Calidum Bibere* (quod apud Antiquos in vsu fuit) in Desuetudinem abierit. Nouimus certè *Medicum* admodùm celebrem, qui in Prandio, & Cœnâ, *Iusculum* etiam *præcalidum* auidè ingerere solebat, & paulò post optare vt regestum esset; Neque enim
[V4ʳ] mihi *Iusculo* opus est (inquit) sed *Calido* tantùm. |

25 **6.** Omninò vtile arbitramur, primam *Potionem,* siue *Vini,* siue *Ceruisiæ,* siue *Potûs* alterius (cui quis insueuit) in Cœnâ semper calidam exhiberi.

**7.** *Vinum Extinctionis Auri,* vtile arbitramur semel in Mensâ; Non quod *Aurum,* aliquid Virtutis ad hoc largiri credamus, sed quia *Extinc-*
30 *tionem* omnem *Metallicam,* in aliquo Liquore, Astrictionem potentem indere nouimus. *Aurum* autem deligimus, quia præter illam (quam optamus) *Astrictionem,* nil aliud *Metallicæ Impressionis* post se relinquit.

**8.** *Offas Panis* in *Vino,* mediâ Mensâ, vtiliores, quam ipsum *Vinum*
[V4ᵛ] esse iudicamus; præsertìm | si *Vino* cui *Offa* intingatur, *Ros-Marinus,* &
35 *Cortex Citri,* fuerint infusi; idque cum *Saccharo,* vt tardiùs labatur.

---

2 incurrunt.] ~:     9 poterint.] ~:     11 est.] ~:     16 cùm] cum / some copies (e.g. BNF) with grave, others (e.g. Folger copy 2) without     20 *Bibere*] ~, 21 abierit.] ~;     24 est] ~,     26 alterius] ~,     31 nouimus.] ~:

invade the principal ones, yet for the prolongation of life, repair by aliment, and putting off the atrophy of age, if the concoctions and principal viscera work well, the others for the most part work as one might wish.

3. And from the volumes of the physicians, | which have the comfort and convenience of the four principal members as their theme, we should each of us choose, according to the particular habit of our body, the things which can be applied to our diet and routines. For health generally requires short-term medicines; but we should hope for longevity only from sensible eating, and an unbroken run of supporting medicines. And now indeed I will put down a few of the choicest.

4. The stomach (which is (as they say) the paterfamilias whose strength is the foundation of all the other concoctions) should be so buttressed and secured, as to be devoid of | excessive heat; tight not loose; clean and not burdened with disgusting residues of the humours; and yet (since it is fed by itself rather than the veins) not absolutely empty or starved; lastly its appetite should be kept keen, for appetite sharpens digestion.

5. I marvel how taking drinks hot (which was customary with the ancients) has fallen out of use. I knew a very famous physician who used eagerly to down very hot soup at supper or dinner, and not long after wished it thrown up again, 'for', he declared, 'I had no use for the soup, only for its warmth'. |

6. In my judgement, it is very helpful always to take the first drink at supper hot, whether it be of wine, beer, or whatever other drink one is used to taking.

7. Wine in which gold has been quenched should, in my judgement, be taken once during a meal: not because I believe that gold bestows any virtue but because I know that all quenching of metal in any liquor gives it strong astringency. And I choose gold because besides the astringency that I want, it leaves no metallic impression behind it.

8. In the middle of the meal, I judge that it is better to take bread soaked in wine rather than straight, especially | if rosemary or citron rind were infused in the wine absorbed in the bread, and if sugar were added to make it less runny.

9. Vsum *Cotoneorum,* ad *Stomachi* Robur, vtilem esse certum est. Meliùs tamen adhiberi iudicamus in Succis depuratis, cum *Saccharo* (quos *Myuas* vocant) quam in *Carnibus* ipsorum, quia *Stomachum* nimis grauant. Illæ verò *Myuæ* post Mensam, *simplices,* at ante Mensam, cum
5 *Aceto,* vtilissimè sumuntur.

10. Vtilia *Stomacho* sunt præ cæteris Simplicibus *Ros-marinus, Enula, Mastyx, Absynthium, Saluia, Mentha.*

[V5ᵛ] 11. *Pillulas* ex *Aloe,* & *Mastice,* & *Croco,* præsertìm Temporibus Hyemalibus, ante Prandium sumptas, probamus; ita tamen, vt *Aloe* non
10 tantum Succo *Rosarum* multis vicibus abluta sit, sed etiam in Aceto (in quo dissolutum fuerit *Tragaganthum*) & posteà in *Oleo Amygdalino* Dulci, & Recenti, ad aliquot horas macerata sit, antequàm formetur in *Pillulas.*

12. *Vinum* aut *Ceruisia Infusionis Absynthij,* cum modico *Enulæ,*
15 & *Santali Cytrini,* rectè per vices adhibetur; atque hoc Hyeme potiùs.

13. At Æstate, Haustus ex *Vino Albo,* cum Aquâ *Fragariæ* diluto, in quo *Vino,* Pulueres exquisiti *Perlarum,* & *Testarum Cancrorum Fluuiatil-*
[V5ᵛ] *ium,* & (quod mirum fortassè videatur) parum *Cretæ* fuerint infusa,
20 *Stomachum* optimè recreat, & roborat.

14. At generalitèr, omnis *Haustus Matutinus* (quales frequentèr in vsu sunt) *Refrigerantium* (*Succorum, Decoctionum, Seri Lactis, Hordeatorum,* & similium) fugiendus est; nihílque prorsùs immittendum *Stomacho ieiuno,* quod sit *Frigidum purum.* Meliùs exhibebuntur Res huiusmodi
25 (si necessitas postulet) vel *Horâ quintâ* post *Prandium,* vel *Horâ vnâ* post leue *Ientaculum.*

15. *Ieiunia* frequentia mala sunt ad Longæuitatem; quinetiam *Sitis* quæcunque euitanda, & seruandus *Stomachus* satis *Mundus,* sed
[V6ʳ] perpetuò quasi *Humidus.*

30 16. *Oleum Oliuarum* Recens, & Bonum, in quo *Mithridatij* nonnihil dissolutum fuerit, inunctum *Spinæ Dorsi,* ex aduerso *Oris Stomachi, Stomachum* mirum in modum confortat.

17. Sacculus ex Floccis Carmosinis, infusis in *Vinum Austerum,* in quod infusa fuerint *Myrtus,* & *Cortex Citri,* & parùm *Croci,*
35 super *Stomachum* perpetuò gestari potest. Atque de *Stomachum*

---

1 est.] ~:        3 *Myuas*] / *SEH* (II, p. 187 n. 1) glosses as 'conserves', but see *cmt,* p. 450
below        4 grauant.] ~:        8 *Pillulas*] / *SEH* (II, p. 187) reads as Pilulas
10 Aceto] ~,        19 fortassè] / see Introduction, p. lxxiii above        21 *Matutinus*] / see
Introduction, p. lxxiii above)        22 *Refrigerantium*] ~,        24 *purum.*] ~:
huiusmodi] ~,

9. Taking quinces for strengthening the stomach is certainly helpful. Yet in my judgement they would be better taken in with sugar in refined juices (which they call *Myuæ*) than in the meat itself because the latter load the stomach too much. These *Myuæ* are best taken neat after dinner, but before dinner with vinegar.

10. Before all other simples, these are good for the stomach: rosemary, horse-heel, mastic oil, wormwood, sage, and mint.

11. I recommend pills of aloes, mastic, and saffron taken before dinner, | especially in winter time; yet in such a way that the aloe be washed repeatedly not just in rose-oil but also soaked for some hours in vinegar (in which tragacanth has been dissolved), and lastly in fresh oil of sweet almonds before it is shaped into pills.

12. Wine or beer with an infusion of wormwood, and a dash of horse-heel and yellow sandalwood, is worthwhile at times, and particularly in winter.

13. But in summer a glass of white wine diluted with strawberry water in which fine powders of pearl and crayfish shells, and (strange as | it may seem) a little chalk have been infused, refreshes and strengthens the stomach no end.

14. Yet in general, all morning drinks (of the kinds taken frequently) of cooling things (as juices, decoctions, whey, barley water, and the like) should be avoided; and absolutely nothing that is completely cold should be taken on an empty stomach. Things of that kind (if necessary) are better taken five hours after dinner, or one hour after a light breakfast.

15. Frequent fasts are bad for longevity; all thirsts should likewise be avoided, and the stomach should be kept clean but more or less always moist. |

16. Good, fresh olive oil, in which some mithridate has been dissolved, rubbed on the spine opposite the mouth of the stomach, comforts the stomach wonderfully.

17. A little bag of crimson wool, steeped in dry wine in which myrtle, citron rind, and a little saffron have been infused, may always be worn on the stomach. So much then for comforting the stomach; to

*Confortantibus* hactenùs; cùm etiam haud pauca ex his, quæ alijs Operationibus inseruiunt, ad hoc etiàm iuuent.

**18.** *Iecori*, si à *Torrefactione*, siue *Desiccatione*, atque ab *Obstructione*, [V6ʳ] immune seruetur, nil vltrà ¦ opus est: Etenim *Exolutio* illa, quæ *Aquositates* generat, *Morbus* prorsùs est; At reliqua duo etiam *Senectus* obrepens inducit.

**19.** Hùc pertinent, vel maximè ea, quæ in Operatione super *Sanguinem*, descripta sunt; Iis adjiciemus Pauca admodùm, sed Electa.

**20.** Præcipuè in vsu sit, *Vinum Granatorum* Dulcium, aut si illud haberi non possit, *Succus ipsorum* recens expressus; *Manè* sumendus, cum aliquanto *Sacchari*, & immisso in Vitrum (in quod fit Expressio) modico *Corticis Citri* recentis, & *Gariophyllis* tribus, aut quatuor [V7ʳ] integris: Hocque vsurpetur à *Februario* ad Finem *Aprilis*. ¦

**21.** In vsum adducatur, ante alias omnes *Herbas*, *Nasturtium*, sed tamen *Pubescens*, non *Vetus*: vsurpetur siue *Crudum*, siue in *Iusculis*, siue in *Potu*; Et post hanc *Cochlearia*.

**22.** *Aloe*, quocunque modo abluta, aut correcta, *Hepati* noxia; Itaque nunquàm familiaritèr sumenda est. *Rhubarbarum* contrà vitale *Hepati*, modò tres adhibeantur *Cautiones*: Primò, vt sumatur *ante Cibum*, ne desiccet nimis, aut Vestigium *Stypticitatis* relinquat; Secundò, vt *maceretur* ad Horam vnam, aut duas, in *Oleo Amygdalino* recenti, cum *Aquâ Rosaceâ*, antequàm aliàs infundatur, aut detur in Substantiâ; Tertiò, vt [V7ᵛ] vicibus alternis suma¦tur, aliàs *simplex*, aliàs cum *Tartaro*, aut parùm *Salis nigri*, ne leuiora tantùm asportet, & reddat Massam Humorum magis obstinatam.

**23.** *Vinum*, aut *Decoctum* aliquod *chalybeatum*, ter aut quater in Anno sumi probo, ad *Obstructiones* potentiores soluendas; ita tamen vt sempèr præcedat *Haustus* duorum, aut trium Cochlearium *Olei Amygdalini*, dulcis & recentis, & sequatur *Motus Corporis*, præsertìm *Brachiorum*, & *Hypochondriorum*.

**24.** *Liquores dulcorati*, idque cum Pinguedine quadam, ad arcendam *Arefactionem*, & *Salsedinem*, & *Torrefactionem*, & denique *Senilitatem* [V8ʳ] *Iecoris*, præcipuè & plu¦rimùm possunt; præsertìm si per Vetustatem benè incorporentur; Tales fiant ex *Fructibus*, & *Radicibus* dulcibus; Scilicet *Vina* & *Potus* ex *Vuis passis* recentibus, *Iuiubis, Carycis, Dactilis, Pastinacis, Bulbis* siue *Potadis*, & huiusmodi, cum admistione *Lycoritiæ*

9 Dulcium,] ~;      11 Vitrum] ~,      13 *Februario*] ~,      18 *Hepati*] / e.g. BNF copy; other copies (e.g. Folger copy 2) *Epati*      20 relinquat;] ~:      22 Substantiâ;] ~: 26 *chalybeatum*] / see Introduction, p. lxxiii above      34 *Fructibus*,] / see Introduction, p. lxxiv above

which end not a few of the things which assist the other operations help
with this too.

18. Nothing more is needed for the liver | if it be kept from parching
or drying out, and from obstruction. As for its dissolution, which generates dropsies, it is an illness in every respect. But the other two, the
onset of old age brings them on.

19. The things relevant here are most of all the ones described in the
operation on the blood; but I will add a very few choice ones now.

20. In particular let wine of sweet pomegranates, or if that is
unobtainable, its juice freshly squeezed, be taken in the morning, with
some sugar and, into the glass with the juice, a little fresh citron rind,
and three or four whole cloves; and let this be taken repeatedly from
February until the end of April. |

21. Before all other herbs, let watercress be brought into use; but
growing to maturity, not old. Let it be taken plain, or in soups or
drinks. The next best thing is scurvy-grass.

22. Aloes, however washed or rectified, are harmful to the liver, and
so they should never be taken routinely. Rhubarb, on the other hand,
sustains it, provided that we take three precautions: first, that we take it
before meals in case it desiccates too much, or leaves a styptic residue;
secondly that we soak it for an hour or two in fresh almond oil with rose
water, before it is infused subsequently or is given in substance; thirdly,
that we take it alternately, | i.e. at times by itself, at others with tartar or a
little black salt, in case it only carries off light matter, and makes the
mass of humours more obstinate.

23. I recommend that wine or a decoction of steel be taken three or
four times a year, to dissolve the stronger obstructions, but yet in such a
way that we take a dose of two or three spoonfuls of fresh oil of sweet
almonds first, and that afterwards we get the body moving, especially
the arms and abdomen.

24. Sweetened liquors, and with a certain fatness, especially if age has
incorporated them well, | are very good for staving off arefaction, salti-
ness, and parching, and in short the effects of age on the liver. Such can
be made from fruits and sweet roots, like wines and drinks of fresh
raisins, jujubes, figs, dates, carrots, bulbs, or potatoes and the like,
sometimes with liquorice admixed. Also very effective is a drink made

quandoque; Etiàm *Potus*, ex granis *Indicis* (quæ *Mayz* vocant) cum *Mixturâ Dulcium*, plurimùm confert. Notandum est autem, Intentionem *Præseruationis Iecoris* in Mollicie quâdam, & Pinguedine, longè potentiorem esse illâ alterâ, quæ pertinet ad *Apertionem Iecoris*, quæ
5 potius innuit ad *Sanitatem*, quam ad *Diuturnitatem vitæ*; nisi quod
[V8ᵛ] *Obstructio* ea, quæ inducit *Torrefactionem*, ᴵ æquè maliciosa est, ac aliæ *Arefactiones*.

**25.** *Radices Cichoreæ, Spinachiæ, Betæ*, à Medullis purgatas, atque ad Teneritudinem coctas in Aquâ, cum tertiâ parte *Vini Albi*, pro *Condi-*
10 *mentis* Familiaribus cum *Oleo*, & *Aceto*, laudo; Etiam Gemmas, siue Caules *Asparagi, Pulpas Atriplicis*, & *Radices Bardanæ*, debitis modis elixas, & conditas; & *Iuscula* (tempore *Veris*) ex Folijs pubescentibus *Vitium*, & herbâ viridi *Tritici*. Atq*ue* de *Iecore* muniendo hactenùs.

**26.** *Cor* Iuuamentum suscipit maximè, atque Nocumentum ex *Aere*,
[Xɪʳ] quem spiramus; ex *vaporibus*; atque ex *Affectibus*. Atque ᴵ complura ex
16 ijs, quæ de *Spiritibus*, suprà dicta sunt, hùc transferri possunt; Indigesta autem *Moles Cordialium* apud *Medicos*, ad Intentionem nostram parùm valet: Attamen, quæ *Venenorum Malignitati* occurrere deprehenduntur, ea demùm ad muniendas *Cordis* vires, sano cum Iudicio adhiberi
20 possunt; præsertìm si sint ex eo genere, quod non tam propriam *Veneni Naturam* frangat, quàm *Cor* & *Spiritus* in *Venenum* insurgere faciat. Atque de *Cordialibus* consule *Tabulam* superiùs positam.

**27.** *Aeris Bonitas* in *Locis*, *Experientiâ* potiùs dignoscitur, quàm *Sig-*
[Xɪʳ] *nis*. Optimum iudicamus *Aerem* spirare in *Locis Æquis*, & ᴵ Planis, atque
25 ex omni parte *Perflatilibus*; si fuerit *Terra Sicca*, neque tamen prorsùs *Arida*, aut *Arenosa*; Quæque emittat *Serpillum*, & *Amaraci* genus, & hic inde *caules Mentæ Campestris*; Quæque sit non prorsus *Rasa*, sed *Arboribus* nonnullis (ad *vmbram*) sparsìm consita; Atque vbi *Rosa Rubi* spiret aliquid *Muscatellum*, & *Aromaticum*. *Flumina* si adsint, nocere
30 potiùs arbitramur, nisi fuerint *Exigua* admodùm, & *limpida*, & *glareosa*.

**28.** *Aerem Matutinum* certum est *Vespertino* esse magis Vitalem, licet ad Delicias alter magis ametur.

**29.** *Aerem* à *Vento* agitatum paulò leniore, *Aere Cœli Sereni* fœli-
[X2ʳ] cio|rem esse arbitramur. Optimus autem est *Zephyrus Matutinus*, &
35 *Boreas Post-meridianus*.

---

1 *Indicis*] ~,      3 Mollicie] / *nld* as mollitie in *SEH* (II, p. 189)      10 laudo;] ~:
11 *Atriplicis*] Artiplicis / see *tns* to Vɪʳ; emended thus in *SEH* (II, p. 189 n. 2).      25 si] Si
29 *Aromaticum*.] ~:      34 arbitramur.] ~:

from Indian grain (which they call maize) with sweet things intermixed. However, we should note that the intention of keeping the liver in a certain soft fatty condition is much more powerful than that other intention which has to do with opening it, for that tends rather to health than long life; except that obstruction which brings on parching | is just as bad as any of the other kinds of arefactions.

**25.** I recommend chicory, spinach, and beetroots stripped of their marrow and boiled in water until tender, with a third part of white wine, for an ordinary relish with oil and vinegar. I also recommend the buds and spears of asparagus, pulps of artichokes, and burdock roots properly boiled and seasoned; and in the spring soups made from fresh vine leaves and green shoots of wheat. And so much for fortifying the liver.

**26.** The heart receives the most help and harm from the air we breathe, from vapours, and from our feelings. And | many of the things which I said above about the spirits can be carried over here. Now the undigested mass of cordials in the physicians' writings has little bearing on our intention. However, things which are known to resist the malignity of poisons can be used with discretion to strengthen the heart's forces, especially if they belong to the class which does not so much destroy the specific nature of a poison as make the heart and spirits rise up against poisons in general. On the subject of cordials consult the table set out above.

**27.** The goodness of air in particular places is identified rather by experience than by signs. I judge that the best air breathes in flat | and level places open in every respect to the winds; and if the land be dry but nevertheless not completely arid or sandy, and puts out thyme, marjoram-kind, and here and there shoots of field mint; and is not absolutely bare but with scattered spinneys of trees for shade's sake; and where the wild rose breathes out a musky or aromatic smell. But if rivers are there, I think that is a bad thing, unless they are very narrow, clear, and gravelly.

**28.** The morning air is certainly more life-giving than the evening, although the latter is preferred for the pleasure it provides.

**29.** In my judgement air wafted gently by the wind is more favourable | than calm weather: best is a westerly in the morning, and a northerly in the afternoon.

30. *Odores* ad *Confortationem Cordis* præcipuè vtiles sunt; neque tamen, ac si *Odor bonus* esset *Aeris boni* Prærogatiua. Certum enim est, quemadmodùm inueniuntur *Aeres* prorsùs *Pestilentes,* qui non tantùm *fœtent,* quantum alij *minùs noxij;* similitèr inueniri è contrà *Aeres salu-*
5 *berrimos,* & *Spiritibus Amicissimos,* qui aut prorsùs sint *inodori,* aut ad sensum minùs *Grati* & *Fragrantes.* Atque omninò, vbi degitur in *Aere bono, Odores* per vices tantùm repeti debent. *Odor* enim *Continuus*
[X2ᵛ] (licet optimus) *Spiritus* nonnihil onerat.

31. Laudamus ante omnes alios (vt etiam superiùs innuimus) *Odores*
10 ex *Plantis Vegetantibus,* & non *Auulsis,* in *Aere aperto* exceptis; quales sunt ex *Violis, Floribus Gariophylli* (tam *Maioris* quam *Minoris*), *Floribus Fabarum, Floribus Tiliæ, Floribus* siue *Puluisculo Vitium, Floribus Madre-selue, Floribus Parietariæ luteæ, Rosâ Muscatellâ* (nam cæteræ *Rosæ germinantes* parcè emittunt *Odores*), *Fragariâ* (præsertìm *Moriente*),
15 *Rubo suaui* (præcipuè ineunte *Vere*), *Menthâ Campestri, Lauendulâ Florente,* atque in Regionibus Calidioribus, *Malo Arantio, Cytrio, Myrto, Lauro.* Itaque *Ambulatio,* aut *Sessio* inter huiusmodi *Auras,* in Vsu esse
[X3ʳ] debet.

32. Ad *Cordis* iuuamentum, *Odores Refrigerantes Calidioribus* ante-
20 ponimus; *Suffitus* itaque *Matutinus,* aut sub *Calores Meridiei* optimus fuerit, ex æquis Portionibus *Aceti, Aquæ Rosaceæ,* & *Vini Generosi,* super *Laminam Ferri* quasi candentem, fusorum.

33. Neque verò *Matri Telluri* libare nos quis existimet, si præcipi-amus inter *Fodiendum,* aut *Terram vertendam, vinum Generosum*
25 super-infundi.

34. *Aquam* è *Floribus Arantiorum* bonam, cum modicâ Parte *Aquæ Rosaceæ,* & *vini Fragrantis,* etiam per *Nares* attrahi, aut per *Syringam Errhini* more immitti (sed rariùs) bonum est.

[X3ᵛ] 35. At *Masticatio* (quamuis non habeamus *Betel*) & *Detentio* in *Ore*
30 eorum quæ *Spiritus* fouent (licet assidua) vtilis admodùm est. Fiant itaque *Grana,* aut pusilli *Pastilli* ex *Ambrâ,* & *Musco,* & *Ligno Aloes,* & *Ligno Rhodio,* & *Radice Iridis,* & *Rosâ;* atque formentur illa *Grana,* aut *Pastilli,* per *Aquam Rosaceam,* quæ per paululùm *Balsami Indi* transierit.

36. *Vapores* verò, qui ex *Rebus intrò sumptis, Cor* muniunt, & fouent,
35 hæc tria habere debent; vt sint *Amici, Clari,* & *Refrigerantes. Caliditas* enim *Vaporum,* mala; Atque ipsum *Vinum,* quod putatur habere

---

1 neque] Neque          7 debent.] ~:          9 alios] ~,          11 *Minoris*,] ~),
13 *luteæ,*] ~;          14 *Odores*),] ~),          *Moriente*),] ~),          15 *suaui*] ~,          *Vere*),] ~,)
16 *Cytrio,*] ~;          17 *Lauro.*] ~:          28 immitti] ~,          30 fouent] ~,

30. Smells are especially useful for comforting the heart; not however that a good smell is a prerogative of good air. For it is certain that just as we find very pestilential airs which do not stink as much as others less harmful, so we find very healthful airs sympathetic to the spirits which either do not smell or have little perceptible goodness or fragrance. And generally when one lives in good air, odours should be used only from time to time. For an uninterrupted odour (however good) to a certain extent overloads the spirits. |

31. I recommend above all else (as I have hinted above) the smells of growing plants, not ones pulled up, inhaled in the open air, such as are the smells of violets, greater and lesser gillyflowers, bean flowers, lime flowers, grape-bloom or vine flowers, clary flowers, yellow wallflowers, musk-rose (for other roses when they sprout give off little smell), strawberry (especially as they die), sweet-briar (especially as spring comes in), field mint, flowering lavender, and, in hotter climes, oranges, citron, myrtle, and laurel. We should therefore take up walking or sitting among the scents of such plants. |

32. I place cooling smells before hot for helping the heart. Thus the best fumigant in the morning or in the heat of the day is made by pouring equal parts of vinegar, rose-water, and heady wine on a hot iron plate.

33. But let no one think I offer a libation to Mother Earth if I recommend that during digging or ploughing strong wine be poured on top of the furrows.

34. Good water of orange blossom, with a dash of rose-water, and fragrant wine, inhaled through the nostrils or introduced by a syringe in the manner of a sternutative medicine, is good if not done too often.

35. But chewing (despite our | lack of betel) and keeping things in the mouth which cherish the spirits is very useful (even if done persistently). Make therefore grains or little pastilles of ambergris, musk, lignum aloes, Rhodium wood, iris root, and roses. And let these grains or pastilles be made with rose-water which has passed through a little Indian balsam.

36. Now vapours emanating from things taken internally fortify and cheer the heart. And these should have three attributes: that they be

*vaporem* solummodò *calefacientem,* non expers est prorsùs *Qualitatis*
[X4ʳ] *Opiatæ. Claros* autem *Vapores* vocamus eos, qui ᐧ plus habent ex *Vapore,*
quam ex *Exhalatione,* neque sunt omninò *Fumei,* aut *Fuliginosi,* aut
*Vnctuosi,* sed *Humidi,* & *Æquales.*

5      37. Inter Turbam inutilem *Cordialium,* pauca ad *Diætam* in vsu esse
debent; Loco omnium *Ambra-grisia,* & *Crocus,* & *Granum Kermes,* ex
*Calidioribus;* Atque *Radices Buglossi,* & *Boraginis,* atque *Mala Cytria,* &
*Limones dulces,* & *Poma Fragrantia,* ex *Frigidioribus.* Etiam eo (quo
diximus) modo, & *Aurum,* & *Margaritæ,* non tantùm intra *Venas,* sed
10   etiam in *Transitu,* & circa *Præcordia* aliquid possunt; per *Refrigerium*
scilicet, absque aliquâ *noxiâ Qualitate.*

[X4ᵛ]    38. De *Lapide Bezoar,* ob multas ᐧ Probationes, Virtuti eius fidem
non prorsùs derogamus; Sed omninò *Modus* eius sumptionis talis esse
debet, vt facillimè *Virtus* eius communicetur *Spiritibus.* Itaque nec in
15   *Iusculis,* nec in *Syrupis,* nec in *Aquâ Rosaceâ,* aut huiusmodi, Vsum eius
probamus; sed tantùm in *Vino,* aut *Aquâ Cynamomi,* aut huiusmodi
*Distillato,* sed *Tenui,* non *Calido* aut *Forti.*

39. De *Affectibus* iam superius inquisitum est; Illud tantùm adi-
icimus, omne *Desiderium magnum,* & *constans,* & (vt loquuntur)
20   *Heroicum, Cordis Virtutes* roborare, & ampliare. Atque de *Corde*
[X5ʳ] hactenùs. ᐧ

40. Ad *Cerebrum* quod attinet (vbi *Cathedra,* & *Vniuersitas
Spirituum Animalium* residet) quæ superiùs inquisita sunt, de *Opio,* &
*Nitro,* & *Subordinatis* ad ipsa, & de *Conciliatione Somni placidi,* etiam
25   hùc aliquatenùs spectant. Illud quoque certum, *Cerebrum* tanquam in
*Tutelâ Stomachi* esse; ideóque quæ *Stomachum* confortant, & muniunt,
*Cerebrum* per *Consensum* iuuant, atque hùc similitèr transferri debent.
Adijciemus pauca, tria *Externa, Internum* vnum.

41. *Balneationem Pedum* omninò in vsu esse volumus, ad minùs,
30   semel in *Septimanâ; Balneumque* fieri ex *Lixiuio,* cum *Sale nigro,* &
[X5ᵛ] *Saluiâ, Camomillâ, Fœniculo, Sam|suco,* & *Costo,* cum Folijs *Angelicæ
Viridis.*

42. *Suffitum* laudamus etiam *Quotidianum Manè,* ex *Rore-marino*
arido, Ramulis *Lauri* siccis, & *Ligno Aloes;* nam *Gummi suauia Caput*
35   grauant.

---

3 *Fuliginosi*] / see Introduction, p. lxxiv above      6 *Kermes*] / see Introduction, p. lxxiv
above      8 *Fragrantia*] / see Introduction, p. lxxiv above

friendly, clear, and cooling. For vapours with heat are bad; and wine itself, which people think has only a vapour which heats, is not utterly devoid of an opiate quality. We call those vapours clear which $^|$ have more vapour than exhalation to them, and are not in the least bit smoky, sooty, or unctuous, but are moist and equable.

37. Amid the swarm of useless cordials, there are a few which ought to be adoped into diet; above all ambergris, saffron, and Kermes grain among the hotter remedies; and bugloss and borage roots, citrons, sweet lemons, and fragrant apples among the colder. Gold and pearls too can help when used in the way I mentioned, and not just in the veins but also in transit and around about the heart, i.e. for refrigeration, and without any noxious quality.

38. As for the bezoar stone, it has $^|$ survived many tests and so I do not entirely lack faith in it. But it should surely be taken in such a way that it most easily transmits its virtue to the spirits. Therefore I do not approve of its use in gravy, syrups, rose-water, or the like, but only in wine, or cinnamon-water, or distillates of that kind, and not hot or strong either, but dilute.

39. I have already inquired above concerning the affections. This only will I add: that every great, steady and (as they say) heroic desire strengthens and enlarges the virtues of the heart. So much then for the heart. $^|$

40. As for the brain (where the cathedral and university of the animal spirits is set), what I investigated above in connection with opium and nitre, and their subordinates, and about getting a good night's sleep, is relevant here up to a point. It is certain too that the brain is so to speak in the care of the stomach, and accordingly that what comforts and fortifies the stomach helps the brain by consent, and so should be carried over to this place. I will add a few extra things here: two external and one internal.

41. I very much want bathing of the feet to be used, at least once a week in a bath made up of lye, with black salt, sage, camomile, fennel, $^|$ sweet marjoram, costum, and fresh angelica leaves.

42. I also recommend a fumigation every morning of dried rosemary, dry laurel twigs, and lignum aloes; for nice gums weigh heavy on the head.

43. Cauendum prorsùs, ne *Capiti* per *Exteriùs* admoueantur *Calida*; qualia sunt *Aromata*, non exceptâ *Nuce Muscatâ*. Etenim *Calida* illa ad *Plantas pedum* præcipitamus, ibique solùm applicari volumus. *Vnctionem* verò *Capitis* leuem ex *Oleo*, cum *Rosâ*, & *Myrto*, & parùm 5 *Salis*, & *Croci*, laudamus.

44. Memores eorum, quæ de *Opiatis*, & *Nitro*, & Similibus, antè [X6ʳ] proposuerimus, quæ *Spiritus*| tantoperè densant, non existimamus ab re fore, si semel *Diebus* quatuordecim, accipiantur in *Brodio Matutino*, *Grana* tria, vel quatuor *Castorij*, cum modico *Seminis Angelicæ*, & 10 *Calami Aromatici*; quæ & ipsa *Cerebrum* roborant, & in *Densitate Substantiæ Spirituum* (quæ ad vitæ Longæuitatem tam necessaria est), *Motûs Viuacitatem* & *Vigorem* excitant.

45. In *Confortatiuis* quatuor *Viscerum Principalium* ea proposuimus, quæ & *Propria* sunt, atque *Electa*, atque in *Diætam*, & *Regimen vitæ* 15 transferri, tutò & commodè possunt. *Varietas* enim *Medicamentorum*, [X6ʳ] *ignorantiæ Filia* est; neque *Multa Fercula* | (quod aiunt) *tam multos Morbos* fecere, quam *multa Medicamenta paucas Curas*. Atque de *Operatione* super *Viscera principalia*, ad *Extrusionem Alimenti*, hæc inquisita sunt.

20 ## OPERATIO
super partes exteriores ad Attractionem
Alimenti. VI.

### Historia.

24 1. Licet *Concoctio* bona, per *Partes Interiores* facta, primas Partes ad [X7ʳ] probam *Ali*|*mentationem* teneat, tamen concurrere etiam debent *Actiones Partium Exteriorum*; vt sicut *Facultas Interior Alimentum* emittit, & extrudit; ita *Facultas Partium Exteriorum* idem arripiat, et attrahat; Quóque imbecillior fuerit *Facultas Concoctionis*, eò magis opus est Auxilio concurrente *Facultatis Attractiuæ*.

30 2. *Attractio valida Partium Exteriorum*, excitatur præcipuè per *Motum Corporis*, per quem *Partes calefactæ*, et *confortatæ*, *Alimentum* ad se alacriùs vocant, et attrahunt.

---

1 *Calida*;] / see Introduction, p. lxxiv above  2 *Muscatâ*.] ~:  3 volumus.] ~:
6 *Nitro*] / corrected forme (e.g. BNF copy); uncorrected inner forme (e.g. Huntington
56659) *Nitra* (for the same error see P7ʳ)  11 *Spirituum*] ~,  est),] ~,)
12 *Viuacitatem*] / see Introduction, p. lxxiv above  15 possunt.] ~:
16 *Fercula*] ~,  25 teneat,] ~;

43. Beware especially of bringing hot things to the outside of the head, like aromatic plants, nutmeg not excepted. For I want them brought down to the feet and only applied there. But I recommend light anointing of the head with oil together with roses, myrtle, and a little salt and saffron.

44. Bearing in mind what I set out before about opiates, nitre, and the like, which so ⌐ strongly condense the spirits, I do not believe it would be beside the point if once a fortnight we took three or four grains of castor in a morning broth with a dash of angelica seed and sweet flag, for these both strengthen the brain and excite liveliness and vigour in the density of the spirits' substance (which is so essential to long life).

45. In relation to remedies that comfort the four chief viscera, I have proposed things which are both fitting and selected, and which can be transferred with safety and convenience into diet and life's regimen. For variety of remedies is the daughter of ignorance; nor do many dishes ⌐ (as they say) make for many diseases so much as many remedies make few cures. So much then for the operation on the principal viscera to force out the food.

## VI. The Operation
### on the external parts to attract
### the food

### *History*

1. Although good digestion, carried out by the internal parts, takes first place in effective ⌐ alimentation, yet the actions of the exterior parts should harmonize with them too, so that as the interior faculty sends out and extrudes the nourishment, so the faculty of the exterior parts should lay hold of and attract it. The weaker too the digestive faculty may be, the more it needs the concurrent aid of the attractive faculty.

2. The exterior parts' strong attraction is excited in the main by bodily motion, which warms and comforts the parts so that they summon and draw aliment to them more eagerly.

3. Illud verò maximè cauendum, et prohibendum; Ne idem *Motus,*
[X7$^v$] et *Calor,* qui ad *Membra Nouum Succum* euocat, *Membrum* | simul eo
*Succo,* quo anteà perfusum erat, nimiùm exoluat.

4. *Fricationes* huic Intentioni optimè subseruiunt, factæ præcipuè
5 *Manè*; sed hoc perpetuò comitetur, vt post *Fricationem,* fiat leuis *Inunctio* cum *Oleo,* ne *Attritio Partium Exteriorum,* eas per *Perspirationem*
reddat effœtas.

5. Proxima est *Exercitatio,* per quam *Partes* ipsæ se *confricant,* et *con-
cutiunt*; modò sit *Moderata,* et quæ (vt superiùs notatum est) nec sit
10 *Celeris,* nec ad *Vltimas Vires,* nec ad *Lassitudinem*; Verùm in hâc ipsâ,
atque *Fricatione,* eadem est ratio, et *Cautio*; Ne Corpus nimiùm
[X8$^r$] perspiret. Itaque *Exercitatio* melior est sub | *Dîo,* quam sub *Tecto,* et
*Hyeme* quam *Æstate*; Atque insupèr, *Exercitatio Inunctione* non tantùm
*claudi* debet, vt *Fricatio,* sed etiam in *Exercitationibus* vehementioribus,
15 adhibenda est *Vnctio,* et in *Principio,* et sub *finem,* more *Athletarum.*

6. Ad *Exercitationem,* vt quam minimùm aut *Spiritus,* aut *Succos*
exoluat, vtile est, vt vsurpetur *Stomacho* non prorsùs *ieiuno.* Itaque vt
*Exercitatio,* nec *Stomacho Repleto* (quod plurimùm interest *Sanitatis*) nec
*Ieiuno* (quod non minùs interest *Longitudinis vitæ*) vsurpetur; in vsum
20 adduci debet *Ientaculum* Manè, non ex *Medicamentis,* aut *Haustibus*
[X8$^v$] *Matutinis,* aut *Vuis passis,* aut *Fi*|*cubus,* aut huiusmodi; sed planè ex
*Cibo,* & *Potu*; at *Leui* admodùm, & *Modica Quantitate.*

7. *Exercitationes* ad *Irrigationem Membrorum,* debent esse *Membris*
omnibus quasi æquales; non vt (quemadmodùm ait *Socrates*) *Tibiæ*
25 *moueant, Brachia quiescant,* nec è contra; sed vt *Partes vniuersæ* ex Motu
participent; Atque omninò ad vitam prodest, vt *Corpus* nunquàm diù in
eâdem *Positurâ* permaneat; sed singulis Semi-horis, ad minùs, *Posituram*
mutet, præterquàm in *Somno.*

29 8. Quæ ad *Mortificationem* vsurpantur, ad *Viuificationem* traduci
[Yr$^r$] possunt; Nam & *Indusia setosa,* & *Flagellationes,* & omnis *Ex*|*teriorum*
*Vexatio,* Vim eorum *Attractiuam* roborat.

9. *Vrticationem* commendat *Cardanus,* etiàm ad *Melancholiam*;
verùm de hâc parùm nobis compertum est; Et suspecta nobis est illa,
ne propter *Venenatam* nonnullam *Qualitatem Vrticæ, Serpigines* vsu
35 frequenti inducat, & *Mala Cutis.* Atque de *Operatione* super *Partes Exteriores,* ad *Attractionem Alimenti,* hæc inquisita sunt.

---

12 perspiret.] ~;     14 *Fricatio,*] / see Introduction, p. lxxiv above     19 *Ieiuno*] ~,

3. But we should watch out for and prevent that this same heat and motion, which calls forth new juice to the parts, does not simultaneously | do away too much with the juice with which the member was suffused before.

4. Massage, especially when given in the morning, answers this intention very well; but let it always be followed by a light application of oil, in case the rubbing of the outer parts makes them weak through perspiration.

5. The next best thing is exercise, which causes the parts to rub and stroke each other, provided (as I said above) that this be moderate and not fast or furious, or taken to the point of lassitude. But in this as well as in massage the same principle and precaution applies, that the body does not sweat too much. Thus exercise is better outside | than in, and in winter rather than summer. Again, oil ought to be applied not just when exercise comes to a close, as is the case with massage, but also at the beginning and end in more violent exercise, as is the practice with athletes.

6. To stop exercise doing away with the spirits or juices, it is important that it is not taken on a completely empty stomach. Thus, so that exercise may neither be taken on a full stomach (which is bad for health), nor on an empty one (which is no less bad for longevity), breakfast should be such that no medicines, morning drafts, or raisins or figs, or the like should be consumed | but just food and drink, and that very light and in moderate quantity.

7. Exercises for irrigating the members should affect them all equally and not (as Socrates says) so that the legs move while the arms stay still, or vice versa, but that all parts share in motion. And it is very important for longevity that the body should not stay in the same posture for long, but should change every half hour at least, except during sleep.

8. What people use for mortification can be transferred to vivification. For hair shirts and flagellation, and all agitation of the outer parts, | strengthens their attractive force.

9. Cardan recommends stinging nettles, even for melancholy; but I have little experience of this, and suspect that the poisonous quality of nettles would with frequent use cause itching and skin complaints. So much then for the operation on the external parts to attract food.

| OPERATIO
super Alimentum ipsum
ad Insinuationem eiusdem. VII.

### Historia.

5 **1.** Reprehensio vulgaris de *Multis Ferculis, Censorem* potiùs decet, quàm *Medicum*; aut vtcúnque constantiæ *Sanitatis* vtilis esse potest, ad *Longitudinem vitæ,* noxia est; Proptereà quod *Mistura Alimentorum varia,* & aliquantùm *Heterogenea,* Exitum reperit in *Venas,* & *Succos,* meliùs & [Y2ʳ] alacriùs, quàm | *Simplex,* & *Homogenea*; Cùm insupèr ad *Appetitum* 10 excitandum (qui *Acies* est *Digestionis*) plurimùm possit. Itaque & *Mensam variam,* & *Mutationes* subindè *Ciborum,* pro *Temporibus Anni,* aut aliàs, probamus.

**2.** Etiàm illud de *Simplicitate Ciborum* absque *Condimentis, Simplicitas Iudicij* est; cùm *Condimenta* bona, & benè electa, sint *Præparationes* 15 *Ciborum* saluberrimæ, atque tùm ad Sanitatem, tùm ad Vitam conferant.

**3.** Videndum est, vt cum *Cibis Durioribus,* coniungantur *Potus Fortiores,* & *Condimenta* quæ *penetrent,* & *incidant*; Cum *Cibis* contrà [Y2ᵛ] *Facilioribus, Potus Tenues,* & *Condimenta Pinguia.* |

**4.** Cùm paulò antè monuerimus, vt prima *Potio* in cœnâ excipiatur 20 *Calida*; nunc addimus, quod ad *Præparationem Stomachi,* etiam *Semihorâ* ante *Cibum,* bonus *Haustus Potûs* (cui quisque maximè insueuit) *calidus* vsurpetur; sed parùm *Aromatizatus* ad gratiam Saporis.

**5.** *Præparatio* Ciborum, & *Panis,* & *Potuum,* si benè, & in ordine ad *Intentionem* instituatur, magni est prorsùs Momenti; licet sit *Res* 25 *Mechanica,* & sapiat *Culinam,* & *Cellam*; cùm tamen longè præstet *Fabellis* de *Auro,* & *Gemmis,* & huiusmodi.

**6.** *Humectatio Succorum Corporis* per *Præparationem Alimentorum* [Y3ʳ] *Humidam,* puerilis Res est; Ad | *Feruores Morborum* nonnihil facit; ad *Alimentationem* verò *roscidam,* omninò contraria est; itaque *Elixatio* 30 *Ciborum* longè inferior est, ad Intentionem nostram, *Assatione,* & *Coctione* in *Furno,* & similibus.

**7.** *Assatio* debet fieri *Igne viuido,* & celeriùs perfici; non *Igne lento,* & nimiâ Morâ.

**8.** *Carnes* omnes *Solidiores* in vsu esse debent non prorsùs *Recentes,* 35 sed nonnihil *Salis Expertæ*; ex *Sale* ipso autem in *Mensâ,* eò minus sumi

---

10 *excitandum*] ~,    17 *incidant*] *incîdant*

## | VII. The Operation
## on the Aliment itself
## to insinuate the same

### *History*

1. The common criticism of many dishes better befits a moralist than a physician, or at any rate it can be useful for maintaining health, but it is bad for prolonging life because mixture of various and heterogeneous aliments finds better and swifter access to the veins and juices than | does a simple and homogeneous one. In addition it does much to whet the appetite (which is the spur of digestion). I therefore approve of a varied diet and frequent changes of menu with the seasons or other circumstances.

2. That food should be simple and without seasoning is simple-minded, since good, well-chosen seasoning is food's healthiest preparation, and it contributes as much to health as to length of life.

3. We should note that with heavy foods, stronger drinks and more penetrating and piercing sauces should be used; but with lighter foods, weaker drinks and fatty condiments. |

4. Following my advice of a while ago to take the first drink of supper warm, I now add that to make the stomach ready, you should take hot a good swig of whatever drink you are used to half an hour before dining, with a little spice according to taste.

5. The preparation of food, bread, and drinks, if it be well and systematically established according to the present intention, is a matter of the greatest moment; although it is a mechanical business and smacks of the kitchen and cellar, it is still far more important than stories about gold, gems, and so on.

6. The moisturizing of bodily juices by moist preparation of aliment is a childish idea; | it helps with the heats of illness, but is altogether at odds with dewy alimentation. Therefore the boiling of food is far inferior in respect of our intention than roasting, baking, and the like.

7. Roasting should be done with a strong fire and finished fast; not with a slow fire, and long cooking.

8. All the more solid meats should not be consumed too fresh, but salted; and salt should not be taken at meals in any quantity or indeed

debet, aut nihil omninò. *Sal* enim *Alimento Incorporatus* magis valet ad *Distributionem*, quàm per se sumptus.

9. Debent in vsum adduci *Macerationes*, & *Infusiones Carnium*
[Y3$^v$] va⌐riæ, & bonæ, in *Liquoribus idoneis*, ante *Assationes*; quemadmodùm
5 quandóque in vsu sunt Similia, ante *Coctiones* in *Furno*, & in *Murijs* aliquorum *Piscium*.

10. At *Pulsationes*, & tanquam *Verberationes Carnium*, antequàm coquantur, haud paruam rem præstant. Certè in confesso est, & *Perdices*, & *Phasianos*, in *Aucupio*; & *Damas*, & *Ceruos* in *Venatione*
10 occisos (nisi fuerit ea Fuga longior) gratiores esse etiàm ad Gustum. *Pisces* autem nonnulli *Flagellati*, & *Verberati*, euadunt meliores. Etiam *Pyra Duriora*, & *Austera*, atque alij nonnulli *Fructus*, *Compressione* dulcescunt. Bonum esset in Vsum adduci, *Carnium duriorum* non-
[Y4$^r$] nullam *Pul⌐sationem*, & *Contusionem*, antequàm Ignem patiantur; Idque
15 ex optimis *Præparationibus* erit.

11. *Panis* modicè *Fermentatus*, & valdè parùm *Salitus*, optimus est; quíque etiam in *Furno Feruenti* satis, nec admodùm *Elanguido*, coctus est.

12. *Potûs Præparatio* ad vitam longam simplici ferè Præcepto constat.
20 Atque de *Aquæ-potoribus* nihil attinet dicere; Potest huiusmodi *Diæta* (vt alibi diximus) vitam *aliquandiù remorari*, sed nunquam *maiorem* in *Modum prolongare*. At in alijs *Potibus Spirituosis*, qualia sunt *Vinum*, *Ceruisia*, *Hydromel*, & huiusmodi, id tanquam Summa Summarum
[Y4$^v$] affectari, & obseruari debet; Vt ⌐ *Partes Liquoris* sint subtilissimæ, &
25 *Spiritus* lenissimus. Hoc *Vetustate simplici* difficile erit efficere, quæ gignit *Partes* paulò *Subtiliores*, *Spiritus* verò multò *Acriores*; Itaque de *Infusione* in *Dolijs Substantiæ* alicuius *Pinguis*, quæ *Spirituum Acrimo-niam* compescat, iam anteà præceptum est. Est & alius modus absque *Infusione*, aut *Mixturâ*; Is est, vt *Liquor Potûs* continuò agitetur, siue per
30 *Vecturam* in *Mari*, siue per *Vecturam* in *Carris*, siue suspendendo *vtres* ex *Funibus*, eosque quotidiè agitando, aut alijs huiusmodi Modis. Certum enim est, *Motum* illum *localem*, *Partes subtilizare*, ac *Spiritus* in partibus
[Y5$^r$] interìm ita *fermentare*, vt *Acedini* ⌐ (quod *Putrefactionis* genus est) non vacent.
35 13. *Vergente* autem *Senectute*, etiam talis *Præparatio Ciborum* instituenda est, quæ sit tanquam in Mediâ Viâ ad *Chylum*. Atque de

---

8 præstant.] ~:      10 occisos] ~,      19 constat.] ~:      22 *prolongare*.] ~:
23 huiusmodi,] ~;      25 lenissimus.] ~:      28 est.] ~:      31 Modis.] ~:
33 *Acedini*] ~,      est)] ~,)

taken at all; for salt is better distributed with food than when taken by itself.

9. Various and good macerations and infusions of meats ˈ in appropriate liquors before roasting should be put in practice, just as similar practices are sometimes used before baking in the oven and in the pickling of some fish.

10. But the pounding and as it were thrashing of meats before they are cooked is no small matter. Certainly it is generally agreed that pheasants and partridges taken by hawks, and bucks and stags killed in the chase (provided that it has not gone on too long) taste better. Some fish too turn out better when beaten and thrashed. Even some hard and sour pears and some other fruits become sweeter with squeezing. It would be good too to bring into use the pounding and bruising ˈ of harder meats before they are put in the fire. And this will be one of the best kinds of preparation.

11. Lightly leavened bread with very little salt is best, and it is best baked in an oven sufficiently hot, and not for too long.

12. The preparation of drink for long life can be summed up in a simple precept. There is nothing to be said of water-drinkers for, as I said above, a diet of this kind may keep life going for a bit but cannot prolong it appreciably. But in other spirituous drinks—such as wine, beer, mead, and so on—the one thing to be grasped and noted as the highest priority is ˈ that the parts of the liquor be as subtle as can be, and the spirit as gentle. This is difficult to achieve by ageing alone; for that makes the parts a bit subtler but the spirits much keener. Therefore have I already instructed that an infusion of some fatty substance be made in the barrels to temper the acrimony of the spirits. There is also another way to do this but without infusion or mixture, and that is to shake up the liquor constantly either by sea voyages or by jolting it in carts or by hanging the vessels from ropes and giving them a daily shaking, or other such means. For it is certain that this local motion makes the parts subtle, and meanwhile so ferments the spirits in the parts that they ˈ have no scope for acidity (which is a species of putrefaction).

13. In advanced age, such a way of preparing food should be adopted that it be half turned into chyle. As for distillations of foods, they are a

*Distillationibus Ciborum,* meræ Nugæ sunt; Etenim *Portio Nutritiua,* vel optima, non ascendit in *Vaporem.*

14. *Incorporatio Cibi* & *Potûs,* antequàm concurrant in *Stomacho,* gradus est ad *Chylum;* Itaque sumantur vel *Pulli,* vel *Perdices,* & *Phasi-*
5 *ani,* & similia; Et coquantur in *Aquâ,* cum parùm *Salis;* deinde mundentur, & siccentur; posteà siue in *Musto,* siue in *Ceruisiâ feruescente,*
[Y5ᵛ] infundantur, cum parum *Sacchari.* |

15. Etiàm *Expressiones Ciborum,* & *Concisiones minutæ,* benè conditæ, *Senibus* vtiles sunt; eò magis quod *Officio Dentium* in *Manduca-*
10 *tione* (quod *Præparationis* præcipuum Genus est) ferè destituantur.

16. Atque de *Iuuamentis* eius *Defectûs* (*Dentium* scilicet *Roboris,* ad *Cibum molendum*) tria sunt quæ conferre possint. Primum, vt alij *Dentes* renascantur; Id quod difficile omninò esse videtur, nec posse perfici absque *Instauratione Corporis Intimâ,* et *Potenti.* Secundum est vt
15 *Mandibula* per *Astringentia debita,* ita firmentur, vt *Officio Dentium,* aliquâ ex parte, sufficere possint; quod non malè cedere posse videtur.
[Y6ʳ] Tertium, | vt *Cibus* sit ita *præparatus,* vt istâ *Masticatione* non egeat; quod promptum est, & expeditum.

17. Subit etiam cogitatio de *Quantitate Cibi,* & *Potûs;* Eam in *Excessu*
20 nonnullo, quandóque ad *Irrigationem* Corporis, vtilem esse. Itaque & *Epulæ Profusæ,* & *Perpotationes,* non omninò inhibendæ sunt. Atque de *Operatione* super *Alimenta,* & eorundem *Præparationem,* hæc inquisita sunt.

[Y6ᵛ]
## | OPERATIO
25
### super Actum vltimum
### Assimilationis.
### VIII.

**Connexio.** *De* Actu Vltimo Assimilationis (*quem* Operationes *tres proximè præcedentes intuentur*) breuis, *& simplex erit* Præceptio;
30 *Resque magis* Explicatione *indiget, quam* Præceptione *aliquâ variâ.*

[Y7ʳ]
## | Commentatio.

1. *Certum est Corpora omnia* Assimilandi, *quæ in Contiguo sunt,* Desiderio *nonnullo indui. Id faciunt generosè, & alacritèr*

9–10 *Manducatione*] ~,     10 est)] ~,)     11 *Defectûs*] ~,     20 esse.] ~:
28 Assimilationis] ~,     33 *indui.*] ~:

waste of time, for the nutritious or best portion does not rise in the vapour.

14. The incorporation of food and drink before they come together in the stomach is a step towards chyle. Therefore let chickens, partridges, pheasants, and the like be taken, and boiled in water with a little salt, then let them be cleaned and dried, and afterwards steep them with a little sugar in new wine or beer still fermenting. |

15. Also useful for old men are food extracts and food finely ground and well seasoned; and the more so because the teeth are no longer up to the job of chewing (which is the chief kind of preparation).

16. Now to remedy this defect (namely of the teeth for grinding food) three things may help. The first is that one acquire new teeth, but that it seems to be extremely difficult and not to be effected without a thorough and powerful renewal of the body. The second is to make the gums so strong with suitable astringents that they can to some degree do the job of teeth, a proposal which does not seem implausible. The third |️ is that food be so prepared that it does not need chewing, and this can be done promptly and expeditiously.

17. It also springs to mind concerning the quantity of food and drink that some excess is sometimes useful for irrigating the body. Therefore bingeing on food and drink should not be entirely ruled out. And so much then for the inquiry concerning the operation on aliments and their preparation.

| VIII. The Operation
on the last Act
of Assimilation

*Connection.* Guidance concerning the last act of assimilation (to which the previous three operations aim) will be short and sweet; for the matter requires explanation rather than any variety of instruction.

| Speculation

1. It is certain that all bodies are endowed with some desire to assimilate things in contact with them. Thin and pneumatic

Tenuia, & Pneumatica; *veluti* Flamma, Spiritus, Aer. *At contrà, quæ* Molem *habent* Crassam, & Tangibilem, *debilitèr admodùm; eò quod* Desiderium *illud* Assimilandi, *à fortiori* Desiderio [Y7$^v$] Quie¦tis, & *se* non Mouendi, *ligetur.*

5    **2.** *Certum est itidem,* Desiderium *illud* Assimilandi, *in Mole corporeâ, ligatum vt diximus,* & *inutile redditum, à* Calore *aut* Spiritu *in Proximo, liberari nonnihil,* & *excitari, vt tum demùm actuetur; Quæ vnica est Caussa, cur* Inanimata *non assimilent,* Animata *assimilent.*

10    **3.** *Certum* & *hoc quoque est: quò durior sit* Corporis Consisten-[Y8$^r$] tia, *eò illud* | *indigere* Maiore Calore *ad* Stimulum Assimilationis; *Quod in* Senibus, *malè omnino cedit; quia* Partes *sunt* obstinati-ores, Calor imbecillior. *Itaque aut* Obstinatio Partium *mollienda, aut* Calor *intendendus; Atque de* Malacissatione Membrorum 15 *posteà dicemus, cùm iam antè etiam plura, quæ ad* Duritiem *huiusmodi prohibendam,* & *præueniendam pertinent, proposuerimus.* [Y8$^v$] *De* Calore *autem* intenden¦do, *iam simplici* Præcepto *vtemur, si priùs etiam alterum* Axiôma *assumpserimus.*

    **4.** Actus Assimilationis (*qui à* Calore (*vt diximus*) circumfuso 20 *excitatur*), *est* Motus *admodùm* Accuratus, & Subtilis, & *in* Minimis. *Omnes autem huiusmodi* Motus, *tum demùm sunt in* Vigore, *cum omnis* Localis Motus *cesset, qui eum obturbet. Etenim* [Z1$^r$] Motus Separationis *in* Homogenea, *qui in* Lacte | *est, vt* Flos *supernatet,* Serum *subsidat, nunquàm fiet, si* Lac *lenitèr agitetur.* 25 *Neque* Putrefactio *vlla in* Aquâ, *aut* Mistis *procedet, si illa continuò* localitèr *moueantur. Ex His itaque, quæ* Assumpta *sunt, hoc iam ad* Inquisitionem præsentem *concludemus.*

    **5.** Actus *ipse* Assimilationis *perficitur præcipuè in* Somno & 29 Quiete, *præsertìm versus* Auroram, *factâ iam* Distributione. *Non* [Z1$^v$] *habemus igitur aliud,* | *quod ad præcipiendum occurrit, nisi vt* Homines *dormiant in* Calido; *Atque insupèr, vt versus* Auroram *sumatur aliqua* Inunctio, *vel* Indusium *intinctum, excitans moderatè* Calorem, *atque post illud sumptum redintegretur* Somnus. *Atque de* Actu vltimo Assimilationis, *hæc inquisita sunt.*

---

1 Aer.] ~:      8 *Caussa*] / this spelling occurs only here and on 2B7$^r$, 2C3$^r$ and 2F5$^v$
10 *est.*] ~;      20 *excitatur*),] ~,)      24 *agitetur.*] ~:      29 Distributione.] ~:

bodies, like flame, spirit, and air, do that lavishly and eagerly; on the other hand, ones whose mass is gross and tangible do so extremely weakly, because their desire to assimilate is curbed by the stronger desire | for rest and immobility.

2. It is also certain that the desire to assimilate, which is, as I have said, curbed, and made ineffective in a corporeal mass, is somewhat freed, stimulated, and eventually actuated by heat or spirit near by—which is the only reason why inanimate bodies do not assimilate whereas animate ones do.

3. This too is certain: that the harder a consistent body is, the | greater is the heat needed to stimulate assimilation; and this is bad news for old men, for the parts are less yielding, and the heat feebler. Therefore the stubbornness of the parts must be softened, or the heat be increased. Now I shall speak of the softening of the members later for I have already set out many proposals concerning the halting and prevention of this kind of hardness. But as for increasing the heat | I will now afford but one precept, after first adopting this other axiom.

4. The act of assimilation (which is, as I have said, provoked by the surrounding heat) is a very fine and subtle motion which goes on in the smallest particles. Now all motions of this kind flourish only when every local motion which can confuse them stops. For motion of separation where like moves to like, which happens in milk | as the cream swims above and the whey sinks beneath, never happens at all if the milk be gently stirred. Nor will any putrefaction occur in water or mixtures if they be kept in constant local motion. These things being granted, I will draw the following conclusion in relation to the present inquiry.

5. The act itself of assimilation works through, especially during sleep and rest, and particularly towards dawn, when distribution is accomplished. Thus I have no other | recommendation to make except that men sleep in the warm; and more, that towards dawn they apply some ointment or anointed shirt, to stimulate a moderate heat, and after that go off to sleep again. And so much for the final act of assimilation.

[Z2ʳ]                          | OPERATIO
super Intenerationem eius quod arefieri cœpit, siue
Malacissatio Corporis.
IX.

5  **Connexio.** *De* Inteneratione *per* Interius, *quæ per multas* Ambages,
*& Circuitus fit, tam* Alimentationis, *quam* Detentionis Spiritûs
[Z2ᵛ] (*ideóque sensìm perficitur*) *superiùs inquisi* tum est; *De eâ autem,
quæ fit per* Exterius, *& quasi subitò, siue de* Corpore Malacissando,
*iam videndum est.*

10                              *Historia.*

1. In *Fabulâ* de *Restitutione Peliæ* in *Iuuentutem, Medea* cùm id se moliri
fingeret, eam proposuit Rationem Rei conficiendæ; vt *Corpus Senis* in
*Frusta* concideretur; deinde in *Lebete* cum *Medicamentis* quibusdam
14  decoqueretur. *Coctio* fortassè aliqua ad hoc requiretur, *Concisione in*
[Z3ʳ] *Frusta* scilicet non est Opus. |
2. Attamèn etiam *Concisio in Frusta* adhibenda aliquatenùs videtur,
non *Ferro,* sed *Iudicio*; Cùm enim *Viscerum,* & *Partium* sit *Consistentia*
multùm diuersa, necesse est vt *Inteneratio* ipsorum, non iisdem Modis
absoluatur, sed vt instituatur *Cura singulorum*; præter ea quæ pertinent
20  ad *Intenerationem* totius *Massæ Corporis*; de quâ tamen primùm.
3. Huic Operationi, per *Balnea, Vnctiones,* & similia (si modò sit eius
Rei aliqua Potestas) satisfieri verisimile est; Circa quæ obseruanda sunt
ea, quæ sequuntur.
24  4. Non nimis indulgendum est spei, quod hæc Res confici posset,
[Z3ᵛ] propter ea quæ fieri cerni mus, in *Imbibitionibus* & *Macerationibus*
*Inanimatorum,* per quas illa intenerantur; Cuius aliqua Exempla
superiùs adduximus; Facilior enim est Operatio huiusmodi super
*Inanimata,* quia *attrahunt,* & *sugunt Liquores*; At in *Corpore Animali*
difficilior, quia *Motus* in iis fertur potiùs, ad *Circumferentiam.*
30  5. Ideò *Balnea,* quæ in vsu sunt, *Emollientia,* parùm prosunt, sed
obsunt potiùs; quia *extrahunt* magis, quam *imprimunt*; & *soluunt*
Compagem Corporis, potiùs quam *consolidant.*

---

2  arefieri] / thus in BNF copy; other copies (e.g. Folger copy 2) have arifieri      5  Interius] /
see Introduction, p. lxxiv above        6  Spiritûs] ~,

## <sup>|</sup> IX. The Operation
## to Tenderize the Parts which have Dried up,
## or to Soften the Body

*Connection.* I have already looked into tenderizing from within, which is accomplished by indirect and twisting paths both of alimentation and by detention of the spirit (and accordingly takes place gradually); so I now come to <sup>|</sup> tenderizing or softening imposed from outside and in no time at all.

### *History*

1. In the fable of Pelias' restoration to youth, Medea, when she pretended to be trying to accomplish it, put forward her scheme of cutting the old man's body to pieces and then of cooking it up with certain drugs in a cauldron. Perhaps some cooking helps, but the cutting to pieces is not necessary. <sup>|</sup>

2. Yet again maybe we should up to a point go in for cutting to pieces but not with a blade but with discernment. For since the consistency of the viscera and parts is very diverse, it necessarily follows that they cannot be tenderized in the same ways, but there should be a treatment for each of the particular parts besides that applied to tenderizing the whole mass of the body; and I shall take this last first.

3. Now if we have any power to achieve it, this operation can likely be satisfied by baths, anointings, and the like; concerning which pay attention to the points which follow.

4. We must not build too much hope of achieving our goal from the fact that we see the job <sup>|</sup> being done when inanimate things are steeped and macerated and so become tender—of which I have given examples above. For this kind of operation is easier on inanimate things because they draw and suck in the liquors; but it is more difficult in animal bodies because motion inside them is directed towards the circumference.

5. Therefore the emollient baths currently in use do more harm than good because they draw out rather than push in, and they loosen the body's structure rather than consolidate it.

**6.** *Balnea*, & *Vnctiones*, quæ Operationi præsenti (*Corporis* scilicet [Z4ʳ] benè & solidè *Malacissandi*) inseruire possint, | tres debent habere *Proprietates*.

**7.** Prima & præcipua est, vt constent ex iis, quæ totâ *Substantiâ* 5 *Similia* sunt *Corpori*, & *Carni Humanis*, quæque sint tanquam *Alma*, & *Nutricantia* per *Exteriùs*.

**8.** Secunda est, vt habeant *Admista ea*, quæ *subtilitate* nonnullâ *imprimant*, vt *Vim Nutritiuam* eorum, quibus admiscentur, insinuent, & inculcent.

10 **9.** Tertia, vt recipiant nonnullam *Mixturam* (licet reliquis longè minorem) eorum, quæ sunt *Astringentia*; Non *Austera*, aut *Acerba*, sed *Vnctuosa*, & *Confortantia*; Vt dum reliqua duo operentur, interìm [Z4ᵛ] prohibeatur | (quantùm fieri potest) *Exhalatio* è *Corpore*, quæ *Virtutem Malacissantium* perdat; sed potiùs vt per *Astrictionem Cutis*, et *Clau-* 15 *suram meatuum*, *Motus* ad *Intrà* promoueatur, et iuuetur.

**10.** *Consubstantiale* maximè *Corpori Humano*, est *Sanguis tepidus*, vel ex *Homine*, vel ex alijs *Animalibus*. At *Ficini* illud *Commentum*, ad Instaurationem Virium in Senibus, de *Exuctione Sanguinis Humani*, ex *Brachio Adolescentis sani*, leue admodùm est: Etenim, quod per *Interius* 20 nutrit, nullo modo debet esse *Æquale*, aut planè *Homogeneum Corpori*, quod nutritur; sed aliquatenus *Inferius*, & *Subordinatum*, vt subigi [Z5ʳ] possit. At in *exteriùs Applicatis*, | quantò *Substantia* est *similior*, tantò *Consensus melior*.

**11.** Ab Antiquo receptum est, *Balneum ex Sanguine Infantium* sanare 25 *Lepram*, & *Carnes* iam *corruptas* restituere; Adeò vt hoc ipsum fuerit *Regibus* quibusdam Inuidiæ apud *Plebem*.

**12.** Proditum est *Heraclitum Hydrope laborantem*, se in *Ventre calido Bouis*, nupèr occisi, immersisse.

**13.** In vsu est *Sanguis* tepidus *Catulorum Felis*, ad *Erysipelata*, & 30 instaurandas *Carnes* et *Cutem*.

**14.** *Brachium*, aut *Membrum* aliquod *Abscissum*, aut ex quo Sanguis aliàs nimiùm profluit, vtilitèr inseritur in *ventrem* alicuius *Animalis* [Z5ᵛ] nuper dissectum; | nam potentèr operatur ad *sistendum Sanguinem*; *Sanguine Membri Abscissi*, *Sanguinem Recentem Animalis*, per *Consensum* 35 sorbente, & ad se vehementèr trahente; Vnde & ipse sistitur, & refluit.

**15.** Multùm in vsu est, in *Morbis Extremis*, et quasi *Desperatis*, vt *Columbæ Scissæ*, aliæ post alias mutatæ, ad *Plantas Pedum Ægroti*

5 quæque] Quæque   10 *Mixturam*] ~,   17 *Animalibus*.] ~:   22 possit.] ~;
35 vehementèr] vehementrèr / some copies, see Introduction, p. lxxiv above
36 Multùm] Multum / some copies, see Introduction, p. lxxiv above

6. Baths and anointings which subserve the present operation (of softening the body firmly and well) | ought to have three properties.

7. The first and main one is that they should consist in things which in their entire substance are similar to the body and flesh of human beings, and whatever are as external wet nurses and nourishers.

8. The second is that they have those things admixed which give them with some subtlety, so that they insinuate and impart the nutritive power of the things with which they are admixed.

9. The third is that they take on some mixture (though far inferior to the former ones) of those things which are astringent, not harsh nor acerbic but unctuous and comforting; so that while the two factors just mentioned are in play, exhalation from the body | (which destroys the softening virtue) is in the meantime (as far as possible) prevented, and instead that motion to the innards is set going, and is assisted by astriction of the skin and closing of the passages.

10. That which is most consubstantial to the human body is warm blood, either from man or from animals. But that figment of Ficino's, i.e. that the vital powers of old men can be renewed by sucking human blood from the arm of a fit young man, is extremely shallow. For that which nourishes from the inside should not in any way be equal to or quite like the body nourished, but to some extent inferior or subordinate so that it can be worked up. But with things applied externally | the more the substance is like the body the better its consent with it.

11. It is anciently received that bathing in the blood of infants cures leprosy and restores flesh already rotten. But instances of this brought down popular hatred on the heads of certain kings.

12. They say that Heraclitus, struggling with dropsy, sunk himself in the warm belly of a freshly killed bull.

13. They use the warm blood of house kittens for erysipelas, and to renew the flesh and skin.

14. When an arm or limb has been cut away, and too much blood still flows, it is a good idea to put the stump inside the belly of some animal just cut open; | for this has a powerful effect in stopping the flow, since the blood of the cut absorbs fresh animal blood by consent, and draws it energetically to itself, whence the blood itself stops and flows back.

15. In widespread use in extreme and pretty well desperate illnesses, is the practice of cutting up pigeons and applying them alternately to the soles of the patient's feet, and from this comes wonderful succour

apponantur; vnde sequitur interdum Auxilium mirabile; Id vulgò imputatur, quasi *Maligna Morbi traherent;* Sed vtcúnque, Caput petit ista *Medicatio,* et *Spiritus Animales* confortat.

4    **16.** Verùm *Balnea* ista, et *Vnctiones sanguinolentæ,* nobis videntur
[Z6ʳ] sordidæ, et odiosæ; videndum | de alijs, quæ minùs fortassè habent *Fastidij,* neque tamen minus *Iuuamenti.*

**17.** Post *Sanguinem* igitur *Recentem, similia Substantiæ Corporis Humani,* sunt *Alimentosa; Carnes* pinguiores, *Bouinæ, Suillæ, Ceruinæ; Ostrea* inter *Pisces; Lac, Butyrum; Vitella Ouorum; Pollen Tritici; Vinum*
10 *Dulce,* aut *Saccharatum,* aut *Mulsum.*

**18.** Quæ admisceri debent ad *Impressionem,* sunt, loco omnium, *Sales,* præsertìm *Niger;* Etiàm *vinum* (cùm *Spiritu* turgeat) *imprimit,* & vtile est *Vehiculum.*

14    **19.** *Astringentia* eius generis, quæ descripsimus, *Vnctuosa* scilicet, &
[Z6ᵛ] *Confortantia* sunt, *Crocus, Mastyx,* & *Myrrha,* & *Baccæ Myrti.* |

**20.** Ex his, pro nostro Iudicio, optimè fiet *Balneum,* quale desideramus. *Medici* & *Posteri* Meliora reperient.

**21.** Longè autèm potentior fiet operatio, si *Balneum* quale proposuimus (quod Caput Rei esse arbitramur), comitetur quadruplex
20 *Operationis Series,* siue *Ordo.*

**22.** Primò, vt *Balneum* præcedat *Fricatio Corporis,* & *Inunctio ex Oleo,* cum aliquo *Spissamento;* vt Virtus, & Calor humectans *Balnei* potiùs subintret *Corpus,* quam *Aquea pars Liquoris.* Deinde, sequatur *Balneum* ipsum, ad horas fortè duas. A *Balneo* autem *Emplastretur*
25 *Corpus* ex *Mastice, Myrrhâ, Tragagantho, Diapalmâ, Croco,* vt cohibeatur
[Z7ʳ] (quantùm fieri | potest) *Perspiratio,* donec *Malacum* paulatìm vertatur in *Solidum;* idque per viginti quatuor Horas, vel amplius. Postremò, amotâ *Emplastratione,* fiat *Vnctio* cum *Oleo,* addito *Sale,* & *Croco;* Et renouetur *Balneum* post Quatriduum, cum *Emplastratione,* & *Vnctione* (vt priùs)
30 & continuetur huiusmodi *Malacissatio* per Mensem vnum.

**23.** Etiàm durante tempore *Malacissationis,* Vtile iudicamus, & Proprium, & secundum Intentionem nostram, vt *Corpus* benè nutriatur, & ab *Aere Frigido* abstineatur, & nil nisi *Calidum* bibatur.

34    **24.** Hoc verò (vt Initio in Genere monuimus) est ex ijs, quæ
[Z7ᵛ] no|bis *Experimento* probata non sunt, sed descripta tantummodò ex *Collimatione* ad *Finem.* Etenim *Metâ* positâ, alijs *Lampada* tradimus.

---

14 *Vnctuosa* scilicet,] some copies (see Introduction, p. lxxiv above) *Vnctuosa,* scilicet
18–19 proposuimus] ~,     19 arbitramur),] ~)ₐ     34 verò] ~,

which is commonly ascribed to the fact that they draw the sting from the disease. But in any event this treatment goes to the head, and comforts the animal spirits.

16. But to me these blood-bolstered baths and anointings seem filthy and disgusting; so we must seek out ⌐ alternatives which are perhaps less unpleasant but no less effective.

17. Thus after fresh blood, things similar to the substance of the human body are alimentary substances—fat flesh, beef, pork, venison; among fish, oysters; milk, butter, yolks of eggs, wheat meal, sweet wine either sugared or mixed with honey.

18. For things which should be mixed in to make an impression, salts and especially sea-salt are best of all. Wine too (as it swells with spirit) makes an impression, and is a useful vehicle.

19. Astringents of the kind I have described, namely unctuous and comforting, are saffron, gum mastic, myrrh, and myrtle berry. ⌐

20. Of these (in my judgement) a bath of the kind I desire can best be made. Physicians and future generations will find better ingredients.

21. The operation would become much more powerful if a bath of the kind I proposed (a thing which I judge to be indispensable) were accompanied by a fourfold series or sequence of operations.

22. The first is that the bath be preceded by massage and anointing with oil mixed with some thickening agent so that the virtue and moistening heat of the bath, rather than the watery part of the liquor, may enter the body. Next comes the bath for two hours or so. After the bath apply to the body plasters of gum mastic, myrrh, tragacanth, palm oil, and saffron so that (as far as ⌐ possible) the sweat can be kept back, and the soft be gradually turned into solid, and do that for twenty-four hours or more. Lastly, with the plasters removed anoint the body with oil to which salt and saffron have been added. Take the bath again with the plasters and anointing (as before) four days later, and continue this softening up routine for a month.

23. Now while the softening is underway, I judge that it would be useful, appropriate, and in line with my intention to nourish the body well, to keep it from cold air, and to drink only warm liquids.

24. But this (as I warned in general at the outset) is one of those things which I have ⌐ not tested by experiment, but have described only as a goal to be aimed at. For having set up the target, I hand on the lamp to others.

**25.** Neque negligenda sunt *Fomenta* ex *Corporibus Viuis*. *Ficinus* ait (neque id per Iocum) *Dauidem contubernio Puellæ, aliàs salubritèr, sed nimis serò vsum fuisse*; Debuerat autem addere, quod *Puellam* illam, more *Virginum Persiæ*, oportuisset inungi *Myrrhâ*, & Similibus, non ad
5 *Delicias*, sed ad *augendam Virtutem Fomenti*, ex *Corpore viuo*.

**26.** *Barbarossa Ætate Extremâ*, ex Consilio *Medici Iudæi*, *Puerulos*
[Z8ᵛ] continuè *Stomacho*, & *Ilijs* applicabat ad *Fomenta*; Etiam *Senes* | nonnulli, *Caniculas* (Animalia scilicet inter calidissima) *Stomacho* noctù applicare consueuerunt.

10 **27.** De *Hominibus* quibusdam *Nasonibus* (qui Irrisionis pertæsi, *Nasorum Tuberes*, & quasi *Surculos* amputarunt, atque in *Brachiorum Vlnas*, Incisione nonnullâ ad-apertis, ad tempus insuerunt, atque inde *Nasos* magis decentes efformarunt) increbuit Relatio quasi certa: idque in multis Nominibus; Ea si vera sit, *Consensum Carnis* ad *Carnem*,
15 præsertìm *Viuarum*, planè testatur.

**28.** De *Inteneratione* particulari *Viscerum Principalium*, *Stomachi*,
[Z8ᵛ] *Pulmonum*, *Iecoris*, *Cordis*, *Cerebri*, *Spi*|*nalis Medullæ*, *Renum*, *Fellis*, *Iliorum*, *Venarum*, *Arteriarum*, *Neruorum*, *Cartilaginum*, *Ossium*, nimis longa foret Inquisitio, & *Præscriptio*; Cùm iam non *Praxim* instruamus,
20 sed *Indicationes* ad *Praxim*.

[2Arʳ] | OPERATIO
super Expurgatione*m* Succi veteris, &
Restitutionem Succi noui,
siue Renouationem per Vices. X.

25 *Historia.*

Licet, quæ hîc ponemus, superiùs ferè præoccupata sint, tamen quia ista *Operatio* est ex *Principalibus*, retractabimus ea paulò fusiùs.

**1.** Certum est, *Boues Aratores*, atque *Laboribus exhaustos*, in *Pascua*
[2Arᵛ] noua, & *læta* admissos, *Carnes* re|cipere *Teneras*, & *Iuueniles*; Idque Esu,
30 & Palato comprobari; vt manifestum sit *Carnium Intenerationem* non esse difficilem; verùm & *Carnis Intenerationem* sæpiùs *repetitam*, etiàm ad *Ossa*, & *Membranas*, & Similia peruenire posse, verisimile est.

**2.** Certum est, *Diætas*, quæ in vsu sunt, ex *Guaiaco* præcipuè, atque ex *Sarsa-perillâ*, & *Chinâ*, & *Sassafras*, præsertìm longiùs continuatas,

---

7 *Fomenta*,] ~:    10 *Nasonibus*] ~,    14 Nominibus] ~:    29 *Iuueniles*,] ~:
31 difficilem;] ~:

**25.** Nor should we neglect soothing compresses made from living bodies. Ficino says (and he was not joking) that David would have felt a lot better cohabiting with a girl, but that he took the treatment too late. Now he should have added that the girl ought to have been anointed with myrrh and the like, in the manner of Persian virgins, not for the thrill of it but to increase the virtue of the living compress.

**26.** On the advice of a Jewish physician, Barbarossa in extreme old age continually applied little boys as compresses to his stomach. Some old | men were also accustomed to applying puppies (evidently among the hottest animals) to their stomachs at night.

**27.** A tolerably certain report has spread on the authority of many concerning certain men who, fed up with the mockery to which their noses exposed them, have amputated their nasal protuberances and outgrowths. They then made an incision in their forearms, in which they sewed their noses for a time, and so fashioned better ones. If this be true, it testifies to the consent of flesh with flesh, especially living flesh.

**28.** Concerning the individual tenderizing of the main viscera—the stomach, lung, liver, heart, brain, spinal | marrow, kidneys, gall, guts, veins, arteries, nerves, cartilages, bones—it would take too long to look into them and to make recommendations since at the moment I am providing suggestions for practice, not practice itself.

| X. The Operation
on the Elimination of old Juice and
its Replacement with new,
or Renovation by Turns

### History

Although the things I set down here have, for the most part, been anticipated above, nevertheless I shall deal with them again a little more thoroughly because this is one of the chief operations.

**1.** It is certain that plough oxen, exhausted by their labours, when they are released into pastures fresh and new, regain flesh | tender and young. This is confirmed by its taste when you eat it; and so it is plain to see that this softening of the flesh oft repeated may likely reach to bones, membranes, and so on.

**2.** It is certain that the diets in current use—especially those based on guaiac, and on sarsaparilla, china root, and sassafras—above all if

& secundum Regulas rigidiores, *Vniuersum Corporis Succum*, primò *Attenuare*, deinde *Consumere*, atque *Sorbere*; Quod manifestissimum est, quia *Morbum Gallicum* vsque ad *Gummositates* prouectum, quíque [2A2ʳ] intimos *Corporis Succos*, occupauerit, & | deprauauerit, ex illis *Diætis*, 5 posse curari probatum est: Atque insupèr quia æquè manifestum est, per huiusmodi *Diætas*, Homines factos *Macilentos, Pallidos*, & quasi *Cadauerosos*, paulò post *Impinguari, Colorari*, & manifestò *Renouari*. Quamobrem huiusmodi *Diætas*, vergente Ætate, semel Biennio, ad Intentionem nostram vtiles esse, omninò existimamus, tanquam 10 *Exuuias*, & *Spolia Serpentum*.

3. Fidentèr dicimus (neque verò quis rogo nos inter *Hæreticos Catharos* reponat) *Purgationes repetitas*, atque factas *Familiares*, longè magis ad Diuturnitatem vitæ facere, quàm *Exercitia* & *Sudores*. Id autem fieri [2A2ᵛ] necesse | est, si teneatur, quod positum est, *Vnctiones Corporis*, & *Meat-* 15 *uum* ab *Extrà Oppletiones*, & *Aeris Exclusiones*, & *Spiritûs* in Massâ Corporis, *Detentiones*, plurimum conducere ad vitam longæuam. Etenìm certissimum est, per *Sudores*, & *Perspirationes Exteriores*, non solùm *Humores*, & *Vapores Excrementitios*, exhalari, & absumi; sed vnà etiàm *Succos*, & *Spiritus bonos*, qui non tam facilè reparantur; In 20 *Purgationibus* autem (nisi fuerint admodùm immoderatæ) non item, cùm super *Humores* præcipuè operentur. *Purgationes* autem, ad hanc *Intentionem*, optimæ sunt, quæ paulò ante *Cibum* sumuntur, quia *desic-* [2A3ʳ] *cant minùs*; Ideóque de|bent esse ex ijs *Catharticis*, quæ *Ventriculum* minimè turbant.

25 Intentiones Operationum, *quas proposuimus* (*vt arbitramur*) *verissimæ sunt*; Remedia Intentionibus *fida. Neque credibile est dictu* (*licet haud pauca ex ipsis velut* Plebeia *videri possint*) *quantâ cum Curâ, & Delectu, ea à nobis examinata fuerint; vt sint* (*saluâ* [2A3ᵛ] *semper* Intentione) *& Tuta, & Efficacia. Rem ipsam, Expe|rimen-* 30 *tum & comprobabit, & promouebit. Talia autem in omnibus Rebus sunt, Opera Consilij cuiusque Prudentioris; quæ sunt Effectu admiranda, Ordine quoque egregia, Modis faciendi tanquam vulgaria.*

---

they are carried on for a long time and according to strict rules, first thin out and then consume and absorb all the body's juice. And this is very obvious in that the French disease, carried to the stage of gummosities, and which has occupied and ¹ degraded the inner juices of the body, can still be cured by diets of this kind. And it is just as obvious that men made thin, pale, and almost cadaverous by such diets, soon after put on weight, regain colour, and evidently get a new lease on life. And so with the onset of old age diets of this kind, practised at two-year intervals, are, I believe, altogether useful for my intention, acting as they do like the sloughing off of snakes' skins.

3. I say with confidence (but, I declare, no one should align me with the Cathar heretics because of it) that repeated purges, and made famil- iar, contribute far more to length of life than exercise and sweats. Now this must occur of necessity, ¹ if my former position holds true, i.e. that anointings of the body, stoppings up of passages from the outside, and both exclusions of the air and detentions of the spirit in the bodily mass, are very advantageous to long life. For it is very certain that by outward sweats and perspirations not only are the humours and excrementitious vapours exhaled and used up, but that with them the good juices and spirits are too, which are not so easily repaired. But this is not so with purges (unless they be extremely immoderate) since they work in the main on the humours. The very best purges for this intention are those taken a little before meals because they desiccate less. And for that reason ¹ they should be chosen from among those cathartics which least upset the stomach.

The intentions of these operations which I have proposed are (in my judgement) very true, and the remedies right for the inten- tions. For although many of them may seem commonplace yet it is hard to credit with what care and discrimination I have exam- ined them in order (always keeping the intention unimpaired) to ensure their safety and effectiveness. But experiment ¹ will firm up and carry forward this business. But such things in all matters are products of wiser counsels: admirable in their effects, excellent in their disposition, yet in a way ordinary in the means used to carry them out.

[2A4$^{r-v}$: blank]

[2A5$^r$]                         | Atriola Mortis.

*Ad Artic.* **15.** *Connexio.* *De* Atriolis Mortis *iam inquirendum*; *id est de ijs, quæ accidant* Morientibus *in* Articulo Mortis, *& paulò Antè, & Post. Vt cùm multis vijs perueniatur ad* Mortem, *intelligi* 5 *possit in quæ* Communia; *illæ desinant*; *Præcipuè in* Mortibus, *quæ* [2A5$^v$] *inferuntur per* Indigentiam Naturæ, *potiùs* | *quàm per* Violentiam; *tametsi etiam aliquid ex his, propter Rerum Connexionem inspergendum sit.*

### Historia.

10      1. *Spiritus Viuus* videtur tribus indigere, vt subsistat: *Motu commodo*; *Refrigerio temperato*; & *Alimento idoneo. Flamma* verò duobus ex his tantùm indigere videtur; *Motu* nimirùm, & *Alimento*; proptereà quod *Flamma* simplex sit Substantia, *Spiritus* Composita; ita vt si transeat 14 paulo propiùs in *Naturam Flammeam*, se perdat.

[2A6$^r$]      2. Etiam *Flamma*, maiore *Flammâ* & potentiore resoluitur, & | necatur, vt benè notauit *Aristoteles*; multò magis *Spiritus.*

3. *Flamma*, si comprimatur nimiùm, extinguitur; vt cernere est in *Candelâ*, super-imposito *Vitro*; Etenim *Aer* per Calorem dilatatus contrudit *Flammam*, eamque minuit, & extinguit. Neque in *Caminis* 20 concipitur *Flamma*, si Materies, absque Spatio aliquo interiecto, compingatur.

4. Etiam *Ignita Compressione* extinguuntur; veluti si *Carbonem* ignitum *Ferro*, aut *Pede*, fortitèr comprimas, extinguitur statim *Ignis.*

25      5. At vt ad *Spiritum* veniamus; Si *Sanguis* aut *Phlegma* irruat in *ven-* [2A6$^v$] *triculos Cerebri*, fit Mors su|bitò, cùm *Spiritus* non habeat vbi se moueat.

6. *Contusio* etiam *Capitis* vehemens inducit subitam Mortem, *Spiritibus* in *ventriculis Cerebri* angustiatis.

7. *Opium*, & alia *Narcotica* Fortiora, coagulant *Spiritum*, eúmque 30 priuant Motu.

8. *Vapor Venenatus, Spiritui* totalitèr odiosus, infert Mortem subitam, vt in *venenis Mortiferis*, quæ operantur per *Malignitatem*

---

5 Communia;] / some copies (e.g. BNF) with semicolon, others (e.g. Huntington 601102) with a comma          18 *Vitro*;] ~:          19 extinguit.] ~:          26 su|bitò,] ~;
29 coagulant] Coagulant

|

## | Death's Anterooms

*To Article* 15. **Connection.** Now we must inquire into the ante-rooms of death, i.e. into the things which befall the moribund during the critical instants just before and just after death; so that, since the approaches to death are many, one may understand the common destination at which they end, especially in deaths occurring by defect of nature rather | than by violence; though I shall be obliged to say something here and there about the latter by reason of the connection of things.

### *History*

1. The living spirit seems to need three things to survive: suitable motion, moderate coolness, and appropriate aliment. But flame seems only to need two of these, namely motion and aliment; for flame is a simple substance, spirit a composite one, such that if it gets a little nearer to the flamy nature it destroys itself.

2. Indeed as Aristotle has rightly noted, flame falls | back and gets suppressed by a larger and stronger flame, and that is even truer of spirit.

3. Flame squeezed hard goes out, as we see when we put a glass over a candle, for the air, dilated by the heat, cramps the flame, diminishes it, and puts it out. Neither is flame generated in ovens, if the fuel is compacted, and without any gaps in it.

4. Burning bodies are also quenched by compression, so that if you compress burning charcoal with iron or under foot, the burning is immediately put out.

5. Let us now come to the spirit: if blood or phlegm bursts into the ventricles of the brains, sudden death results | since the spirit has no room to move.

6. A severe contusion to the head also causes sudden death, the spirits in the ventricles of the brain being restricted.

7. Opium and other stronger narcotics coagulate the spirit and take away its motion.

8. A toxic vapour which is utterly obnoxious to the spirit causes sudden death, as in deadly poisons which work by what they call *specific*

(vt loquuntur) *specificam.* Incutit enim Fastidium *spiritui,* vt ampliùs mouere, aut Rei tam inimicæ occurrere nolit.

9. Etiam *Extrema Ebrietas,* aut *Crapula,* quandóque inferunt Mortem subitam; cùm *Spiritus* | non tam *Densitate,* aut *Malignitate Vaporis* (vt in *Opio,* & *venenis malignis*) quam ipsâ *Copiâ,* obruatur.

10. *Extremus Mœror,* & *Metus,* præsertìm *subitus* (vt fit in *Nuncio malo,* & improuiso) quandóque dant subitam Mortem.

11. At non solùm nimia *Compressio,* sed etiam nimia *Dilatatio Spiritûs,* Mortifera.

12. *Gaudia Ingentia* & *Repentina* Complures exanimârunt.

13. In magnis *Euacuationibus,* quales fiunt in secandis *Hydropicis,* exeuntibus confertìm Aquis; multò magis in *Ingentibus,* & *Repentinis Profluuijs Sanguinis,* sequitur sæpiùs Mors subita; Idque per meram *Fugam* | *Vacui* in Corpore, omnibus affatìm mouentibus ad Spatia implenda, quæ exinaniuntur; atque inter alia *Spiritu* ipso. Nam quoad *Profluuia Sanguinis* tardiora, res spectat ad *Indigentiam Alimenti,* non ad *Refusionem Spiritûs.* Atque de *Motu Spiritûs,* in tantum vel *Compresso,* vel *Effuso,* vt Mortem inferat, hæc inquisita sunt.

14. Veniendum ad *Indigentiam Refrigerij. Cohibitio Respirationis* Mortem infert subitam, vt in omni *Suffocatione* aut *Strangulatione.* Neque tamen videtur Res referri debere, tam ad *Impedimentum Motûs,* quam ad *Impedimentum Refrigerij;* quia *Aer* ni|mis *Calidus,* licet liberè attractus, non minùs suffocat, quàm si inhibeatur *Respiratio;* Vt fit in iis, qui suffocati aliquandò sunt ex *Carbonibus* incensis, aut *Lithanthracibus,* aut *Parietibus Recentèr Dealbatis,* in Cubiculis Clausis, Igne etiam accenso; quod Genus mortis traditur fuisse *Imperatoris Iouiniani.* Aut etiam ex *Balneis* siccis *super-calefactis,* quod vsurpatum fuit in *Nece Faustæ, Constantini Magni* vxoris.

15. Valdè pusillum est *Tempus,* quo Natura *Anhelitum* repetit, atque expelli *Fuliginem Aeris* in *pulmones* attracti, & recentem intro-recipi desiderat; vix certè ad tertiam partem Minutæ.

16. Rursùs *Pulsus Arteriarum,* & | *Motus Cordis, Systoles* & *Diastoles,* triplò velocior quàm *Respiratio;* adeò vt si fieri posset, vt ille *Motus* in *Corde,* absque inhibitâ *Respiratione* sisti posset, sequeretur Mors etiam celeriùs, quam ex *Strangulatione.*

---

1 *specificam.*] ~:     4 *Vaporis*] ~,     15 ipso.] ~:     17 *Spiritûs.*] ~:
26 accenso;] ~:     quod] Quod     34 sequeretur] Sequeretur

*malignity.* For it hits the spirit with such revulsion that it no longer wants to move or face such a terrible foe.

9. Sometimes too extreme drunkenness or overeating cause sudden death, as the spirit | is overwhelmed not so much by the density or malignity of the vapour (as in opium and malign poisons) as by its abundance.

10. Extreme grief and fear, especially if sudden (as happens with unexpected bad news), sometimes cause sudden death.

11. But it is not only severe compression, but also severe dilatation of the spirit that is lethal.

12. Great and sudden joys have finished off many.

13. In great evacuations of the kind which happen in sections for dropsy when the waters gush out *en masse*, and much more in great and sudden effusions of blood, sudden death often results; and this is caused by simple avoidance | of a vacuum in the body when everything, spirit included, moves in bulk to fill the empty spaces. As for slower effusions of blood, this has to do with lack of aliment, not with back-flow of spirit. So much then for the motion of the spirit when it causes death by compression or effusion.

14. Now we must go on to lack of coolness. Stopping respiration causes sudden death, as in all suffocation or strangulation. But it seems that this should not be ascribed so much to obstruction of motion, as to obstruction of cooling, because overheated air, | though freely drawn in, suffocates no less than prevention of respiration, as in those who have sometimes been suffocated by burning charcoal, coal, or walls freshly whitewashed in closed rooms with a lighted fire—which kind of death is said to have overtaken the emperor Jovinian. This also happens in overheated dry baths, which was what was going on in the death of Fausta, wife of Constantine the Great.

15. The time which nature lets pass between each breath, and during which it wants to expel the dirty air drawn into the lungs, and to take in fresh, is exceedingly short, barely twenty seconds.

16. Again the arterial pulse and | the systolic and diastolic motions of the heart are three times quicker than breathing, so that if this motion of the heart could be stopped without preventing breathing, death would follow more rapidly than by strangulation.

17. *Vsus* tamen & *Consuetudo*, in hâc Naturali Actione *Respirationis* nonnihil valet; vt in *Vrinatoribus Delijs*, & *Piscatoribus Perlarum*, qui perpetuo Vsu, Decuplum Temporis ad minimum, retinere *Anhelitum* possunt, plusquam pro Ratione aliorum Hominum.

5    18. Sunt ex *Animalibus*, etiam ex iis, quæ *Pulmones* habent, alia quæ
[2B1ʳ] ad longius Tempus, alia quæ ad breuius, *Anhelitum* co'hibere possunt; prout maiore scilicet, aut minore indigent *Refrigerio*.

19. *Pisces* minore indigent *Refrigerio*, quàm *Animalia Terrestria*; Indigent tamen, atque refrigerantur per *Branchias*; Atque quemadmodùm
10 *Terrestria* Aerem nimis feruidum, aut occlusum non ferunt, ita & *Pisces*, in Aquâ Glacie totalitèr, & diutiùs Coopertâ, suffocantur.

20. Si *Spiritus* Insultum patiatur, ab alio *Calore*, Proprio longè Vehementiore, dissipatur, & perditur. Si enim *Proprium Calorem* non
14 sustineat, absque *Refrigerio*, multò minùs *Alienum intensiorem* tolerare
[2B1ᵛ] potest. Id Cernitur in *Febribus Ardentibus*, vbi | *Calor Humorum putrefactorum, Calorem Natiuum* superat, vsque ad *Extinctionem*, siue *Dissipationem*.

21. *Somni* quoque Indigentia, & Vsus, refertur ad *Refrigerium. Motus* enim, *Spiritum* attenuat & rarefacit, & *Calorem* eius acutit, & intendit.
20 *Somnus* contrà *Motum*, & *Discursum* eius Sedat, & compescit. Etsi enim *Somnus* Actiones *Partium*, & *Spirituum Mortualium*, & omnem Motum ad Circumferentiam Corporis, roboret, & promoueat; tamen Motum proprium *Spiritûs viui*, magnâ ex parte consopit, & tranquillat. At
24 *Somnus* regularitèr, semel intra 24 Horas, Naturæ humanæ debetur,
[2B2ʳ] idque | ad 6 aut 5 Horas ad minimum; Licet sint etiam in hac parte quandóque Naturæ Miracula, vt refertur de *Mæcænate*, quod longo tempore ante Obitum non dormisset. Atque de *Indigentiâ Refrigerij*, ad *Spiritum conseruandum*, hæc inquisita sint.

22. Quod verò ad Tertiam *Indigentiam* attinet (*Alimenti* scilicet)
30 videtur illa ad *Partes* potiùs, quàm ad *Spiritum Viuum* pertinere. Facilè enim quis credat, *Spiritum Viuum* subsistere in *Identitate*, non per *Successionem*, aut *Renouationem*. Atque quoad *Animam Rationalem* in Homine, certo certius est, eam nec ex *Traduce* esse, nec *reparari*, nec
[2B2ᵛ] *interire*. | Loquuntur de *Spiritu Naturali* Animalium, atque etiam
35 Vegetabilium, qui ab illâ alterâ essentialitèr, & formalitèr differt; Ex

---

9 *Branchias*;] ~:    10 ferunt,] ~;    15 potest.] ~;    24 24] ~.    25 6] ~.
5] ~.    34 *interire*.] ~:

17. Custom and practice have no little influence over the natural action of breathing, as we see in the Delian divers and pearl fishers, who, by constant practice, can hold their breath for at least ten times longer than other men.

18. Among animals, and indeed among those with lungs, some can hold their breath ¹ longer than others, according as they need more or less cooling.

19. Fish need less cooling than land animals, but they still need it and breathe through gills; and as land animals cannot bear air too hot or close, so fish are suffocated in water if it be completely frozen for a long time.

20. If the spirit endures an insult from another heat far stronger than its own it is dissipated and destroyed. For if it cannot put up with its own heat without cooling, so much less can it bear a more intense alien heat. We see this in burning fevers where ¹ the heat of putrefied humours surpasses the native heat to the point of extinguishing or dissipating it.

21. Sleep and lack of it have to do with cooling. For motion attenuates and rarefies the spirit, and sharpens and intensifies its heat. Sleep, on the other hand, calms and subdues its motion and bustling. For although sleep strengthens and promotes the actions of the parts, of the non-living spirits, and all motion to the circumference of the body, it still to a large extent quietens and makes tranquil the living spirit. But human nature needs regular sleep once in twenty-four hours, and ¹ for at least five or six hours; though in this connection there are sometimes miracles of nature too, as Mæcenas is said not to have slept for a long period before his death. So much then for lack of cooling to conserve the spirit.

22. As for the third lack (i.e. aliment), this seems to relate rather to the parts than to the living spirit. For one may readily believe that the living spirit persists in its essential integrity and not successively or by renovation. And as for man's rational soul, it is perfectly certain that it is not a legacy of the flesh, and that it neither suffers repair nor death. ¹ People also speak of the natural spirit of animals and, indeed, of vegetables, saying that it differs from the soul *essentialiter* and *formaliter*; for

horum enim Confusione, *Metempsuchôsis* illa, & innumera tam *Ethnicorum,* quam *Hæreticorum* Commenta emanârunt.

**23.** *Renouatio* per *Alimentum,* in Corpore Humano, regularitèr singulis Diebus requiritur. *Triduanum* autem *Ieiunium Sanis* vix toleratur:
5 Vsus tamen & Consuetudo, etiam in hac Parte, haud parùm valet; At *Morbo languentibus Inedia* minùs grauis est. Etiàm *Somnus Alimentationi* nonnihil parat, quemadmodùm contrà *Exercitatio* eam efflagitat
[2B3ʳ] magis. Inuenti etiam sunt (sed rarò) ali|qui, qui quodam Naturæ Miraculo, sine *Cibo* & *Potu,* ad Tempus non Mediocre vixerunt.

10 **24.** *Corpora Mortua,* si non intercipiantur à Putredine, diutiùs sine notabili Absumptione subsistunt. At *Corpora viua* non multùm vltra *Triduum* (vt dictum est) nisi reparentur per *Alimentationem;* Id quod indicat citam illam Absumptionem, esse Opus *Spiritûs viui,* qui aut se reparat; aut Partes ponit in Necessitate se reparandi, aut vtrunque:
15 Quam Rem etiam illud astruit (quod paulo antè notatum est) nempe, quod possint *Animalia,* sine *Alimento,* paulò diutiùs durare, si *dormiant.*
[2B3ʳ] At *Somnus* om|ninò nil aliud est, quàm *Receptio Spiritûs viui* in se.

**25.** Copiosa nimis & continua *Effluxio Sanguinis,* qualis aliquandò fit ab *Hæmorrhoidibus;* interdùm à *Vomitu Sanguineo,* Venis interioribus
20 reseratis aut fractis; interdùm ex *Vulneribus,* Mortem infert festinam; cùm *Sanguis Venarum, Sanguini Arteriarum* ministret, *Sanguis Arteriarum, Spiritui.*

**26.** Haud paruum est *Quantum Cibi* & *Potûs,* quod Homo bis in die pastus, intra Corpus recipit; Longè plus, quàm aut per Sellam, aut per
25 Vrinam, aut per Sudores egerit: Nil mirum (inquis) cum reliquum in
[2B4ʳ] Succos, & substantiam Corporis mute|tur. Rectè: Sed cogita paulispèr, quod ea Accessio fit bis in Die, neque tamen Corpus exundat; Similitèr, licet *Spiritus* reparetur, tamen *Quanto* suo non enormitèr excrescit.

**27.** Nil attinet adesse *Alimentum* in *Gradu remoto,* sed eius Generis,
30 & ita præparatum, & ministratum, vt *Spiritus* in illud agere possit. Neque enim Baculus Cerei sufficiet ad Flammam continuandam, nisi adsit Cera; Neque Homines Herbis solis pasci possunt: Atque inde fit *Atrophia Senilis,* quod licèt adsit Caro & Sanguis, tamen Spiritus est
34 factus tam paucus & rarus, & Succi & Sanguis tam effœti, & obstinati,
[2B4ʳ] vt non teneat ¦ Proportio ad Alimentandum.

---

5 valet;] ~:   11 subsistunt.] ~:   15 astruit] ~,   20 festinam;] ~:
21 cùm] Cùm

actually muddling up the two has given birth to the notion of trans-
migration of souls, and no end of heathenish and heretical fabrications.

**23.** In the human body renovation by aliment is needed regularly
every day. For a healthy body can hardly put up with three days of
starvation. Yet habit and custom have no little influence in this matter.
But not eating is less serious to those enfeebled by sickness. And just as
sleep to some degree stands in for aliment, so by contrast exercise
demands more of it. On the other hand, we come across people (though
not often) ⎮ who, by some miracle of nature, have lived for an appreciable time without food or drink.

**24.** If they are not overtaken by putrefaction, dead bodies last for
some time without obvious wasting. But, as I have said, living bodies do
not manage for much more than three days unless they are repaired by
aliment. This indicates that wasting is the work of the living spirit
which either repairs itself, or puts the parts in a position where they
must repair themselves, or both. This fact (as I noted a moment ago)
also contributes to that conclusion, namely that animals without aliment last a little longer if they sleep. For sleep, to be sure, is really
nothing other than ⎮ the living spirit's withdrawal into itself.

**25.** Loss of blood which is too great and goes on too long, as sometimes happens with piles, sometimes with throwing up of blood from
the opening or bursting of the inner veins, and sometimes with wounds,
brings quick death; for venous blood ministers to the blood of the
arteries, while the blood of the arteries ministers to the spirit.

**26.** The amount of food and drink that a man dining twice a day
takes into his body is not small. Indeed, it is a great deal more than he
loses in excrement, urine, and sweat. And no wonder (you may say)
since the rest is turned into the juices and substance of ⎮ the body. You
would be right; but think for a moment that this twice-daily addition
takes place but without swamping the body. And in the same way,
although the spirit gets repaired, its amount does not suffer undue
increase.

**27.** It does not help to have aliment available in a distant degree, but
it must be of a kind so prepared and furnished that the spirit can work
on it. For a candlewick cannot keep a flame going in the absence of
wax, and men cannot live on salad alone. And this is what causes the
atrophy of old age: that although flesh and blood are not absent, the
spirit has been made so scarce and sparse, and both the juices and blood
so worn out and intractable, that they ⎮ cannot do the job of
alimentation.

335

28. Subducamus *Calculos Indigentiæ,* secundum Cursum Naturæ Ordinarium, & Consuetum. *Explicatione Motûs sui,* in ventriculis Cerebri, & Neruis, indiget *Spiritus* perpetuò; *Motu Cordis,* tertiâ Parte Momenti; *Respiratione,* singulis Momentis; *Somno* & *Alimento,* intra
5 Triduum; *Potentiâ* ad *Alimentandum* quasi post Octoginta Annos. Atque si alicui ex his Indigentijs non succurratur, sequitur Mors. Atque tria planè esse videntur *Atriola Mortis*; Destitutio Spiritûs, in *Motu* suo; in *Refrigerio*; in *Alimento.*

9 **Monita.** 1. *Errauerit, qui existimet* Spiritum Viuum, *Exemplo*
[2B5ʳ] Flammæ *perpe|tuò generari, & extingui, nec ad Tempus aliquod notabile durare. Neque enim hoc facit* Flamma *ipsa ex naturâ suâ, sed quia inter Inimica versatur.* Nam Flamma *intra* Flammam *durat. At* Spiritus Viuus *inter Amica degit, & Obsequia plurima. Itaque cùm* Flamma *sit substantia Momentanea,* Aer *autem substantia Fixa*; Spiritûs Viui *media est Ratio.*
15 2. *De* Interitu Spiritûs, *per* Destructionem Organorum (*qualis fit per* Morbos, *&* Violentiam) *non est Inquisitio præsens (vt ab initio diximus) tametsi & ille in eadem tria* Atriola *desinat. Atque de ipsâ* Formâ Mortis
[2B5ᵛ] *hæc inquisita sint.* |

29. Duo sunt magni *Præcursores Mortis.* Alter à *Capite,* alter à *Corde*
20 missus, *Conuulsio,* & *Extremus Labor Pulsûs*; nam etiam *Singultus* ille *læthalis* est *Conuulsionis* genus; *Labor* autem *Pulsûs læthalis,* habet Velocitatem insignem, quandoquidem *Cor* sub ipsâ Morte ita trepidet, vt *Systole,* & *Diastole* ferè confundantur. Habet etiam coniunctam *Debilitatem,* & *Humilitatem,* & sæpiùs *Intermissionem* magnam,
25 labascente motu Cordis, nec fortitèr, aut constantèr insurgere valente.

30. Præcedunt etiàm *Mortem* in Propinquo summa *Inquietudo,* &
[2B6ʳ] *Iactatio*; *Motus Manuum* tanquam Floccos colligendo; *Nix|us Prehensionis,* & *Tentionis* fortis; *Dentes* etiam *fortitèr comprimere*; *Glutire vocem*; *Tremor Labij inferioris*; *Pallor oris*; *Memoria confusa*; *Sermonis Priuatio*;
30 *Sudores frigidi*; *Corporis Elongatio*; *Sublatio Albuginis Oculorum*; *Faciei totius Alteratio* (Naso acuto, Oculis concauis, Genis labantibus); *Linguæ Contractio* & *Conuolutio*; *Frigus Extremitatum*; In aliquibus *Emissio Sanguinis,* aut *Spermatis*; *Clamor acutus*; *Anhelitus Creber*; *Inferioris Maxillæ Lapsus,* & similia.

35 31. *Mortem* sequuntur, *Sensûs* omnis, & *Motûs,* tam Cordis & Arteriarum, quàm Neruorum & Artuum, *Priuatio*; Impotentia *Corporis*

---

15 Organorum] ~,     16 *præsens*] ~,     17 *desinat*.] ~:     19 *Mortis.*] ~;
20 missus,] ~:     23 confundantur.] ~;     31 *Alteratio*] ~,     labantibus);] ~;)

336

28. Let me now reckon up what we need for life according to the ordinary and common course of nature. The spirit always needs to express its motion in the ventricles of the brain and in the nerves; motion of the heart three times a moment; respiration once a moment, sleep and aliment within three days; the capacity for alimentation beyond the age of 80 or so. And if any of these needs are not met, death results. And so there clearly seem to be three anterooms of death: failure of the spirit in its motion, in its cooling, and in its aliment.

*Advice.* 1. That man is wrong who thinks that the living spirit is, |
like flame, perpetually generated and extinguished, and does not last for an appreciable interval. For even flame does not do that of its own nature but because it exists among hostile bodies. For one flame situated within another lasts. But the living spirit dwells among a lot of friendly and deferential ones. Thus since flame is an evanescent substance, and air a permanent one, the living spirit is a principle poised between those two.

2. Death of the spirit by destruction of the organs (as happens through illness and violence) does not (as I said at the start) belong to the present inquiry, but this too ends up in those same three anterooms. So much then for the form of death.|

29. Death has two great forerunners, one dispatched from the head, the other from the heart: respectively, convulsion, and desperate labour of the pulse; for even that fatal catching of the breath is a kind of convulsion. Now that deadly labour of the pulse goes at great speed, because at death's door the heart trembles so much that you can hardly tell systole and diastole apart. It is also associated with weakness and abjection, and often great stalling when the motion of the heart is giving out and no longer capable or rousing itself steadily or strongly.

30. The two forerunners of imminent death are terrible anxiety and shaking; the hands move as if gathering wool; | they strive strongly to clutch and grasp; the teeth are tight shut; the voice chokes; the lower lip trembles; the face grows pale and the memory confused; speech is lost and cold sweats come; the body lengthens; the whites of the eyes roll up; the face changes completely (with the nose sharp, eyes hollow, and cheeks fallen); the tongue contracts and rolls back; the extremities grow cold; in some, blood or sperm is discharged; the cries grow shrill, the breathing rapid; the lower jaw falls; and so on.

31. Death is followed by loss of all sensation, and of motion both of the heart and arteries, and of the nerves and members; the body cannot

[2B6ᵛ] *se sustentandi erectum;* ¹ *Rigor Neruorum & Partium; Depositio* omnis *Caloris;* paulò post *Putrefactio* & *Fœtor.*

32. *Anguillæ, Serpentes,* & *Insecta* diù mouentur, singulis Partibus, post *Concisionem;* vt etiam *Rustici* putent, Partes singulas, ad se rursùs
5 vniendum expedire. Etiàm *Aues* Capitibus auulsis ad Tempus subsultant; quin et *Corda Animalium* auulsa diù palpitant. Equidem meminimus ipsi vidisse *Hominis Cor,* qui euisceratus erat (Supplicij Genere apud nos versus *Proditores* recepto), quod in Ignem de more
9 iniectum, saltabat in altum, primò ad Sesquipedem, & deinde gradatìm
[2B7ʳ] ad minus; durante spatio (vt meminimus) septem, ¹ aut octo Minutarum. Etiàm Vetus, & Fide digna Traditio est, de *Boue* sub *Euiscerationem* Mugiente. At magis certa de *Homine,* qui eo Supplicij genere (quod diximus) euisceratus, postquàm *Cor* auulsum penitùs esset, & in *Carnificis* Manu, tria aut quatuor Verba Precum, auditus est proferre;
15 Quod idcircò magis credibile esse diximus, quàm illud de *Sacrificio;* quia solent Amici huiusmodi *Reorum,* Mercedem dare *Carnifici,* vt Officium suum pernicissimè expediat, quò illi celeriùs à Doloribus liberentur. In *Sacrificijs* verò non videmus am, cur similis præstetur à
[2B7ᵛ] *Sacerdote* diligentia. ¹
20 33. Ad *Resuscitandum* eos, qui *Deliquia Animi,* aut *Catalepses* subitas patiuntur (quorum haud Pauci, absque Ope, etiàm expiraturi fuissent), hæc sunt in Vsu. *Exhibitio Aquarum* ex *Vino distillatarum* (quas *Aquas* vocamus *Calidas,* & *Cordiales*); *Inflexio Corporis in Pronum; Obturatio* fortis *Oris* & *Narium; Flexio Digitorum cum Torturâ quâdam; Euulsio*
25 *Pilorum Barbæ,* aut *Capitis; Frictio Partium,* præsertim *Faciei,* & *Extremorum; Subita Inspersio Aquæ Frigidæ* in *Faciem; Strepitus Acuti, & subiti; Appositio ad Nares Aquæ rosaceæ,* cum *Aceto,* in languorcaussibus; *Incensio Plumarum, Pannorum,* in *Suffocatione Vteri;* At maximè *Sartago*
[2B8ʳ] feruefacta ¹ vtilis est *Apoplecticis;* Etiàm *Fotus arctus Corporum viuorum,*
30 aliquibus profuit.
34. Complura fuerunt Exempla Hominum tanquam Mortuorum, Aut *Expositorum è Lecto,* Aut *Delatorum ad Funus,* quinetiam *nonnullorum in Terrâ Conditorum,* qui nihilominùs reuixerunt; Id quod in ijs, qui Conditi sunt, repertum est (terrâ aliquantò post apertâ), per Obtu-
35 sionem & Vulnerationem Capitis, ex iactatione & Nixu Cadaueris intra Feretrum: Cuius Exemplum recentissimum, & maximè memorabile

stay upright; ' nerves and parts become stiff; all warmth is lost; and soon after come putrefaction and stench.

32. Eels, snakes, and insects cut up go on moving for some time in all their parts so that country folk think that the bits are trying to get back together again. Birds too jump about for a while after decapitation, and the hearts of animals beat for a long time after being torn out. As for myself, I remember seeing the heart of a man who had been disembowelled (the punishment traditionally inflicted on traitors here with us), which when, as custom decrees, it was thrown into the fire, jumped up at first a foot and a half into the air, and then gradually less and less over an interval (as I recall) of seven ' or eight minutes. There is also an old and trustworthy tradition of a bull lowing after being gutted. But more certain is the story of the man who, disembowelled in the punishment just mentioned, and his heart quite ripped out and in the executioner's grasp, was heard to offer up three or four words of prayer. This is more credible I say than the one about the sacrifice because friends of criminals of this kind usually pay the executioner to do his job as fast as possible, and put the condemned out of their misery more quickly. But in sacrifices I do not see any reason why a priest would do the same. '

33. To bring round people who have passed out or suffered sudden catalepsy (of whom without help not a few would have perished), these treatments are available: the production of waters (which we call hot or cordial waters) distilled from wine; bending the body downwards; stopping up the mouth and nostrils; bending and twisting the fingers; pulling out the hairs of the beard or head; massaging the parts, chiefly the face and extremities; sudden splashing of cold water on the face; a sudden and penetrating din; applying rose-water with vinegar to the nostrils in fainting fits; the burning of feathers or rags in hysteria; but most helpful of all is a heated ' pan in apoplexies. Also being snuggled up to living bodies has worked for some people.

34. Many are the cases of men left for dead, or laid out on the deathbed; or carried off for burial, or indeed who have been actually buried, who have none the less come back to life again. This is apparent (after exhumation) in the case of the ones buried, from the bruising and injuries to the head caused by the struggles and violent travails of the body in its box. The latest and most famous case of this was of Duns

fuit, *Ioannis Scoti, Subtilis* illius & *Scholastici*; qui à Seruo, cùm sepultus
[2B8ᵛ] esset, absente (quíque, vt videtur, huiusmodi *Catalepsium* eius *Symp-
tomata* nouerat) aliquantò post effossus, in tali Statu repertus est; Et
simile quiddam accidit nostrâ Ætate, in personâ *Histrionis, Sepulti*
5 *Cantebrigiæ*; Memini me accepisse de *Generoso* quodam, qui ludi-
bundus, ex Curiositate, desiderabat scire, qualia paterentur in *Patibulo*
Suspensi; seséque suspendit, super Scabellum se alleuans, & deinde se
demittens; putans etiàm penes se futurum, vt Scabellum, pro arbitrio
suo, recuperaret, Id quod facere non potuit; sed tamen ab Amico
10 præsente adiutus est; Ille interrogatus, quid passus esset? Retulit se
[2C1ʳ] *Dolorem* non sensisse; sed primò obuersatam sibi fuisse circa oculos
*Speciem Ignis, & Incendij*; deinde *Extremæ Nigredinis*, siue *Tenebrarum*;
postremò *Coloris* cuiusdam *Cærulei pallidioris*, siue *Thalassini*; qualis
etiam conspicitur sæpè, *Animo Linquentibus*. Audiui etiam de *Medico*
15 adhùc viuente, qui Hominem, qui se suspenderat, atque per Horam
dimidiam suspensus manserat, in vitam, *Fricationibus*, & *Balneis*
*Calidis*, reduxerat; Quíq*ue* etiam profiteri soleat, se non dubitare, quin
Suspensum quemcunque ad tempus prædictum, reuocare posset, modò
19 Ceruices ei, per Impetum primæ Demissionis, non fuerint effractæ.

[2C1ᵛ] <sub></sub>

|Discrimina Iuuentutis
& Senectutis.

**Ad Artic. 16. 1.** *Scala Humani Corporis* talis est: *Concipi*; *Viuificari in*
*Vtero*; *Nasci*; *Mamilla*; *Depulsio à Mamillâ*; *Vsus Cibi* & *Potûs*, ab initio,
qualis *Infantibus* conuenit; *Dentire* primò, circa Annum secundum;
25 *Incipere Gradiri*; *Incipere Loqui*; *Dentire* secundò, circa Annum septimum; *Pubescere*, circa Annum duodecimum, aut decimum quartum;
*Potentem esse* ad *Generandum*, & *Fluxus Menstruorum*; *Pili* circa *Tibias*,
[2C2ʳ] & *Axillas*; *Barbescere*; atque hùc vsque, & quandóque vlteriùs, |*Gran-*
*descere*; Deindè *Roboris Artuum Status* & *vltimitas*, etiàm *Agilitatis*;
30 *Canescere*, & *Caluescere*; *Cessatio Menstrui*, & *Potentiæ Generationis*;
*Vergere* ad *Decrepitum*, & *Animal Tripes*; *Mori*. Interim *Animus* quoque
suas habet Periodos, sed per Annos non possunt describi, vt *Memoriam*
*labilem*, & *Similia*, de quibus posteà.

2. *Discrimina Iuuentutis*, & *Senectutis* hæc sunt. *Cutis* Iuueni *lœuis*, &

---

2 absente] ~,    3 est;] ~:    5 *Cantebrigiæ*] / silently emended in *SEH* (II, p. 209) to
*Cantabrigiæ*, though the *c-t* form seems to have been common in the seventeenth century
32 describi,] ~:    vt] Vt

Scotus, known as the Subtle Doctor, and a scholastic; he was buried
when the servant (who, it seems, knew the symptoms of his master's
cataleptic ¹ fits) was away; and the servant later dug him up and found
him in that condition. Something similar happened in my time, in the
shape of an actor buried at Cambridge. I recall being told about a
certain gentleman who, in playful mood and out of curiosity, wanting
to know what people suffered when they were hanged, hanged himself.
He did this by getting up on a stool and then jumping off, thinking to
get back on it again when he pleased. But this he was unable to do but
for the help of a friend who happened to be present. Asked what he had
experienced, the gentleman replied that he had felt no pain but at first
noticed a kind of fire ¹ and burning about his eyes, then a profound
blackness or shadows, and then a pale blue or aquamarine colour, which
is of a kind often seen by people in fainting fits. I also heard from a
physician who is still with us that he had with massage and warm baths
revived a man who had hanged himself and been dangling for half an
hour. This physician is also given to saying that he is sure that he could
bring round anyone who has been hanging for that length of time, so
long as his neck had not been broken by the jolt of the initial drop.

## ¹ The Differences between Youth and Old Age

*To Article* 16. 1. The stages through which the human body passes are
these: conception; quickening in the womb; birth; suckling; weaning
and the consumption of food and drink suitable for babies; cutting the
first teeth around the second year; beginning to walk and to speak;
cutting the second teeth around the age of 7; puberty around ages of 12
to 14; becoming sexually potent, and the start of menstruation; hair
springing on legs and arms; growth of the beard; and all this while, and
sometimes beyond, ¹ increasing stature; then reaching the height of
bodily strength and agility; grey hair and baldness; the menopause, and
loss of generative power; decline into decrepitude, and wretchedness on
three legs; death. In the meantime the mind also has its periods, but
these cannot be marked out by years—like memory loss and the like, of
which more later.

2. These are the differences between youth and old age: in youth the

*Explicata, Seni Arida,* & *Rugosa,* præsertìm circa *Frontem,* & *Oculos; Carnes* Iuueni *Teneræ,* & *Molles,* Seni *Duriores; Robur* Iuueni, & *Agilitas,* Seni *Diminutio Virium,* & *Motuum Tarditas;* Iuueni *Coctionum Validi-*
[2C2ᵛ] *tas,* Seni <sup>|</sup> *Debilitas;* Iuueni *Viscera Mollia,* & *Succulenta,* Seni *Salsa,* &
5 *Retorrida;* Iuueni *Corpus Erectius,* Seni *Inclinatio in Curuum;* Iuueni *Constantia Artuum,* Seni *Debilitas,* & *Tremor;* Iuueni *Humores Biliosi,* & *Sanguis Feruidior,* Seni *Humores Phlegmatici,* & *Melancholici,* & *Sanguis Frigidior,* Iuueni *Venus in promptu,* Seni *Tardior;* Iuueni *Succi Corporis* magis *Roscidi,* Seni magis *Crudi,* & *Aquei;* Iuueni *Spiritus Multus,* &
10 *Turgescens,* Seni *Paucus,* & *Ieiunus;* Iuueni *Spiritus Densus, & Viridis,* Seni *Acris,* & *Rarus;* Iuueni *Sensus Viuaces,* & *Integri;* Seni *Hebetiores,* & *Deficientes;* Iuueni *Dentes Robusti,* & *Integri,* Seni *Debiles, Attriti,* &
[2C3ʳ] *Decidui;* <sup>|</sup> Iuueni *Pili colorati,* Seni, cuiuscunque fuerint Coloris, *Cani;* Iuueni *Coma,* Seni *Caluities;* Iuueni *Pulsus Grandior,* & *Incitatior,* Seni
15 *Obscurior,* & *Tardior;* Iuueni *Morbi* magis *Acuti,* & *Curabiles,* Seni magis *Chronici,* & *Curatu Difficiles;* Iuueni *Vulnera citiùs Coalescentia,* Seni *Tardiùs;* Iuueni *Genæ florentes Colore,* Seni aut *Pallidæ,* aut *Rubi-cundæ,* atque *Sanguine Spisso;* Iuueni *minor Molestia ex Catarrhis,* Seni *Maior;* Neque scimus in quæ proficiant Senes (quoad corpus) nisi
20 quandóque in *Obesitatem;* Cuius Caussa præstò est; quia Corpora Senum, nec benè perspirant, nec benè assimilant; *Pinguedo* autem, nihil
[2C3ᵛ] aliud est, quàm *Exu*<sup>|</sup>*berantia Alimenti,* vltra id, quod excernitur, aut perfectè assimilatur. Etiam in quibusdam Senibus in *Edacitatem* proficitur, propter *Acidos Humores,* licèt Senes digerant minùs. Ac vniuersa,
25 quæ iam diximus, *Medici* quasi Feriantes, referent ad *Caloris Naturalis,* & *Humoris Radicalis Diminutionem,* quæ Res nihili sunt ad Vsum; Illud certum, *Siccitatem* in Decursu Ætatis, *Frigiditatem* præcedere; Atque Corpora cùm sint in *Statu,* & *Acme Caloris,* ad *Siccitatem* declinare, *Frigiditatem* autem posteà sequi.

30     3. Iam verò etiam de *Affectibus Animi* videndum. Equidem memini,
[2C4ʳ] cùm Adolescens essem *Pictauij* in *Galliâ,* me consueuisse <sup>|</sup> familiaritèr cum *Gallo* quodam, Iuuene Ingeniosissimo, sed paululùm Loquaci; qui

---

1 *Explicata,*] ~;    *Oculos,*] ~:    2 *Molles,*] ~;    *Duriores,*] ~:    *Agilitas,*] ~;
3 *Tarditas,*] ~:    3–4 *Validitas,*] ~;    4 *Debilitas,*] ~:    *Succulenta,*] ~;
5 *Retorrida,*] ~:    *Erectius,*] ~;    *Curuum,*] ~:    6 *Artuum,*] ~;    *Tremor,*] ~:
7 *Feruidior,*] ~;    8 *Frigidior,*] ~:    *promptu,*] ~;    *Tardior,*] ~:    9 *Roscidi,*] ~;
*Aquei,*] ~:    10 *Turgescens,*] ~;    *Ieiunus,*] ~:    *Viridis,*] ~;    11 *Rarus,*] ~:
12 *Deficientes,*] ~:    *Integri,*] ~;    13 *Decidui,*] ~:    *colorati,*] ~;    *Cani,*] ~:
14 Iuueni] Iuneni / first occurrence    *Caluities,*] ~:    *Incitatior,*] ~;    15 *Tardior,*] ~:
*Curabiles,*] ~;    16 *Difficiles,*] ~:    *Coalescentia,*] ~;    17 *Tardiùs,*] ~:
*Colore,*] ~;    18 *Spisso,*] ~:    *Catarrhis,*] ~;    19 *Maior,*] ~:    *Senes*] ~,

skin is smooth and clear, in age it is dry and wrinkled, especially round the forehead and eyes; in youth the flesh is soft and tender, in age harder; in youth you are strong and agile, in age you lose strength and mobility; in youth concoction is strong, in age ⎮ weak; in youth the innards are soft and succulent, in age salty and parched; in youth the body is upright, in age bent double; in youth strength of limbs, in age weakness and tremor; in youth bilious humours and hot blood, in age cold blood and phlegmatic, melancholy humours; in youth quick sexual arousal, in age slow; in youth the bodily juices are more dewy, in age more crude and watery; in youth abundant and swelling spirit, in age meagre and weak; in youth a spirit dense and fresh, in age biting and scanty; in youth the sense is keen and complete, in age dull and defective; in youth the teeth are strong and whole, in age weak, worn down, and rotten; ⎮ in youth the hair has colour, in age it goes grey regardless of its original colour; in youth you have hair, in age baldness; in youth a strong and rapid pulse, in age a fainter and slower one; in youth illnesses more acute and curable, in age more chronic and intractable; in youth wounds heal quickly, in age slowly; in youth a fresh face, in age a pale or ruddy one, with thick blood; in youth less trouble from catarrhs, in age more. In fact I do not know that age confers any benefit on the body other than in fatness sometimes; and the reason for that is not hard to find, namely that old bodies neither sweat nor assimilate well. For fatness is nothing other than ⎮ a surplus of aliment beyond that which is excreted or completely assimilated. In some old men too the appetite grows because of acid humours, though their digestion is less efficient. But everything I have spoken of just now the physicians, as if their brains were on holiday, refer to the dwindling of natural heat and radical moisture, things which have no practical use. Yet this is certain: that as the years advance dryness comes before coldness, and that bodies in the highest condition of heat decline into dryness, and that coldness follows after.

3.  Now I must come to the affections of the mind. For my own part I remember when I was in my youth at Poitiers in France, I was on very good terms ⎮ with a certain young Frenchman who was very brilliant but

posteà in Virum Eminentissimum euasit. Ille in Mores *Senum* inuehere solitus est; atque dicere, si daretur conspici *Animos Senum*, quemadmodùm cernuntur *Corpora*, non minores apparituras in ijsdem *Deformitates.* Quinetiam Ingenio suo indulgens, contendebat *Vitia*
5 *Animorum* in Senibus, *Vitijs Corporum* esse quodammodò Consentientia, & *Parallela.* Pro *Ariditate Cutis*, substituebat *Impudentiam;* Pro *Duritie Viscerum, Immisericordiam;* Pro *Lippitudine Oculorum, Oculum Malum*, & *Inuidiam;* Pro *Immersione Oculorum*, & *Curuatione Corporis*
[2C4$^v$] versus Terram, | *Atheismum* (*Neque enim cælum*, inquit, *respiciunt, vt*
10 *priùs*); Pro *Tremore Membrorum, Vacillationem Decretorum*, & *fluxam Inconstantiam;* Pro *Inflexione Digitorum*, tanquam ad Prehensionem, *Rapacitatem* & *Auaritiam;* Pro *Labascentia Genuum, Timiditatem;* Pro *Rugis, Calliditatem*, & *Obliquitatem;* & Alia quæ non occurrunt. Sed vt Serij simus, Iuueni adest *Pudor*, & *Verecundia;* Seni paululùm *obduruit;*
15 Iuueni *Benignitas*, & *Misericordia*, Seni *Occalluit;* Iuueni *Æmulatio laudabilis*, Seni *Inuidia Maligna;* Iuueni *Inclinatio ad Religionem*, & *Deuotionem*, ob Feruorem, & Inexperientiam *Mali*, Seni *Deferuescentia*
[2C5$^r$] *in Pietate*, ob Charita|tis Teporem, & diutinam Conuersationem inter Mala, nec-non ob Credendi Difficultatem: Iuueni *Valdè velle*, Seni
20 *Moderatio;* Iuueni *leuitas* quædam, & *Mobilitas;* Seni *Grauitas* maior, & *Constantia;* Iuueni *Liberalitas*, & *Beneficentia*, & *Philanthropia*, Seni *Auaritia*, & *Sibi sapere*, & *consulere;* Iuueni *Confidentia*, & *benè sperare*, Seni *Diffidentia*, & *Plurima habere pro suspectis;* Iuueni *Facilitas*, & *Obsequium*, Seni *Morositas*, & *Fastidium;* Iuueni *Sinceritas*, & *Animus*
25 *Apertus*, Seni *Cautio*, & *Animus Tectus;* Iuueni *Magna appetere*, Seni *Necessaria curare;* Iuueni *Præsentibus Rebus fauere*, Seni *Anteacta potiora*
[2C5$^v$] *habere;* Iuueni *Superiores* | *reuereri*, Seni *Censurâ in illos vti;* & complura alia, quæ ad *Mores* potiùs pertinent, quam ad *Inquisitionem præsentem.* Attamen, quemadmodùm in *Corpore*, ita in *Animo*, in nonnulla pro-
30 ficiunt Senes, nisi fuerint admodùm Emeriti; Nempe, vt cùm ad *Excogitandum* minùs sint prompti, *Iudicio* tamen valeant; & *Tutiora*, &

1 euasit.] ~;      6 *Impudentiam;*] ~:      7 *Immisericordiam;*] ~:
8 *Inuidiam;*] ~:      9 *Atheismum*] ~,      10 *priùs*);] ~.)      11 *Inconstantiam;*] ~:
12 *Auaritiam;*] ~:      *Timiditatem;*] ~:      13 *Obliquitatem;*] ~:      14 simus,] ~;
*obduruit;*] ~:      15 *Misericordia,*] ~;      *Occalluit;*] ~:      16 *laudabilis,*] ~;
*Maligna;*]:      17 *Mali,*] ~;      19 *velle,*] ~;      20 *Moderatio;*] ~:      *Mobilitas,*] ~:
21 *Constantia;*] ~:      *Philanthropia,*] ~;      22 *consulere;*] ~:      *sperare,*] ~;
23 *suspectis;*] ~:      24 *Obsequium,*] ~;      *Fastidium;*] ~:      25 *Apertus,*] ~;
*Tectus;*] ~:      *appetere,*] ~;      26 *curare;*] ~:      *fauere,*] ~;      27 *habere;*] ~:
*reuereri,*] ~;      *vti;*] ~:      28 *præsentem.*] ~:

a bit of a chatterbox, who later turned out to be very distinguished. He was accustomed to inveigh against the manners of the old, and to say that if their minds could be seen as their bodies could, the former would seem no less deformed than the latter. Moreover, giving his wit free rein, he argued that the vices of their minds had in a way some agreement or parallel with the defects of their bodies. He aligned their dry skin with their effrontery; their tough guts with their pitilessness; their rheumy eyes with the evil eye and envy; their sunken eyes and earth-bent bodies | with their atheism (because, he said, they could no longer gaze upon the heavens as before); their tremulous limbs with their wavering and inconstant judgement; their arthritic, grasping fingers with their rapacity and avarice; their knocking knees with timidity; their wrinkles with their crafty deviousness; and other things which slip my mind. But in all seriousness, youth is diffident and has a sense of shame, whereas old age is rather brazen; youth is well-disposed and merciful, old age is thick-skinned; youth has an admirable desire to emulate, old age malign envy; youth an inclination to religion and devotion born of fervour and inexperience of evil, old age sinking piety born of lukewarm | charity, and long habituation to evil, as well as difficulty of believing; youth has strong desires, old age moderation; youth has a certain lightness of mind and capriciousness, old age greater gravity and constancy; youth is open-handed, beneficent, and philanthropic, old age avaricious, self-seeking, and self-interested; youth is confident and optimistic, old age distrustful and suspicious; youth is easy-going and compliant, old age captious and scornful; youth sincere and straightforward, old age cautious and secretive; youth yearns for great things, old age worries about practical necessities; youth lives for the present, old age in the past; youth admires | superior persons, old age criticizes them; and so on for many other things which have more to do with custom than with the present inquiry. Yet as with the body so with the mind: old men do get better in one or two respects, unless they are completely superannuated. For instance, though they are not so quick at thinking things up, they have sounder judgement and settle rather on safer, more reasonable

*Saniora,* quam *Speciosiora* malint; Etiam, in *Garrulitatem* proficiunt,
& *Ostentationem*; Fructum enim *Sermonis* petunt, cum *Rebus* minùs
valeant; vt non absurdè *Tithonum* in *Cicadam* versum fuisse, *Poetæ*
fingant.

[2C6<sup>r–v</sup>: blank]

[2C7<sup>r</sup>] CANONES
6 mobiles de Duratione Vitæ, & Formâ Mortis.

## CANON. I.

*Non fit* Consumptio, *nisi quod deperditum sit de Corpore, trans-
migret in* Corpus Aliud.

10 EXPLICATIO.

*Nvllus est Rerum* Interitus; *Itaque quod absumitur, aut euolat in* Aerem,
[2C7<sup>v</sup>] *aut reci*|*pitur in* Corpus *aliquod* Adiacens. *Quare videmus* Aaneam,
*aut* Muscam, *aut* Formicam, *in* Electro, *Monumento plus quam Regio,
sepultas, Æternizari; cùm tamen sint Res Teneræ, & Dissipabiles. Verùm*
15 *non adest* Aer, *in quem aliquid euolet; atque Substantia Electri est tam
Heterogenea, vt nihil ex illis recipiat. Simile etiam fore arbitramur, misso*
Ligno, *aut* Radice, *aut eiusmodi, in* Argentum Viuum. *At* Cera, *&* Mel,
*&* Gummi, *habent similem Operationem, sed ex Parte tantùm.*

## CANON. II.

[2C8<sup>r</sup>] *Inest Omni Tangibili* Spiritus, *Corpore crassi*|*ore obtectus, &*
21 *obsessus; Atque ex eo Originem habet* Consumptio, *&* Dissolutio.

## EXPLICATIO.

*Nvllum* Corpus *nobis notum, hîc in Superiore Parte Terræ,* Spiritu *vacat;
siue per Attenuationem, & Concoctionem Caloris Celestium, siue aliàs.*
25 *Neque enim Caua Rerum Tangibilium* Vacuum *recipiunt; sed aut* Aerem,
*aut* Spiritum *Rei proprium.* Spiritus *autem ille (de quo loquimur) non est
Virtus aliqua, aut* Energia, *aut* Entelechia, *aut* Nugæ; *sed planè* Corpus
[2C8<sup>v</sup>] *Tenue, Inuisibile;* | *attamen Locatum, Dimensum, Reale. Neque rursùs*

11 Interitus;] ~:    12 Adiacens.] ~:    28 *Reale.*] ~:

346

conclusions than outwardly impressive ones. They also become more garrulous and ostentatious, since being less a match for things, they seek satisfaction in words, so that the poets' fiction that Tithonus was turned into a cricket is not completely absurd.

| Variable rules concerning
life's duration, and the form of death

### RULE 1

Consumption does not happen unless what is lost from one body takes up residence in another.

### EXPLANATION

Things are never destroyed. Therefore whatever is consumed either escapes into the air, or | is absorbed by some other body nearby. Accordingly we see a spider, fly, or ant buried in amber, a mausoleum more than royal, and eternized though they are fragile things and perishable. But the truth is that no air is present into which anything can escape, and the amber's substance is so heterogeneous that it can absorb nothing from them. In my judgement the same would happen if a piece of wood, or a root or the like were immersed in quicksilver. But wax, honey, and gum work in the same way, but only up to a point.

### RULE 2

In every tangible substance there exists a spirit hidden | and invested in the grosser body; and from this consumption and dissolution originate.

### EXPLANATION

No body known to us here in the upper part of the Earth lacks a spirit infused by the attenuation and concoction wrought by the heat of the heavenly bodies, or by other means. For the hollows in tangible things do not admit of a vacuum but only of air or the particular spirit of the thing. Now this spirit which I speak of is neither some virtue or energy nor an entelechy or silly trifle; it is plainly a body thin, and invisible, |

347

*Spiritus ille* Aer *est* (*quemadmodùm nec Succus Vuæ est Aqua*) *sed Corpus Tenue, cognatum Aeri, at multùm ab eo diuersum.* Partes autem Rei Crassiores (*cùm sint Naturæ Pigræ, nec admodùm Mobilis*) *per periodos longas duraturæ forent; sed* Spiritus *ille est, qui turbat, & illas fodicat, &*
5 *subruit, atque Humidum Corporis, & quicquid digerere potest, in nouum* Spiritum, *deprædatur; deinde tam* Spiritus *Corporis præ-inexistens, quàm nouitèr Factus, simul sensìm euolant. Id optimè ostenditur, in* Diminutione Ponderis, *Corporum Arefactorum per Perspirationem. Neque enim*
[2D1ʳ] *quicquid emitti|tur, erat* Spiritus, *quandò ponderauerat; neque non* Spiri-
10 tus, *quandò euolauerat.*

## CANON. III.

Spiritus *Emissus* desiccat; *Detentus, & moliens intùs, aut* colliquat; *aut* putrefacit; *aut* viuificat.

## EXPLICATIO.

15 *Qvatuor sunt* Processus Spiritûs; *Ad* Arefactionem; *ad* Colliquationem;
[2D1ᵛ] *ad* Putrefactionem; *ad* | Generationem *Corporum.* Arefactio *non est Opus Proprium* Spiritûs, *sed* Partium Crassiorum, *post emissum* Spiritum: *tùm enim illæ se contrahunt, partìm per Fugam Vacui, partìm per vnionem Homogeneorum: Vt liquet in omnibus, quæ arefiunt per Ætatem, et in*
20 *Siccioribus Corporibus, quæ desiccantur per Ignem, vt* Lateribus, Carboni-bus, Panibus. Colliquatio *est merum Opus* Spirituum; *Neque fit nisi* Calore *excitentur; tùm enim* Spiritus *se dilatantes, neque tamen exeuntes, se insinuant, & perfundunt inter Partes crassiores; Easque ipsas reddunt*
24 *Molles, & Fusiles; vt in* Metallis, *&* Cerâ: *Etenim* Metalla, *& alia Tenacia,*
[2D2ʳ] *apta sunt* | *ad cohibendum* Spiritum, *ne excitatus euolet.* Putrefactio *est Opus Mixtum,* Spiritûs, *&* Partium Crassiorum: *Etenim* Spiritu (*qui Partes rei continebat, & frænabat*) *partìm emisso, partìm languescente, omnia soluuntur, & redeunt in* Heterogenias *suas, siue* (*si placet*) Elementa *sua; Quod* Spiritûs *inerat Rei, congregatur ad se* (*vnde* Putre-
30 facta *incipiunt esse grauis Odoris*); Oleosa *ad se* (*vnde* Putrefacta *habent nonnihil Læuoris, & Vnctuositatis*); Aquea *itidem ad se;* Fæces *ad se* (*vnde fit Confusio illa in* Putrefactis). *At* Generatio, *siue* Viuificatio, *est Opus itidem Mixtum* Spiritûs, *&* Partium crassiorum; *sed longè alio*

---

1 *est*] ~,     2 *diuersum.*] ~:     4 *forent;*] ~:     19 *Ætatem,*] ~;   *et*] Et
30 *Odoris;*] ~:)    *ad se*] ~;    31 *Vnctuositatis;*] ~:)    *ad se*] ~; / first occurrence   *se*
(*vnde*] / some copies (e.g. Bibliotheca Regia Monacensis) *se;* (*vnde*    32 Putrefactis).] ~.)

yet something real with place, and extension. Nor again is this spirit air (any more than grape juice is water) but a thin body related to air but very different from it. Now the grosser parts of a thing (since they are torpid, and not very lively) would endure for long periods, were it not that this spirit stirred up, undermined, and destroyed them; and moreover preyed on the moisture of the body and whatever else it could digest into new spirit; after which the pre-existing and new-made spirit gradually escape together. We see this very well in the weight loss of bodies dried out by perspiration. For whatsoever is given out ᴵ was not spirit when it had weight, nor was it other than spirit when it escaped.

## RULE 3

Spirit given out desiccates; spirit held in and labouring within either liquefies, putrefies, or vivifies.

## EXPLANATION

The spirit has four processes: they are directed towards arefaction, liquefaction, putrefaction, and ᴵ the generation of bodies. Arefaction is not properly speaking a work of the spirit but of the grosser parts once the spirit has been given off; for then the parts contract in on themselves, partly to avoid a vacuum and partly by the coming together of homogeneous substances, as is evident in all things which dry out with age, and in drier bodies desiccated by fire, like bricks, coals, and loaves. Liquefaction is the work of the spirits alone, and it happens only when stimulated by heat, for then the spirits, expanding but yet not going out, insinuate and spread themselves among the grosser parts, and make them soft and molten, as in metals and wax. For metals and other tough bodies are good at ᴵ constraining the spirit by stopping it getting out when it is stimulated. Putrefaction is the combined work of the spirits and grosser parts. For when the spirit (which contained and curbed the parts of the thing) has in part escaped and in part grown weak, all things are dissolved, and are reduced to their heterogeneous parts or (if you like) their elements; for whatever spirit that existed in the body gathers itself together (whence the putrefied bodies begin to smell nasty), and the oily parts come together (whence putrefied bodies have a degree of smoothness and greasiness); the watery parts also come together, as do the dregs (whence comes the confusion in putrefied bodies). But generation or vivification is also the combined work of the spirit and

[2D2ᵛ] *modo*: Spiritus *enim* ᐟ *totalitèr detinetur, sed tumet, & mouetur localitèr*;
Partes *autem* Crassiores *non soluuntur, sed sequuntur Motum* Spiritûs,
*atque ab eo, quasi difflantur, & extruduntur in varias Figuras; vnde fit illa*
Generatio, *& Organizatio. Itaque sempèr fit* Viuificatio *in Materiâ*
5 Tenaci, *& Lenta; atque etiam* Sequaci, *& Molli; vt simul & Spiritûs
fiat Detentio; atque etiam* Cessio lenis Partium, *prout eas effingit*
Spiritus. *Atque hoc cernitur in Materiâ, omnium tam* Vegetabilium,
*quam* Animalium; *siue generentur ex* Putrefactione, *siue ex* Spermate;
*In his enim omnibus, manifestissimè cernitur esse* Materia, *difficilis ad*
10 Abrumpendum, *facilis ad* Cedendum.

[2D3ʳ]                                    ᐟ CANON. IV.

*In omnibus* Animatis *duo sunt Genera* Spirituum; Spiritus
Mortuales, *quales insunt* Inanimatis; *& Super-additus* Spiritus
vitalis.

15                                      EXPLICATIO.

*Iam antè dictum est, ad Longæuitatem procurandam, debere considerari*
Corpus Humanum, *primò, vt* Inanimatum, *& Inalimentatum; Secundò,*
[2D3ᵛ] *vt* Animatum, *& Alimentatum:* ᐟ *Nam Prior Consideratio dat Leges de*
Consumptione; *Secunda de* Reparatione. *Itaque nosse debemus, inesse*
20 *Humanis* Carnibus, Ossibus, Membranis, Organis, *denique* Partibus
*singulis, dum viuunt, in Substantiâ earum perfusos, tales* Spiritus, *quales
insunt in huiusmodi Rebus,* Carne, Osse, Membranâ; *& cæteris, Separatis,
& Mortuis; quales etiam manent in* Cadauere. At Spiritus Vitalis, *tametsi
eos regat, & quendam habeat cum illis Consensum, longè alius est ab ipsis;*
25 Integralis, *& per se* Constans. *Sunt autem duo Discrimina præcipua,
inter* Spiritus Mortuales, *&* Spiritus Vitales; *Alterum, quod* Spiritus
[2D4ʳ] Mortuales, *minimè sibi continuentur, sed sint tanquam* ᐟ *abscissi, &
circundati Corpore Crassiore, quod eos intercipit; quemadmodùm* Aer *per-
mixtus est in* Niue, *aut* Spuma. At Spiritus Vitalis *omnis sibi continuatur,*
30 *per quosdam* Canales, *per quos permeat, nec totalitèr intercipitur. Atque
hic* Spiritus, *etiam duplex est: Alter* Ramosus *tantùm, permeans per paruos*
Ductus, *& tanquam* Lineas; *Alter habet etiam* Cellam, *vt non tantùm sibi
continuetur, sed etiam congregetur in Spatio aliquo cauo, in benè magnâ*

---

4 Organizatio.] ~:        7 Spiritus.] ~;        11 IV.] 4.        23 Cadauere.] ~:
31 *duplex est*.] ~;        32 Lineas;] ~:

grosser parts, but in a quite different way. For the spirit | is completely kept in, and swells up and moves about locally, while the grosser parts are not dissolved but follow the motion of the spirit, and that drives them as if by the breeze, and forces them out into various shapes, and that causes this generation and organization. Thus vivification always happens in matter tenacious and sticky, and at the same time pliant and soft, so that there is at once a keeping back of the spirit, and a gentle yielding of the parts according as the spirit fashions them. And this we see in the matter of all things, vegetable as well as animal, be they generated from putrefaction or from seed; for in all these things we most plainly see matter which is hard to break through but easy to yield.

## | RULE 4

In all living things there are two kinds of spirits: non-living ones of the kind found in inanimate substances, and the superadded vital spirits.

## EXPLANATION

As I said earlier, to procure longevity the human body ought first to be considered as something inanimate and not needing aliment, and secondly as an animate being requiring aliment. | For the first consideration gives laws concerning consumption; the second laws concerning reparation. Thus we should know that there exists in flesh, bones, membranes, organs, and every single part of the human body, spirits which pervade them while they live, and which are identical to those which exist in those parts—flesh, bone, membrane, and the rest—when they are separate and dead, and identical to the ones remaining in the corpse. But the vital spirit, though it rules and has some consent with them, is very different from them, as it is integral and self-consistent. Now the non-living and vital spirits differ in two main ways: the first is that the non-living spirits are not in the least self-continuous, but are as it were | cut off and surrounded by the grosser body which intercepts them rather as air is intermixed in snow or froth. But all vital spirit is self-continuous through certain channels which it pervades, without being completely intercepted. This spirit too is of two kinds: the one is just branched and runs through little thread-like tubes; the other has in addition a cell so that it is not just self-continuous but is also gathered together in some hollow space and, relative to the body, in an

*Quantitate, pro Analogiâ Corporis; atque in illâ* Cellâ *est* Fons Riuulorum,
*qui inde diducantur. Ea* Cella *præcipuè est in* Ventriculis Cerebri, *qui in*
[2D4ᵛ] Animalibus *magis* Ignobilibus *angusti sunt; adeò vt vide|antur* Spiritus
*per vniuersum* Corpus Fusi, *potiùs quam* Cellulati: *Vt cernere est, in*
5 Serpentibus, Anguillis, Muscis, *quorum singulæ* Portiones *abscissæ
mouentur diù; Etiàm* Aues *diutiùs Capitibus auulsis subsulta*nt; *quonia*m
*parua habeant Capita, et paruas* Cellas. *At* Animalia Nobiliora Ventricu-
los *eos habent ampliores; Et maximè omnium* Homo. *Alterum Discrimen
inter* Spiritus *est; quod* Spiritus vitalis *nonnullam habeat* Incensionem;
10 *atque sit tanquam* Aura Composita *ex* Flammâ, *& Aere; quemadmodum*
Succi Animalium *habeant &* Oleum, *&* Aquam. *At illa* Incensio, *pecu-*
[2D5ʳ] *liares præbet* Motus, *&* Facultates; *Etenim &* Fumus Inflammabilis, |
*etiam ante* Flammam *conceptam, Calidus est, Tenuis, Mobilis; Et tamen
alia Res est, postquàm facta sit* Flamma. *At* Incensio Spirituum Vitalium,
15 *multis partibus lenior est, quam mollissima* Flamma, *ex* Spiritu Vini, *aut
aliàs; Atque insupèr Mixta est ex magnâ parte, cum* Substantiâ Aereâ; *vt sit
& Flammeæ, & Aereæ* Naturæ Mysterium.


## CANON. V.

Actiones Naturales *sunt propriæ* Partium *singularum, sed* Spiritus
20 vitalis *eas excitat, & acuit.*


[2D5ᵛ] | EXPLICATIO.

Actiones *siue* Functiones, *quæ sunt in singulis* Membris, *Naturam ipsorum* Membrorum *sequuntur* (Attractio, Retentio, Digestio, Assimilatio,
Separatio, Excretio, Perspiratio, *etiàm* Sensus *ipse*), *pro Proprietate*
25 Organorum *singulorum* (Stomachi, Iecoris, Cordis, Splenis, Fellis,
Cerebri, Oculi, Auris, *& cæterorum*). *Neque tamen vlla, ex ipsis* Actionibus, *vnquam actuata foret, nisi ex vigore, & Præsentiâ* Spiritûs vitalis, *&*
Caloris *eius; Quemadmodùm nec* Ferrum *aliud* Ferrum *attracturum foret,*
[2D6ʳ] *nisi exci|taretur à* Magnete; *Neque* Ouum *vnquam fœcundum foret, nisi*
30 Substantia Fœmellæ *actuata fuisset ab* Initu Maris.


6 *diù,*] ~:          7 Cellas.] ~:          14 Flamma.] ~;          23 *sequuntur*] ~;
24 *ipse*),] ~;)      25 *singulorum*] ~,     26 *cæterorum*).] ~.)

appreciable quantity; and in this cell is the source of the rivulets which go their separate ways from there. This cell is mainly in the cerebral ventricles, which in humbler creatures are narrow, such that ' the spirits seem to be diffused through the whole body rather than concentrated in cells, as we see in snakes, eels, and flies whose individual parts still move after being cut away. Birds too jump about for a while after decapitation because they have small heads and small cells. But nobler animals have larger cells, men most of all. The other difference between the spirits is this: that the vital spirit has some inflammation to it, and is like a breath made up of flame and air, just as the juices of animals contain both oil and water. But that inflammation provides special motions and faculties; for an inflammable fume ' before it has caught fire is hot, thin, and mobile; but it is another thing once it has been turned into flame. But the inflammation of the vital spirits is by many degrees gentler than the softest flame from spirit of wine or anything else; and it is in addition mixed to a great extent with an airy substance, so that it is a mysterious union of the airy and the flamy nature.

## RULE 5

Natural actions belong to the particular parts, but the vital spirit whets and stimulates them.

### ' EXPLANATION

Actions or functions belonging to particular members flow from the nature of the members themselves (as attraction, retention, digestion, assimilation, separation, excretion, perspiration, and indeed the very sense itself) act according to the character of the particular organs (stomach, liver, heart, spleen, gall, brain, eye, ear, and the rest). Yet not one of these actions would ever be sparked off but for the vigour and presence of the vital spirit, as well as its heat—just as one piece of iron would not attract another unless it had been ' stimulated by a loadstone, nor an egg ever become fertile if the female substance had not been actuated by the virility of the male.

## CANON. VI.

Spiritus Mortuales *Aeri proximè consubstantiales sunt*; Spiritus vitales *magis accedunt ad Substantiam Flammæ.*

### EXPLICATIO.

5 *Explicatio* Canonis quarti *præcedentis, est etiam Declaratio* Canonis
[2D6ᵛ] Præsentis; ᴵ *Verùm insupèr hinc fit, vt quæcunque sint* Pinguia, *&* Oleosa, *diu maneant in Esse suo; Neque enim* Aer *illa multùm vellicat, neque illa etiam ipsa, cum* Aere *coniungi multùm desiderant. Illud autem prorsùs vanum est, quod* Flamma *sit* Aer Accensus, *cùm* Flamma *&* Aer, *non* 10 *minus Heterogenea sint, quam* Oleum *&* Aqua. *Quod verò dicitur in* Canone, *quod* Spiritus Vitales *magis accedant ad* Substantiam Flammæ; *Illud intelligendum est, quod magis hoc faciant, quam* Spiritus Mortuales; *non quod magis sint* Flammei, *quam* Aerij.

ᴵ CANON. VII.

15 Spiritûs Desideria *duo sunt; vnum se* Multiplicandi; *Alterum* Exeundi, *& se* Congregandi, *cum suis Connaturalibus.*

### EXPLICATIO.

*Intelligitur* Canon *de* Spiritibus Mortualibus; *Etenim quoad* Desiderium secundum, Spiritus Vitalis *Exitum è Corpore suo, maximè exhorret; neque* 20 *enim inuenit* Connaturalia *hîc in Proximo: Ruit fortè in Occursum Rei*
[2D7ᵛ] *Desiderabilis, ad Extima Corporis sui;* ᴵ *sed Egressum, vt dictum est, fugit. Verùm de* Spiritibus Mortualibus, *vtrunque* Desiderium *tenet: Quoad primum enim attinet,* Omnis Spiritus inter Crassiora locatus, *non* fœlicitèr habitat; *Itaque cum* Simile sui *non inueniat, eò magis* Simile sui 25 *creat, & facit, in tali Solitudine positus, & strenuè laborat, vt se multiplicet, & Volatile Crassiorum deprædetur, vt augeatur suo* Quanto. *Quod verò ad* Secundum Desiderium Euolandi, *& se in Aerem recipiendi, Certum est omnia* Tenuia *(quæ sempèr sunt Mobilia) ad sui Similia in Proximo* 29 *libentèr ferri; vt* Bulla Aquæ *fertur ad* Bullam, Flamma *ad* Flammam. *At*
[2D8ʳ] *multò magis hoc fit, in* Euolatione Spiritûs *in* Aerem ᴵ Ambientem; *quia non fertur ad* Particulam *sui similem; Sed etiam tanquam ad* Globum

---

8 *desiderant.*] ~:     18 *quoad*] qnoad / i.e. turned letter     21 *fugit.*] ~:
27 *recipiendi,*] ~;     28 *Tenuia*] ~,     29 *Flammam.*] ~:

354

## RULE 6

The non-living spirits are nearly consubstantial to air; the vital spirits come closer to the substance of flame.

### EXPLANATION

The explanation of Rule 4 above is also an elucidation of this one; | but going on from there, it so happens that all fatty or oily substances stay true to their nature for a long time. For the air does not eat away at them much, nor do they in their turn much want to ally themselves with the air. Now the notion that flame is kindled air is an absolutely futile idea, seeing that flame and air are no less heterogeneous than oil and water. Thus, when this rule states that vital spirits come closer to the substance of flame, it must be taken to mean that they do so more than the non-living ones, and not that they are more flamy than airy.

### | RULE 7

The desires of the spirit are two: one is to multiply itself; the other is to go out and get together with its connaturals.

### EXPLANATION

This rule applies to the non-living spirits. For in connection with the second desire, the vital spirit is absolutely terrified of leaving its body for it has no connaturals nearby. Perhaps it will rush to the far reaches of its body to come into contact with something it desires, | but as I have said it avoids going out. But the non-living spirits are gripped by both desires, and as far as the first is concerned, all spirit, situated among the grosser parts, is housed wretchedly; and therefore, as it finds itself friendless and with nothing similar at hand, it labours all the harder to make something like itself, to multiply itself, and prey on the volatile material in the grosser parts. But as for the second desire, to escape and to give itself to the air, it is certain that all tenuous bodies (which are always mobile) gladly take themselves off to similar substances nearby, as one bubble of water moves towards another, or one flame to another; but this happens much more in the spirit's escape into | the surrounding air, because this is not movement to a particle like itself but to the very globe of its connaturals. But in the meantime it should be noted that

Connaturalium *suorum. At illud interìm notandum; quod Exitus, &* Euolatio Spiritûs *in Aerem, est duplicata Actio: partìm ex* Appetitu Spiritûs, *partìm ex* Appetitu Aeris; Aer *enim communis tanquam Res indigens est, atque omnia auidè arripit;* Spiritus, Odores, Radios, Sonos, 5 *& alia.*

## CANON. VIII.

Spiritus Detentus, *si alium* Spiritum *gignendi Copiam non habeat, etiam* Crassiora *intenerat.*

| EXPLICATIO.

10 Generatio Noui Spiritûs *non fit nisi super ea, quæ sunt in Gradu ad* Spiritum *propiore: qualia sunt* Humida. *Itaque si* Partes Crassiores (*inter quas versatur* Spiritus) *sint in Gradu remotiore, licet* Spiritus *eas conficere non possit, tamen (quod potest) eas labefactat, & emollit, & fundit; vt cùm Quantum suum augere non possit, tamen habitet laxiùs, &* 15 *inter ea degat, quæ sint magis Amica: Iste autem* Aphorismus, *ad Finem nostrum, admodùm vtilis est; quia innuit ad* Intenerationem Partium *Obstinatarum, per* Detentionem Spiritûs.

| CANON. IX.

Inteneratio Partium *Duriorum benè procedit, cùm* Spiritus *nec* 20 *euolet, nec generet.*

## EXPLICATIO.

*Iste* Canon *soluit Nodum, & Difficultatem in* Operatione Intenerandi, *per* Detentionem Spiritûs. *Si enim* Spiritus *non Emissus deprædetur omnia intùs, nil fit Lucri ad* Intenerationem Partium, *in Esse suo; sed* 25 *potiùs soluuntur illæ, & corrumpuntur. Itaque vnà cum* Detentione *refrigerari debent* Spiritus, *& astringi, ne sint nimis* Actiui.

| CANON. X.

Calor Spiritus, *ad* Viriditatem *Corporis, debet esse* Robustus, *non* Acris.

---

2 *Actio:*] ~;     6 CANON.] ~ₐ     18 CANON.] ~ₐ     23 Spiritûs.] ~:

the spirit's going out and escape into the air is a twofold action arising in part from the spirit's appetite, and partly from the air's. For ordinary air is a starveling, and greedily swallows them all up, spirits, smells, rays, sounds, and the rest.

## RULE 8

Spirit kept in, if it does not have the resources for generating new spirit, actually softens the grosser parts.

### EXPLANATION

Generation of new spirit takes place only on things which are a degree closer to the spirit, like moist bodies. Therefore if the grosser parts (in which the spirit busies itself) are in a degree further off, the spirit, although it cannot digest them, still (as far as it can) weakens, softens, and loosens them, so that since it cannot increase its quantity, it can nevertheless have more room, and live among things which are better disposed to it. Now this aphorism is very important for what we aim at because it tends to the softening of the stubborn parts by keeping back the spirit.

### RULE 9

Softening of the harder parts goes well when the spirit neither escapes nor generates.

### EXPLANATION

This rule solves a knotty problem in the operation of softening by keeping in the spirit. For if spirit not given off preys on everything within, nothing is gained in the way of softening the parts in their proper state, but rather they are dissolved and corrupted. Thus the spirits as well as being kept in ought to be cooled and compressed, in case they are too active.

### RULE 10

To keep the body youthful the spirit's heat should be robust but not fierce.

## EXPLICATIO.

*Etiam iste* Canon *pertinet ad soluendum Nodum Supradictum, sed longè latiùs patet; Describit enim, qualis debeat esse Temperamenti* Calor, *in Corpore ad Longæuitate*m. *Hoc verò vtile est, siue* Spiritus *detineantur, siue* [2E2ʳ] *non; Vtcunque enim talis debet esse* Calor Spirituum, *vt vertat se potiùs | in* 6 Dura, *quam deprædetur* Mollia; *Alterum enim desiccat, alterum intenerat. Quinetiam, eadem Res valet ad* Alimentationem *benè perficiendam; talis enim* Calor *optimè excitat* Facultatem Assimilandi, *atque vnà optimè præparat* Materiam *ad* Assimilandum. *Proprietates autem huiusmodi* 10 Caloris, *tales esse debent: Primò vt* Tardus *sit, nec* Subitò *calefaciat; Secundò, vt non sit admodùm* Intensus, *sed* Mediocris; *Tertiò, vt sit* Æqualis, *non* Incompositus, *scilicet se Intendens, & Remittens; Quartò, vt si inueniat* Calor *iste, quod ei resistat, non facilè* suffocetur, *aut* langueat. *Subtilis admodùm hæc* Operatio; *sed cùm sit ex Vtilissimis, non deserenda* [2E2ᵛ] *est. Nos | verò in* Remedijs *(quæ ad indendum* Spiritibus Calorem Robu-16 stum, *siue eum, quem vocamus* Fabrilem, *non* Prædatorium, *proposuimus) huic Rei aliquâ ex Parte satisfecimus.*

## CANON. XI.

Spirituum Densatio *in Substantiâ suâ, valet ad Longæuitatem.*

20                    EXPLICATIO.

*Svbordinatus est* Canon *ad Præcedentem; Etenim* Spiritus Densior *suscipit omnes illas quatuor* Caloris Proprietates, *quas diximus. Modi autem* [2E3ʳ] Densationis, *| in Primâ ex decem* Operationibus *habentur.*

## CANON. XII.

25 Spiritus *in* Magnâ Copiâ *& magis festinat ad Exitum, & magis deprædatur, quam in* Exiguâ.

## EXPLICATIO.

*Clarus est per se* Canon *iste, cùm* Quantum *ipsum regularitèr augeat* Virtutem; *Atque cernere est in* Flammis, *quòd quantò fuerint maiores,*

4 *Longæuitate*m.] ~;        10 *debent:*] ~;     *calefaciat;*] ~:        11 Mediocris;] ~:
12 *Remittens;*] ~:        15 Remedijs] ~,

358

## EXPLANATION

This rule also relates to solving the knotty problem mentioned above, but its scope is wider. For it tells us what kind of temperament the heat in the body should have to promote longevity. And this is useful whether or not the spirits are kept in; for in any event the heat of the spirits should be such as to work rather $^|$ on the hard parts than to prey on the soft, for the former softens while the latter dries things up. Besides, this same thing helps to round off alimentation well, as such a heat best stimulates the faculty of assimilation, and at the same time works up the material to be assimilated. Now heat of this kind should have these properties: first, that it should warm slowly, not in a trice; secondly, that it not be very intense but moderate; thirdly, that it be uniform and not irregular, i.e. not flaring up and then dying down; fourthly, that, if this heat comes across anything that opposes it, it will not easily be smothered or weakened. For this operation is extremely subtle, but since it is among the most useful we should not neglect it. Indeed, I $^|$ in my remedies (which I put forward to endow the spirits with a robust heat or that which I call constructive rather than predatory) have to some extent accomplished this objective.

## RULE 11

Condensation of the spirits in their substance is good for longevity.

## EXPLANATION

This rule is subordinate to the preceding one; for spirit made denser acquires all four properties of heat just mentioned. But means of condensation $^|$ can be obtained in the first of the ten operations.

## RULE 12

Spirit in abundance is both readier to escape, and more predatory than when it is in short supply.

## EXPLANATION

This rule is transparent in itself, since as a general rule quantity increases virtue of itself, as we see in flames which, the bigger they are, the more

*tantò & erumpant fortiùs, & absumant celeriùs. Itaque nimia* Copia, *aut* Tur'gescentia Spirituûs, *prorsùs nocet Longæuitati. Neque amplior est optanda* Copia Spirituum, *quam quæ Munijs Vitæ, & bonæ Reparationis Ministerio sufficiat.*

5

## CANON. XIII.

Spiritus æqualitèr Perfusus, *minùs festinat ad exitum, & minus deprædatur, quam* imparitèr Locatus.

## EXPLICATIO.

9
[2E4ʳ]
*Non solùm* Copia Spirituum *secundum Totum, Durationi Rerum obest, sed etiàm eadem* Copia, *minùs refracta,* | *similitèr obest. Itaque quò magis fuerit* Spiritus comminutus, *& per Minima insinuatus, eò deprædatur minùs.* Dissolutio *enim incipit à Parte, vbi* Spiritus *est laxior, Itaque &* Exercitatio, *&* Fricationes, *Longæuitati multùm conferunt.* Agitatio *enim optimè comminuit, & commiscet Res per Minima.*

15

## CANON. XIV.

Motus Spirituum Inordinatus, *&* Subsultorius *magis properat ad Exitum, & magis deprædatur, quam* Constans, *&* Æqualis.

[2E4ᵛ]
## | EXPLICATIO.

*In* Inanimatis *tenet iste* Canon *certò;* Inæqualitas *enim* Dissolutionis
20 Mater; *In* Animatis *verò (quia non solùm spectatur* Consumptio, *sed* Reparatio; Reparatio *autem procedit per Rerum* Appetitus; Appetitus *rursùs acuitur per* Varietatem) *non tenet rigidè, sed eousque tam*en *recipiendus est, vt* Varietas *ista potiùs sit* Alternatio, *quam* Confusio, *& tanquam Constans in Inconstantiâ.*

25

## CANON. XV.

Spiritus *in* Corpore Compagis solidæ, *detinetur licet inuitus.*

2 *Longæuitati.*] ~:  20 *verò*] ~,  21 Appetitus;] ~: / first occurrence
25 CANON.] ~ₐ

powerfully they burst out and more quickly they swallow things up. Thus too great an abundance or swelling | of spirit is a great enemy of long life; and abundance of spirits should be no greater than necessary for life's functions and the business of maintaining them.

## RULE 13

Spirit evenly distributed is in less of a hurry to get out and less predatory than when it is distributed unevenly.

## EXPLANATION

Not only is abundance of spirits in relation to the whole body bad for the durability of things; abundance when it is not | well split up is bad too. Therefore the more finely divided the spirit and more diffused through the smallest portions, the less predatory it is. For dissolution starts at the point where the spirit is unrestricted. Thus exercise and massage do a lot for longevity. For agitation is excellent for achieving fine division and mixing things in with the smallest portions.

## RULE 14

The motion of the spirits when it is disorderly and fitful makes them more liable to escape and more predacious than when it is steady and settled.

## | EXPLANATION

In inanimate things this rule holds for certain; for inequality is the mother of dissolution. But in animate bodies (because we consider not just consumption but also repair, and repair stems from the appetites of things, appetites stimulated by variety) it holds less rigidly but can nevertheless be accepted to the extent that this variety be rather alternation than confusion, and so to speak uniform in its variability.

## RULE 15

Spirit in a body of solid structure is kept in, albeit reluctantly.

| EXPLICATIO.

*Omnia* Solutionem Continuitatis *suæ exhorrent; attamen pro Modo Densitatis, aut Tenuitatis suæ. Etenim, quò Corpora sunt magis Tenuia, eò in minores, et angustiores Meatus se compelli patiuntur.* Itaque Aqua
5 *subintrabit Meatum, quem non subintrabit* Puluis; Aer *etiam, quem non subintrabit* Aqua; Quin Flamma, & Spiritus, *quem non subintrabit* Aer. *Veruntamèn est huiusce rei, aliquis Terminus; Neque enim* Spiritus *in tantùm, Desiderio Exeundi laborat, vt patiatur se discontinuari nimis, &*
9 *in nimis arctos Poros, aut Meatus agi.* Itaque si Spiritus Corpore Duro,
[2E5ᵛ] *aut etiam* Vnctuoso, & Tenaci | (*quod non facile diuiditur*) *circumdetur, planè constringitur, et tanquam incarceratur, et* Appetitum exeundi *posthabet; Quare videmus* Metalla, & Lapides *longo Æuo egere, vt exeat* Spiritus; *nisi aut* Spiritus *Igne excitetur, aut Partes crassiores* Aquis Corrodentibus, & Fortibus *disiungantur. Similis est Ratio* Tenacium,
15 *qualia sunt* Gummi, *nisi quod leniore Calore soluantur. Itaque* Succi Corporis Duri, Cutis Constricta, *et similia* (*quæ procurantur ab* Alimentorum Siccitate, *et* Exercitatione, *et* Aeris Frigore) *vtilia sunt ad Longæuitatem; quia Claustra circundant* Spiritui *arcta, ne exeat.*

| CANON. XVI.

20 *In* Oleosis, & Pinguibus *detinetur* Spiritus *libentèr, licet non sint Tenacia.*

EXPLICATIO.

Spiritus, *si nec à Corporis circundati* Antipathiâ *irritetur, nec à Corporis nimiâ* Similitudine *pascatur, nec à* Corpore Externo *sollicitetur, aut*
25 *prouocetur, non tumultuatur multùm ad Exeundum. Quæ omnia* Oleosis *desunt; Nam nec tam* Spiritui Infesta *sunt, quam* Dura; *nec tam* Propinqua, *quam* Aquea; *nec cum* Aere ambiente *benè consentiunt.*

| CANON. XVII.

Evolatio cita Humoris Aquei, *conseruat diutiùs* Oleosum *in Esse*
30 *suo.*

---

4 *patiuntur.*] ~:        9 *agi.*] ~;        10 Tenaci] ~,        19 CANON.] ~‸
28 CANON.] ~‸

## ᴵ EXPLANATION

All things dread solution of their continuity, but this varies according to their density or tenuity; for the more tenuous bodies are, the smaller and tighter are the passages into which they will allow themselves to be driven. Thus water will seep into passages where powder will not; air where water will not; and flame and spirit where air will not. Yet there are limits to this, for spirit does not labour so much with the desire to get out that it puts up with becoming too bitty, and being forced into pores and passages which are too narrow. Therefore if the spirit be surrounded by a hard body, ᴵ or indeed a greasy or tenacious one (which is not easy to part), it is plainly curbed, and as it were pent up, and its desire to get out slackens; and so we see that it takes ages for the spirit of metals and stones to get out, unless the spirit is either stimulated by fire or the grosser parts are put asunder by strong and corrosive waters. Much the same is true of tenacious bodies like gums, unless they are dissolved by a gentler heat. Thus the juices of a hard body, a firm skin, and so on (which you get from dryness of aliment, exercise, and cold-ness of air) are useful for longevity, because they place the spirit in close confinement, and stop it getting out.

## ᴵ RULE 16

In oily and fatty things, even though they are not tenacious, the spirit is kept in willingly.

## EXPLANATION

Spirit, if it is not annoyed by the antipathy of a surrounding body, nor grazed down by being too much like that body, nor tempted and stimu-lated to go out by a body outside, does not make a great fuss to get out. Now in all oily bodies these conditions are absent, for those bodies are neither as troublesome to the spirit as hard ones, nor as akin as watery ones, nor do they consent well with the surrounding air.

## ᴵ RULE 17

Rapid escape of watery humour keeps the oily humour in being for a long time.

## EXPLICATIO.

*Diximus* Aquea, *vtpote* Aeri *Consubstantialia, citiùs euolare,* Oleosa *tardiùs, vt cum* Aere *minùs consentientia.* At *cùm* Humidum vtrunque *plerisque Corporibus insit, euenit vt* Aqueum *veluti prodat* Oleosum; 5 *Nam illud sensìm exiens, hoc etiam asportat. Itaque nil magis iuuat ad Corporum Conseruationem, quam* Siccatio Lenis, *quæ* Humorem [2E7ʳ] Aqueum *expirare faciat, nec* oleosum *sollici|tet, tum enim* Oleosum *fruitur Naturâ suâ; Neque hoc spectat ad inhibendam* Putredinem (*licet etiam & illud sequatur*) *sed ad conseruandam* Viriditatem. *Hinc fit vt* 10 Fricationes molles, *&* Exercitationes Moderatæ, *ad* Perspirationem *potius quam ad* Sudorem, *Longæuitati plurimùm conferant.*

## CANON. XVIII.

Aer Exclusus *confert ad Longæuitatem, si alijs* Incommodis *caueas.*

## EXPLICATIO.

15 *Diximus paulò antè,* Euolationem Spiritûs *esse Actionem duplicatam, ex* [2E7ʳ] Appetitu Spiritûs, *& Aeris. Quare si al|tera tollatur, haud parum proficitur, Id quod ex* Inunctionibus *præcipuè expectari debet. Attamen hoc sequuntur varia* Incommoda; *quibus, quomodò subueniatur, in* Operatione Secundâ *ex decem, annotauimus.*

20 ## CANON. XIX.

Spiritus Iuueniles Senili Corpori *inditi, Naturam compendiò retrouertere possint.*

## EXPLICATIO.

24 *Natura* Spirituum *est quasi* Rota *suprema, quæ alias* Rotas *in* Corpore [2E8ʳ] Humano circumagit. *Itaque illa in* Intentio|ne *Longæuitatis, prima poni debet. Hùc accedit, quod facilior, & magis expedita* Via *patet ad alterandos* Spiritus, *quàm ad alia. Etenim duplex est* Operatio super Spiritus: *Altera*

---

3 *consentientia.*] ~:     4 *euenit*] Euenit     8 Putredinem] ~,     12 CANON.] ~ₐ
20 CANON.] ~ₐ     27 Spiritus:] ~;

## EXPLANATION

I have said that watery bodies escape more quickly because they are more consubstantial to air whereas oily ones escape more slowly since their consent with air is less. But since both humours exist in most bodies, it turns out that watery matter as it were betrays the oily, for the former gradually going out carries off the latter with it. Therefore nothing contributes so much to the conservation of bodies as that mild dryness which makes the watery humour breathe out without stirring up | the oily, for then the oily delights in its own nature. And this tends not to the curbing of putrefaction (though that too follows) but to the conservation of youthful vigour. Hence it happens that gentle massage and exercise that encourage perspiration rather than sweat make for a long life.

## RULE 18

Air kept out makes for a long life, provided that you watch out for other disadvantages.

## EXPLANATION

I said a moment ago that escape of spirit is a twofold action springing from the appetite of the spirit and of the air. Thus if you remove one or other | of these, you make not a little progress; and this is chiefly to be looked for from anointings. But various disadvantages follow from this, which can be relieved by the means I noted in the second of the ten operations.

## RULE 19

Young spirits introduced into an old body can reverse the course of nature in short order.

## EXPLANATION

The nature of the spirits is as it were the chief cog which keeps all the other cogs in the human body turning. Thus it ought | to stand first in relation to the intention of lengthening life. Now it is better to proceed by the easier and quicker route, which is to alter the spirits, than to take

*per* Alimenta, *quæ est Tarda, & tanquam per Circuitum; Altera (& illa Gemina) quæ est Subita, &* Spiritus *rectà petit. Nempe per* Vapores; *Aut per* Affectus.

## CANON. XX.

5 Svcci *Corporis* subduri, *& Roscidi, faciunt ad Longæuitatem.*

### EXPLICATIO.

*Ratio perspicua est, cùm anteà posuerimus* Dura, *& Oleosa, siue* Roscida, [2E8$^v$] *ægriùs dis|sipari. Illud tamen interest (sicut etiam in* Operatione *Decimâ notauimus) quod* Succus Subdurus *minùs dissipabilis est, sed est simul* 10 *minùs* Reparabilis. *Itaque* Commodum *cum* Incommodo *coniunctum est. Neque possit propterea aliquod* Magnale *per hoc præstari; At* Succus Roscidus *vtrique Rei satisfacit; Itaque diligentiùs huic incumbendum.*

## CANON. XXI.

*Qvicquid* Tenuitate *penetrat, neque tamen* Acrimoniâ *rodit, gignit* 15 Succos Roscidos.

[2F1$^r$]

Canon *iste magis difficilis est Practicâ, quam Intellectu. Manifestum est enim, quicquid benè* penetrat, *sed tamen cùm* Stimulo, *aut* Dente *(qualia sunt omnia* Acria, *& Acida) relinquere, vbicunque transit, Vestigium non-* 20 *nullum* Siccitatis, *& Diuulsionis; vt* Succos *induret,* Partes *conuellat; At contrà, quæ penetrant merâ* Tenuitate, *tanquam furtìm, & insinuatiuè, absque violentiâ, irrorare, & irrigare in transitu. De his in* Operationibus *quartâ, & septimâ, haud pauca descripsimus.*

## CANON. XXII.

25 Assimilatio *optimè fit cessante omni* Motu locali.

another path. For the operation on the spirits is twofold: the one by aliment which is slow and roundabout; the other (also bipartite) which has immediate effect and goes straight to the spirits, namely, by vapours or the affections.

## RULE 20

Somewhat hard and dewy bodily juices make for long life.

### EXPLANATION

The grounds for this are obvious since I laid it down before that hard and oily or dewy bodies can hardly ¹ be dissipated. Yet this too makes a difference (as I have already noted in the tenth operation) that juice a bit on the hard side is less susceptible of dissipation but at the same time less reparable; so an advantage is linked with a disadvantage, and accordingly no miracle can be wrought by that. But a dewy juice does the trick in both respects; so we must devote our energies more diligently to this matter.

## RULE 21

Whatever penetrates by its tenuity, and yet does not gnaw by its acrimony, engenders dewy juices.

### ¹ EXPLANATION

This rule is more difficult to put into practice than to understand. For it is plain that whatever penetrates well, but yet with pricking or bite (as all acrid and acid things are) leaves in its wake wherever it goes a trace of dryness or disruption, as it hardens the juices and undermines the parts. But on the other hand, things that penetrate by sheer tenuity, as if by stealth and guile, and without violence, moisten and irrigate on their way. I have said quite a lot about these in the fourth and seventh operations.

## RULE 22

Assimilation works best when all local motion has stopped.

ᴵ EXPLICATIO.

*Hvnc* Canonem, *in* Commentatione *ad* Operationem *octauam, satis explicauimus.*

## CANON. XXIII.

5 Alimentatio *per Exterius, aut saltem non per* Stomachum, *Longæui-tati vtilissima, si fieri possit.*

### EXPLICATIO.

*Videmus omnia, quæ per* Nutritionem *peraguntur, fieri per longas Ambages; quæ verò per* Amplexus similiu*m* (*vt fit in* Infusionibus) *non*
10 *longam requirere Moram. Itaque vtilissima foret* Alimentatio *per Exterius,*
[2F2ʳ] at*q*ue *eò* ma|*gis, quod deciduæ sint* Facultates Concoctionum *sub* Senectute; *Quamobrèm si possint esse* Nutritiones *aliæ Auxiliares, per* Balneationes, Vnctiones, *aut etiam per* Clysteria, Coniuncta *possint proficere, quæ* Singula *minùs valeant.*

15 ## CANON. XXIV.

*Vbi* Concoctio *debilis est ad* Extrusione*m* Alimenti, *ibi* Exteriora *confortari debent ad* Euocationem Alimenti.

### EXPLICATIO.

*Non est hoc, quod in isto* Canone *proponitur, eadem Res cum Præcedente;*
[2F2ᵛ] *Aliud enim est, si* Alimentum Exterius *intrò* tra|*hatur, aliud, si* Alimen-
21 tum Interius *extrà trahatur; At in hoc concurrunt, quod Debilitati* Concoctionu*m Interiorum, aliâ Viâ subueniant.*

## CANON. XXV.

*Omnis subita* Renouatio *Corporis fit, aut per* Spiritus; *aut per*
25 Malacissationes.

---

10 *Moram.*] ~:     23 CANON.] ~‸

## ' EXPLANATION

I have said enough to explain this rule in the commentary on the eighth operation.

## RULE 23

Alimentation from the outside or at any rate not by the stomach is, if it can be done, very good for longevity.

## EXPLANATION

We see that all things done by nutrition go by way of long meanderings, whereas those which take place by embracing similar substances (as happens in infusions) need no long lapse of time. Therefore alimentation from outside would be very useful, and all the more ' because in old age the faculties of concoction fall into decay, so that if there could be supplementary nutrition by bathing, anointings, or indeed by clysters, and if these things were combined, they could improve matters more than if taken singly.

## RULE 24

Where concoction is not strong enough to send out aliment, there the outer parts should be invigorated to call forth the aliment.

## EXPLANATION

What I put forward in this rule is not the same as what was advanced in the previous one, for it is one thing to draw aliment ' from outside to in, and another to draw it from inside to out. But they agree in this, that by different routes they support weakness of concoction.

## RULE 25

All immediate renovation of the body is brought about either by the spirits or by emollients.

## EXPLICATIO.

*Dvo sunt in Corpore,* Spiritus, *& Partes; Ad vtrunque, longâ viâ peruenitur per* Nutritionem; *At viæ breues ad* Spiritus *per* Vapores, *& Affectus;*
4 *& ad* Partes, *per* Malacissationes. *Illud autem paulò attentiùs notandum,*
[2F3$^r$] *quod nullo modo confundimus* Ali'mentationem *per* Exterius, *cum* Malacissatione; *neque enim Intentio est* Malacissationis, *vt nutriat Partes, sed tantùm vt eas reddat, magis idoneas ad Nutriendum.*

## CANON. XXVI.

Malacissatio *fit, per* Consubstantialia, Imprimentia, *&*
10 Occludentia.

## EXPLICATIO.

*Manifesta Ratio est, quòd* Consubstantialia *propriè malacissent,* Imprimentia *deducant,* Occludentia *retineant, &* Perspirationem, *quæ est*
14 *Motus* Malacissationi *oppositus, cohibeant. Itaque (vt in* Operatione *Nonâ*
[2F3$^v$] | *descripsimus)* Malacissatio *simul benè fieri non potest, sed per Seriem, & Ordinem: Primùm excludendo* Liquorem, *per* Spissamenta; *quia Extranea, & crassa* Infusio *non benè coagmentet Corpus, Subtile debet esse, & ex Vaporis genere, quod intrat. Secundò* Intenerando, *per* Consensum Consubstantialiu*m: Corpora enim ad Tactum eorum, quæ valdè consen-*
20 *tiunt, se aperiunt, &* Poros *laxant. Tertio,* Imprimentia Vehicula *sunt, & nonnihil* Consubstantialia *inculcant, &* Mixtura *lenitèr* Astringentium, Perspirationem *interìm paululùm cohibet. At sequitur quarto loco Magna illa* Astrictio, *&* Clausura *per* Emplastrationem; *& posteà gradatim per*
[2F4$^r$] Inunctione*m; donec* Malacu*m* verta'*tur in* Solidum, *vt suo loco diximus.*

25 ## CANON. XXVII.

Crebra Renouatio Reparabilium, *irrigat etiàm minùs* Reparabilia.

8 CANON.] ~. 25 CANON.] ~.

## EXPLANATION

There are two things in the body: spirits and parts; to both nutrition takes the long way round. But there are short ways to the spirits by way of vapours and the affections; and to the parts by way of emollients. Now we should be careful to note that I do not muddle up in any way $^|$ alimentation from the outside with use of emollients. For the intention of emollients is not to nourish the parts but only to make them readier to be nourished.

## RULE 26

Emollience is effected by consubstantial things, by things which imprint themselves, and by things which close up.

## EXPLANATION

The reason for this is obvious, for consubstantial things properly have an emollient effect; things which imprint themselves take things forward, and things which close up retain them, and curb perspiration, whose effect is the opposite of the emollient motion. Therefore (as described in the ninth $^|$ operation) the emollient effect cannot be well achieved at once but by series and in proper order. In the first place we keep out liquor by using sealants because an external and gross infusion will not consolidate the body properly; and what enters should be subtle and of the vaporous sort. Secondly, we soften by consent of consubstantial bodies. For by contact with things which consent powerfully with them, bodies open themselves up and relax their pores. Thirdly, things which imprint themselves are carriers which to a degree impress consubstantial bodies, and in the mean time a mixture of astringents gently curbs perspiration somewhat. But in the fourth place stands that great astriction and closing up achieved by applying plasters, and after that by gradual application of ointments, until the soft is turned $^|$ into the solid, as I pointed out in the proper place.

## RULE 27

Frequent renovation of the reparable also irrigates the less reparable.

## EXPLICATIO.

*Diximus in* Aditu *ipso* Historiæ *huius, Eam esse Viam Mortis, quod magis* Reparabilia, *in consortio minùs* Reparabilium *intereant; vt totis viribus, in* Reparatione *huiusmodi* Partium *minùs* Reparabilium, *sit exudandum.*
5 *Itaque admoniti* Aristotelis *Obseruatione de* Plantis, *quod scilicet,* Nouitas Ramorum Truncum ipsum in transitu reficiat; *similem Rationem fore*
[2F4$^v$] *arbitrati sumus, si sæ$^|$pè reparentur* Carnes, *&* Sanguis *in Corpore Humano; vt inde ipsa* Ossa, *&* Membranæ, *& reliqua, quæ Naturâ minùs sunt* Reparabilia, *partìm per Transitum alacrem* Succorum, *partìm per*
10 *Vestitum illum nouum* Carnium, *et* Sanguinis *Recentiorum, irrigentur, & renouentur.*

## CANON. XXVIII.

Refrigeratio, *quæ non transit per* Stomachum, *vtilis ad Longæuitatem.*

15 ## EXPLICATIO.

*Ratio præstò est, quia cùm* Refrigeratio *non Temperata, sed Potens (præ-*
[2F5$^r$] *sertim Sanguinis)* $^|$ *ad vitam Longam sit præcipuè necessaria; omninò hoc non fieri possit per Intùs, quantum opus est, absq*ue *Destructione* Stomachi, *&* Viscerum.

20 ## CANON. XXIX.

Complicatio *illa, quod tam* Consumptio, *qua*m Reparatio, *sint* Caloris Opera, *maximu*m est *Obstaculum ad Longæuitatem.*

## EXPLICATIO.

*Destruuntur ferè omnia Magna Opera à* Naturis Complicatis; *cùm quod*
25 *aliâ Ratione iuuet, aliâ noceat; Atq*ue *hîc librato Iudicio, & sagaci Practicâ*
[2F5$^v$] *opus est. Id nos, quantu*m *res permittit, & in præsen$^|$tiâ occurrit, fecimus; Separando* Calores Benignos, *à* Nociuis; *& ea, quæ ad vtrunque faciunt.*

---

4 *exudandum.*] ~;　　　7 *reparentur*] reparentnr / turned letter　　　12 CANON.] ~‸
20 CANON.] ~‸

## EXPLANATION

Actually, in the preface to this history, I declared that this was death's way: that the more reparable parts perished in partnership with the less reparable, so that we must pour every ounce of energy into the business of repairing the latter. Thus, prompted by Aristotle's observation concerning plants, namely that newly sprouted branches make good the actual trunk as they grow, I judged that the same principle might apply if ˡ the flesh and blood of the human body were often repaired, with the result that the very bones, membranes, and other parts whose nature is less reparable might be irrigated and renewed partly by vigorous transit of juices, and partly by putting on a new outfit of fresh flesh and blood.

## RULE 28

Cooling which does not pass via the stomach is good for longevity.

## EXPLANATION

The reason for this is clear, for since cooling (especially of the blood) which is not temperate but powerful ˡ is above all necessary to long life; this absolutely cannot be done to the necessary degree from inside without destroying the stomach and innards.

## RULE 29

The complication that both consumption and repair are the works of heat is the greatest obstacle to longevity.

## EXPLANATION

Almost all great works perish from complication of natures, since what helps in one way is harmful in another; and this calls for balanced judgement and skilful practice, which is what I have achieved, as far as circumstances currently permit, ˡ by parting benign from harmful heats, and those things which contribute to both.

## CANON. XXX.

Cvratio Morboru*m* te*m*porarijs *eget* Medicinis; *at* Longæuitas vitæ *expectanda est à* Diætis.

## EXPLICATIO.

5 *Qvæ ex* Accidente *superueniunt, sublatis* Caussis *desinunt; At* Cursus Naturæ *Continuus, instar* Fluuij *labentis, etiàm continuâ indiget* Remigatione, *aut* Velificatione *in aduersum: Itaq*ue *operandu*m *est, regularitèr* [2F6ʳ] *per* Diætas. Diætæ *autem genere duplices sunt;* ⏐ Diætæ Statæ, *quæ certis temporibus, &* Diæta Familiaris, *quæ in victu quotidiano vsurpari debet.*
10 *Potentiores autem sunt,* Diætæ Statæ, *id est, Series Remediorum ad Tempus. Etenim quæ tanta Virtute pollent, vt naturam retrò vertere valeant, fortiora sunt plerunque, & magis subitò alterantia, quam quæ familiaritèr in Vsum recipi tutò possint. Atque in* Remedijs *nostris* Intentionalibus, *tres tantum* Diætas Statas *reperias:* Diætam Opiatam; Diætam Malacis-
15 santem; *& Diætam* Emaciantem, *& Renouantem. At inter ea, quæ ad* Diætam Familiarem, *& Victum quotidianum præscripta à nobis sunt, efficacissima sunt hæc, quæ sequuntur; quæ etiam validitatem Diætarum* [2F6ʳ] Stataru*m ferè æquant:* ⏐ Nitrum, *& Subordinata ad* Nitrum; Regimen Affectuum, *& Studiorum Genus; Refrigeria, quæ non transeunt per*
20 Stomachu*m; Potus* Roscidantes; Perspersio Sanguinis *cum* Materiâ Firmiore, *vt Margaritis,* Lignis; Inunctiones *debitæ, ad cohibendum* Aerem, *& Detentionem* Spirituum; Calefactoria per Exterius, *tempore* Assimilationis *post* somnum; Cautio *de ijs, quæ incendunt* Spiritum, *induntque ei* Calorem acrem, *vt de* Vinis, *&* Aromatibus; *&* Vsus Mode-
25 ratus, *& Tempestiuus eorum, quæ indunt* Spiritibus *Calorem Robustum, vt* Croci, Nasturtij, Allij, Enulæ, Opiatoru*m* compositoru*m.*

[2F7ʳ] ⏐ CANON. XXXI.

Spiritus viuus Interitu*m patitur immediatè, cùm destituitur aut* Motu; *aut* Refrigerio; *aut* Alimento.

---

1 CANON.] ~ₐ     5 Caussis] / *nld* in *SEH* (II, p. 224) as causis     10–11 *Tempus.*] ~:

## RULE 30

The cure of illnesses requires temporary medicines, but we must look for length of life from diets.

## EXPLANATION

What comes by accident stops when the causes are cancelled; but the ceaseless course of nature, like a rolling river, requires us ceaselessly to row or sail against the flow. Thus we are obliged to work as a general rule by diets. Now diets are of two kinds: | diets adopted for fixed periods, and the usual one which we have day by day. The former, i.e. a series of remedies taken for a time, are more powerful; for things which have virtue enough to be able to turn nature back are mostly stronger and cause more peremptory alterations than those which can safely be taken into normal use. Now in the remedies related to my intentions, you will find only three of these diets: the opiate diet, the softening diet, and the emaciating and renovating diet. But among the things that I have prescribed for the normal day-to-day diet, the most effective are the ones which almost match the force of diets taken for fixed periods, and they are the following: | nitre and its subordinates; management of the affections; the variety of one's studies; cooling which does not go via the stomach; drinks which foster dewiness; the suffusing of the blood with firmer matter like pearl or wood; suitable ointments to restrain the air and keep in the spirits; external calefactions when assimilation is taking place after sleep; guarding against things which inflame the spirit and imbue it with violent heat, such as wine and seasonings; and the moderate and timely use of things which give the spirits a robust heat, such as saffron, watercress, garlic, horse-heel, and compound opiates.

## | RULE 31

The living spirit suffers sudden death when it is deprived either of motion, or coolness, or aliment.

## EXPLICATIO.

*Svnt hæc scilicet illa tria, quæ superiùs vocauimus* Atriola Mortis; *Suntque* Passiones Spiritûs *propriæ, & immediatæ. Etenim Organa omnia partium Principalium seruiunt, vt hæc tria* Officia *præstentur. Et rursùs, Omnis*
5 Destructio Organorum, *quæ est Læthifera, eò rem deducit, vt vnum, aut*
[2F7ᵛ] *plura ex his tribus deficiant. Itaque alia omnia sunt diuer|sæ viæ ad Mortem; sed in hæc desinunt.* Fabrica *autem* Partium, Organum Spiritus *est; quemadmodùm & Ille,* Animæ rationalis; *quæ Incorporea est, & Diuina.*

## CANON. XXXII.

10 Flamma *Substantia Momentanea est.* Aer *Fixa;* Spiritûs Viui *in* Animalibus, *media est Ratio.*

## EXPLICATIO.

*Res est hæc & altioris Indagationis, & longioris Explicationis, quàm faciat*
14 *ad* Inquisitionem *præsentem. Sciendu*m *interìm* Flammam *continentèr*
[2F8ʳ] *generari, & extingui; vt per* Successi|onem *tantùm continuetur.* Aer *autem Corpus fixum est, nec soluitur, licet enim* Aer *ex* Humido Aqueo *Nouum* Aerem *gignat; tamen vetus* Aer *nihilominùs manet; vnde fit* Superoneratio *illa* Aeris, *de quo diximus in* Titulo *de* Ventis. *At* Spiritus, *vtriusque Naturæ particeps est, &* Flammeæ, *&* Aereæ; *quemadmodùm &*
20 Fomites *eius, sunt* Oleum, *quod est Homogeneum* Flammæ; *&* Aer, *qui est Homogeneus* Aquæ. Spiritus *enim non nutritur ex* Oleoso Simplici, *neque ex* Aqueo Simplici, *sed ex vtroque: atque licet nec* Aer *cum* Flammâ, *nec* Oleum *cum* Aquâ *benè componantur, tamen satis conueniunt in* Misto.
[2F8ᵛ] *Etiàm,* Spiritus *habet ex* Aere, *faciles suas & delicatas* Im|pressiones,
25 *& Receptiones; A* Flammâ *autem, Nobiles suos, & Potentes Motus, & Actiuitates. Similitèr etiam* Duratio Spiritûs, *Res composita est, nec tam Momentanea, quam* Flammæ, *nec tamen tam fixa, quam* Aeris; *Atque eò magis non sequitur rationes* Flammæ, *quod* Flamma *etiàm ipsa extinguitur per* Accidens; *nempè à* Contrarijs, *& Destruentibus circumfusis; quam*
30 *causam & necessitate*m *non habet paritèr* Spiritus. *Reparatur autem* Spiritus *ex* Sanguine *viuido, &* Florido Arteriarum *exiliu*m, *quæ insinuantur in* Cerebrum; *Sed fit* Reparatio *ista suo modo, de quo nunc non est* Sermo.

## FINIS.

## EXPLANATION

These are the three things which earlier I called Death's Anterooms, and they are the proper and immediate passions of the spirit. For all the organs of the principal parts serve to keep these three functions going; and, again, all fatal destruction of these organs amounts to this: that one or more of these three fails. Therefore all other things are just different ways to death which end up in these three. For the structure of the parts is the instrument of the spirit, just as the spirit is the instrument of the rational soul which is incorporeal and divine.

## RULE 32

Flame is an evanescent substance; air is permanent; the living spirit in animals is a principle poised between those two.

## EXPLANATION

This subject needs deeper investigation and longer explanation than is appropriate to the present inquiry. In the mean time it should be understood that flame is continually generated and extinguished so that its continuity comes only from one flame succeeding another. But air is a permanent body and cannot fall into dissolution, for although air generates new air from watery moisture, the original air none the less remains, whence arises that overburdening of the air which I spoke of in the title concerning the winds. But the spirit partakes of both natures, the flamy one and the airy, in the same way as what kindles it are oil, which is akin to flame, and air, which is akin to water. For spirit is not nourished by pure oily matter or simple airy stuff but by both together; and although air does not well consort with flame nor water with oil, yet they get along tolerably well in mixture. Indeed, spirit gets its easy and delicate impressions and receptions from the air, and from flame its noble and powerful motions, and capacity for action. Likewise too the spirit's duration is a compound thing, for it is not as evanescent as flame, nor yet as permanent as air; and it diverges the more from the attributes of flame because flame itself is also put out by accident, i.e. by contraries and by destructive ambient bodies, and the spirit is not affected by that cause and necessity to the same extent. In fact, the spirit is repaired by the fresh and bright blood of the narrow arteries which work their way into the brain. But this repair takes place in its own way, but I am not concerned with that just now.

## THE END

# THE COMMENTARIES

# COMMENTARY ON THE
# PRELIMINARIES TO THE
# *HISTORIA NATURALIS*
# *ET EXPERIMENTALIS*

Note: all citations of and quotations from Pliny's *Historia naturalis* are referenced by book and chapter number as given in editions of the Latin. These numbers are followed by a page number which refers to *The historie of the world. Commonly called, the natvrall historie of C. Plinivs Secvndvs. Translated into English by Philemon Holland Doctor in Physicke*, Adam Islip: London, 1601. English quotations from this edition are supported where necessary with quotations (in whole or part) from the Latin.

## A2ᵛ

Page 4, l. 8: veluti *Granum Sinapis*—implicitly contrasting the scale of the history presented in the 1622 volume with the bulk of conventional natural histories (see p. xxvi above). Bacon alludes to Matthew 13: 31–2: 'The kingdom of heaven is like to a grain of mustard seed [*grano sinapis*], which a man took, and sowed in his field: which indeed is the least of all seeds: but when it is grown, it is the greatest among herbs, and becometh a tree: so that the birds of the air come and lodge in the branches thereof.' There may be a secondary allusion to Luke 17: 6: 'And the Lord said, "If ye had faith as a grain of mustard seed, ye might say unto this sycamine tree, 'Be thou plucked up by the root, and be thou planted in the sea', and it should obey you".'

l. 11: tanquam in Cantico nouo—cf. Psalms 98: 1: 'O sing unto the LORD a new song, for he hath done marvellous things.'

## A3ʳ

Page 4, ll. 16–17: *Historiâ Naturali* bonâ—cf., for instance, *Novum organum* in *OFB*, XI, p. 214, ll. 24–6.

## A4ʳ

Page 6, ll. 1–10: TITVLI Historiarum—for the prefaces to these histories see pp. xlv, 132–8 above, and *OFB*, XIII, pp. 36–9. For the reasons why Bacon chose these titles see *HNE*, C2ᵛ–C3ʳ, pp. 12–14 above, and *cmts* thereon, p. 384 below.

## B1ʳ⁻ᵛ

Page 6, ll. 17–22: *Monendi vtique sunt Homines*—this idea is connected with the criticism made a little later (see *cmts* on B2ᵛ–B3ʳ below).

**B1ᵛ–B2ᵛ**

Page 6, ll. 24–6: *Pythagoræ . . . Zenonis*—see *NO* in *OFB*, XI, pp. 98, 102, 106–8, 114, and *cmts* thereon pp. 516–18, 521–2. The play simile echoes, of course, the name given to one class of idol in *NO*, viz. Idols of the Theatre; see *OFB*, XI, pp. 80–3, 94–103.

**B2ᵛ**

Page 8, ll. 4–5: *Patricius . . . Campanella*—for Francesco Patrizi da Cherso (1529–97) and Bacon's use of his ideas, see *OFB*, VI, pp. xxiv, 158–60, 402–5; XI, pp. 174, 532. Bernardino Telesio (1509–88) is mentioned in many places in Bacon's philosophical works, and was the only philosopher whose ideas Bacon subjected to large-scale and systematic examination; see *DPAO* in *OFB*, VI, pp. 224–67. Giordano Bruno (1548–1600) is not mentioned elsewhere in Bacon's works. Petrus Severinus, i.e. Peder Sørensen (1540/2–1602), was credited by Bacon with making the braying of Paracelsus harmonious; see *OFB*, VI, p. xlv; XI, pp. 532–3. William Gilbert (1544–1603) was the modern natural philosopher most cited in *NO*; for references see *OFB*, XI, p. 513; also see pp. 132–4 above and *cmts* thereon p. 416 below (S2ʳ⁻ᵛ). Tommaso Campanella (1568–1639) may have been a source for some of Bacon's earliest ideas about the structure of the cosmos; see *OFB*, VI, pp. xxxviii–xxxix.

**B2ᵛ–B3ʳ**

Page 8, ll. 9–11: *nihil est luminis sicᵈci . . . ex specu Platonis*—for the idea that the mind is not a dry light see *OFB*, XI, p. 86. For the cells of imagination cf. *ibid.*, p. 24. For Plato's cave and Idols of the Cave see *ibid.*, pp. 80, 509.

**B3ʳ**

Page 8, ll. 13–16: *ex quorundam Virorum*—here Bacon is thinking of university scholasticism and its reliance on Aristotle and Aristotelian commentary.

ll. 17–18: *Cicero in Cæsaris annum*—see Plutarch, *The lives of the noble Grecians and Romaines . . . translated out of Greeke into French by Iames Amiot . . . and out of French into English, by Sir Thomas North*, Richard Field: London, 1603, p. 738: 'Cæsar . . . did set forth an excellent and perfect kalendar, more exactly calculated, then any other that was before: the which the Romaines do vse vntill this present day . . . But his enemies notwithstanding that enuied his greatnesse, did not sticke to find fault withall. As Cicero the Orator, when one said, to morrow the starre Lyra will rise: Yea, said he, at the commandement of Cæsar, as if men were compelled so to say and think, by Cæsars edict.'

**B3ᵛ–B4ʳ**

Page 8, ll. 22–8: *Etenim Mundos creamus*—for similar sentiments see the preface to *IM* in *OFB*, XI, pp. 20, 36. This is still in Bacon's mind when he speaks of the Babel and the confusion of tongues on B5ᵛ (p. 10 above) and *cmt* thereon (p. 383 below).

**B5ʳ**

Page 10, l. 6: *Theses Hypothesibus anteposuerunt*—this contrast is not self-evident. Thesis is a very rare word in Bacon, and I find it in only two other places in his writings: in a legal context (*DAS*, 3P1ᵛ (*SEH*, I, p. 825) ), and in a cosmological one (*OFB*, VI, p. 132, l. 22), in connection with the very limits of the heavens, where the term refers to assertions which lack any empirical support whatever. The term may mean that here. As for hypothesis, Bacon almost invariably uses it to refer (often disparagingly) to the convenient fictions of geometrical astronomy; see, for instance, *OFB*, VI, p. 134, ll. 16, 18; p. 186, l. 15; p. 188, l. 36. That is not what is meant here, where it may mean something like a reasonable inference from evidence.

**B5ᵛ**

Page 10, ll. 10–12: *Hic est ille Sermo*—Vulgate, Psalms 18: 2: 'Caeli enarrant gloriam Dei, Et opera manuum eius annuntiat firmamentum. [18: 3] Dies diei eructat verbum, Et nox nocti indicat scientiam. [18: 4] Non sunt loquelae, neque sermones, Quorum non audiantur voces eorum. [18: 5] In omnem terram exivit sonus eorum, Et in fines orbis terrae verba eorum.' For Babel and the confusion of tongues, cf. *DAS*, G1ʳ⁻ᵛ (*SEH*, I, p. 466): 'Sic ante Diluuium, Sacri Fasti, inter paucissima, quæ de eo seculo memorantur, dignati sunt memoriæ prodere, Inuentores Musicæ, atque Operum metallicorum. Sequenti Seculo post Diluuium, grauissima pœna, quâ Deus Humanam Superbiam vltus est, fuit Confusio Linguarum, quâ Doctrinæ liberum Commercium, & literarum ad inuicem Communicatio maximè interclusa est.' Also see *DAS*, 2N1ʳ (*SEH*, I, p. 653).

**B5ᵛ–B6ʳ**

Page 10, ll. 12–15: *repuerascentes*—for becoming again as a little child, and the abecedarium metaphor, see *ANN* in *OFB*, XIII, pp. 172–3, and *cmts* thereon pp. 305–6.

**B6ʳ**

Page 10, ll. 16–18: *in* Instauratione *nostrâ*—Bacon takes it for granted that the reader knows what the *Instauratio* and its parts are, and what the *Novum organum* is, i.e. he assumes that the reader is fully acquainted with the contents of *DO*, a piece published in 1620; see *OFB*, XI, pp. 26–47. He makes the same assumption at the beginning of *ANN*, a preparative for Part IV of the *Instauratio*; see *OFB*, XIII, p. 172.

**B6ᵛ**

Page 10, l. 19: *supersint in* Organo—for what remained to be completed in *NO* see *OFB*, XI, pp. xcii–xcv.

    ll. 20–1: *promouere . . . paucis*—cf. *ANN*, *OFB*, XIII, p. 172, l. 21.

**B6ᵛ–B8ʳ**

Page 10, ll. 24–7: *Spargi proculdubiò*—cf. the first paragraph of *ANN* in *OFB*, XIII, p. 172.

**B8ʳ⁻ᵛ**

Page 12, ll. 5–6: Nauiculas *quasdam*—for similar ship imagery see *NO* in *OFB*, XI, p. 344; *DAS* 2Y1ʳ, 3P4ʳ⁻ᵛ (*SEH*, I, pp. 714, 829).

ll. 10–13: *vt* Organum—this is one of Bacon's strongest statements of the priority that he gave to natural history in the last years of his life.

**B8ᵛ–C1ʳ**

Page 12, ll. 14–18: *Deus Vniuersi Conditor*—this prayer appears in the same words at the end of *ANN* (*OFB*, XIII, p. 224); cf. the prayer at the end of *DO* (*OFB*, XI, pp. 44–6).

**C2ʳ**

Page 12, ll. 19–23: Norma Historiæ præsentis—the whole of this closely resembles *Norma abecedarij* at the end of *ANN*—which was written just before *HNE*—and fulfils a similar function for, just as *Norma abecedarij* was meant to prepare for inquiries to be presented in Part IV of the *IM*, so Norma Historiæ served the same purpose for Part III; see *OFB*, XIII, pp. xxii–xxiii, 220–4.

ll. 20–1: *Qvamuis sub finem eius partis* Organi *nostri*—this refers to *PAH*; see *OFB*, XI, pp. 454–73.

**C2ʳ⁻ᵛ**

Page 12, ll. 23–9: Titulis *in* Catalogo—these titles are the ones listed after *PAH*; see *OFB*, XI, pp. 474–84. These are histories of concrete bodies in the sense that the list does not embrace the dozens of the all-important abstract (or cardinal) natures (dense, rare; heavy, light; hot, cold; etc.) discussed synoptically in order of their classes (schematisms of matters, simple motions, etc.) in *ANN*. Bacon believed that knowledge of the forms of such natures was the key to theoretical and operative power over nature (*NO, OFB*, XI, pp. 204–6). Bacon had promised in *DO* to produce a separate history of these 'Naturæ primordia', and in *PAH*, had already talked of reserving them for himself (*OFB*, XI, pp. 38–40, 472). The new abecedarium (i.e. *ANN*) did not appear at the end of the volume containing *HV* probably because Bacon had come to see it as properly belonging to Part IV of *IM* and so perhaps decided at the last moment not to have it printed with *HV*; see *OFB*, XIII, pp. xxi–xxii.

Page 14, ll. 1–6: Titulos—while the reasons Bacon gives for choosing the titles may fairly represent those chosen for the first six months (see A4ʳ, p. 6 above), only two titles were taken from *CHP*: '*Historia Ventorum, & Flatuum repentinorum, & Vndulationum Aeris*' and '*Historia Vitæ & Mortis*' (items 6 and 58 in *OFB*, XI, pp. 474, 480).

**C3ʳ⁻ᵛ**

Page 14, ll. 7–12: *In* Titulis—for the function of particular topics, see *DAS*, 2K1ʳ (*SEH*, I, p. 635): '*Verùm missum facientes Vitium illud & Fastum, quæ nimiùm diù regnarunt in Scholis; Videlicet, vt quæ præstò sint, infinitâ*

Subtilitate persequantur; quæ paulò remotiora, ne attingant quidèm; Nos sanè *Topicam Particularem*, tanquam Rem apprimè vtilem amplectimur; hoc est, *Locos Inquisitionis* & *Inuentionis, Particularibus Subiectis* & *Scientijs* appropriatos. Illi autem, Mixturæ quædam sunt, ex *Logicâ,* & *Materiâ ipsâ propriâ* singularum *Scientiarum.* Futilem enim esse constat, & angusti cuiusdam Animi, qui existimet *Artem,* de *Scientijs* inueniendis, perfectam iam à Principio, excogitari & proponi posse; eandemque posteà in Opere poni, & exerceri debere.'

**C3ᵛ–C4ᵛ**

Page 14, ll. 13–15: Historia *&* Experimenta—the wording at points is almost identical to that in *ANN* (*OFB*, XIII, pp. 220–2). For experiments of light see *OFB*, VI, pp. 363–4; XI, p. 338, 565. For crucial instances see *NO, OFB*, XI, pp. 318–38, 560–1. For the meaning of Mandata and Historia designata see *OFB*, VI, pp. xxiv–xxv.

**C4ᵛ–C5ᵛ**

Page 14, l. 22–p. 16, l. 3: Experimenti *alicuius subtilioris* Modum—for a shorter version of this see *ANN* (*OFB*, XIII, p. 222, ll. 6–24).

**C5ᵛ–C6ʳ**

Page 16, ll. 4–8: *Vtilitatis humanæ*—cf. *ANN* (*OFB*, XIII, p. 222, ll. 25–9).

**C6ʳ⁻ᵛ**

Page 16, ll. 14–20: *Patet ex antedictis*—once again Bacon assumes that the reader is entirely *au fait* with *DO* (the plan of *IM*), for which see *cmts* p. 383 above, and *OFB*, XI, pp. 26–46.

# COMMENTARY ON
# *HISTORIA VENTORUM*

**C7ʳ–C8ʳ**

Page 18, ll. 1–17: *Aditus, siue Præfatio*—see p. 14 (Norma Historiæ) above.

ll. 10–12: *& ad Nauigandum, & ad Molendum*—about a quarter of the text of *HV* is devoted to practical matters of this sort, i.e. to ships, mills, and weather forecasting; see above, pp. 90–106.

l. 15: *Æolus Iunoni*—according to Virgil, Juno, in her determination to prevent the fleeing Aeneas and his companions reaching Italy to found a new Troy, asks Aeolus to raise a storm to disperse their fleet. Aeolus (*Aeneid*, I, ll. 76–80) agrees to do so: 'tuus, o regina, quid optes, | explorare labor; mihi iussa capessere fas est. | tu mihi quodcumque hoc regni, tu sceptra Iovemque | concilias, tu das epulis accumbere divum, | nimborumque facis tempestatumque potentem.'

ll. 15–17: *Primariæ Creaturæ non sunt*—meteorological phenomena (and by these Bacon means the wide range of effects traditionally assigned to that category) were not created during the six days but arose from the subsequent interactions of the material substances that were primary creations.

**C8ᵛ**

Page 18, ll. 22–3: *Nomina Ventorum*—see *cmts* on E1ʳ–E3ʳ, pp. 388–9 below.

**C8ᵛ–D1ʳ**

Page 18, ll. 27–9: *Venti generales*—see E3ᵛ–E6ᵛ, pp. 32–6 above, and *cmts* thereon p. 389 below.

**D1ʳ**

Page 20, ll. 1–3: *Venti Stati*—see E7ʳ–F2ʳ, pp. 36–40 above, and *cmts* thereon p. 390 below.

ll. 4–7: *Venti Asseclæ*—see F2ʳ–F7ʳ, pp. 40–2 above, and *cmts* thereon pp. 390–1 below.

**D1ʳ⁻ᵛ**

Page 20, ll. 8–10: [*Venti Marini.*]—these are later treated as species of *venti asseclæ*; see F4ʳ–F7ʳ, p. 42 above, and *cmts* thereon p. 391 below.

**D1ᵛ**

Page 20, ll. 11–16: *Venti Liberi*—notice that these are dealt with *before* the winds mentioned immediately above; see E3ʳ⁻ᵛ, p. 32 above, and *cmts* thereon p. 389 below. As Bacon said (p. 14 above), he did not consider himself bound to treat the investigations arising from the list of particular topics in list order.

**D1ʳ–D2ʳ**

Page 20, ll. 17–19: *Qualitates Ventorum*—see F7ʳ–G8ʳ, pp. 44–54 above, and *cmts* thereon pp. 391–5 below.

**D2ʳ⁻ᵛ**

Page 20, ll. 20–30: *Originales locales*—see G8ᵛ–I1ᵛ, pp. 54–64 above, and *cmts* thereon pp. 395–8 below.

**D2ᵛ**

Page 20, ll. 31–2: *Generationes Accidentales*—see I1ᵛ–I3ᵛ, pp. 64–6 above, and *cmts* thereon p. 398 below.

Page 22, ll. 3–4: *Venti Rari, & Prodigiosi*—see I3ᵛ–I5ʳ above, and *cmts* thereon p. 398–9.

ll. 5–7: *Vaporosi, & Mercuriales*—for mercurial and sulphurous winds, also see F8ʳ, pp. 44–6 above, and *cmts* thereon p. 391 below.

**D3ʳ–D4ʳ**

Page 22, ll. 15–35: *Confacientia ad Ventos*—see I5ᵛ–K6ʳ, pp. 68–76, above, and *cmts* thereon pp. 399–401 below.

**D3ʳ⁻ᵛ**

Page 22, ll. 15–20: Circa *Astrologica*—see I7ᵛ–I8ʳ, p. 70, ll. 11–12 above. For Bacon's general views on astrology and astrological influences, see *DAS*, V4ʳ–X4ʳ (*SEH*, I, pp. 554–60).

**D4ᵛ–D5ᵛ**

Page 24, ll. 3–16: *Limites Ventorum*—see K6ᵛ–K8ʳ, pp. 76–8 above.

**D5ᵛ**

Page 24, ll. 20–7: *Successiones Ventorum*—see K8ᵛ–L2ʳ, p. 80 above.

**D6ʳ–D7ʳ**

Page 26, ll. 3–30: *Motus diuersi Ventorum*—see L2ʳ–M2ʳ, pp. 82–90 above.

**D7ʳ⁻ᵛ**

Page 26, ll. 25–30: *Motûs* | *Ventorum, in impulsu Nauium*—see M2ᵛ–N7ʳ, pp. 90–102 above.

**D7ᵛ**

Page 26, ll. 27–30: *Motûs Ventorum in velis Molendinorum*—see N7ᵛ–O2ʳ, pp. 102–6 above.

**D7ᵛ–D8ʳ**

Page 26, l. 33: *Potestates Ventorum*—see F7ʳ–G8ʳ, pp. 44–54 above.

**D8ʳ⁻ᵛ**

Page 28, ll. 13–15: *Prognostica Ventorum*—see O2ʳ–P7ᵛ, pp. 106–120 above.

**D8ᵛ–E1ʳ**

Page 28, l. 18: *Imitamenta Ventorum*—see P7ᵛ–Q3ᵛ, pp. 120–4 above.

E1ʳ–E3ʳ

Page 28, l. 30 ff.: Nomina *Ventis*—Bacon had plenty of ancient and modern sources available to him on the names of winds and, in some cases, the quarters from which they blew. See, for instance, Aristotle, *The situations and names of winds*, in *Minor works* (Loeb Classical Library), trans. W. S. Hett, London and Cambridge, Mass., 1936, repr. 1993, pp. 452–7; Lucretius, *De rerum natura*, V. 737–50; Pliny, *Historia naturalis*, II. 45 ff., pp. 21–4; Seneca, *Naturales quaestiones* (Loeb Classical Library), trans. T. H. Corcoran, London and Cambridge, Mass., 1972, V. 1–17, pp. 74–113 (this edition contains a useful note (pp. 311–12), with a fold-out of wind directions, on classical wind names). Among the moderns, Bacon would have known Agricola's 24-point classification of the winds; see *De re metallica*, lib. 3. *CCC met.* (cols. 69–71) contains a useful summary of early seventeenth-century opinion and presents a 32-point classification not very different from Bacon's. *CCC met.* notes that everyone agrees that there are four principal winds: 'è 4. mundi partibus seu præcipuis angulis spirantes, quos Aeolus primum notasse perhibetur . . . videlicet ab Oriente Aequinoctiali [Subsolanum:] a Meridie [Austrum:] ab Occasu Aequinoctiali. [Fauonium] à Polo Arctico [Septentrionem.].' But, it adds, others of the ancients placed collateral winds between each of these, so bringing the number up to eight. Aristotle elects for eleven: viz., Favonius, Vulturnus, Africus, Aquilo, Auster, Cæcias, Argestes, Subsolanus, Meses, Thrascias, and Phœnicias. The moderns have, however, opted for thirty-two winds: 'Secto in 32. minutiores partes Horizonte, sub ea parte, quæ est ad polum Arcticum, collocant Septentrionem: & in parte per diametrum opposita, Austrum. Tum hac diametro ad angulos rectos intersecta, ab ortu Aequinoctiali ad Aquinoctialem occasum, sub ortu Aequinoctiali ponunt Subsolanum; sub occasu Aequinoctiali Zephirum, seu Fauonium . . . Deinde in puncto medio inter quoslibet duos horum principum ventorum ventum vnum constituunt: nimirum inter Septentrionem & Subsolanum ponunt Mesem: inter Subsolanum & Austrum, Euronotum: inter Austrum & Zephirum, Notezephirum: inter Zephyrum & Septentrionem, Traciozephyrum: sicque octo ventos descriptos habent. Deinde in puncto medio inter quosuis duos horum octo ventorum constituunt ventum alium: vt inter Septentrionem & Mesem, Boream: inter Mesem & Subsolanum, Cæciam: inter Subsolanum, & Euronotum, Eureum seu Vulturnum: inter Euronotum & Austrum, Phœnicam: inter Austrum & Notozephyrum, Libonotum: inter Notozephyrum & Zephyrum, Africum; inter Zephyrum & Thraciozephyrum, Corum: denique inter Thraciozephyrum, & Septentrionem, Thraciam, seu Circium. Atque ita constitutos habent sexdecim ventos. Tum rursus in puncto medio inter quoslibet duos constituunt alium ventum, numerantque 32.'

E3ᵛ–E6ᵛ

Page 32, l. 21–p. 36, l. 13: De *Ventis Generalibus*—in Bacon's theory of the universe (here kept firmly in the background) this wind is an expression (as are consilient tidal phenomena) of a universal east-to-west motion which also embraces the stellar and planetary heavens. In Bacon's theory wind and water have to move with the heavens because of the sympathetic alliance that they have with them in virtue of their membership of the two families or quaternions of substances (see *cmts* on S5ʳ–S8ʳ, pp. 416–18 below) which fill most of the cosmos. This wind and related phenomena are discussed often and at length in Bacon's philosophical works; see, for instance, *TC* and *DGI* (*OFB*, VI, pp. li, 78, 379); *NO* (*OFB*, XI, pp. 314–16, 560–3), and *ANN* (*OFB*, XIII, pp. 188–90). Also see I6ᵛ, p. 68, ll. 24–6 above. It should be noted that José de Acosta, one of Bacon's sources of information in *HV*, speaks of the east to west wind in the tropics at some length; see *The natvrall and morall historie of the East and West Indies. Intreating of the remarkeable things of heaven, of the elements, mettalls, plants and beasts which are proper to that country* . . . *Written in Spanish by Ioseph Acosta, and translated into English by E. G.*, V. Sims: London, 1604, pp. 129–30: 'for that within the Tropickes, the Easterne winds continually blow, the which are fittest to go from, *Spaine* to the West *Indies*, for that their course is, from Easte to west . . . The like discourse is of the Navigation made into the South sea, going from new *Spaine*, or *Peru*, to the *Philippines* or *China*, and returning from the *Philippines* or *China* to new *Spaine*, the which is easie, for that they saile alwaies from East to West, neere the line, where they finde the Easterly windes to blow in their poope.' Bacon's theoretical preoccupations guided his choice of the six natural-historical topics announced in *HNE*; see p. xxxvii above. For a brief and ageing discussion of this issue, see Rees, 'Francis Bacon's semi-Paracelsian cosmology and the *Great Instauration*'.

**E5ᵛ**

Page 34, l. 29: *Quod* Briza—see Acosta, *The natvrall and morall historie of the* . . . *Indies*, pp. 128, 132.

**E6ᵛ**

Page 36, l. 13: *Euro-aquilo*—*SEH*, II, p. 28 notes that elsewhere this wind is called *Euro-boreas* (see, for instance, G5ʳ, p. 50 above), and wonders if Bacon picked up the term from the Vulgate, Acts 27: 14: 'Non post multum autem misit se contra ipsam ventus typhonicus, qui vocatur Euroaquilo.'

**E7ʳ**

Page 36, ll. 14–29: Venti Stati—see *OFB*, VI, pp. 64–5, 374. Also see *CCC met.*, cols. 73–4.

**E7ᵛ**

Page 36, ll. 23–5: Apud Antiquos—see Pliny, V. 9, p. 97: 'Many and divers causes of this rising and increase of his, men have given: but those which carrie the most probabilitie, are either the rebounding of the water, driven backe by the winds Etesiæ, at that time blowing against it, and driving the sea withall upon the mouths of Nilus: or else the Summer raine in Æthyopia, by reason that the same Etesiæ bring clouds thither from other parts of the world.'

ll. 26–9: Inueniuntur in Mari—for Bacon on currents see *DFRM* in *OFB*, VI, pp. 64–6, 376.

### E7ᵛ–E8ʳ

Page 36, ll. 30–4: *Columbum*—also see *HNE*, F4ʳ, p. 42 above.

### F1ʳ–F2ʳ

Page 38, ll. 23–6: In partibus *Europæ*—cf. Pliny, II. 47, p. 23: 'Some call *Favonius* (which beginneth to blow about the seventh day before the Calends of March) by the name of *Chelidonius*, upon the sight of the first swallowes: but many name it *Orinthias*, comming the 71 day after the shortest day in Winter; by occasion of the comming of birds: which wind bloweth for nine daies . . . in the hotest season of the Summer, the Dog-starre ariseth [*exoritur Caniculæ sydus*], at what time as the Sun entreth into the first degree of Leo [*primam partem Leonis ingrediente*], which commonly is the fifteenth day before the Calends of August. Before the rising of this Starre for eight daies space or thereabout, the Northeast winds [*Aquilones*] are aloft, which the Greekes call *Prodromi, i.* forerunners. And two daies after it is risen, the same winds hold still more stiffely, and blow for the space of fortie daies, which they name *Etesiæ*.'

### F2ʳ

Page 38, l. 14: Præcisus reditus *Ventorum*—see *cmt* immediately above.

### F3ᵛ–F4ʳ

Page 40, ll. 25–35: *Memini me à Mercatore*—this could have been any merchant associated with the London and Bristol Company's attempt to found colonies in Newfoundland. A good candidate may be John Guy (d. 1629), who in 1610 led the first formal attempt to set up a colony. Guy returned to England twice, in 1611 and 1614, but remained in England thereafter. Bacon himself was a founder member of the Company, and may have derived much of his knowledge of Canadian conditions from associates in the venture; see Patent Roll 8 James I. Part VIII, No. 6: Charter of the London and Bristol Company.

### F4ʳ⁻ᵛ

Page 42, ll. 6–7: Ea præcipuè, quæ *Columbo*—see *cmts* on E7ᵛ–E8ʳ, p. 390 above.

### F4ᵛ–F5ʳ

Page 42, ll. 15–24: Non satis constat sibi *Acosta*—see Acosta, *The natvrall and morall historie of the . . . Indies*, p. 113: 'at *Peru* . . . every morning the winde from the sea doth cease, and the Sunne beginnes to cast his beames . . . vntill the returne of the same windes, which otherwise they call the tide or winde of the sea, which makes them first to feele cold . . . at noone we felt a fresh aire; for that then, a North easterly wind which is fresh and coole, doth commonly blow'; cf. *ibid.*, p. 183: 'vpon all that coast it blowes continually with one onely winde, which is South and Southweast, contrary to that which dooth vsually blow vnder the burning Zone.' *SEH* (II, n. p. 32) remarks that Acosta's two statements are not inconsistent because the coast of Peru runs north-west to south-east.

### F7ᵛ–F8ʳ

Page 44, l. 31–p. 46, l. 3: *Paracelsi* Schola—this is a reference to a Paracelsian whom Bacon admired (see *TPM* (*Sc*, V6ʳ⁻ᵛ, *SEH*, III, p. 533)), namely Petrus Severinus (i.e. Peder Sørensen), to whose book the poem by Johannes Pratensis quoted here is appended; see *Idea medicinæ philosophicæ, fvndamenta continens totius doctrinæ Paracelsicæ*, Basle, 1571, 2G3ʳ [*sic* for 3G3ʳ].

### F8ʳ

Page 46, ll. 4–5: vt in prouerbio sit—see the old rhyme: 'When the wind is in the East, 'tis neither good to man or beast. | When the wind is in the North, the skilful fisher goes not forth. | When the wind is in the South, it blows the bait in the fish's mouth. | When the wind is in the West, then it is at its very best.'

### F8ᵛ

Page 46, ll. 12–17: *Auster* minùs Anniuersarius—see Aristotle, *Problems*, Books 32–38 (Loeb Classical Library), trans. W. S. Hett, London and Cambridge, Mass., 1937, rev. edn., 1957, repr. 2001, XXVI. 2, pp. 72–3: 'Why are the north winds periodic, whereas the south winds are not? Or are the south winds also periodic, but not continuous, because the origin of the south wind is far away from us, but we live under the north wind?' Bacon made extensive use of the *Problems* from this point onwards in *HV*. He recorded his respect for it (which he aligned with the *Historia animalium* as a praiseworthy Aristotelian text) in several places; see *OFB*, XI, p. 517.

### F8ᵛ–G1ʳ

Page 46, ll. 18–21: *Auster* nobis pluuiosus—cf. Aristotle, *Problems*, XXVI. 50, pp. 104–5: 'Why are the south winds which are dry and not rainy liable to produce fever? Is it because they produce in the body an unnatural warm moisture? For these winds are naturally moist and warm, and this produces fever. For fever is due to an excess of these two things. So when the winds blow under the influence of the sun without bringing rain they produce this condition in us, but when they come in conjunction with rain, the water cools us.'

### G1ʳ⁻ᵛ

Page 46, ll. 30–3: *Auster* & *Boreas*—ibid., XXVI. 35, pp. 94–7: 'The reason why the north and south winds are the commonest is that, when one contrary is overpowered by another, it is least able to continue in a straight line, but it is better able to resist a wind blowing across it. So the south and north winds blow from districts which are on either side of the course of the sun, but the others blow in the exactly contrary direction.'

**G1ʳ**

Page 46, ll. 34–6: *Auster* Saluberrimus—*ibid.*, XXVI. 17, pp. 82–3: 'Why does the south wind produce an unpleasant smell? Is it because it makes bodies moist and hot, and in this condition they are more liable to decomposition? But south winds from the sea are good for plants; for they fall upon them from the sea. This is the case in Attica on the Thriasian Plain, for the south wind arrives cool. Now red blight is due to moisture which is hot, but comes from elsewhere.'

**G1ʳ–G2ʳ**

Page 48, ll. 1–4: *Auster* lenior—cf. *ibid.*, XXVI. 20, pp. 82–5, but esp. 38, pp. 96–9: 'Why do the south winds not make the sky overcast, when they blow gently, but when they are violent they do? Is it because when they blow gently they cannot produce many clouds? So they only affect a small area. But when they grow strong, they drive many clouds before them, and so they seem to make the sky more overcast.'

**G2ʳ**

Page 48, ll. 8–11: Post pruinas—*ibid.*, XXVI. 3, pp. 72–3: 'Why does the south wind blow after a frost? Is it because frost occurs after concoction [πέψεως] has taken place, and after absorption and concoction a change to the contrary occurs? And the south wind is contrary to the north. For the same reason the south wind also blows after snow. Speaking generally, snow and hail and rain and all such purgation are proof of concoction. So after rain and storms of a like nature the winds blow.'

**G2ʳ⁻ᵛ**

Page 48, ll. 12–14: *Auster*, & frequentiùs—cf. *ibid.*, XXVI. 9, 14, pp. 76–7, 80–1: 'Why is there a saying "Boreas at night does not survive the third day"? Is it because the winds blowing from the north are weak when they occur at night? The proof that the quantity of air moved is not great is the fact that the wind blows from that quarter when there was but little heat; and a small amount of heat moves a small quantity of air . . . Why does the nightly north wind cease on the third day? Is it because it arises from a small and weak beginning, and the third day is its critical moment? Or is it because it rushes on all at once, like the hurricane winds [ἐκνεφιῶν]? So its cessation is equally swift.'

**G2ᵛ**

Page 48, ll. 17–18: Spirante *Austro*—cf. *ibid.*, XXVI. 37, pp. 96–7: 'Why is the sea blue when the south wind blows, and black when the north wind blows? Is it because the north wind disturbs the sea less, and everything which is less disturbed appears black?'

**G3ʳ**

Page 48, ll. 27–8: *Boreas* sementi—Pliny, XVIII. 33, p. 608: 'They also that graffe sions in the stocke by cleft, or set bud in the scutcheon by way of inoculation, must take heed how they meddle in this wind [*hunc caveat insitor calamis, gemmisque inoculator*].'

ll. 29–30: A parte *Austri*—Pliny, XVIII. 33, p. 608: 'Let hardly the Vine spread her braunches, and run into this wind, in all places of Italie: but leave not the cuts either of tree or vine looking that way [*In hunc Italiæ palmites spectent, sed non plagæ arborum vitiumve*].'

### G3ʳ⁻ᵛ

Page 48, ll. 31–6: In latis pascuis—Pliny, XVIII. 33, pp. 608–9: 'As charie also and heedfull must thou be to drive thy cattaile Northward from the Sunne, and there to let them graze: for marke what I say, In so doing, they will not bee able to hold open their eies; this wind will make them bleared & bloudshotten [*Clodunt ita, lippiuntve ab afflatu*]; nay, it will drive them into a gurrie or flux of the bellie, which will soone make an end of them. Howbeit, if thou wouldest have the beasts conceive and bring forth females, force them when they be leaped and covered, to stand with their heads into this wind, and thou shalt see the proofe hereof.' *SEH* (II, p. 35, n. 6) notes (correctly) that Bacon reads *claudicant* instead of Pliny's *clodunt*. As for Pliny's inconsistency, *SEH* (*loc. cit.*, n. 7) refers to Pliny, VIII. 72, but there are only fifty-nine chapters in that book. Pliny, VIII. 47, p. 227 reads thus: 'They say, that if the North winds [*Aquilonis flatu*] blow when they take the ramme, they will bring forth males; but if the South winds [*Austri*] be up, females.'

### G3ᵛ

Page 50, ll. 1–4: *Venti* tribus temporibus—cf. Pliny, XVIII. 44, p. 574 (the Latin and Holland's English differ, so only the Latin is given here): 'Venti autem tribus temporibus nocent frumento et hordeo: in flore aut protinus cum defloruere vel maturescere incipientibus; tum enim exinaniunt grana, prioribus causis nasci prohibent.'

### G3ᵛ–G4ʳ

Page 50, ll. 5–9: Flante *Austro*—see Aristotle, *Problems*, XXVI. 42–3, pp. 100–1: 'Why are men heavier and more feeble when the wind is in the south? Is it because moisture becomes abundant instead of scanty, permeating through the warmth, and heavy moisture replaces light air? So men's strength is relaxed. Why are men more inclined to eat when the wind is in the north than when it is in the south? Is it because the north winds are colder?' See also *ibid.*, XXVI. 17, pp. 82–3, quoted in *cmt* (p. 392 above) on G1ᵛ, p. 46 above.

### G4ᵛ

Page 50, ll. 23–4: Flante *Euro*—ibid., XXVI. 53, pp. 106–7: 'Why, when the east wind blows, do all the things seem larger? Is it because it makes the air very gloomy?'

**G4ᵛ–G5ʳ**

Page 50, ll. 25–9: *Cœciam nubes—ibid.*, XXVI. 1, pp. 70–1: 'Why does Caecias alone of the winds attract clouds towards itself? . . . Falling then, as has been said, upon regions of the earth which are towards the west, and collecting the clouds because of the shape of its path, on its return journey it thrusts the clouds towards itself.' See also *ibid.*, XXVI. 29, pp. 88–91: 'Why is the north-easter [*καικίας*] the only wind which attracts clouds to itself, as the proverb says, "Drawing clouds to itself like the north-easter"? . . . but does the wind naturally travel along the circumference of a circle? The other winds all blow round the earth; but the concave side of this circle faces the heavens and not the earth, so that it blows back in the direction of its starting-point, and draws the clouds towards itself.' Cf. Erasmus, Adagia 1. 5. 62: 'Mala attrahens ad sese, ut caecias nubes.'

**G5ᵛ**

Page 52, ll. 4–6: Tonitrua, & fulgura—Pliny, *Historia naturalis*, II. 48–9, quoted in *cmts* on I3ᵛ–I5ʳ, p. 398 below.

**G6ʳ**

Page 52, ll. 15–17: *Venti* omnes—Aristotle, *Problems*, XXVI. 28, pp. 88–9: 'Why do winds which are cold have a drying effect? Is it because the colder winds produce evaporation? Why do they cause more evaporation than the sun? Is it because they drive off the vapour, but the sun leaves it? So it produces more moisture and less dryness.'

**G6ᵛ**

Page 52, ll. 20–3: Martij magis—cf. Gilbert, *DM*, pp. 260–1: 'Ineunte vere, postquam desiere imbres hiemales apud nos, Venti longe plus corpora exsiccant celeriusque quam media æstate, quare artifices nostri musici & materiales [i.e. martiales], ventos expetunt [i.e. expectant?], ad materiam & opificia varia confirmanda.' Bacon clearly had a better manuscript text of *DM* than the 1651 printed text.

**G6ᵛ–G7ʳ**

Page 52, ll. 28–34: Certè *Venti*—Acosta, *The natvrall and morall historie of the . . . Indies*, pp. 110–12.

**G7ᵛ**

Page 54, ll. 5–8: Niues quandóque—see Richard Knolles, *The generall historie of the Turkes, from the first beginning of that nation to the rising of the Othoman familie . . .*, Adam Islip: London, 1603, p. 650: 'The countrey neere vnto the citie of Sᴠʟᴛᴀɴɪᴀ wherein *Solyman* lay encamped at large, is on euerie side enuironed with huge mountaines . . . Whilest *Solyman* in those plaine fields most fit to fight a battell in, expected the comming of *Tamas*, such a horrible and cruell tempest . . . fell downe from those mountaines . . . with such abundance of raine, which frose so eagerly as it fell, that it seemed the depth of Winter had euen then of a sudden been come in: for such was the rage of the blustring winds, striuing with themselues as if it had beene for victorie, that they swept the snow from off the tops of those high mountaines, and cast it downe into the plaines in such abundance, that the Turkes lay as men buried aliue in deepe snow . . . wherein a wonderfull number of sicke souldiours and others of the baser sort which followed the campe perished; and many others were so benummed, some their hands, some their feet, that they lost the vse of them for euer: most part of their beasts which they vsed for carriage, but especially their camels, were frozen to death.'

### G8ᵛ

Page 54, l. 24: *in* Scripturis *notata*—John 3: 8: 'The wind bloweth where it listeth, and thou hearest the sound thereof, but canst not tell whence it cometh, and whither it goeth: so is every one that is born of the Spirit.'

### H1ᵛ

Page 56, ll. 4–5: Finxerunt *Poetæ*—see *Aeneid,* I, ll. 50–5: 'Talia flammato secum dea corde volutans | nimborum in patriam, loca feta furentibus Austris, | Aeoliam venit. hic vasto rex Aeolus antro | luctantis ventos tempestatesque sonoras | imperio premit ac vinculis et carcere frenat.'

### H2ʳ

Page 56, ll. 7–9: *Qui producit Ventos*—Ecclesiasticus 43: 15: 'Propterea aperti sunt thesauri, Et evolaverunt nebulae sicut aves'; cf. Acosta, *The natvrall and morall historie of the . . . Indies,* p. 123.

ll. 9–10: de Thesauris Niuis, & grandinis—Job 38: 22: 'Hast thou entered into the treasures of the snow? or hast thou seen the treasures (Vulgate: *thesauros*) of the hail.'

### H2ᵛ

Page 56, ll. 18–20: Si expirat Aer è terrâ—cf. Gilbert, *DM,* pp. 259–60.

### H3ʳ

Page 56, ll. 32–5: Spiritus subterraneus—Gilbert, *DM,* pp. 259–60.

### H3ᵛ–H4ʳ

Page 58, ll. 5–11: In Comitatu—see Gilbert, *DM,* p. 260: 'in Comitatu Derbiæ, montosa regione & lapidosa, ex cavernis quibusdam tam vehementes ventorum eruptiones existunt, ut injecta vestimenta, pannique, rursus magna vi efflentur, & altius in aërem efferantur. In Aber Barry juxta Sabrinam fluvium, in saxoso quodam clivo, apertis foraminibus qui aurem apposuerit, sonitus varios & flatus in terræ penetralibus exaudiet.'

### H4ʳ

Page 58, ll. 12–17: *Notauit* Acosta—see Acosta, *The natvrall and morall historie of the . . . Indies,* pp. 112–13: 'why are not the nightes in summer at *Peru,* as hotte and troublesome as in *Spaine*? . . . Why is all the coast of

*Peru*, being ful of sands, very temperate? And why is *Potozi* (distant from the silver Citie but eighteene leagues, and in the same degree) of so divers a temperature, that the Countrie being extreamely colde, it is wonderfully barren and drie? And contrariwise, the silver Citie is temperate, inclining vnto heat, and hath a pleasant and fertil soile? It is more certaine, that the winde is the principall cause of these strange diversities; for without the benefite of these coole windes, the heate of the Sunne is such, as (although it bee in the midst of the snow,) it burnes and sets all on fire: but when the coolenes of the aire returnes suddenly, the heat is qualified how great soever it be: and whereas this coole winde rains ordinarie, it keepes the grosse vapours and exhalations of the earth from gathering together, which cause a heavie and troublesome heat.'

### H4$^{r-v}$

Page 58, ll. 18–21: *vt voluit* Parmenides—see Aristotle, *Metaphysics*, I. v, 986$^b$–987$^a$. For a full note on Parmenides as viewed through the eyes of Bernardino Telesio and other sources of Bacon's knowledge of Parmenides, see *OFB*, VI, pp. 422–3.

### H4$^v$

Page 58, ll. 22–5: *Sunt quidam putei*—see Pliny, II. 45, p. 21: 'Now, there be certaine caves and holes which breed winds continually without end like as that is one which we see in the edge of Dalmatia, with a wide mouth gaping, and leading to a deepe downefall [*hiatu*]: into which if you cast any matter of light weight, be the day never so calme otherwise, there ariseth presently a stormie tempest like a whirlepuffe [*turbini similis emicat procella*]. The places name is Senta. Moreover, in the province Cyrenaica there is reported to bee a rocke consecrated to the South-wind, which without prophanation may not be touched with mans hand; but if it be, presently the South wind doth arise and cast up heapes of sand.'

ll. 26–8: *Flammas euomunt* Ætna—see Pliny, II. 106, p. 47: 'Bvt amongst the wonderfull mountaines, the hill Ætna burneth alwaies in the nights: and for so long continuance of time yeeldeth sufficient matter to maintaine those fires: in winter it is full of Snow, & covereth the ashes cast up, with frosts.' Pliny goes on to give other instances of fires rising from the ground.

### H5$^v$

Page 60, ll. 10–13: *Increbescere murmur*—Pliny, XVIII. 86, p. 613: 'Let us come aland againe, and marke the disposition of woods and hills: you shall heare the mountains and forests both, keep a sounding and rumbling noise, and then do they foretell some change of weather: nay you shall marke the leaves of trees to move, flicker & play themselves, and yet no wind at all stirring [*sine aura quæ sentiatur folia ludentia*].'

**H5ᵛ–H6ʳ**

Page 60, ll. 14–16: Stellas sagittantes—cf. Aristotle, *Problems*, XXVI. 23, pp. 84–5: 'Why is it a sign of wind when there are shooting stars? Is it because they are carried by the wind, and the wind reaches them before us? So also the wind rises in the quarter from which the stars are travelling.' Also cf. Pliny, XVIII. 80, p. 612: 'If the starres make semblance as if they flew up and downe many togither, and in their flying seeme whitish, they denounce winds from that coast where they thus do shoot. Now if it seeme to the eye, as if they ran and kept one certaine place, those winds will hold and sit long in one corner: but in case they do so in many quarters of the heaven, they betoken variable and inconstant winds, going and comming, and never at rest [*si volitare plures stellæ videbuntur, quo ferentur albescentes ventos ex is partibus nuntiabunt, si coruscabunt, certos, si id in pluribus partibus fiet, inconstantes ventos et undique*].'

**H6ʳ**

Page 60, ll. 19–21: Stellæ exiguæ—cf. Pliny, XVIII. 80, p. 612.

**H6ʳ⁻ᵛ**

Page 60, ll. 22–6: Circuli apparent—the substance of items 20 and 21 originates with Pliny; for quotations see *cmts* pp. 106–20 below, on prognostics of winds.

**H7ʳ**

Page 60, l. 36–p. 62, l. 2: in mediâ regione—for antiperistasis see *OFB*, VI, pp. 174, 407; XI, pp. 294, 398–400, 541.

**H7ᵛ–H8ʳ**

Page 62, ll. 8–18: *Aer nouitèr factus*—for Bacon's highly artificial quantitative experimental work on the volume of air produced from a given volume of water and his consequent denial (central to his thinking about matter) of the Peripatetic theory of the decuple proportionality of the elements, see *PhU* in *OFB*, VI, pp. 48–50; *NO* in *OFB*, XI; p. 82, ll. 18–21; *HDR* in *OFB*, XIII, pp. 38, 70, 277. The fact is that Bacon thought that his denial of the decuple theory had led him to discover a main cause of the wind phenomena. Also see *HV*, Q4ᵛ–Q6ᵛ, pp. 124–6 above.

**H8ᵛ–I1ʳ**

Page 62, ll. 27–30: Notatum est—according to *SEH* (II, p. 44), this is based on Gilbert, *DM*. However, Gilbert (pp. 254–66) says nothing as definite as this.

**I1ʳ⁻ᵛ**

Page 62, l. 34–p. 64, l. 6: *Quicunque nôrit*—see *cmt* (p. 397 above) on H7ᵛ–H8ʳ, p. 62 above.

**I1ᵛ–I3ᵛ**

Page 64, l. 7–p. 66, l. 5: Accidentales generationes—most of these observations seem to be Bacon's own, except for the final item, for which see Gilbert, *DM*, p. 259.

I3$^v$–I5$^r$

Page 66, l. 6–p. 68, l. 4: Venti extraordinarij—cf. Pliny, *Historia naturalis*, II. 48–9, pp. 24–5: 'Now will we speake of suddaine blasts: which being risen . . . by exhalations of the earth, and cast downe againe; in the meanwhile appeare of many fashions [*multiformes*], enclosed within a thin course of clouds newly overcast. For such as be unconstant, wandering, and rushing in manner of land flouds [*torrentium modo*] . . . bring forth thunder and lightening. But if they come with a greater force, sway, and violence, and withal burst and cleave a drie cloud asunder al abroad, they breed a storm, which of the Greekes is called *Ecnephias*: but if the clift or breach bee not great, so that the wind be constrained to turn round, to rol and whirle in his discent, without fire, that is to say lightening, it makes a whirlepuffe [*vorticem*] or ghust called *Typhon*, that is to say, the storm *Ecnephias* aforesaid, sent out with a winding violence [*id est vibratus Ecnephias*]. This takes with it a peece broken out of a congealed cold cloud, turning winding, and rolling it round, and with that weight maketh the [d]ownefall more heavie, and changeth from place to place with a vehement and suddaine whirling [*rapida vertigine*]. The greatest danger and mischeefe that poore sailers have at sea, breaking not onely their crossesaile-yards [*antennas*], but also writhing and bursting in peeces the very ships: and yet a smal matter is the remedy for it, namely, the casting of vinegre out against it as it commeth, which is of nature most cold. The same storme beating upon a thing, is it selfe smitten backe againe with a violence, and snatcheth up whatsoever it meeteth in the way aloft into the skie, carrying it backe, and swallowing it up on high. But if it breake out from a greater hole of the said cloud, by it so borne downe, and yet not altogether so broad, as the abovenamed storme *Procella* doth, nor without a cracke; they call this boisterous wind *Turbo*, casting downe and overthrowing all that is next it. That same, if it be more hote and catching a fire as it rageth, is named *Prester*, burning, and withall laying along, whatsoeuer it toucheth and encountreth . . . No *Typhon* commeth from the North, ne yet any *Ecnephias* with Snow, or while Snow lieth on the ground. This tempestuous wind, if when it brake the cloud burned light withall, hauing fire of the owne before, and catched it not afterward, it is verie lightning; and differeth from *Prester*, as the flame from a cole of fire. Againe, *Prester* spreadeth broad with a flash and blast; the other gathereth round with forcible violence. *Typhon* moreouer or *Vortex*, differeth from *Turben* in flying backe: and as much as a crash from a cracke. The storme *Procella* from them both, in breadth: and to speake more truly, rather scattereth than breaketh the cloud. There riseth also upon the Sea, a darke mist, resembling a monstrous beast; and this is euer a terrible cloud to the sailers. Another likewise called a Columne or Pillar, when the humour and water ingendred, is so thicke and stiffe congealed, that it standeth compact of it selfe. Of the same sort also is that cloud which draweth water to it, as it were into a long pipe [*in longam veluti fistulam*].' Also see *CCC met.*, cols. 74–6.

**I5ᵛ–I6ᵛ**

Page 68, ll. 8–23: *Quæ à veteribus*—for a concise, contemporary expression of the Aristotelian position, see *CCC met.*, cols. 1–7 (the association of hot/dry exhalations with the generation of wind, and of hot/moist vapours with the production of rain). On the three regions of the air and antiperistasis, see col. 9: 'suprema regio, vltra quam eius natura postulet, calida est. Primum ob ignis vicinitatem. Secundò ob flammulas ab igni, vt quidam volunt, nonnunquam decide*n*tes . . . Infima verò regio calida est, plusquam natiua eius conditio poscat: minus tamen, quam suprema . . . At media regio perfrigida est multis de causis. Primum propter vapores, qui cum eò perueniunt, sublatis iam impedimentis vltro sese ad natiuum frigus reuocant, aeremq*ue* circumfusum refrigerant. Secundo, quia aer ita refrigeratus cum ex superiori & inferiori parte calido obsideatur, per antiperistasim algescit (compertum namq*ue* est, etiam in alieno subiecto intendi qualitatem circumstante aduersario, vt calorem in aquis cisternarum hyemali tempore . . .) Tertiò, quia nec repercussio solarium radiorum, nec elementaris ignis feruor; nec impetus, quo superior pars in orbem versatur eò pertingunt.'

**I6ᵛ**

Page 68, ll. 24–6: *Rotatio naturalis*—this is the general wind described above at E3ᵛ–E6ᵛ (pp. 32–6) and see *cmts* thereon p. 389 above.

**I7ʳ⁻ᵛ**

Page 70, ll. 1–2: In *Peruviâ*—Acosta, *The natvrall and morall historie of the . . . Indies*, p. 96, says that rain not wind is abundant 'when the Moone is at the full.'

**I7ᵛ**

Page 70, ll. 8–10: *Aquam & Aerem esse corpora valdè homogenea*—for Bacon air and water are convertible, and in any case belong to the same family of substances, namely the mercury quaternion; see *cmts* (pp. 416–17 below) on *HNE*, S5ʳ–S8ʳ (pp. 136–8 above).

**I7ᵛ–I8ʳ**

Page 70, ll. 13–18: Exortu *Orionis*—cf. Aristotle, *Problems*, XXVI. 13, pp. 80–1: 'Why is there more variety in the day and more frequent change of wind at the rising of Orion than at other times? Is it because everything is less fixed in time at the moment of change, and Orion rises at the beginning of autumn, and sets in the winter, so that as one season is not yet established, but one is arriving and another dying, for this reason the winds also must be unsettled because the conditions are intermedial between the two seasons? In fact, Orion is described as an unpleasant season both rising and setting because of the lack of fixity of the season; it must necessarily be a time of confusion and inconsistency.' Also see Pliny, XVIII. 28, p. 599: 'Tempests, which comprehend hailes, stormes of wind and raine . . . and these for the most part proceed from some of the horrible and dreadful Stars, as . . . Arcturus, Orion, and the Kids [*haedis*].'

**I8$^v$**

Page 70, ll. 27–9: in *Peruviâ*—see Acosta, *The natvrall and morall historie of the . . . Indies*, p. 112; for the quotation see *cmts* (pp. 395–6 above) on H4$^r$, p. 58 above.

ll. 30–2: In *vitro calendari*—the calendar glass became one of Bacon's favourite instruments in works belonging to *IM*; see, for instance, *OFB*, XI, pp. xl–xli, lxxiv, and, on the construction of the glass, *ibid.*, pp. 248–50.

**K1$^r$**

Page 70, l. 33–p. 72, l. 2: *Experimentum fecimus*—this is an instance, though not the best one, of Bacon's willingness to conduct actual and artificial experiments; for this aspect of his work, see *OFB*, XI, pp. xli–xlii, xliii, lxxxii–lxxxiii.

**K1$^{r–v}$**

Page 72, ll. 3–7: Etiam receptio Aeris—see Acosta, *The natvrall and morall historie of the . . . Indies*, p. 112. For antiperistasis see *OFB*, VI, pp. 174, 407; XI, pp. 294, 407, 541.

**K1$^v$–K2$^r$**

Page 72, ll. 11–19: de *ventis Vaporarijs*—this is again an implicit denial of the Aristotelian notion that only rain was generated from vapours; see *cmt* (p. 399 above) on I5$^v$–I6$^v$ (p. 68 above). Behind this is Bacon's idea that a primary cause of wind was overloading of the air by the conversion of water into new air. For Bacon a given volume of water was not converted into ten volumes of air (the late Aristotelian view) but into over a hundred; hence the wind-creating overload; see *cmts* (p. 397 above and p. 414 below) on H7$^v$–H8$^r$ (p. 62 above) and Q4$^v$–Q6$^v$ (pp. 124–6 above).

**K2$^v$–K3$^r$**

Page 72, ll. 32–5: nauigatione ad *Russiam*—cf. Gilbert, *DM*, pp. 256–7: 'Quæritur, cur apud septentrionem, ultimis scilicet navigationibus, quibus mercatores nostri Angli in Moscoviam negotiantur, per Warhusam, & divi Nicolai fanum, sub altitudine 75 graduum navigant, ventorum procellis minus longe afflictantur, quam in nostro Britannico pelago?'; cf. Acosta, *The natvrall and morall historie of the . . . Indies*, pp. 140–1.

**K3$^{r–v}$**

Page 74, ll. 6–10: opinio Veterum—see *cmt* on I5$^v$–I6$^v$, p. 399 above.

**K3$^v$–K4$^r$**

Page 74, ll. 15–17: Etiam *Anniuersarij Aquilones*—cf. Pliny, II. 47, p. 23 (quoted in *cmts* on *HV*, F1$^r$–F2$^r$, p. 390 above.

**K6$^r$**

Page 76, ll. 28–9: Narrat *Plinius Turbinis*—see *cmt* on *HV*, R3$^r$ (p. 415 below).

**K6ᵛ–K7ʳ**

Page 76, l. 31–p. 78, l. 2: Traditur de Monte—Mount Athos excepted, all the same phenomena are reported in *NO*; see *OFB*, XI, pp. 220–2. Also see Acosta, *The natvrall and morall historie of the . . . Indies*, pp. 146–7: 'There is in *Peru*, a high mountaine which they call *Pariacaca* . . . [at] the top of this mountaine, I was suddenly surprized with so mortall and strange a pang, that I was ready to fall from the top to the ground . . . I was surprised with such pangs of straining & casting, as I thought to cast vp my heart too . . . and not onely the passage of *Pariacaca* hath this propertie, but also all this ridge of the mountaine, which runnes above five hundred leagues long . . . the best remedy (and all they finde) is to stoppe their noses, their eares, and their mouthes, as much as may be . . . for that the ayre is subtile and piercing, going into the entrailes.' Also see Aristotle, *Problems*, XXVI. 36, pp. 96–7: 'on some very high mountains there is no wind—for instance on Mount Athos and in other similar places. We have proof of this; for whatever is left behind by those who sacrifice in one year, is said to be found still remaining the year after.' For Olympus see Julius Solinus, *The excellent and pleasant work of Iulius Solinus Polyhistor . . . translated . . . by Arthur Golding*, I. Charlewoode for Thomas Hacket: London, 1587, I3ᵛ: '*Olympus* . . . ryseth so bigge, with so hygh a toppe, that the dwellers by doo call the knappe of it heauen. Ther is on the top of it an Altar dedicated to *Iupiter*, where vpon if any part of the inwards be layd, they are neyther blowne a sunder wyth blastes of wynde, nor washed away with rayne: but when the yeere comes about againe, they are founde the selfe same that they were left. And whatsoever is once consecrated there vnto the God, it is priuiledged for euer from corruption of the aire. Letters written in the ashes continue tyll the Ceremonies of the next yeere.'

**K8ʳ**

Page 78, ll. 29–30: Pusilli (vt dictum est) *Turbines*—see I5ʳ, pp. 66–8 above.

**K8ᵛ**

Page 80, ll. 1–3: Non solum *Eurus*—see F3ᵛ–F4ʳ, p. 40 above.

**L1ᵛ**

Page 80, ll. 20–3: Si *Auster* cœperit—cf. Aristotle, *Problems*, XXVI. 47, pp. 102–3: 'Why does the north wind follow swiftly on the south, but the south wind does not follow swiftly on the north? Is it because the arrival of the former is from near by, but of the latter from a distance? For we live near the north.'

ll. 24–6: Cum Annus inclinârit—cf. *ibid.*, XXVI. 46, pp. 100–3: 'Why is it said that "if the south wind summons the north wind, winter is upon us"? Is it because it is the nature of the south wind to collect clouds and heavy rain? So when in these conditions the north wind blows as well, as it carries much matter with it, it freezes and produces winter.'

**L1$^v$–L2$^r$**

Page 80, ll. 27–9: *Plinius* citat *Eudoxum*—Pliny, II. 47, p. 24: 'And verily *Eudoxus* is of opinion (if wee list to observe the least revolutions) that after the end of every fourth yeere, not onely all winds, but other tempests and constitutions also of the weather, returne again to the same course as before. And alwaies the Lustrum or computation of the five yeers, beginneth at the leape year, when the Dog star doth arise.' For Eudoxus of Cnidus (*c.*390–*c.*340 BC) see *OCD*. I think this is the only reference to him in Bacon's writings.

**L2$^r$**

Page 80, ll. 29–32: Illud ex aliquorum—Bacon elaborates on this in the essay *Of Vicissitude of Things*, and adds that a 35-year cycle has been reported of the Low Countries; see *OFB*, XV, p. 173, ll. 61–9 and *cmt* thereon p. 313.

**L3$^r$**

Page 82, ll. 17–18: vt in Commentatione—see H7$^v$–H8$^v$, p. 62 above.

**L3$^{r-v}$**

Page 82, ll. 19–29: Etiam huius rei Imaginem—see K1$^r$, pp. 70–2 above. For a formal expression of Bacon's idea of variation of experiment, and of the idea of *experientia literata*, of which it forms one branch, see *DAS*, 2H2$^v$ (*SEH*, I, pp. 623–4).

**L7$^r$**

Page 86, ll. 22–4: Si verum sit, *Columbum*—see E7$^v$–E8$^r$, p. 36 above and *cmts* thereon p. 390 above.

**L8$^r$**

Page 88, ll. 7–8: quicquid dicant *Poetæ*—almost certainly an allusion to Ovid, *Tristia*, I. 2, ll. 19–25: 'me miserum, quanti montes uoluuntur aquarum! | iam iam tacturos sidera summa putes. | quantae diducto subsidunt aequore ualles! | iam iam tacturas Tartara nigra putes'.

**M2$^v$–M3$^r$**

Page 90, ll. 18–20: In *Nauibus* maioribus—in Willaert's paintings of the Prince Royal (see page xliii, n. 112 above), the masts are disposed as Bacon reports, with the inclined bowsprit with a spritsail yard terminating in a short vertical post (a spritsail topmast) with a spritsail topsail yard.

**M3$^r$**

Page 90, ll. 27–31: His *Malis* superimpendent *Vela* decem—pictures of the Prince Royal (see *cmt* above) show only two sails to foremast and mainmast. There are no topgallant sails or yards to suspend them from. However, it was not uncommon for the topgallant topmasts to be left ashore; see Alan Moore, 'Rigging in the seventeenth century', *The Mariner's Mirror*, 2, 1912, pp. 267–74; continued *ibid.*, pp. 301–8, and continued *ibid.*, 3, 1913, pp. 7–13. The pictures also show mizzenmasts designed for two sails: the upper suspended from a horizontal sail-yard, the lower, wider triangular one suspended from an inclined sail-yard. The bowsprit is set up for two sails. For an early seventeenth-century account of masts and the sails they carry, see *A treatise on shipbuilding and a treatise on rigging written about 1620–1625*, ed. W. Salisbury and R. C. Anderson, The Society for Nautical Research: London, 1958, pp. 47, 62. Also see John Smith, *A sea grammar*, ed. Kermit Goell, Michael Joseph: London, 1970, pp. 21, 39–40.

## M3ᵛ–M4ʳ

Page 92, ll. 12–15: In *Naui*—for relationships between tonnage, length, and width of ships see *A treatise on shipbuilding*, pp. 14–15. The Prince Royal is referred to in this treatise (p. 16), but the example the anonymous writer worked from was a ship of 550 tons burthen, and his rules for scaling up or down to ships of other sizes do not seem (with their difficult mathematics) to produce length and width measurements which agree with Bacon's figures. However, a survey of the Prince Royal undertaken in 1632 gives figures which are almost identical to Bacon's: 1186.80 tons; length of keel 115 feet; breadth 43 feet. The results of the survey are given in a short vellum roll (BL Add. MS 18037); also see *The autobiography of Phineas Pett*, pp. lxxviii, 217.

## M4ʳ–M5ᵛ

Page 92, ll. 29–36: *Velum à suprà*—I have found no source that corroborates Bacon's account of sail dimensions, but they seem far too circumstantial not to have been based on accurate information about, no doubt, the canvas carried by the Prince Royal.

## M6ᵛ–M7ʳ

Page 94, ll. 34–7: Motus *Ventorum* in *velis*—Bacon is right that the greater 'leverage' of the topsails would, if they were used in high winds, endanger the ship's stability.

## M8ᵛ

Page 96, ll. 28–30: *motus* Directionis—Smith, *A sea grammar*, p. 40: 'All *after sailes*, that is, all the sailes belonging to the maine Mast and Miszen, keepes her to wind ward; therefore few ships will steare upon quarter winds with one saile, but must have one after saile and one head saile.'

## M8ᵛ–N1ʳ

Page 96, l. 31–p. 98, l. 6: Cum Pyxis nautica—cf. Smith, *A sea grammar*, p. 50: 'a crosse saile. [square-rigged ship] cannot come neerer the wind than six points; but a Carvell [caravel], whose sailes stand like a paire of Tailer's sheeres, will goe much neerer.' *SEH* notes (II, p. 61) that 'a square-rigged vessel will lie within six points of the wind . . . so that there is no change in this respect since his [Bacon's] time.'

**N1<sup>r–v</sup>**

Page 98, ll. 7–12: Omnis *ventus* in *velis*—I have found no early seventeenth-century source for this but it sounds sensible.

**N1<sup>v</sup>–N2<sup>r</sup>**

Page 98, ll. 13–23: Potest fieri per motum—there are seventy Italian miles to a degree of latitude, as against sixty English statute miles per degree. On serpent motion see *OFB*, XI, p. 292, ll. 26–8, p. 556.

**N7<sup>v</sup>–N8<sup>r</sup>**

Page 102, l. 25–p. 104, l. 2: Motus *Molendinorum* ad *Ventum*—see Introduction, p. xliv above.

**O2<sup>r</sup>**

Page 106, ll. 4–5: Traditur alicubi esse *Rhedas*—the Dutch engineer and scientist Simon Stevin (1548–1620) built (*c*.1600), a wind-driven carriage which he sailed (with Prince Maurice of Nassau among others) on the beach at Scheveningen. The carriage reached, with twenty-six passengers and a favourable wind, speeds faster than a horse's.

**O2<sup>r</sup>–O3<sup>v</sup>**

Page 106, ll. 12–17: Diuinatio *quò magis pollui*—for Bacon on superstitious, natural, and artificial divinations see *DAS*, M1<sup>r</sup>, 2E4<sup>v</sup>–2F1<sup>r</sup> (*SEH*, I, pp. 498, 607–8). A history of natural divinations is no. 79 in *CHP*; see *OFB*, XI, p. 480.

**O3<sup>v</sup>**

Page 106, ll. 28–30: 1. *Sol* si oriens—most of the material on prognostics was taken, with changes to the order and wording, from Pliny, *Historia naturalis*, XVIII. 78–90. In the case of this first item cf. *ibid.*, XVIII. 78, p. 611: 'If the Sun in rising seeme hollow, he foretelleth raine [*concavus oriens pluvias praedicit*].' The amplification of this statement seems to be Bacon's own.

ll. 31–2: 2. Si *Sol* oriatur pallidus—cf. Pliny, *loc. cit.*, where a pale sun rising announces hail: 'When he riseth cleare and not fiery red, it is a signe that the day will be faire; but if he shew pale and wan, it presageth a cold winter-like haile-storme that very day [*Purus oriens atque non fervens serenum diem nuntiat, at hibernum pallidus grandine*].'

**O3<sup>v</sup>–O4<sup>r</sup>**

Page 108, ll. 1–2: 3. Si corpus ipsum *Solis*—cf. Pliny, *loc. cit.*

**O4<sup>r</sup>**

Page 108, ll. 3–4: 4. Si in exortu *Solis*—this item does not appear in Pliny, *loc. cit.*

ll. 5–7: 5. Si in ortu—cf. Pliny, *loc. cit.*, 'Marke at his rising, or going downe, if his beames be short and as it were drawne in, be sure of a good showre [*si in ortu aut in occasu contracti cernentur radii, imbrem*].'

ll. 8–9: 6. Si ante Ortum *Solis*—cf. Pliny, *loc. cit.*, 'If he spread his beams before he be up and appear in our Horizon, looke for wind and water both [*si ante ortum radii se ostendent, aquam et ventum*].'

## O4$^{r-v}$

Page 108, ll. 10–14: **7. Si in Exortu *Solis*—cf. Pliny, *loc. cit.*, 'If at his rising you see him to cast his beams afar off among the clowds, and the mids between be void thereof, it signifieth raine [*si in exortu longe radios per nubes porriget et medius erit inanis, pluviam significabit*].'

## O4$^{v}$

Page 108, ll. 15–17: **8. Si *Sol* oriens—cf. Pliny, *loc. cit.*, 'If as the Sun riseth he be compassed with a circle, marke on what side the same breaketh and openeth first, and from thence looke for wind without faile: but if the said circle passe and vanish away all at once equally, as well as of one part as another, you shall have faire weather upon it [*Si oriens cingetur orbe, ex qua parte is se ruperit, expectetur ventus. Si totus defluxerit æqualiter, serenitatem dabit*].'

ll. 18–20: **9. Si sub Occasum *Solis*—Pliny, *loc.cit.*, 'If about him toward his going downe there be seene a white circle, there will bee some little tempest and troublesome weather that night ensuing: but if in stead thereof hee be over-cast with a thicke mist, the tempest will be the greater and more violent. If the Sun couchant appeare fiery and ardent, there is like to bee wind. Finally, if the circle aforesaid bee blacke, marke on which side the same breaketh, from hence shall you have blustering winds [*Si circa occidentem candidus circulus erit, noctis levem tempestatem. Si nebula, vehementiorem. Si candente Sole, ventum. Si ater circulus fuerit, ex qua regione is ruperit se, ventum magnum*].'

## O5$^{r}$

Page 108, ll. 21–2: **10. Si *Nubes* rubescant—cf. Pliny, *loc. cit.*, 'and when before his rising the clowds be red, the winds will bee aloft that day . . . Are the clowds red about the Sun as he goeth downe? you shall have a faire day the morrow after [*idem ventos cum ante exorientem eum nubes rubescunt . . . si circa occidentem rubescunt nubes, serenitatem futuri diei spondent*].'

ll. 23–5: **11. Si sub Exortum *Solis*—Pliny, *loc. cit.*, 'If before he rise, the clowds gather round togither like globes, they threaten sharpe, cold, and winter weather: but in case hee drive them before him out of the East so as they retire into the West, we have a promise thereby of a faire time [*Si ante exortum nubes globabuntur, hiemem asperam denuntiabunt. Si ab ortu repellentur et ad occasum abibunt, serenitatem*].'

ll. 26–8: **12. Si in Exortu—Pliny, *loc. cit.*, 'If when the Sun doth rise you see flying clowds dispearsed, some to the South and others Northward (say all be cleare and faire otherwise about him) make reckoning that day of wind and raine both [*si in exortu spargentur partim ad Austrum, partim ad Aquilonem, pura circa eum serenitas sit licet, pluviam tamen ventosque significabunt*].'

ll. 29–31: **13. Si *Sol* sub Nube—Pliny, *loc. cit.*, 'If at the Suns setting it raine, or that his raies either looke darke and blew, or gather a banke of clowds, surely these be great tokens of tempestuous weather & storms the morrow after [*si in occasu eius pluet aut radii nubem in se trahent, asperam in proximum diem tempestatem significabunt*].'

## O5ʳ⁻ᵛ

Page 108, l. 32–p. 110, l. 2: **14.** and **15.** Si *Nubes*—cf. Pliny, *loc. cit.*, 'If there appeare about the bodie of the Sun, a circle of clowds compassing it round [*Si nubes solem circumcludent*], the nearer they come about him and the lesse light that they leave him, the more troubled and tempestuous weather will follow: but in case he be environed with a double circle [*duplex orbis*], so much more outragious and terrible will the tempest be. If peradventure this happen at his rising, so as the said clowds be red againe which compasse the Sunne, looke for a mightie tempest one time or other of that day. If haply these clowds enclose him not round, but confront and seeme as if they charged upon him, looke from whence they come, from that quarter they portend great wind: and if they encounter him from the South, there will be raine good store and wind both.'

## O5ᵛ–O6ʳ

Page 110, ll. 3–5: **16.** *Nouilunia* dispositionum—the general point made here is simply a preparation for the rest that Bacon says (or rehearses) concerning the Moon.

## O6ʳ

Page 110, ll. 6–7: **17.** Diuturnâ obseruatione—Pliny, *loc. cit.*, does not mention this.

ll. 8–9: **18.** Si *Luna* à Nouilunio—Pliny, XVIII. 79, p. 612: 'If shee shew not in our Horizon before the prime or fourth day after the chaunge, and the West wind blow withall, then that moone throughout (*toto mense*) threatneth cold and winter weather.'

ll. 10–14: **19.** Si *Luna* nascens—cf. Pliny, *loc. cit.*, 'The new moone whiles shee is croissant, if shee rise with the upper tip or horne blackish [*obatrato*], telleth beforehand that there will be store of raine after the full, and when she is in the wane: but if the nether tip be so affected, the rain will fall before she be at the full.'

## O6ʳ–O7ᵛ

Page 110, ll. 15–37: **20 to 28.** Si Ortu in quarto—Pliny, XVIII. 79, pp. 611–12: 'the Ægyptians observe most her Prime, or the fourth day after the chaunge: for if she appeare then, pure, faire, and shining bright, they are verily persuaded that it will bee faire weather: if red, they make no other reckoning but of winds: if dim and blackish [*nigra*], they looke for no better than a foule and rainie moneth. Marke the tips of her hornes when she is five daies old, if they be blunt, they foreshew raine; if pricking upright and sharpe pointed withall, they alwaies tell of winds toward: but upon the fourth day especially, this rule faileth not, for that day telleth truest. Now if that upper horne of hers only which bendeth Northward, appeare sharpe pointed and stiffe withall, it presageth wind from that coast: if the nether horne alone seeme so, the wind will come from the South: if both stand streight and pricking at the point, the night following will be windie. If the fourth day after her chaunge, she have a red circle or Halo about her, the same giveth warning of wind and raine. As for *Varro*, hee (treating of the presages gathered from the Moone) writeth thus: If (quoth he) the new moone when she is just foure daies old, put her horns direct and streight forth, she presageth thereby some great tempest at sea presently to follow, unlesse it be so that she have a guirland or circle about her, and the same clear and pure; for then there is good hope that there will be no foule nor rough weather before the full. If at the full, one halfe of her seeme pure and neat, a signe it is of a faire season; if it be red, the wind will be busie; if enclined to blacke, what else but raine, raine. Doe you see at any time a darke mist or clowd round about the body of the moone [*si caligo orbisve nubium incluserit*]? it betokeneth winds from that part where it first breaketh: and in case there bee two such clowdie and mistie circles environing her, the tempest will bee the greater: but how if there bee three of them forfailing, and those either blacke, or interrupted, distracted and not united? surely then there will be more stormes & more.'

**O7<sup>r–v</sup>**

Page 112, ll. 1–8: **29 to 32.** *Luna*—these items do not appear in Pliny's material on prognostics.

**O7<sup>v</sup>–O8<sup>r</sup>**

Page 112, ll. 9–11: **33.** *Stellæ* (vt loquimur)—cf. Pliny, XVIII. 80, p. 612: 'If the starres make semblance as if they flew up and downe many togither, and in their flying seeme whitish, they denounce winds from that coast where they thus do shoot. Now if it seeme to the eye, as if they ran and kept one certaine place, those winds will hold and sit long in one corner: but in case they do so in many quarters of the heaven, they betoken variable and inconstant winds, going and comming, and never at rest [*si volitare plures stellae videbuntur, quo ferentur albescentes ventos ex is partibus nuntiabunt, si coruscabunt, certos, si id in pluribus partibus fiet, inconstantes ventos et undique*].'

**O8<sup>r</sup>**

Page 112, ll. 12–15: **34.** Cum non conspiciantur—cf. Pliny, *loc. cit.*: 'Within the signe Cancer, there be two pretie stars which the Mathematicians call Aselli, [*i.* little Asses] betweene which there seemeth to be a small clowd taking up some little roome, and this they name in Latine Præsepia, [*i.* a Crib, Crarch, Bowzey, or Manger:] now if it chaunce that this Racke or Crib appeare not, and yet the aire be faire and cleare otherwise, a signe it is of cold, foule, and winter weather. Also if the one of those two little stars, to wit, that which standeth Northerly, be hidden with a mist, then shall you have the South wind to rage; but in case the other which is more Southerly, be out of sight, then the Northeast wind will play his part [*sunt in signo cancri duae stellae parvae aselli appellatae, exiguum inter illas spatium obtinente nubecula quam praesepia appellant; haec cum caelo sereno apparere desiit, atrox hiems sequitur; si vero alteram earum aquiloniam caligo abstulit, auster saevit, si austrinam, aquilo].'

O8$^{r-v}$

Page 112, ll. 16–20: **35. Cœlum æqualitèr**—cf. the rather different formulation in Pliny, *loc. cit.*: 'The starrie skie, if it shew cleare and bright all over, and in every part alike [*aequaliter totum erit splendidum*] . . . it is a fore-token of a faire and drie Autumne, but yet cold.

If the Spring and Summer both, passed not cleare without some raine and wet weather, it will bee an occasion that the Autumne following shall be drie, and lesse disposed to wind; howbeit, thicke, muddie, and enclined to mists. A faire and drie Autumne, bringeth in alwaies a windie winter.

When all on a sodaine the stars loose their brightnesse and looke dim, and that neither upon a clowd nor a mist in the aire, it signifieth either raine, or grievous tempests [*cum repente stellarum fulgor obscuratur et id neque nubilo nec caligine, pluvia aut graves denuntiantur tempestates*].'

O8$^{v}$

Page 112, ll. 21–3: **36. Si Planetarum**—cf. Pliny, *loc. cit.*: 'when a man seeth new circles still about any planets, there will be much raine soone after [*circulus nubis circa sidera aliqua pluviam*].'

ll. 24–5: **37. Cum tonat vehementiùs**—cf. Pliny, XVIII. 81, pp. 612–13: 'In Summer time, if there chaunce to bee more thunder than lightning, it threatneth winds from that coast where it thundred: contrariwise, if it lighten much & thunder little, looke for rain plentie [*Cum aestate vehementius tonuit quam fulsit, ventos ex ea parte denuntiat, contra si minus tonuit, imbrem*].'

l. 26: **38. Tonitrua *Matutina***—cf. Pliny, XVIII. 81, p. 613: 'morning thunders foreshew winds; but if they be heard at noone, they presage store of raine [*tonitrua matutina ventum significant, imbrem meridiana*].'

O8$^{v}$–P1$^{r}$

Page 112, ll. 27–9: **39. Tonitrua mugientia**—this is not in Pliny.

P1$^{r}$

Page 112, ll. 30–5: **40 to 41. Cum cœlo sereno**—cf. Pliny, XVIII. 81, p. 613: 'when you see it lighten, and the skie otherwise cleare and faire, it is a token that rain and thunder will follow thereupon, yea and rigorous cold weather besides: but the cruellest and most bitter impressions of the aire, ensue upon such lightnings as come from all foure quarters of heaven at once: if it lighten from the Northwest [*ab aquilone*] onely, it betokeneth raine the day following; if from North [*a septentrione*], it is a signe of wind from thence: if from the South, Northwest, or full West [*ab austro vel coro aut favonio*], it happen to

lighten in the night & the same be faire, it sheweth wind and rain from out of those coasts.'

Page 112, ll. 36–8: **42.** Magni feruores—not apparently in Pliny.

**P1ʳ⁻ᵛ**

Page 114, ll. 1–9: **43.** Globus flammæ, quem *Castorem*—cf. Pliny, II. 37, p. 18: 'I have seene myselfe in the campe, from the souldiours sentinels in the nightwatch, the resemblance of lightening to sticke fast upon the speares and pikes set before the rampiar. They settle also upon the crossesaile-yards, and other parts of the ship, as men doe saile in the sea: making a kind of vocall sound [*vocali quodam sono insistunt*], leaping too and fro, and shifting their places as birds doe which flie from bough to bough. Daungerous they be and unluckie, when they come one by one without a companion: and they drown those ships on which they light, and threaten shipwrack, yea, and they set them on fire if haply they fall upon the bottome of the Keele. But if they appeare two and two together, they bring comfort with them, and foretell a prosperous course in the voiage, as by whose comming, they say, that dreadfull, cursed, and threatening Meteor called Helena; is chased and driven away. And thereupon it is, that men assigne this mightie power to *Castor* and *Pollux*, and invocate them at sea, no lesse than gods. Mens heads also in the even-tide are seene many times to shine round about, and to be of a light fire, which presageth some great matter. Of all these things there is no certain reason to be given, but secret these be, hidden with the majestie of Nature, and reserved within her Cabinet [*Omnia incerta ratione, & in naturæ majestate abdita*].' The phenomenon is also known as St Elmo's fire. Castor and Pollux were twin sons of Tyndareos (or Zeus in some traditions) and Leda, and brothers of Helen and Clytemnestra. These were regarded as special patrons of sailors, to whom they appeared in this fiery form. Also see Giovan Battista della Porta, *De aeris transmutationibus*, ed. Alfonso Paolella, Edizioni Scientifiche Italiane: Naples, 2000, pp. 131–5; *CCC met.*, cols. 18–19; *OFB*, XI, p. 226.

**P1ᵛ–P2ʳ**

Page 114, ll. 10–14: **44.** Si conspiciantur Nubes—Pliny, XVIII. 82, p. 613: 'As touching clouds, if you see the racke ride apace in the aire, the weather beeing faire and drie, looke for wind from that quarter whence those clouds do come, and if they seeme to gather thicke in that place, dispearsed they will bee and scattered when the Sun approacheth: but more particularly, if this happen from the Northeast, they portend raine [*ventos*, in fact, in the Latin]; if from the South, storme and tempest [merely *imbres* in the Latin].'

**P2ʳ**

Page 114, ll. 15–17: **45.** Si occidente Sole—cf. Pliny, *loc. cit.*: 'If at the Suns setting the racke seeme to ride from both sides of him into the open aire, they shew of tempests toward: if the clouds be exceeding black, flying out of the East, they threaten raine against night; but if they come out of the West, it will

surely raine the morrow after [*sole occidente si ex utraque parte eius caelum petent, tempestatem significabunt: vehementius atrae ab oriente in noctem aquam minantur, ab occidente in posterum diem*].'

ll. 18–20: **46.** Liquidatio—not in Pliny.

**P2ᵛ**

Page 114, ll. 21–7: **47.** Conspiciuntur—see Gilbert, *DM*, p. 258: 'Vidimus namque sæpissime Austrum nubes ducentem in inferioribus aëris partibus, Aquilonem vero superne alias in contrarium ferentem. Vidi egomet, indicantibus nubium globis, quinque simul dispares ventos altitudine in aëre, & positione ab horizonte.'

ll. 28–30: **48.** *Nubes,* si vt vellera—Pliny, XVIII. 82, p. 613, does not quite say that: 'if the clouds be disparkled many togither out of the East, and flie like fleeces or flockes of wooll, they shew raine for three daies after [*si nubes ut vellera lanae spargentur multae ab oriente, aquam in triduum praesagient*].' Pliny also says nothing about clouds stacked up.

**P3ʳ**

Page 114, ll. 31–2: **49.** *Nubes* plumatæ—not in Pliny, apparently.

ll. 33–5: **50.** Cùm *Montes*—cf. Pliny, *Historia*, XVIII. 82, p. 613: 'when clouds flie low, and seeme to settle upon the tops of the hills, looke shortly for cold weather [*cum in cacuminibus montium nubes consident, hiemabit*].'

**P3ʳ⁻ᵛ**

Page 116, ll. 6–8: **53.** *Nubecula* aliqua—cf. Pliny, XVIII. 82, p. 613: 'bee the skie never so cleare, the least cloud appearing therein, is enough to engender and foreshew wind and storme [*caelo sereno nubecula quamvis parva flatum procellosum dabit*].'

**P3ᵛ**

Page 116, ll. 9–11: **54.** *Nebulæ*—for part of this cf. Pliny, XVIII. 83, p. 613: 'mists if they come downe and fall from the mountains, or otherwise descend from heaven and settle upon the vallies, promise a faire and drie season [*Nebulae montibus descendentes aut caelo cadentes vel in vallibus sidentes serenitatem promittent*].'

ll. 12–13: **55.** *Nube* grauidâ—cf. Pliny, XVIII. 82, p. 613: 'when the clouds seeme to be heavily charged and full, and yet looke white withall (which constitution of the aire is called commonly the white weather) there is an haile-storme at hand [*nube gravida candicante, quod vocant tempestatem albam, grando imminebit*].'

**P3ᵛ–P4ʳ**

Page 116, ll. 14–18: **56.** Autumnus serenus—this does not come from Pliny. I have not discovered a source for the proverb.

ll. 19–21: **57.** *Ignes* in focis—cf. Pliny, XVIII. 84, p. 613: 'If the fire then burne in the chimney pale, and keepe therewith a buzzing noise, wee finde by experience that it foresheweth tempest and stormie weather: as also we may be sure of raine, in case wee see a fungous substance or soot gathered about lamps and candle snuffs [*pallidi namque murmurantesque temepestatum nuntii sentiuntur, pluviae etiam si in lucernis fungi, si flexuose volitet flamma*].'

ll. 22–3: **58**. *Carbones*—cf. Pliny, *loc. cit.*: 'if you see the flame either of fire or candle mount winding and waving as it were, long you shall not be without wind ... or when the fire being raised in embers, keepeth a spitting and sparkling from it: also, if the ashes lying upon the hearth grow togither: and last of all, when the live-cole shineth brighter or scorcheth more than ordinarie: all these be signs of raine [*ventum nuntiant lumina cum ex sese flammas elidunt aut vix accenduntur ... aut cum contectus ignis e se favillam discutit scintillamve emittit, vel cum cinis in foco concrescit et cum carbo vehementer perlucet*].'

ll. 24–5: **59**. *Mare* cum conspicitur—cf. Pliny, XVIII. 85, p. 613: 'if you see the sea within the haven, after the floud is gone, in a low and ebb water to bee calme, and yet heare it keepe a rumbling noise within, it foresheweth wind: if it doe thus by times and fits one after another, resting still and quiet betweene-whiles, it presageth cold weather & rain [*Mare si tranquillum in portu cursitabit murmurabitve intra se, ventum prædicit, si idem hieme, et imbrem*].'

## P4ʳ–P5ʳ

Page 116, ll. 26–36: **60** to **63**. *Littora* in tranquillo—cf. Pliny, *loc. cit.*: 'if in calme and faire weather the sea strond or water banks [*litora ripæque*] resound or make a noise, it is a token of a bitter tempest: so it fareth also with the very sea itselfe; for if it be calme, & yet make a roaring; or if the fome therof be seen to scatter too & fro, or the verie water to boile & buble, you may be bold to foretell of tempests: the Puffins [*pulmones marini* sea lungs or jellyfish] also of the sea ... if they appeare swimming above water, do foresignifie cold wether for many daies togither: oftentimes the sea being otherwise calme, swelleth, & by hooving higher than ordinarie, sheweth that she hath wind good store enclosed within her, which soon after will breake out to a tempest.' For sea lungs also see *NO* in *OFB*, XI, p. 226 and *cmts* thereon p. 543. Philemon Holland's 'Puffins', see *OED*, **Puffin**[2] 2.

## P5ʳ

Page 118, ll. 1–6: **64** to **65**. Sonitus à *Montibus*—cf. Pliny, XVIII. 86, p. 613: 'you shall heare the mountains and the forests both, keep a sounding and a rumbling noise, and then do they foretell some change of weather: nay you shall marke the leaves of trees to move, flicker & play themselves, and yet no wind at all stirring; but be sure then that you shall not be long without. The like prediction is to be gathered by the light downe [*lanugo*] either of poplars or thistles flying too and fro in the aire; also of plumes and feathers floting upon the water. Goe downe lower to the vales and plaines: if a man chaunce to heare a bustling there, he may make account that a tempest will follow. As for the rumbling in the aire, it is an undoubted signe and token thereof.'

## P5ʳ–P6ʳ

Page 118, ll. 7–28: **66** to **72**. *Aues aquaticæ*—cf. Pliny, XVIII. 87, p. 614: 'The like [approaching tempests] may bee said of Froggs, when they crie more than their custome is; and of Seamews [*fulicæ*] also, when they gaggle in a morning betimes extraordinarily: semblably, the Cormorants, Gulls, Mallards, and Ducks [*mergi anatesque*], when they keepe a proining of their feathers with their bills, foreshew wind: and generally, when you see other water-foule to gather and assemble togither and then combat one with another, or Cranes [*grues*] make hast to flie into the midland parts of the maine. The Cormorants and Gulls flying from the sea and standing lakes, and Cranes soaring aloft in the aire still, without any noise, doe put in comfort of a faire and drie season: so doth the Howlat [*noctua*] also, when shee cries chuit in rainie weather: but if it be then faire and drie, we shall be sure to have foule tempests for it afterwards: Ravens [*corviquæ*] crying one to another as if they sobbed or yexed therewith, and besides clapping themselves with their wings, if they continue this note, doe portend winds; but if they give over between-whiles, and cut their crie short as if they swallow it backe againe, they presage raine and wind both. Iacke-dawes [*graculi*], if it be late ere they returne from their reliefe abroad, foretoken cold and hard weather; so doe the white-birds [*albæ aves*] when they assemble and flock togither, as also when land-foule (and the crow [*cornix*] especially) keepe a crying against the water, clapping their wings, washing also and bathing themselves. If the swallow [*hirundo*] flie low and so neare the water, that she flap the same oftentimes with her wings, it is a signe of raine and foule weather. Semblably, all other birds that nestle in trees, if they seeme to make many flights out, but returne quickely againe to their nests. Moreover, if Geese [*anseres*] hold on a continuall gaggling out of all order untunably, a man may guesse no better by them, no more than he can of the Heron [*ardea*] which he seeth heavie and sad upon the sands.'

## P6ʳ⁻ᵛ

Page 118, ll. 29–32: **73**. *Delphini* tranquillo—cf. Pliny, *loc. cit.*: 'the dolphins playing and disporting themselves in a calme water, doe certainely fore-shew wind comming from that coast whence they fetch these friskes and gambols: contrariwise, if they fling and dash water this way and that way, the sea at that time being rough and troubled, it is an infallible signe of a calme and of faire weather toward [*Delphini tranquillo mari lascivientes flatum, ex qua veniunt parte: item spargentes aquam iidem turbato, tranquillitatem*].'

## P6ᵛ

Page 118, ll. 33–5: **74**. Ingruente *vento*—cf. Pliny, XVIII. 88, p. 614: 'the dull and heavie oxen holding up their nose and muzzles, snuffe and smell into the aire, yea and keepe a licking against the haire [toward raine] [*Et boves cœlum olfactantes, seque lambentes, contra pilum*].'

ll. 36–7: **75**. Paulò ante *Ventum*—cf. Pliny, XI. 24, p. 324, who does not concur: 'there bee many presages and prognostications depend upon these Spiders . . . In faire and cleare weather, they neither spin nor weave: upon thicke and cloudie daies, they be harde at worke: and therefore many cobwebs be a signe of raine [*Sunt ex eo & auguria . . . iidem sereno non texunt, nubilo texunt: ideoque multæ araneæ imbrium sunt*].'

l. 38: **76**. Ante pluuiam—cf. *SS*, Gι<sup>r</sup>, Iι<sup>v</sup> (*SEH*, II, pp. 395, 418).

**P7<sup>r</sup>**

Page 120, ll. 3–6: **77** and **78**. *Trifolium* inhorrescere—cf. Pliny, *Historia*, XVIII. 89, p. 614: 'It is knowne for certaine, that the Claver-grasse or hearbe Trefoile [*Trifolium*] will looke rough against a tempest, yea and the leaves thereof will stand staring up as if it were afraid thereof . . . whensoever you see at any feast the dishes and platters wherein your meat is served up to the bourd, sweat or stand of a dew, and leaving that sweat which is resolved from them, either upon dresser, cupbourd, or table, be assured that it is a token of terrible tempests approaching.'

**P7<sup>v</sup>**

Page 120, l. 13: Prognostica Pluuiarum, *sub titulo suo*—the history of rains is item 10 in *CHP*; see *OFB*, XI, p. 474.

**P7<sup>v</sup>–P8<sup>r</sup>**

Page 120, ll. 15–22: *Si animum homines*—what Bacon is recommending here is what he elsewhere calls *experientia literata*; see *DAS*, 2H2<sup>v</sup>–2I3<sup>v</sup> (*SEH*, I, pp. 623–33).

**P8<sup>r–v</sup>**

Page 120, ll. 23–8: *Folles*—for Aeolus see *cmts* on C7<sup>r</sup>–C8<sup>r</sup>, p. 386 above. For Bacon's views on the vacuum see, for example, *OFB*, XI, 574–5, 579–80.

**P8<sup>v</sup>**

Page 120, ll. 31–5: De cœnaculorum—cf. I2<sup>v</sup>, p. 64 above.

**P8<sup>v</sup>–Qι<sup>r</sup>**

Page 122, ll. 1–7: *Flatus* in *Microcosmo*—for Bacon on the macrocosm-microcosm analogy, see *DAS*, F2<sup>v</sup>, 2B4<sup>r–v</sup> (*SEH*, I, pp. 460, 587). Also see *HDR* in *OFB*, XIII, pp. 76, 120.

**Qι<sup>r</sup>**

Page 122, ll. 8–10: In destillatione—cf. *OFB*, XIII, pp. 98–100.

**Qι<sup>v</sup>**

Page 122, ll. 11–18: *Ventus* factus ex *Nitro*—for Bacon on gunpowder, see *OFB*, XI, pp. 332–4, 378–80, 388, 564–5.

**Q1ᵛ–Q2ʳ**

Page 122, ll. 19–23: Latet spiritus flatuosus—Bacon mentions two violent primary explosives here: mercury fulminate and fulminating gold. Mercury fulminate, $Hg(ONC)_2$, can be prepared by mixing alcohol with a solution of mercury in concentrated nitric acid. It is very sensitive to friction and shock. Fulminating gold (gold hydrazine ($AuHN.NH_2$)), a mixture of gold dissolved in aqua regia (itself a mixture of nitric and hydrochloric acids) with an ammoniacal solution, will, when dried as an olive-green powder, explode on heating, scratching, and percussion.

**Q2ʳ**

Page 122, ll. 25–6: Motus Ventorum—cf. Aristotle, *Problems*, XXVI. 36, pp. 96–7: 'There are indeed some similarities to that which appears to happen in the case of water; for water when it is travelling down a steep slope flows more quickly than when it is stagnant in a plain and on the level, and it is somewhat similar with winds; for on peaks and in high places the air is always in motion, but in hollow places it is often calm and there is no wind.'

**Q2ʳ–Q3ᵛ**

Page 122, ll. 27–34: Venti magni—cf. *HV*, I4ᵛ–I5ᵛ, L4ʳ⁻ᵛ, L7ᵛ–M1ʳ (pp. 66–8, 84, 86–8 above).

**Q3ᵛ–Q4ʳ**

Page 124, ll. 9–20: Canones mobiles—it is quite possible that these *canones* were in existence before Bacon collected much of the data that went into *HV*, and that the selection of data was guided by the *canones*; see *OFB*, XIII, p. 302. *Canones* in this sense means a general rule or principle; *mobiles* in classical Latin can mean variable or changeable; in neo-Latin, removable or even dismissable. In the seventeenth-century 'official' English translation of *HVM* (Gibson, no. 154), the phrase is translated as moveable canons.

**Q4ᵛ–Q6ᵛ**

Page 124, l. 24–p. 126, l. 16: Ventorum, *qui fiunt*—see H7ʳ–I1ᵛ, pp. 60–2 and *cmt* thereon pp. 397–8 above, for the notion that surcharge of the air is a principal cause of winds.

**Q7ʳ⁻ᵛ**

Page 126, ll. 22–6: Tam Vapores, quam Exhalationes—see *cmt* on *HV*, I5ᵛ–I6ᵛ, p. 399 above.

**Q8ᵛ–R3ᵛ**

Pages 128, l. 9–p. 130, l. 33: Charta humana—this section is not of a kind explicitly mentioned in the *norma* of the *HNE*, which *norma* refers only to 'incentives to practice' and the need to produce records of things 'lying within human power' (*HNE*, C5ᵛ–C6ʳ, p. 17 above). No 'charta humana' appears in *HVM* but there is one (entitled *Optativa cum Proximis*) in *HDR*; see *OFB*, XIII, pp. 166–8.

**R1ʳ**

Page 128, l. 14: *Sed Consule de eo*—see N2ᵛ–N7ʳ, pp. 98–102 above.

l. 17: Experimenta *nostra*—see N8ᵛ–O2ʳ, pp. 104–6 above.

**R1ᵛ**

Page 128, ll. 22–5: *Hùc multa pertinent*—see O2ʳ–P7ᵛ, pp. 106–20 above.

**R2ʳ⁻ᵛ**

Page 130, ll. 1–2: *id quo vsus videtur* Columbus—see E7ᵛ–E8ʳ, F4ʳ⁻ᵛ, L7ʳ, pp. 36–40, 42, 86 above.

Page 130, ll. 3–6: *Itidem de vbertate*—Diogenes Laertius, *Lives of eminent philosophers*, 2 vols. (Loeb Classical Library), London and Cambridge, Mass., 1925, repr. 1972, I, pp. 26–7: 'Hieronymus of Rhodes in the second book of his *Scattered Notes* relates that, in order to show how easy it is to grow rich, Thales, foreseeing that it would be a good season for olives, rented all the oil-mills and thus amassed a fortune.' For Bacon on Thales, see esp. *OFB*, VI, pp. 144–6, 398. For the inquiry into articles 29 and 30, see F7ᵛ–G8ʳ, pp. 44–54 above.

**R2ᵛ**

Page 130, ll. 14–15: Historias Agriculturæ, & Medicinæ—for these histories see *CHP* (nos. 115, 58–66) in *OFB*, XI, pp. 480–2.

**R2ᵛ–R3ʳ**

Page 130, ll. 17–18: *quædam superstitiosa, &* Magica—see, for instance, James I, *Dæmonologie*, in *The workes of the most high and mighty prince, Iames*, Robert Barker and John Bill: London, 1616, pp. 117–18: witches 'can raise stormes and tempests in the aire, either vpon Sea or land, though not vniuersally, but in such a particular place and prescribed bounds, as GOD will permit them so to trouble. Which likewise is very easie to be discerned from any other naturall tempests that are Meteores, in respect of the sudden and violent raising thereof, together with the short induring of the same.'

**R3ʳ**

Page 130, ll. 24–5: *Experimentum* Plinij—see *Historia naturalis*, II. 48, p. 24: whirlwinds at sea may perhaps be calmed 'yet a smal matter is the remedie for it, namely, the casting of vinegre out against it as it commeth, which is of nature most cold [*tenui remedio aceti in advenientem effusi, cui frigidissima est natura*].'

l. 27: Puteo *in* Dalmatiâ—see *ibid.*, II. 45, p. 21; and *cmt* (p. 396 above) on H4ᵛ, p. 58 above.

**R4ʳ–R8ᵛ**

Page 132, ll. 3–11: Historia Densi & Rari—this history exists in a manuscript version, and in a version published by William Rawley in 1658. For editions of these witnesses see *OFB*, XIII. For *cmts* on this preface see *ibid.*, pp. 269–70.

**R8ᵛ–S1ᵛ**

Page 132, l. 12–p. 134, l. 3: Historia Grauis & Leuis—among the Bacon manuscripts inherited by Isaac Gruter from Sir William Boswell, there existed a sketch of this history, now lost; see *OFB*, VI, p. lxxxiv *passim*. Topics of inquiry on this subject appeared in *DAS* (2K1ᵛ–2K3ᵛ (*SEH*, I, pp. 635–9)), no doubt in part as a retrospective preparation for this history. The history was to have been about what the Aristotelians called natural motion, i.e. rectilinear motion in the sublunary realm whereby bodies were carried to the natural places of their dominant element, e.g. heavy, earthy bodies would descend towards the centre of the Earth; light, fiery bodies would ascend to the sphere of fire beneath and contiguous to the inner surface of the Moon's sphere. For the idea of a body falling through a hole bored in the Earth, cf. *OFB*, XI, 316. Bacon seems to have thought that natural motion (of heavy bodies at least) was what the scholastics called uniformly difform, i.e. falling bodies descended at a uniformly accelerating rate. For Bacon's criticisms of natural motion see *OFB*, XI, pp. 104–6, 392–4, 563–4. Hydrostatics—Bacon probably had Archimedes in mind; the modern writer may have been Simon Stevin (1548–1620), or possibly Galileo. As for Gilbert, Bacon alludes to one of his favourite Gilbertian ideas, namely that if a heavy body were projected far enough away from the centre of the Earth it would cease to fall. For this idea and Bacon's criticisms of Gilbert, see *OFB*, VI, pp. 156, 402; XI, pp. 88–90, 316, 513.

**S2ᵛ–S4ᵛ**

Page 134, ll. 4–33: Historia Sympathiæ—for Bacon's criticisms of the natural magicians and their understandings of sympathy and antipathy, see *OFB*, XI, pp. 138–40, 526–7. For *magnalia naturæ* see *OFB*, XI, p. 166; *NA*, g3ʳ (*SEH*, III, pp. 167–8). We are given hints in *SS* (both factual and critical) as to the contents of the proposed history; for instance he attacks the 'vaine Dreames of *Sympathies*' on the subject of raising storms by burning chameleons on the tops of houses (N4ʳ (*SEH*, II, p. 461)); he reports that a vine will grow towards a stake stuck in the ground some way off (Q3ᵛ (*SEH*, II, p. 489) ); he writes at length on the sympathies and antipathies of plants (Q4ᵛ–R3ʳ (*SEH*, II, pp. 493–98)), with a sideswipe at 'Some of the Ancients, and likewise diuers of the Moderne Writers, that haue laboured in *Naturall Magicke*', who note 'a *Sympathy*, between the *Sunne, Moone*, and some Principall *Starres;* And certaine *Herbs*, and *Plants.* And so they haue denominated some *Herbs Solar*, and some *Lunar;* And such like Toyes put into great Words' (R2ᵛ (*SEH*, II, p. 496)); and he gives sceptical attention to the idea of the weapon-salve (2I2ᵛ (*SEH*, II, p. 645)). In *SS* Bacon draws on Porta and Fracastoro for some of his experiments on sympathy and antipathy.

**S5ʳ–S8ʳ**

Page 136, l. 1–p. 138, l. 10: Historia Sulphuris—this preface contains a cardinal doctrine of Bacon's cosmology, a doctrine which he repeats in many other works: that the universe from the Earth's surface to the highest heavens is occupied by two families of things, the sulphur and mercury quaternions. The sulphur quaternion embraces sulphur, oil and oily bodies, terrestrial fire, and sidereal fire. The mercury quaternion comprises mercury, water and watery bodies, air, and celestial ether. Paracelsian in origin, the theory of the two quaternions provides the underpinning for a strikingly anti-Aristotelian conception of the structure of the universe, and of motion (of tide, wind, comets, and planets) within it. The Paracelsian third principle, salt, is here as elsewhere rejected *qua* principle but is taken to be an intermediate combining the natures of mercury and sulphur. Consequently salt is a 'rudiment', first beginning, or primitive analogue of life, for a living organism's defining feature is its possession of animate spirit which itself is an intermediate compounded of higher members of the two quaternions, viz. fire and air; see *OFB*, VI, pp. xlvi–xlviii, liv–lvii; XI, pp. 435–6; XIII, pp. 188–90. For the Paracelsian view of the elements as wombs, see *OFB*, VI, p. xlv. For the Aristotelian *materia prima*, see *ibid.*, pp. 206, 419. For the term 'adiaphora' cf. *ibid.*, p. 214, l. 20; *OFB*, XI, p. 98, l. 31.

S8ᵛ–T7ʳ
Page 138, ll. 11–17: Historia Vitæ & Mortis—for *cmts* see immediately below.

# COMMENTARY ON
## *HISTORIA VITÆ & MORTIS*

**A2$^{r-v}$**

Page 142, ll. 1–18: Cvm *Historiam Vitæ*—Bacon here refers back to the list given at the beginning of *HNE* (A4$^r$, p. 6 above) on which *HVM* stands last among the six natural histories that he meant to publish to illustrate his natural-historical programme.The history which should have appeared after *HV* was *HDR*, but that was not published in Bacon's lifetime; see *OFB*, XIII, pp. xx–xxi, xxx–xxxvi.

**A3$^r$**

Page 142, ll. 14–18: Etsi enim nos *Christiani*—cf. similar sentiments at the beginning of the discussion of prolongation of life in *DAS*, 2D3$^r$ (*SEH*, I, p. 598): 'Licet enìm *Mundus* Homini Christiano, ad *Terram promissionis* contendenti, tanquàm *Eremus* sit; tamen in *Eremo* ipso proficiscentibus, Calceos & Vestes (Corpus scilicet nostrum, quod Animæ loco Tegminis est) minùs atteri, Gratiæ Diuinæ Munus quoddam æstimandu*m*.'

**A4$^v$–A5$^r$**

Page 144, ll. 10–15: *Quinetiam* Discipulus Amatus—for instances of longevity among Christ's disciples and the Church Fathers, see *HVM*, I8$^r$–K2$^r$, pp. 210–12 above. The 'old law' is of course the Old Testament dispensation, superseded (as Christians believe) by the New Testament dispensation.

**A5$^{r-v}$**

Page 144, ll. 18–20: turbâ Medicorum—echoing words used in *DVM* (see *OFB*, VI, p. 272, ll. 10–23), Bacon attacks the common run of physicians who accepted the traditional theory of ageing (see *cmts* on *HVM*, N4$^v$–N7$^r$ (pp. 238–40 below)). For Bacon's reservations about chemical medicines, see *HVM*, N2$^v$, p. 236 above, and *cmt* thereon, p. 442 below. Also see *DAS*, 2D3$^v$ (*SEH*, I, p. 599); *SS*, 2L4$^r$ (*SEH*, II, pp. 670–1).

**A6$^r$**

Page 144, l. 26: Atriolum *commune*—see *HVM*, 2A5$^r$–2C1$^r$, pp. 328–40 above, and *cmts* thereon pp. 452–3 below.

**A6$^r$–A7$^r$**

Page 144, l. 30–p. 146, l. 9: *Quod reparari potest*—this echoes very closely the opening words of *DVM*; see *OFB*, VI, p. 270, ll. 4–21. The vestal flame was the undying fire (*ignis inextinctus*) maintained in the shrine to the goddess Vesta by the Vestal virgins as a guarantee of Rome's permanence; see *OCD*: **Vesta**. For radical moisture and natural heat, see *cmts* (p. 441 below) on *HVM*, N4$^v$–N7$^r$ (pp. 236–8 above).

### A7$^r$–B1$^r$

Page 146, ll. 12–16: *Sed reuerâ hoc fit*—on the questions of reparable and less reparable parts, and of the conspiracy of inanimate spirits with the air, *HVM* once again echoes *DVM*; (see *OFB*, VI, pp. 272–6). For the torture of Mezentius see *DVM* in *OFB*, VI, pp. 352–4. Also see Virgil, *Aeneid*, VIII, ll. 483–8: 'quid memorem infandas cædes, quid facta tyranni, | effera? di capiti ipsius generique reservent! | mortua quin etiam iungebat corpora vivis, | componens manibusque manus atque oribus ora, | tormenti genus, et sanie taboque fluentis | complexu in misero longa sic morte necabat.' The same allusion appears in *SS*, D1$^r$ (*SEH*, II, p. 364).

### B1$^r$–B2$^v$

Page 148, ll. 4–21: *Itaque duplex debet esse* Inquisitio—cf. *DVM*, *OFB*, VI, pp. lvi, 274. The distinction between vital and inanimate spirits is crucial. Bacon claims that although living beings were distinctive in possessing vital spirits, they had to be considered as if they were inorganic things to the extent that their tangible parts embodied inanimate spirits (cf. *DAS*, 2D4$^{r-v}$ (*SEH*, I, p. 600) ). Vital spirits restrain the inanimate, but eventually the latter prevail and destroy their hosts. Possession of vital spirit entails consumption of the body and so the need for nourishment. Thus it is wrong to concentrate on the distinctive qualities of living things (cf. *DVM*, *OFB*, VI, pp. 352–4). Living organisms had to be considered as cradles of the vital spirit *and* as entities undergoing the same processes of decay as lifeless things.

### B3$^r$

Page 150, ll. 4–6: De Naturâ *Durabilis*—see *HVM*, B7$^v$–C7$^v$, pp. 154–9 above, and *cmts* thereon pp. 420–1 below.

### B3$^{r-v}$

Page 150, ll. 7–12: De *Desiccatione*—see *HVM*, C8$^r$–E6$^r$, pp. 162–76 above, and *cmts* thereon pp. 421–3 below.

### B4$^r$

Page 150, ll. 20–1: De *Animalium*—see *HVM*, E6$^v$–G6$^v$, pp. 176–92 above, and *cmts* thereon pp. 423–8 below.

### B4$^{r-v}$

Page 150, ll. 23–7: Quoniam verò—see *HVM*, G7$^r$–H2$^v$, pp. 192–6 above, and *cmts* thereon pp. 428–9 below.

### B4$^v$–B5$^v$

Page 152, ll. 1–19: De *Longæuitate*—see *HVM*, H2$^v$–M7$^v$, pp. 198–232 above, and *cmts* thereon pp. 429–40 below.

**B5ᵛ**

Page 152, ll. 20–1: De *Medicinis*—see *HVM*, M7ᵛ–N3ʳ, pp. 232–7 above, and *cmts* thereon p. 440 below.

**B5ᵛ–B6ʳ**

Page 152, ll. 22–5: De *Signis*—see *HVM*, L1ᵛ–M7ᵛ, pp. 218–32 above, and *cmts* thereon pp. 437–40 below.

**B6ʳ**

Page 152, ll. 29–32: *Particulares* Intentionum—see *HVM*, N4ʳ–2A3ᵛ, pp. 236–326 above, and *cmts* thereon pp. 441–52 below.

**B7ʳ**

Page 154, ll. 13–15: De *Articulo*—see *HVM*, 2A5ʳ–2C1ʳ, pp. 328–40 above, and *cmts* thereon pp. 452–5 below.

**B7ʳ⁻ᵛ**

Page 154, ll. 16–23: *Postremò*—see *HVM*, 2C1ᵛ–2C5ᵛ, pp. 340–6 above, and *cmts* thereon pp. 455–7 below.

**B7ᵛ–C2ʳ**

Page 154, l. 26–p. 158, l. 14: *Metalla*—for other reflections on durability, see *OFB*, XI, pp. 302–5, 418–21; *OFB*, XIII, pp. 52–5.

**C1ᵛ**

Page 156, l. 22: *Dentibus Equi Marini*—not, as *SEH* (V, p. 223) has it, 'teeth of the sea-horse', but 'teeth of the hippopotamus' or possibly 'walrus.'

**C2ʳ⁻ᵛ**

Page 158, ll. 8–14: At *Æqualitas*—for inequality as a source of dissolution see *HVM*, 2E4ᵛ, p. 360 above, and *OFB*, VI, pp. lxi, 320.

**C2ᵛ–C3ʳ**

Page 158, ll. 16–20: *Loco* Assumpti—for these doctrines, see *HVM*, 2C7ᵛ–2D1ʳ, pp. 346–8 above, and *cmt* thereon p. 457–8 below. For the water–air, and fire–oil relationships, which are fundamental assumptions of Bacon's theory of matter, see *cmt* pp. 416–17 above on *HNE*, S5ʳ–S8ʳ.

**C4ʳ–C6ʳ**

Page 160, l. 1–p. 162, l. 9: *Herbæ* quæ habentur—on lasting of plants and trees. Pliny XVI. 44, pp. 494–7.

**C4ᵛ–C5ᵛ**

Page 160, l. 16–p. 162, l. 2: *Frutices* & *Arbores*—for the longevity of some of the trees noticed here, see Pliny, XVI. 44, pp. 494–7.

**C5ᵛ**

Page 160, l. 33: *Ferula*—birch because Bacon has switch in mind (i.e. as in birching).

**C6ᵛ–C7ᵛ**

Page 162, ll. 11–28: *Benè admodùm notauit* Aristoteles—see Aristotle, *On length and shortness of life*, in *On the soul, parva naturalia, on breath* (Loeb Classical Library), trans. W. S. Hett, Cambridge, Mass. and London, 1936, rev. edn. 1957, repr. 1995, pp. 406–9: 'It is among plants that the greatest longevity [μακροβιώτατα] is found, rather than among animals . . . For plants are always being reborn; that is why they last so long. For some shoots are always new, while others grow old. The same is true of their roots . . . and so the tree continues, part dying and part being born . . . for the plant possesses potential [δυνάμει] root and stalk in every part of it. Consequently there is always proceeding from it a new part besides that which is growing old (a case of virtual longevity), just as when slips are taken.' Also see *cmt* (pp. 422–3 below) on D7ʳ–D8ʳ below.

**C8ʳ**

Page 162, l. 32–p. 164, l. 3: *Ignis,* & *Calor*—on the contrary actions of fire see *HDR, OFB,* XIII, p. 162. The quotation, from Virgil, *Eclogues,* VIII. 80, is accurate and also used in *DAS,* V2ᵛ (*SEH,* I, p. 550) and *HDR* (*loc. cit.*) to the same effect.

**C8ᵛ–D1ʳ**

Page 164, ll. 9–14: *Aer,* præcipuè apertus—for air and age as desiccants and their effects on inanimate spirits, see *HDR, OFB,* XIII, p. 118, and *cmts* thereon *ibid.,* p. 294.

**D1ʳ–D2ʳ**

Page 164, ll. 22–9: *Frigus* omnium maximè propriè exiccat—cf. *NO* in *OFB,* XI, pp. 424–6. For March winds see *HV,* G6ᵛ, p. 52 above. 'Olybano': olibanum, says *OED,* is a resin of the trees of the genus *Boswellia,* and used against catarrhs. For spirits of wine 'cooking' eggs and toast, cf. *NO, OFB,* XI, p. 218, ll. 15–18.

**D2ʳ⁻ᵛ**

Page 166, ll. 3–9: *Pulueres* desiccant—cf. *SS,* X2ᵛ, 2I3ᵛ (*SEH,* II, pp. 534, 649–50). For the dispersal of moisture on gems, etc., cf. *NO, OFB,* XI, p. 382.

**D2ᵛ–D3ᵛ**

Page 166, ll. 10–30: *Granaria*—in *NO* (*OFB,* XI, p. 420) the granaries are located in northern, not eastern, Germany. For the granaries of Cappadocia and Thrace see Pliny, XVIII. 73, pp. 603–4: 'the best and most assured way to preserve corne, is in caves or vaults under the ground, which in Latine be called Siri [*siros*], as the practise is in Cappadocia and Thracia. In Ægypt [*sic*] and Barbarie [*Hispania & Africa*], above all things they looke to this, That their garners stand upon a drie ground: and how drie soever the floore be, yet they lay a course of chaffe [*palea*] underneath betweene it and the corne.'

**D3ᵛ–D4ᵛ**

Page 166, l. 31–p. 168, l. 23: *Fructus*—for the preservation of fruit in pots, see *SS*, O2ʳ (*SEH*, II, p. 467): 'It were a profitable *Experiment*, to preserue *Orenges*, *Limons*, and *Pomgranates*, till Summer; For then their Price will be mightily increased. This may be done, if you put them in a Pot or Vessell, well couered, that the *Moisture* of the *Earth* come not at them; Or else by putting them in a *Conseruatorie* of *Snow*. And generally, whosoeuer will make *Experiments* of *Cold*, let him be prouided of three Things; A *Conseruatorie* of *Snow*; A good *large Vault*, twenty foot at least vnder the Ground; And a *Deepe Well.*' Also see Giovan Battista della Porta, *Natvral magick*, for Thomas Young and Samuel Speed: London, 1658 [English trans. of *Magiæ naturalis libri XX*, Naples, 1589], pp. 122–6. Also cf. *SS*, D3ʳ, N1ʳ (*SEH*, II, pp. 370, 454).

**D5ʳ⁻ᵛ**

Page 168, ll. 24–6: In *Vino, Oleo*—cf. *SS*, 2D2ᵛ (*SEH*, II. p. 598): 'Take a *Stocke-Gilly-Flower*, and tye it gently vpon a Sticke, and put them both into a *Stoope Glasse*, full of *Quick-siluer*, so that the *Flower* be couered: Then lay a little *Weight* vpon the Top of the *Glasse*, that may keepe the Sticke downe; And looke vpon them after foure or fiue daies; And you shall finde the *Flower* Fresh, and the *Stalke* Harder, and lesse *Flexible*, than it was. If you compare it with another *Flower*, gathered at the same time, it will be the more manifest. This sheweth, that *Bodies* doe preserue excellently in *Quick-siluer*; And not preserue only, but, by the *Coldnesse* of the *Quick-siluer*, *Indurate.*' Also see Della Porta, *Natvral magick*, pp. 122–39.

**D6ʳ**

Page 170, ll. 3–9: *Fila Candelarum*—cf. *SS*, O1ʳ–O2ʳ (*SEH*, II, pp. 465–6).

**D7ʳ–D8ʳ**

Page 170, l. 19–p. 172, l. 7: *Inteneratio Desiccati*—on softening of desiccated parts cf. *OFB*, VI, pp. 308–10. Also cf. *SS*, C4ᵛ–D1ʳ (*SEH*, II, pp. 363–4); 'There is an Excellent Obseruation of *Aristotle*; That a great Reason, why Plants (some of them) are of greater Age, than *Liuing Creatures*, is, for that they yearely put forth new Leaues, and Boughes; whereas *Liuing Creatures* put forth (after their Period of Growth,) nothing that is young, but Haire and Nailes; which are Excrements, and no Parts. And it is most certaine, that whatsoeuer is young, doth draw Nourishment better, than that which is Old; And then (that which is the Mystery of that Obseruation) young *Boughes*, and *Leaues*, calling the Sap vp to them; the same Nourisheth the *Body*, in the Passage. And this we see notably proued also, in that the oft Cutting, or Polling of *Hedges*, *Trees*, and *Herbs*, doth conduce much to their Lasting. Transferre therefore this Obseruation to the Helping of Nourishment in *Liuing Creatures*. The Noblest and Principall Vse whereof is, for the *Prolongation* of *Life*; *Restauration* of some Degree of *Youth*; and *Inteneration* of the *Parts*. For certaine it is, that there are in *Liuing Creatures* Parts that Nourish, and Repaire easily; And Parts that Nourish and repaire hardly, And you must refresh, and renew those that are easie to Nourish, that the other may be refreshed, and (as it were) Drinke in Nourishment, in the Passage. Now wee see that *Draught Oxen*, put into good Pasture, recouer the Flesh of young Beefe; And Men after long Emaciating Diets, wax plumpe, and fat, and almost New: So that you may surely conclude, that the frequent and wise Vse of those *Emaciating Diets*, and of *Purgings*; And perhaps of some kinde of *Bleeding*; is a principall Meanes of *Prolongation* of *Life;* and *Restoring* some Degree of *Youth.*' Also see *HVM*, 2A1ʳ⁻ᵛ, pp. 324–6 above. Guaiac (guaiacum officinale) was a standard treatment for syphilis; see, for instance, Ulrich von Hutten, *Of the vvood called guaiacum, that healeth the Frenche pockes . . .,* Thomas Berthelet: London, 1539.

### D8ᵛ–E6ʳ
Page 172, l. 16–p. 176, l. 21: Desiccatio—this is a classic statement of Bacon's theory that all disintegration arises from an *actio triplex* of inanimate spirit. This *actio* is discussed at length in *OFB*, VI, pp. lx–lxv. Most of the examples of contraction of parts given in *HVM* appeared in *DVM*; see *OFB*, VI, p. 276, l. 26–p. 278, l. 29. For Bacon's views on the vacuum see, for example, *OFB*, XI, 574–5, 579–80. Bacon thought of putrefaction as a process which could result in the formation of imperfect creatures; see *OFB*, XIII, pp. 231, 281. Also see *SS*, 2H2ᵛ–2H3ʳ (*SEH*, II, pp. 638–9).

### E7ʳ
Page 176, ll. 29–33: *Neque quæ Concomitantia*—Bacon nevertheless regards these concomitants as key variables as far as longevity is concerned; see, for instance, *HVM*, G3ʳ–G4ᵛ, pp. 188–90 above.

### E7ʳ⁻ᵛ
Page 178, ll. 1–6: *Hominis Æuum . . . omnium superat*—for these concomitants of life expectancy see *HVM*, G4ʳ–G6ʳ, pp. 190–2 above. Also see Aristotle, *On length and shortness of life*, pp. 400–3.

### E7ᵛ–E8ʳ
Page 178, ll. 7–13: *Elephas*—Pliny, *Historia naturalis*, VIII. 10, p. 197: 'The common sort of men thinke, that they goe with young ten yeeres: but *Aristotle* saith, that they goe but two yeeres, and that they breed but once . . . also that they live commonly by course of nature 200 yeers, and some of them 300.' Also see Aristotle, *History of animals*, 3 vols. (Loeb Classical Library), ed. and trans. A. L. Peck and D. M. Balme, London and Cambridge, Mass., 1965–91, bk. VIII, p. 391. For elephant's gestation period see *ibid.*, bk. VI, pp. 332–3: 'the female carries her young, as some say, for eighteen months; others say for three years.' For a summary of gestation periods in animals see *SS*, 2C1ᵛ (*SEH*, II, p. 584–5).

### E8ʳ
Page 178, ll. 14–16: *Leones*—Pliny, *Historia naturalis*, VIII. 16, p. 201: 'Mine author *Aristotle* saith moreover, that they live verie long; and he prooveth it by

this argument, That many of them are found toothles for very age [*quod plerique dentibus defecti reperiantur*].' Aristotle does not say anything of the kind in *History of animals*, see bk. VI, p. 341, for instance.

ll. 17–19: *Vrsus*—cf. Aristotle, *History*, bk. VI, p. 339: 'The she-bear is pregnant for thirty days'; Pliny (VIII. 36, p. 215) agrees.

**E8ᵛ**

Page 178, ll. 23–4: *Camelus*—Bacon's figures are the same as in Pliny, VIII. 18, p. 205; and as in Aristotle, *History of animals*, bk. VI, pp. 332–3.

ll. 25–31: *Equi*—Pliny VIII. 42, p. 222: horses take five to six years to reach adulthood, and 'Vixisse equum septuagintaquinque annos proditur.' Pliny VIII. 43, p. 223, gives thirty years as the lifespan of the ass.

**F1ʳ**

Page 178, l. 32–p. 180, l. 2: *Ceruorum*—Aristotle, *History of animals*, bk. VI, pp. 336–7: 'Stories are told of its longevity, but none of them has been established as true; besides the period of gestation and the swift growth of the fawns do not suggest that it is a long-lived animal.' On collared stags see Pliny, VIII. 32, p. 214: 'It is generally held and confessed, that the Stagge or Hind live long: for an hundred yeer after *Alexander* the Great, some were taken with golden collars about their necks, overgrowne now with haire and growne within the skin [*Vita cervis in confesso longa, post centum annos aliquibus captis, cum torquibus aureis, quos Alexander Magnus addiderat, adopertis jam cute, in magna obesitate*].'

**F1ʳ⁻ᵛ**

Page 180, ll. 3–7: *Canis*—Pliny, VIII. 40, p. 220, gives sixty days for gestation of dogs. Aristotle, *History of animals*, bk. VI, pp. 310–11, gives sixty to seventy-two days according to breed. He also notes (pp. 314–15) that the male Laconian hound lives ten years, with 'the bitch about twelve; bitches of other breeds for the most part live about fourteen or fifteen years, some as many as twenty.'

**F1ᵛ–F2ᵛ**

Page 180, ll. 8–12: *Bos*—Aristotle says that cows' gestation period is nine to ten months, their lifespan fifteen years, and 'Some bulls live twenty years or even more, if they are physically sound'; see *History of animals*, bk. VI, pp. 316–17. For sheep, see Pliny, VIII. 47, p. 226: 'The ramme and ewe both, are fit for generation from two yeeres of age upward untill they come to nine, and some also untill they be ten yeers old.' For sheep and goats also see Aristotle *History of animals*, bk. VI, pp. 308–9: 'A goat lives eight years, a sheep ten, though most of them do not live so long, except that bell-wethers (flock-leaders) may live to fifteen.' Bacon agrees with Pliny (VIII. 51, pp. 229–30) about pigs. For cats see Aelian, *On animals*, trans. A. F. Scholfield, 3 vols. (Loeb Classical Library) London and Cambridge, Mass., 1958–9, II, p. 47: 'The Tom-cat is extremely lustful, but the Female cat is devoted to her kittens and

tries to avoid sexual intercourse with the male, because the semen which he ejaculates is exceedingly hot and like fire, and burns the female organ.' For the cat's lifespan see Aristotle, *History of animals*, bk. VI, pp. 344–5: 'they live for about six years.'

### F3<sup>r</sup>

Page 182, ll. 3–4: *Aues* optimè tectæ—in Bacon's view protection from the air offered by feathers arises from their ability to stop inanimate spirit conspiring destructively with the air; see Introduction, p. l above.

### F3<sup>v</sup>

Page 182, ll. 12–15: De *Auium* Generatione—see Aristotle, *Generation of animals* (Loeb Classical Library), trans. A. L. Peck, London and Cambridge, Mass., 1953, pp. 203–5: 'In what sense are we to say that these eggs [i.e. wind-eggs] are alive? We cannot say that they are alive in the same sense as fertile eggs, for in that case an *actual* living creature would hatch out from them . . . and that is why the male is required to take its share in the business . . . wind-eggs become fertile if the male treads the female within a certain period'; also see *ibid.*, p. III.

### F4<sup>r</sup>–F5<sup>v</sup>

Page 182, ll. 20–5: *Aquila*—Aristotle says that eagles are long-lived, but that increasing curvature of the bill with age causes eagles to starve to death; see *History of animals*, VIII, pp. 298–303. He does not mention bill shedding. But see Augustine, *In Psalmum CII. 9*, in *Patrologia Latina*, vol. 37: 'Et quidem renovatur et juventus aquilæ, sed non ad immortalitatem . . . Dicitur aquila, cum senectute corporis pressa fuerit, immoderatione rostri crescentis cibum capere non posse. Pars enim rostri ejus superior, quæ supra partem inferiorem aduncatur . . . Hoc ei facit vetustas . . . Itaque modo quodam naturali in mensura reparandæ quasi juventutis, aquila dicitur collidere et percutere ad petram ipsum quasi labium suum superius, quo nimis crescente edendi aditus clauditur: atque ita conterendo illud ad petram excutit, et caret prioris rostri onere, quo cibus impediebatur. Accedit ad cibum, et omnia reparantur: erit post senectutem tamquam juvenis aquila; redit vigor omnium membrorum, nitor plumarum, gubernacula pennarum, volat excelsa sicut antea, fit in ea quædam resurrectio.' For Pliny on eagles see *Historia naturalis*, X. 3–5, pp. 271–3. Gesner (following Isidore and Cælius) says that vultures live for up to a century; but says nothing about the longevity of hawks and kites; see Gesner, *Historiæ animalivm liber III. Qui est de auium natura. . .* Frankfurt: Andreas Cambierus, 1604, pp. 611–14, 785. For the swan and goose, see *ibid.*, p. 375; 'Cygnus ab Aristotele εὐγήρως dicitur, quod ad senectutem satis commodè & non grauem sibi perueniat, quod & Ælianus scribit. Vitam ei longissimam esse audio: ita vt vulgò vel trecentesimum annum attingere credatur, quod mihi verisimile non est'; p. 143; 'Anseris vita perlonga est. vidimus anserem domesti-cum, qui annos sexaginta excessit, Albertus.' The stork: as *SEH* (II, p. 126, n. 2)

remarks, the story about their dislike of Thebes is told of swallows in Pliny; see *Historia naturalis*, X. 24, p. 283. Pliny speaks of storks in the previous chapter (pp. 281–3), and so Bacon's notes from the *Historia* may have been inadvertently telescoped here.

### F5ᵛ–F6ʳ

Page 184, ll. 12–16: Nam de *Phœnice*—even Pliny is reluctant to credit many of the stories relating to the phoenix; see *Historia naturalis*, X. 2, p. 271.

### F6ʳ

Page 184, ll. 21–2: *Pauo*—Aristotle says that peacocks live for twenty-five years and get their brilliant plumage after three; see *History of animals*, bk. VI, pp. 252–3. Pliny (X. 20, p. 279) follows Aristotle.

### F6ʳ

Page 184, ll. 27–9: *Palumbes*—according to Aristotle wood-pigeons live for twenty-five, thirty, or even forty years; see *History of animals*, bk. VIII, pp. 254–7. Pliny says thirty or forty years; see *Historia naturalis*, X. 35, p. 290. As for doves and turtle-doves, Pliny (*loc. cit.*) gives eight years.

### F6ᵛ

Page 184, ll. 30–2: *Perdices*—Aristotle says that partridges live for up to fifteen years; see *History of animals*, VIII, pp. 256–7; Pliny says sixteen; see *Historia naturalis*, X. 34, p. 290.

ll. 35–7: *Passer*—Pliny, X. 36, p. 290: 'the Sparrow is but short lived, howbeit as leacherous as the best . . . The cocke Sparrow (by report) liveth but one yeare . . . The Hens live somewhat longer.' Cf. Aristotle, *History of animals*, bk. VIII, pp. 256–7. Gesner discusses the sparrow and the linnet; he lists many authorities as to the short life of the former, but says nothing of the lifespan of the latter; see *Historiæ animalivm liber III*, pp. 590–1, 645. Gesner says nothing about the lifespan of the blackbird; see *ibid.*, pp. 602–6.

### F6ᵛ–F7ʳ

Page 186, ll. 1–2: De *Struthionibus*—Pliny speaks of the ostrich but does not say how long-lived it is; see *Historia naturalis*, X. 1, p. 270. Following Apion, Gesner says that the ibis is very long-lived, but says nothing about the longevity of ostriches; see *Historiæ animalivm liber III*, pp. 569, 739–46.

### F7ᵛ

Page 186, ll. 17–19: *Delphini*—Pliny, *Historia naturalis*, IX. 8, p. 238: 'Young Dolphins come very speedily to their growth, for in ten yeeres they are thought to have their full bignesse: but they live thirtie yeeres, as hath been knowne by the experience and triall in many of them, that had their taile cut for a marke when they were young, and let go again.'

### F8ʳ

Page 186, ll. 23–5: Deprehensæ sunt—Pliny (*Historia naturalis*, IX. 53, p. 266) mentions this 60-year-old fish but does not say that it was a lamprey.

Pliny has Hortensius weeping for a lamprey of uncertain age just a little later (IX. 55, p. 267). The story of Crassus is to be found in many classical sources but none mentions the lamprey's age.

ll. 26–8: *Lucius*—Conrad Gesner speaks of the pike's voracity but says nothing about its longevity; see *Historia animalivm liber IV. Qui est de piscium & aquatilium animantium natura*, Frankfurt: Andreas Cambierus, 1604, pp. 502–3.

ll. 29–30: At *Carpio, Abramus*—Gesner discusses these but only cites authorities on the eel's lifespan; see *ibid.*, p. 45: 'Vita anguillis nonnullis vel ad septem octoqu*e* annos protrahitur; Aristoteles. Octonis viuunt annis, Plinius. Platina non rectè scripsit octoginta. Nos seruauimus anguillam in piscina per quindecim annos, Niphus.' Gesner says that the salmon grows quickly but nothing about how long it lives; see *ibid.*, pp. 824–32.

### F8ʳ⁻ᵛ

Page 186, ll. 33–5: moles *Balænarum*—Pliny speaks of these (*Historia naturalis*, IX. 6, pp. 236–7) but says nothing about their lifespan. Gesner is as reticent; see *Historiæ animalivm liber IV*, pp. 114–21, 193–216. Holland keeps *orca* in his English. Did he or Bacon know what it was?

### F8ᵛ

Page 186, l. 34: de *Phocis*—see Gesner, *Historiæ animalivm liber IV*, p. 707: 'Hoc animal triginta annis, quod cognitum est amputatis eorum caudis, viuit, Incertus.' Gesner's authority is 'Incertus.' For nicking of dolphins' tails see *cmts* on F7ᵛ, p. 186 above.

l. 34: *Porcis Marinis*—Pliny, *Historia naturalis*, XXXII. 5, p. 436: 'The Sea-swine or Porpuis, hath prickie finns upon his backe, and those are counted among other venomous things that the sea yeeldeth, putting them to much paine that are wounded or hurt thereby: but what helpe therefore? surely the verie muddie slime that gathereth about the bodie of the same fish, is the onely remedie [*Inter venena piscium sunt porci marini spinae in dorso, cruciatu magno laesorum. remedio est limus ex reliquis piscium eorum corporis*].' Also see Gesner, *Historiæ animalivm liber IV*, p. 756–7.

Page 186, l. 36–p. 188, l. 2: *Crocodili*—Pliny, VIII. 25, pp. 208–9: 'Ordinarily, he is above eighteene cubites in length. The female laieth egs as big as geese doe: and sitteth ever upon them out of the water . . . There is not another creature againe in the world, that of a smaller beginning, groweth to a bigger quantitie . . . he is a great and greedie devourer.' Also see Gesner, *Historiæ animalivm liber IV*, p. 304.

### G1ʳ

Page 188, ll. 4–6: *Normam aliqua*m Longæuitatis—as usual Bacon turns apparent weakness to advantage. He is saying that the deficiencies in his own natural history are nothing other than a minor symptom of the very malaise his programme is designed to cure, namely the lack of reliable data for the reconstruction of the sciences, which is itself an argument for collecting natural history on an unprecedented scale.

### G1ʳ–G4ᵛ

Page 188, l. 7–p. 190, l. 17: *Inueniuntur Plures*—protection from the air is, in fact, protection from the destructive conspiracy between air and inanimate spirit, for which see Introduction, p. l above. See *HVM*, L6ʳ⁻ᵛ, pp. 222–4 above for male and female contributions to the substance of offspring.

### G4ᵛ–G5ʳ

Page 190, ll. 18–26: *Spirituum Sedes*—I think this is the only time that Bacon claims that inanimate as well as animate spirits are concentrated in the head. On the theoretical background to this see *cmts* (p. 458 below) on 2D3ʳ–2D5ʳ.

### G6ʳ

Page 192, ll. 3–4: Bilis *enim*—for bile and curly hair see *HVM*, F1ᵛ–F2ᵛ, p. 180 above; for the 'fact' that youths have more bilious humours than old men, see *HVM*, 2C2ᵛ, p. 342 above.

### G6ʳ⁻ᵛ

Page 192, ll. 8–13: *In omni* Corruptibili—cf. *NO* in *OFB*, XI, pp. 380–3.

### G7ʳ–G8ʳ

Page 192, ll. 26–33: *Alimentum* erga *Alimentatum*—for the Feüillans see *OFB*, XI, pp. 438, 584. Also see *SS*, B3ʳ⁻ᵛ (*SEH*, II, pp. 347–8): 'The *French,* (which put off the Name of the *French Disease,* vnto the Name of the *Disease* of *Naples,*) doe report, that at the Siege of *Naples,* ther were certaine wicked Merchants, that Barrelled vpp *Mans flesh,* (of some that had been, lately slaine in *Barbary*) and sold it for *Tunny;* And that vpon that foule and high Nourishment, was the Originall of that *Disease.* Which may well be; For that it is certaine, that the *Caniballs* in the *West Indies,* eate *Mans flesh;* And the *West Indies* were full of the Pockes when they were first discouered: And at this day the *Mortallest poisons,* practised by the *West Indians,* haue some Mixture of the Bloud, or Fatt, or Flesh of *Man:* And diuers Witches, and Sorceresses, aswell amongst the *Heathen,* as amongst the *Christians,* haue fedd vpon *Mans flesh,* to aid (as it seemeth) their Imagination, with High and foule Vapours.' On cannibals see the famous essay of Michel de Montaigne, *The complete essays,* ed. and trans. M. A. Screech, Penguin Books: London, 1991, p. 235: 'the master of each captive summons a great assembly of his acquaintances . . . they . . . hack at him with their swords and kill him. This done, they roast him and make a common meal of him, sending chunks of his flesh to absent friends. This is not as some think done for food—as the Scythians used to do in antiquity—but to symbolize ultimate revenge.'

### G8ʳ⁻ᵛ

Page 192, l. 34–p. 194, l. 11: Quò *Alimentum*—cf. *DAS*, 2H2ᵛ–2H3ʳ (*SEH*, I, p. 624).

**H1ʳ**

Page 194, ll. 20–21: Atque *Sanguis* in Venis—there is little evidence as to Bacon's views on blood motion, but what little there is suggests that he deviated little (if at all) from conventional Galenic teaching. Blood is made from the fresh and sweet part of the food; the blood of the veins feeds the body's tissues (see *SS*, 2A3ʳ (*SEH*, II, p. 566)); the spirituous blood of the arteries refreshes and repairs the vital spirits (*DAS*, 2E4ʳ (*SEH*, I, p. 606)); the heart's systolic and diastolic actions are examples of motion of trepidation (*NO* in *OFB*, XI, pp. 410–13; *OFB*, XIII, pp. 202–3), but Bacon says nothing about the purpose of these actions, nothing about the heart's function in blood motion, and nothing about the relationship between venous and arterial blood.

**H2ᵛ–H3ʳ**

Page 196, ll. 12–22: Ante *Diluuium*—for Walter Ralegh on the longevity of the Patriarchs, and interpretative problems associated therewith, see *The history of the world*, 2nd edn., William Stansby: London, 1617, H1ᵛ–H3ᵛ, 6T6ʳ⁻ᵛ. Bacon's statements here link up with views presented in *HVM*, L1ᵛ–L3ʳ, pp. 218–20 above, and *cmts* thereon p. 437 below.

**H3ʳ–H4ʳ**

Page 196, l. 24–p. 198, l. 5: *Abraham* Annos—Abraham: Genesis 25: 7: 'these are the days of the years of Abraham's life which he lived, a hundred threescore and fifteen years.' Isaac: Genesis 35: 28: 'the days of Isaac were a hundred and fourscore years.' Jacob: Genesis 47: 28: 'the whole age of Jacob was a hundred forty and seven years.' Ishmael: Genesis 25: 17: 'And these are the years of the life of Ishmael, a hundred and thirty and seven years: and he gave up the ghost and died, and was gathered unto his people.' Sarah: Genesis 23: 1: 'And Sarah was a hundred and seven and twenty years old: these were the years of the life of Sarah.' Joseph: Genesis 50: 26: 'Joseph died, being a hundred and ten years old.' Levi: Exodus 6: 16: 'and the years of the life of Levi were a hundred thirty and seven years.' Levi's son: Exodus 6: 18: 'And the years of the life of Kohath were a hundred thirty and three years.' The grandson of Levi: Exodus: 6: 20: 'and the years of the life of Amram were a hundred and thirty and seven years.'

**H4ʳ–H5ʳ**

Page 198, ll. 6–22: *Moses*—Moses: Deuteronomy 34: 7: 'Moses was a hundred and twenty years old when he died: his eye was not dim, nor his natural force abated.' Aaron: Numbers 33: 39: 'Aaron was a hundred and twenty and three years old when he died in mount Hor.' Phineas: Judges 20: 27–9: 'And the children of Israel inquired of the LORD (for the ark of the covenant of God was there in those days, and Phinehas, the son of Eleazar, the son of Aaron, stood before it in those days), saying, "Shall I yet again go out to battle against the children of Benjamin my brother, or shall I cease?" And the LORD said, "Go up; for tomorrow I will deliver them into thy hand".' Joshua: Joshua 24: 29: 'And it came to pass after these things, that Joshua the son of Nun, the servant of the LORD, died, being a hundred and ten years old.' Caleb: this estimate seems to be based on Numbers 13: 30–1 and 14: 24. Ehud: Judges 3: 26–30: 'And Ehud escaped . . . And it came to pass when he was come, that he blew a trumpet in the mountain of Ephraim, and the children of Israel went down with him from the mount . . . And he said unto them, "Follow after me: for the LORD hath delivered your enemies the Moabites into your hand" . . . And they slew of Moab at that time about ten thousand men . . . So Moab was subdued that day under the hand of Israel. And the land had rest fourscore years.'

## H5<sup>r–v</sup>

Page 198, ll. 23–36: *Iob*—Job: Job 42: 16: 'After this lived Job a hundred and forty years, and saw his sons, and his sons' sons, even four generations.' Eli: 1 Samuel 4: 15: 'Now Eli was ninety and eight years old, and his eyes were dim, that he could not see.' Elisha: 2 Kings 2: 23: 'And he went up from thence unto Beth-el: and as he was going up by the way, there came forth little children out of the city, and mocked him, and said unto him, "Go up, thou bald-head, go up, thou bald-head".' For Isaiah see Isaiah 1: 1: 'The vision of Isaiah the son of Amoz, which he saw concerning Judah and Jerusalem in the days of Uzziah, Jotham, Ahaz, and Hezekiah, kings of Judah.' Walter Ralegh reckoned that the last three of these kings reigned for seventy-one years; see *The history of the world*, pp. 556–97. Isaiah is held to have prophesied the coming of Christ; see, for instance, Matthew 3: 3; 12: 17–21; 15: 7–9; and, above all, Isaiah 7: 14: 'Therefore the Lord himself shall give you a sign: behold, a virgin shall conceive, and bear a son, and shall call his name Immanuel.'

## H5<sup>v</sup>–H6<sup>v</sup>

Page 200, ll. 1–12: *Tobias Senior*—Tobias the Elder: Tobit 14: 11: 'he gave up the ghost in the bed, being a hundred and eight and fifty years old; and he buried him honourably.' Tobit the Younger: Tobit 14: 14: 'And he died at Ecbatane in Media, being a hundred and seven and twenty years old.' For the weeping Jews, see Ezra 3: 12: 'But many of the priests and Levites and chief of the fathers, who were ancient men, that had seen the first house, when the foundation of this house was laid before their eyes, wept with a loud voice.' For Simeon see Luke 2: 25–32; I do not know Bacon's source for Simeon's age. Anna: Luke 2: 36–7: 'And there was one Anna, a prophetess, the daughter of Phanuel, of the tribe of Aser; she was of a great age, and had lived with a husband seven years from her virginity . . . a widow of about fourscore and four years, which departed not from the temple, but served God with fastings and prayers night and day.'

## H7<sup>r–v</sup>

Page 200, ll. 25–9: *Numa*—for Numa see Lucian of Samosata, *ΛΟΥΚΙΑΝΟΥ ΣΑΜΟΣΑΤΕΩΣ ΦΙΛΟΣΟΦΟΥ ΤΑ ΣΩΖΟΜΕΝΑ*. Luciani . . . opera, Paris, 1615, p. 913: 'Numa Pompilius Romanorum Regum longè felicissimus, ac Deorum cultui deditissimus, supra octoginta annos vixisse traditur.' For Corvinus see Pliny, *Historia naturalis*, VII. 48, p. 181: '*M. Valerius Corvinus* lived 100 yeers complet: between his first Consulate and sixt, were 46 yeers: he tooke his seat upon the yvorie chaire of estate, and was created a magistrate Curule 21 times; and no man ever besides him, so often.'

**H7v–H8r**

Page 200, l. 30–p. 202, l. 5: *Solon*—Diogenes Laertius, *Lives*, I, pp. 62–3: 'He died in Cyprus at the age of eighty.' For the seven wise men see *OCD*, and *OFB*, XI, pp. 124, 523. Epimenides: Pliny, *Historia naturalis*, VII. 48, p. 180. Xenophanes: Diogenes Laertius, *Lives*, II, p. 427. Also see *OFB*, XI, pp. 114, 521.

**H8r–v**

Page 202, ll. 6–7: *Anacreon*—Lucian, p. 918: 'Sophocles poëta Tragicus deuorato vuæ acino strangulatus est, cùm egisset annos quinque & nonaginta . . . Porrò Anacreon poëta Lyricus vixit annos quinque & octoginta.'

**H8v**

Page 202, ll. 11–16: *Artaxerxes*—Artaxerxes: Lucian, p. 915: 'Artaxerxes Mnemonis cognomentu*m* sortitus, contra quem Cyrus frater exercitum duxit, imperium Persarum obtinens, morbis consumptus decessit, annos sex & octoginta natus, sed vt Dinon scribit, quatuor & nonaginta.' Agesilaus: see Plutarch, *Lives*, p. 631: 'the sea being rough in the winter quarter, he died by the way . . . after he was foure score and foure yeares old.'

**H8v–I1r**

Page 202, ll. 17–20: *Gorgias*—see Pliny, *Historia naturalis*, VII. 48, p. 181; Diogenes Laertius (*Lives*, II, p. 373) cites Apollodorus to the effect that Gorgias lived 109 years. Also see Cicero, *De senectute*, V. 13: 'Leontinus Gorgias centum et septem complevit annos neque umquam in suo studio atque opere cessavit. Qui, cum ex eo quaereretur, cur tam diu vellet esse in vita, "Nihil habeo," inquit, "quod accusem senectutem." Praeclarum responsum et docto homine dignum.'

**I1r–v**

Page 202, ll. 20–36: *Protagoras*: Diogenes Laertius, *Lives*, II, pp. 466–9: 'According to some his death occurred, when he was on a journey, at nearly ninety years of age, though Apollodorus makes his age seventy.' Isocrates: Lucian, p. 917: 'Isocrates sex & nonaginta annos natus sermonem scripsit Panegyricum.' Democritus: Diogenes Laertius, *Lives*, II, pp. 452–3: 'he let his life go from him without pain, having then, according to Hipparchus, attained his one hundred and ninth year.' Lucian, says (p. 916) that Democritus lived to be 104. Bacon's remarks on Democritus echo words elsewhere in his oeuvre, see for instance *OFB*, VI, pp. 118, 388–9; XI, pp. 512, 570. Diogenes Synopeus: Diogenes Laertius, *Lives*, II, pp. 78–9: 'Diogenes is said to have been nearly ninety years old when he died.' Bacon derived his remarks on him from *ibid.*, II, pp. 72–7. Zeno Cittieus: see *ibid.*, II, pp. 138–41: 'he was ninety-eight when he died and had enjoyed good health without an ailment to the last. Persaeus, however, in his ethical lectures makes him die at the age of seventy-two.' For Zeno's character see *ibid.*, II, pp. 110 ff. As for Seneca, see *OCD*: **Annaeus Seneca** (2).

I1ᵛ–I2ʳ

Page 202, l. 36–p. 204, l. 8: *Plato*—see Diogenes Laertius, *Lives*, I, pp. 276–8: 'Apollodorus in his *Chronology* fixes the date of Plato's birth in the 88th Olympiad, on the seventh day of the month of Thargelion, the same day on which the Delians say that Apollo himself was born. He died, according to Hermippus, at a wedding feast, in the first year of the 108th Olympiad, in his eighty-first year. Neanthes, however, makes him die at the age of eighty-four.' For Bacon on Plato's intellectual style, see *OFB*, VI, p. 208; XI, p. 102. Theophrastus: see Diogenes Laertius, *Lives*, I, p. 487: 'He died at the age of eighty-five, not long after he had relinquished his labours.' For Bacon's sketch of Theophrastus, cf. *ibid.*, I, pp. 482 ff. Carneades: see *ibid.*, I, p. 441: 'According to Apollodorus in his *Chronology*, he departed this life in the fourth year of the 162nd Olympiad at the age of eighty-five years.'

I2ʳ⁻ᵛ

Page 204, ll. 8–11: *Orbilius*—see Suetonius, *De illvstribvs grammaticis*, in *Historiæ Romanæ scriptores Latini veteres*, 2 vols., Yverdon, 1621, II, p. 84: 'Fuit autem naturæ acerbæ, non modò in antisophistas, quos omni sermones lacerauit, sed etiam in discipulos, vt Horatius significat *plagosum* eum appellans . . . Vixit prope ad centesimum ætatis annum, amissa iampridem memoria.'

I2ᵛ–I3ʳ

Page 204, ll. 12–16: *Q. Fabius Maximus*—see Pliny, *Historia naturalis*, VII. 48, p. 181: 'As for *Q. Fabius Maximus* (a Romane) he continued Augure 63 yeeres.' Masinissa: see Lucian, p. 915: 'Masinissa verò rex Mauritaniæ nonaginta annos vixit.'

I3ʳ⁻ᵛ

Page 204, ll. 23–34: *Terentia*—Pliny, *Historia naturalis*, VII. 48, p. 181: '*Ciceroes* wife *Terentia*, out-lived her husband, until she was 103 yeers old . . . *Luceia* a common vice in a play, followed the stage and acted thereupon 100 yeeres. Such another vice that plaied the foole and made sport betweene whiles in enter-ludes, named *Galeria Copiola*, was brought againe to act her feats upon the stage, when *Cn. Pompeius* [*recte C. Poppæo*] and *Q. Sulpitius* were Consuls, at the solemne plaies vowed for the health of *Augustus Cæsar* the Emperor, when she was in the 104 yeere of her age: the first time that ever she entred the stage to shew proofe of her skill in that profession, was 91 yeers before, and then she was brought thither by *M. Pomponius* an Aedile of the Commons, in the yeere that *C. Marius* and *Cn. Carbo* were Consuls. And once againe *Pompeius* the Great, at the solemne dedication of his stately Theatre, trained the old woman to the stage for to make a shew, to the wonder of the world.' *SEH*, II, p. 138, n. 3, explains some of the confusion in the Pliny/Bacon account.

**I3ᵛ–I4ʳ**
Page 204, l. 35–p. 206, l. 8: Fuit & alia *Mima*—Junia Tertulla's death is recorded by Tacitus, but he does not mention her age, see *The annals of imperial Rome*, trans. Michael Grant, Penguin: Harmondsworth, rev. edn. 1971, p. 156.

**I4ᵛ**
Page 206, ll. 9–28: Memorabilis est *Annus*—Pliny, *Historia naturalis*, VII. 49, p. 182, 'in the last taxation, numbring, and review [*census*] . . . that was taken under the *Cæsars Vespasians* the father and the sonne, both Emperours and Censors . . . At Parma, three men were found that lived a hundred and twentie yeeres: at Brixels [*Brixelli*], one that was an hundred twentie and five yeeres old: at Parma moreover two, an hundred and thirtie yeeres of age: at Plaisance [*Placentiæ*], one elder by a yeere: at Faventia, there was one women an hundred thirtie and two yeers old: at Bononie [*Bononiæ*], *L. Terentius* the sonne of *Marcus*, and at Ariminum *M. Aponius*, reckoned each of them an hundred and fiftie yeeres. *Tertulla* was knowne to be an hundred thirtie and seven yeeres old. About Plaisance, there is a towne situate upon the hills, named Velleiacium, wherein six men brought a certificate, that they had lived an hundred yeeres apeece: foure likewise came in with a note of an hundred and twentie yeeres: one, of an hundred and fourteen, namely *M. Mutius* sonne of *Marcus* surnamed *Galerius Fœlix*. But because we will not dwell long in a matter so evident and commonly confessed: in the review taken of the eigth [*sic*] region of Italie, there were found in the toll, foure and fiftie persons of an hundred yeeres of age: 57, of an hundred and ten: two, of an hundred and five and twentie: foure, of an hundred and thirtie: as many that were 135 or 137 yeeres old: and last of all, three men of an hundred and fortie.'

**I5ᵛ**
Page 206, ll. 29–36: *Ne res*—a concise statement of Bacon's method in *HVM*, i.e. to try to correlate longevity with biographical circumstance; see Introduction, pp. lii–liii above.

**I5ᵛ–I8ʳ**
Page 208, ll. 1–37: Inter *Imperatores*—for Augustus see Suetonius, *The historie of tvvelve Cæsars, emperovrs of Rome* . . . trans. Philemon Holland, London, 1606, p. 87: 'Hee died in that very bed-chamber where his father OCTAVIUS left his life before him . . . being 76 yeeres olde wanting five and thirtie daies.' For Tiberius see *ibid.*, p. 120: 'But being kept backe, as well by tempest as the violence of his disease that grew still uppon him hee died soone after in a village bearing the name *Luculliana*, in the 78. yeare of his age' (the passage goes on to recount suspicions regarding the manner of his death). See *ibid.*, pp. 74–5 for Musa's cure. For Tiberius the 'slow-jawed', see *ibid.*, p. 98: 'these words of AUGUSTUS were over heard by the Chamberlaines. *Miserum populum*

*Romanum qui sub tam lentis maxillis erit. O unhappie people of Rome, that shall be under such a slow paire of chawes.'*

**I8ʳ⁻ᵛ**

Page 210, ll. 1–9: Iam à *Sæcularibus*—for Symeon see Eusebius, *History*, in *The avncient ecclesiasticall histories . . . wrytten in the Greeke tongue by . . . Eusebius, Socrates, and Euagrius,* trans. Meredith Hanmer, Thomas Vautroullier: London, 1577, p. 53: '*Symeon* the sonne of *Cleopas . . .* suffered martyrdome being a hundreth, and tvventy yeare olde, vnder *Traian* the Emperour, and *Atticus* the Consul.' For Polycarp see *ibid.,* p. 65.

**I8ᵛ–K2ʳ**

Page 210, ll. 15–25: *Aquila*—for Aquila and Priscilla see Acts 18: 18–19; Romans 16: 3–5; 2 Timothy 4: 19. For Paul the Hermit see Hieronymus (i.e. St Jerome), *Vita Pauli,* in *Catalogvs scriptorvm ecclesiasticorvm,* P. Brubach: Frankfurt, 1549, p. 91: 'Sed ut ad id redeam, unde digressus sum, cum iam centum tredecim annos beatus Paulus uitam coelestem ageret in terris, & nonagenarius in alia solitudine Antonius moraretur (ut ipse asserere solebat): hæc in mentem eius cogitatio incidit, nullum ultra se perfectum monachum in eremo consedisse.'

**K2ʳ–K3ʳ**

Page 212, ll. 3–20: At *Papæ*—Bacon could have found all the facts here from Bartolomeo (Sacchi) de Platina, *Historia B. Platinae de vitis pontificvm Romanorvm,* Bernard Walther: Cologne, 1600. This work has a supplement (3A1ʳ–3D4ᵛ) begun by Onuphrius Panvinius (1529–68), and later brought up to 1592, which lists 235 popes. To these Bacon would have added the six popes who reigned from 1592 to 1623 (Innocent IX to Urban VIII) to bring the total to 241. As *SEH* (II, p. 143, n. 1) points out, Bacon meant to speak of John XXII, not XXIII. For the lives of these two see Platina, P6ᵛ, Q7ᵛ. For Gregory XII, see *ibid.,* S5ʳ. For Paul III who lived to be 81, see *ibid.,* 2C2ᵛ. For Paul IV, who died aged 83, see *ibid.,* 2D8ᵛ. For Gregory X, see *ibid.,* P5ʳ.

**K3ʳ–K4ʳ**

Page 212, l. 21–p. 214, l. 1: Quæ sequentur . . . *Arganthonius*—Pliny, VII. 48, pp. 180–1: '*Anacreon* the poët maketh report, that *Arganthonius* king of the Tartessians, lived 150 yeeres: and *Cynaras* likewise king of the Cyprians, ten yeeres longer . . . *Hellancius* hath written, that among the Epians in Ætolia, there be some that continue ful two hundred yeeres: and with him accordeth *Damases*; adding moreover, that there was one *Pictoreus* among them, a man of exceeding stature, mightie and strong withall, who lived 300 yeeres. *Ephorus* testifieth, that ordinarily the KK. of Arcadia were 300 yeers old ere they died. *Alexander Cornelius* writeth of one *Dando,* a Sclavonian, that lived 500 yeers. *Xenophon* in his treatise of old age, maketh mention of a king of the Latines, or as some say, over a people upon the sea coasts, who continued alive 600 yeeres: and because he had not lied loud enough already, he goeth on still and saith,

That his sonne came to 800 . . . *Mutianus* witnesseth, that in Tempsis (for so they call the crest or pitch of the mountaine Tmolus) folke lived ordinarily 150 yeeres.' It seems that Bacon derived the reading Litorius for Pictoreus from a marginalium (Lictoreus) in Dalechamps's magnificent edition; see C. *Plinii Secvndi historiae mvndi libri XXXVII . . . nouissimè verò laboriosis obseruationibus conquista & solerti iudicio pensitata, Iacobi Dalecampii*, B. Honoré: Lyon, 1587, p. 159. That Xenophon's king ruled over the Latines comes from another marginalium on the same page. For Arganthonius also see Lucian, p. 913.

### K4$^{r-v}$

Page 214, ll. 1–10: *Essæorum* apud *Iudæos*—see Flavius Josephus, *De bello Ivdaico, libri septem*, Sebastian Gryphius: Lyon, 1528, pp. 137–43. For Apollonius of Tyana, see Philostratus, *Historiæ de vita Apollonij Tyanei libri octo*, Egidius Gourbinus: Paris, 1555, p. 566: 'Aliorum verò quidam octogesimum, alij nonagesimum, plerique etiam centesimum ætatis annum excessisse perhibent, & corpus habuisse senio quidem confectum, rectum tamen, & ad omnes vitæ vsus idoneum, ita vt aspectu penè iucundior esset, quàm in iuuenta fuerat. Inest enim, & senibus propter reuerentiam pulchritudo sua, quæ in illo maximè vigebat, idque ex eius imaginibus quæ Tyanis in templo constitutæ sunt, licet intueri.'

### K4$^{v}$–K5$^{r}$

Page 214, ll. 11–24: *Q. Metellum . . . Appium Cæcum*—see Valerius Maximus, *Dictorvm factorvmqve memorabilivm libri nouem*, Wechel: Frankfurt, 1601, pp. 282–3: 'Cvivs vitæ [i.e. M. Valerius] spatium æquauit Metellus, quartoque anno post consularia imperia, senex admodum pontif. max. creatus, tutelam cæremoniarum per duos & viginti annos, neque ore in votis nuncupandis hæsitante, neque in sacrificiis faciendis tremula manu gessit . . . Appii vero æuum clade metirer, quia infinitum numerum annorum orbatus luminibus exegit, nisi quatuor filios, & quinque filias, plurimas clientelas, rem denique publicam hoc casu grauatus fortissime rexisset. Quinetiam fessus iam viuendo, lectica se in Curiam deferri, iussit, vt cum Pyrrho deformem pacem prohiberet. Hunc cæcum aliquis nominet, à quo patria quod honestum erat, per se parum cernens, coacta est peruidere?' Also see *OCD*: **Pyrrhus**. For Appius' speech to the Senate, see Plutarch, *Lives*, p. 407: 'But *Appius Claudius*, a famous man, who came no more to the Senate, nor dealt in matters of state at all by reason of his age, and partly because he was blind . . . caused his seruants to carie him . . . vnto the Senate doore . . . The Senate made silence to honour the comming in of so notable and worthie a personage: and he so soone as they had set him in his seate, began to speake in this sort: Hitherunto with great impatience (my Lords of ROME) haue I borne the losse of my sight, but now, I would I were also as deafe as I am blind, that I might not (as I do) heare the report of your dishonourable consultations determined vpon in Senate, which tend to supuert the glorious fame and reputation of ROME.'

**K5ᵛ**

Page 214, ll. 24–8: *Perpenna*—see Pliny VII. 48, p. 181: '*M. Perpenna*, and of late daies, *L. Volusius Saturninus*, out-lived all those Senators who sat in counsell with them when they were Consuls, & whose opinions they were wont to aske. As for *Perpenna*, when he died, he left but 7 of those Senators alive, whom he had either chosen or re-elected in his Censorship: & he lived himself 98 yeers.'

**K5ᵛ–K6ʳ**

Page 214, ll. 28–33: *Hiero*—see Valerius Maximus, p. 283: 'Siciliæ rector Hiero ad nonagesimum annum peruenit'; also see Lucian, p. 914: 'Hieron quoque Syracusanorum tyrannus duos & nonaginta annos natus, morbis tandem absumptus obiit, cùm regnasset annos septuaginta.' Statilia: see Pliny, *Historia*, VII. 48, p. 181: '*Statilia* a noble ladie of Rome, in the time of *Claudius* the Emperor, was knowne to be 99 yeers of age . . . *Clodia* wife to *Ofilius*, went beyond her, and saw 115 yeers, and yet she had in her youth 15 children.'

**K6ʳ**

Page 214, l. 33–p. 216, l. 11: *Xenophilus*—see Lucian, p. 916: 'Xenophilus Musicus . . . vltra quinque & centum annos Athenis vitam traduxit.' For Demonax see Lucian, pp. 556–7: 'Vixit annos fermè centum, nullo vnquam morbo implicitus, nulla affectus animi ægritudine, neminem vnquam facessendis negotiis molestauit, aut in ius productum accusauit . . . Aliquanto ante verò quàm moriebatur, interrogante quopiam, de sepultura verò quid imperas? Ne vos hæc cura remordeat, inquit, solicitos: etenim fœtor me sepeliet. Respondente autem illo, an non turpe foret & nefarium, si tanti viri corpus volucribus & canibus dilaniandum obiiceretur: Atqui nihil absurdum est, inquit, si mortuus quoque nonnullis animalibus vtilitatem aliquam allaturus sim.'

**K6ᵛ–K7ʳ**

Page 216, ll. 11–18: *Populus Indiæ*—see Pliny, VII. 2, p. 156: '*Isogonus* saith, that certaine Indians named Cyrni, live a hundred and fortie yeeres. The like he thinketh of the Æthyopian Macrobij, and the Seres: as also of them that dwel upon the mount Athos: and of these last rehearsed, the reason verily is rendered to be thus, because they feed of vipers flesh, and therfore is it that neither lice breed in their heads, nor other vermine in their cloths, for to hurt and annoy their bodies.' Also see Lucian, p. 912: 'Iam verò gentes quoque ætate omnium longissima vtentes, ex quibus sunt Seres, ad trecentos annos vsque viuere . . . & Chaldæos vltra centum annos egredi fama est, & eosdem hordeaceo pane vti, tanquam perspicacitatis pharmaco.'

**K7ᵛ–K8ʳ**

Page 216, l. 19–p. 218, l. 1: *Ouidius Senior*—see Ovid, *Tristia*, IV. 10, ll. 77–8: 'et iam complerat genitor sua fata novemque | addiderat lustris altera lustra novem.' Seneca: see *OCD*: **Annaeus Seneca** (1) and (2). For Aretinus' longevity, see J. B. Egnatius, *De exemplis illvstivm virorvm Venetae civitatis atqve aliarvm gentivm*, Venice, 1554. Egnatius, in chapter 12 (De senectute) of Book 8, pp. 280–1, gives the ages at death of Molinus (100 plus), Contarenus (96), and Donatus (84) in that order and without a break. Bacon reverses the order.

### K8ᵛ

Page 218, ll. 1–5: de *Cornaro Veneto*—this instance included, Luigi Cornaro (1464?–1566) is mentioned four times in *HVM*; see pp. 218, 228, 240, 262 above. George Herbert's selective translation of Cornaro's treatise appears at sigs. K6ʳ–M4ʳ of Leonard Lessius, *Hygiasticon: or, the right course of preserving life and health unto extream old age*, Cambridge, 1634. Cornaro got over ill heath caused by riotous living by eating only twelve ounces of food a day accompanied by fourteen ounces of grape juice.

ll. 6–10: *Guilielmus Postellus*—in 1611 Jacques Auguste de Thou said that Postel (1510–81) was nearly a hundred years old when he died; in 1604 Antoine du Verdier claimed that Postel lived for more than a hundred and ten years; see Marion L. Kuntz, *Guillaume Postel: prophet of the restitution of all things. his life and thought*, Martinus Nijhoff Publishers: The Hague, 1981, pp. 4–5. Also see François Secret, *Postel revisité: nouvelles recherches sur Guillaume Postel et son milieu*, S.É.H.A: Paris, and Archè: Milan, 1998.

### L1ʳ

Page 218, ll. 17–19: In *Hospitali Bethleem*—the notorious Bethlem or Bedlam Hospital for the insane which was located just beyond the walls of the City of London to the north of Bishopsgate.

### L1ᵛ

Page 218, ll. 20–4: Ætates . . . *Nympharum*—see *OFB*, XIII, pp. 288–9.

### L1ᵛ–L3ʳ

Page 218, l. 27–p. 220, l. 4: *Decursus Sæculorum*—this passage, which concludes with Bacon's rejection of decline theories, is of course fundamental. The aim of procuring longevity is not such a bright idea if people believe that human life is getting shorter through its implication in the irreversible and general decay of nature. A typical decline theorist was Godfrey Goodman; see *The fall of man, or the corrvption of natvre, proved by the light of our naturall reason*, Felix Kyngston: London, 1616, pp. 353–4, where Goodman even argues that modern humans are weaker than the castle and cathedral builders of the Middle Ages. For a counterblast to such reasoning see George Hakewill's *An apologie of the power and providence of God in the government of the world. Or an examination and censvre of the common errovr tovching natvres perpetvall and vniversall decay*, John Lichfield and William Turner: Oxford, 1627.

### L2ᵛ–3ʳ

Page 220, ll. 4–6: licet *Virgilius*—see *Georgics*, I, ll. 489–97; 'ergo inter sese paribus concurrere telis | Romanas acies iterum videre Philippi; | nec fuit indignum superis, bis sanguine nostro | Emathiam et latos Haemi pinguescere campos. | scilicet et tempus veniet, cum finibus illis | agricola incurvo terram molitus aratro | exesa inveniet scabra robigine pila, | aut gravibus rastris galeas pulsabit inanis, | grandiaque effossis mirabitur ossa sepulcris.'

## L3ᵛ

Page 220, ll. 17–25: *Regionibus Frigidioribus*—see Introduction, p. lvi above on the effect of cold on inanimate spirits. Pliny, VII. 2, p. 157: 'Artemidorus affirmeth, That in the Island Taprobana the people live exceeding long without any maladie or infirmitie of the bodie. *Duris* maketh report, That certaine Indians engender with beasts, of which generation are bred certaine monstrous mungrels, halfe beasts and halfe men. Also, that the Calingian women of India conceive with child at five yeeres of age, & live not above eight.'

## L4ʳ⁻ᵛ

Page 220, l. 32–p. 222, l. 9: *Loca Excelsa*—on Arcadia see *cmts* (pp. 430, 434 above) on H6ᵛ–H7ʳ, K3ʳ (p. 212 above). For Ætolia see Pliny VII. 48, pp. 180–1 and *cmt* on K3ʳ–K4ʳ, p. 434 above.

## L5ʳ

Page 222, ll. 14–17: *Regiones Particulares*—most of the places associated with longevity are mentioned above; see *HVM*, K6ᵛ, L1ʳ, L3ᵛ, L4ʳ, S1ᵛ, S2ʳ⁻ᵛ, pp. 214, 218, 220, 274, 276.

    ll. 18–23: Occulta est res—cf. *SS*, 2E2ᵛ (*SEH*, II, p. 605): 'But for the *Choice* of *Places*, or *Seats*, it is good to make Triall, not only of *Aptnesse* of *Aire* to corrupt, but also of the *Moisture* and *Drinesse* of the *Aire* . . . Wee see that there be some *Houses*, wherein *Sweet Meats* will relent, and *Baked Meats* will mould . . . All which, (no doubt,) are caused chiefly by the *Moistnesse* of the *Aire*, in those *Seats*. But because it is better to know it, before a *Man* buildeth his *House*, than to finde it after, take the *Experiments* following . . . Lay *Wooll*, or a *Sponge*, or *Bread*, in the *Place* you would trie, comparing it with some other *Places*; And see whether it doth not moisten, and make the *Wooll*, or *Sponge*, &c. more Ponderous . . . Because it is certaine, that in some *Places*, either by the *Nature* of the *Earth*, or by the *Situation* of *Woods*, and *Hills*, the *Aire* is more Vnequall, than in Others; And *Inequalitie* of *Aire* is euer an Enemy to *Health*; It were good to take two *Weather-Glasses*, Matches in all things, and to set them, for the same Houres of One day, in seuerall *Places*, where no *Shade* is, nor *Enclosures*; And to marke, when you set them, how farre the *Water* commeth; And to compare them, when you come againe, how the *Water* standeth then: And if you finde them *Vnequall*, you may be sure that the Place where the *Water* is lowest, is in the *Warmer Aire*, and the other in the *Colder*. And the greater the *Inequalitie* be, of the *Ascent*, or *Descent* of the *Water*, the greater is the *Inequalitie* of the *Temper* of the *Aire*.' For Bacon and calendar (weather) glasses, see *cmt* p. 400 above on *HV*, I8ᵛ, p. 72.

**L6ʳ–L8ʳ**

Page 222, l. 34–p. 224, l. 34: Vt *Series*—behind Bacon's ideas about effect of parental condition on the children they produce is his theory of matter, and in particular the action of vital and inanimate spirits. For Plato's opinion concerning the fitness of women, see *Republic*, in *The dialogues of Plato*, III, pp. 330–7. For the ages at which Spartan and Roman women were considered marriageable, see Plutarch, *Lives*, pp. 78–9; and *loc. cit.* for the following: 'the keeping of maidens to be married by *Numaes* order, was much straighter and more honorable for womanhood: and *Lycurgus* order hauing too much scope and liberty, gaue Poets occasion to speake, and to giue them surnames not very honest. As *Ibycus* called them *Phænomeridæ*: to say, thighe shewers: and *Andromanes*: to say, manhood'. 'Manhood' is an eccentric translation of, ultimately, ἀνδρομᾶνής or 'man-mad.'

**L8ʳ–M3ʳ**

Page 224, l. 35–p. 228, l. 17: *Candidiores Genis*—these characteristics are considered, for example, by Aristotle and G. B. della Porta, but neither of these authorities relates them to longevity; see Aristotle, *Physiognomics*, in *Minor works*, pp. 84–137, and Della Porta, *De humana physiognomonia*, Vico Equense, 1586.

**M3ʳ⁻ᵛ**

Page 228, ll. 19–23: De *Temporibus*—see *HVM*, B5ʳ, p. 152 above for Bacon's dismissal of astrological observations on times of nativity and length and shortness of life. He had no time for judicial astrology; see *DAS*, X1ʳ (*SEH*, I, p. 555): '*Fatalia* illa, quod Hora Natiuitatis aut Conceptionis Fortunam Fœtûs regat, Hora Incæptionis Fortunam Incæpti, Hora Quæstionis Fortunam Rei Inquisitæ, atque (vt verbo dicamus) *Doctrinas* de *Natiuitatibus, Electionibus,* & *Quæstionibus,* & istiusmodi Leuitates, maximâ ex Parte, nihil certi aut solidi habere, & Rationibus Physicis planè redargui & euinci, iudicamus.' In the discussion of prologation of life in *DAS*, 2D3ʳ–2E1ᵛ (*SEH*, I, pp. 598–602), he says not a word about astral influences.

**M3ᵛ–M5ʳ**

Page 228, l. 24–p. 230, l. 21: *Victus siue Diæta*—for Pythagorean diets see Diogenes Laertius, *Lives*, II, pp. 332–3, 336, 348–9. Also see *OCD*, **Pythagoras, Pythagoreanism**. For Cornaro's diet see *cmts* (p. 437 above) on *HVM*, K8ᵛ, p. 218 above. For Celsus' recommendations, see A. Cornelius Celsus, *On medicine in eight books, Latin and English*, trans. Alex. Lee, 2 vols., E. Cox: London, 1831, I, p. 25: 'Sanus homo, qui et bene valet, et suae spontis est, nullis obligare se legibus debet; ac neque medico, neque iatralipta egere. Hunc oportet varium habere vitae genus: modo ruri esse, modo in urbe, sæpiusque in agro; navigare, venari, quiescere interdum, sed frequentius se exercere: siquidem ignavia corpus hebetat, labor firmat; illa maturam senectutem, hic longam adolescentiam

reddit.' The first book of *De medicina* is (if the anachronism be allowed) very much in the spirit of *HVM*. For Bacon's good opinion of Celsus, also see *DAS*, 2C4$^v$, 2G3$^{r-v}$ (*SEH*, I, pp. 594, 617).

## M5$^{r-v}$

Page 230, ll. 22–30: *Vita Religiosa*—for Paul the Hermit see *HVM*, I8$^r$–K2$^r$ (p. 210 above) and for Simeon Stylites see *cmts* on R7$^v$–R8$^r$, pp. 446–7 below.

## M5$^v$–M7$^r$

Page 230, l. 31–p. 232, l. 13: *Huic proxima*—for longevity of learned men cf. *HVM*, R1$^v$–R2$^r$ (p. 266 above).

## M7$^{r-v}$

Page 232, ll. 17–24: De *Vitâ Militari*—for Corvinus and Agesilaus see *HVM*, H7$^{r-v}$, H8$^v$, and *cmts* thereon, p. 431 above. For Xenophon see Lucian, p. 917: 'Xenophon Grylli filius vixit vltra nonaginta annos.' For Camillus: see Plutarch, *Lives*, p. 156.

## M7$^v$–M8$^v$

Page 232, l. 27–p. 234, l. 2: Medicina *quæ habetur*—see *cmts* on *HVM*, O2$^r$–O3$^r$, p. 242 above.

## M8$^v$–N1$^v$

Page 234, ll. 3–12: *Aurum* triplici Formâ—cf. *cmts* on *HVM*, T1$^v$–T2$^v$, p. 449 below.

## N1$^r$–N2$^r$

Page 234, ll. 13–18: *Margaritæ*—on pearls, emeralds, and jacinth as medicines, see Johannes Wecker, *Antidotarivm speciale. . . ex opt. avthorvm . . . fideliter congestum*, Eusebius Episcopius and heirs of Nicolaus Episcopius, Basle, 1588, pp. 23–4, 32–3. For dissolutions of pearls (oil of pearls), see *ibid.*, p. 253. R. L. Ellis was the first to recognize that Wecker's systematic compendium could provide valuable contexts for Bacon's pharmacopoeia; see *SEH*, II, p. 155, n. 2. Unicorn horn, hartshorn, bone of stag's heart, and ivory are all considered by Wecker in the same place (p. 148), and without a break, which rather suggests that Bacon owned a copy and dipped into it as he prepared *HVM*. For bezoar, see Wecker, pp. 25–6, and *cmt* on *HVM*, V8$^v$–X4$^v$, p. 450 below. Also see E. Kremers, 'On the bone of the stag's heart', *Isis* **23**, 1935, p. 256.

## N2$^r$

Page 234, l. 33–p. 236, l. 10: Calida—cf. Wecker, *Antidotarivm*, p. 39, where some of Bacon's hot substances are so classified.

## N3$^r$

Page 236, l. 25: vt ait ille—I have not been able to identify this individual, but cf. *DAS*, 2S2$^r$ (*SEH*, I, pp. 689–90).

**N4$^{r-v}$**

Page 236, l. 29–p. 238, l. 2: Intentiones—for the meaning of this term see Introduction, p. liv above. The discussion of intentions accounts for almost 40 per cent of *HVM*.

**N4$^v$–N7$^r$**

Page 238, ll. 2–35: *Nam cùm audiamus*—this is an attack on a good range of traditional prescriptions for and theories of the prolongation of life. For *cmts* on these see below. Also see *DAS*, X4$^r$, Z4$^{r-v}$, 2D3$^v$ (*SEH*, I, pp. 559, 574–5, 599).

**N4$^v$**

Page 238, l. 3: *de confortando* Calore Naturali—this is the standard natural heat and radical moisture theory of ageing. For this and Bacon's response to it see *OFB*, VI, pp. lxv–lxix, 270–2, 304, 437.

ll. 4–5: *de* Cibis—for foods which generate good blood, see Marsilio Ficino, *De vita*, ed. Albano Biondi and Giuliano Pisani, Biblioteca dell'immagine: Pordenone, 1991, pp. 104–6.

**N5$^r$**

Page 238, ll. 8–9: Medicinis ex Auro—for gold medicines see *OFB*, XIII, pp. 14, 92, 285.

**N5$^v$**

Page 238, ll. 15–16: Carnes Aquilarum—see *cmt* p. 425 above on *HVM*, F4$^r$–F5$^v$, pp. 182–4 above.

**N5$^v$–N6$^r$**

Page 238, ll. 16–20: *Quódque quidam, cùm* Vnguentum—see *Epistola fratris Rogerii Baconis, de secretis operibus artis et naturæ*, Frobenius: Hamburg, 1618, pp. 48–9: 'Ac domina de Nemore in Britania majori, quærens cervam albam, invenit unguentum quo custos nemoris se perunxerat in toto corpore, præterquam in plantis: vixit trecentis annis sine corruptione, exceptis pedum passionibus.'

**N6$^r$**

Page 238, ll. 20–3: *Atque de* Artefio—see *Epistola fratris Rogerii Baconis*, p. 49, where we are told that Artefius lived for 1,020 years. Also see William Salmon, *Medicina practica or, practical physick . . . To which is added, the philosophick works of . . . Artefius Longævus . . .*, Thomas Howkins: London, 1692, p. 436 (2F2$^v$): 'the most Learned *Theophrastus Paracelsus in Libro de vita longa . . .* saith, To which term of a Thousand Years, none of the other Philosophers, no nor *Hermes* himself, the Father of them, ever attained, but only *Artephius* our Author: see then whether it be not doubtlesse, that this great Man knew this Stone, and understood the Virtues thereof, better than all others.'

**N6$^{r-v}$**

Page 238, ll. 23–6: *Et de* Horis Fortunatis—on this and planetary seals, see

Ficino, *De vita,* esp. book 3 (*De vita coelitus comparanda*), pp. 196 ff., 216. Also see *DAS,* 2D3ᵛ (*SEH,* I, p. 599).

## N7ʳ–N8ᵛ

Page 238, l. 34–p. 240, ll. 1–18: *Sunt tamen Pauca*—see *HVM,* H7ᵛ, p. 200 above. On Herodicus, see Plato, *Republic,* in *The dialogues of Plato,* trans. B. Jowett, 5 vols., 2nd edn., Clarendon: Oxford, 1875, III, pp. 282–3: 'Herodicus . . . had a mortal disease which he perpetually tended, and as recovery was out of the question, he passed his entire life as a valetudinarian; he could do nothing but attend upon himself, and he was in constant torment whenever he departed in anything from his usual regimen . . . When a carpenter is ill he asks the physician for a rough and ready cure . . . And if any one tells him that he must go through a course of dietetics, and swathe and swaddle his head, and all that sort of thing, he replies at once that he has no time to be ill, and that he sees no good in a life which is spent in nursing his disease to the neglect of his ordinary calling.' For Cornaro see *cmts* on *HVM,* K8ᵛ, p. 437 above.

## N8ᵛ–O1ʳ

Page 240, ll. 19–27: *Secundò*—cf. *DAS,* 2D3ᵛ (*SEH,* I, p. 599): '*monemus,* vt Homines nugari desinant, nec tam faciles sint, vt credant, grande illud opus, quale est Naturæ Cursum remorari & retrouertere, posse Haustu aliquo Matutino, aut Vsu alicuius Pretiosæ Medicinæ, ad exitum perduci; Non Auro potabili, non Margaritarum Essentijs, & similibus Nugis; sed vt pro certo habeant, *Prolongationem Vitæ* esse rem operosam, & quæ ex compluribus Remedijs, atque eorum inter se Connexione idoneâ, constet: Neque enim quisquam ita stupidus esse debet, vt credat, quod nunquàm factum est adhùc, id fieri iam posse, nisi per modos etiàm nunquàm tentatos.'

## O1ᵛ

Page 242, l. 1: *vt ait* Scriptura—Matthew 6: 25, 'Nonne anima plus est quam esca: et corpus plus quam vestimentum?'

## O2ʳ–O3ʳ

Page 242, ll. 4–20: *Quartò*—see *DAS,* 2D3ᵛ–2D4ʳ (*SEH,* I, p. 599): 'monemus, vt Homines ritè animaduertant & distinguant, circa ea, quæ ad *Vitam sanam,* & ea, quæ ad *Vitam longam,* conferre possunt. Sunt enim nonnulla, quæ ad Spirituum Alacritatem, & Functionum Robur, & Morbos arcendos, prosunt; quæ tamen de summâ Vitæ detrahunt, & Atrophiam senilem absque Morbis accelerant: Sunt & alia, quæ ad Prolongationem Vitæ, & Atrophiam senilem longiùs summouendam iuuant; sed tamen non vsurpantur absque periculo Valetudinis; Adeò vt qui ijs vtentur ad Prolongationem Vitæ, debeant simul Incommodis occurrere, quæ alioquìn ex eorum vsu superuenire possint.'

## O4ᵛ–O5ʳ

Page 244, ll. 11–14: *Cum vero hæc*—in fact this part of the work really goes beyond natural history in the strict Baconian sense. There is nothing in the norma set out at the beginning of *HNE* (see pp. 12–16 above) which suggests that the practical materials presented in the discussion of intentions would play so dominant a role.

**O5ʳ–O7ʳ**

Page 244, l. 15–p. 246, l. 20: Operatio super Spiritus—the following twelve points are in fact a concise account of the action of inanimate spirit on the body; see Introduction, pp. xlviii–xlix above.

**O5ᵛ**

Page 244, l. 28: Succos Corporis (vt ait ille)—I have not been able to identify the *ille*, but cf. *cmt* on *HVM*, N3ʳ, p. 440 above.

**O8ʳ⁻ᵛ**

Page 248, ll. 5–10: *Græci* multum posuerunt—Wecker believed that the meconium of the Greeks was opium or a simple like it. He presents a comprehensive list of opiate receipts, among which he lists various theriac and mithridate preparations; see *Antidotarivm*, pp. 323–30.

**O8ᵛ**

Page 248, ll. 16–19: *Turcæ Opium*—Bacon's source may well have been George Sandys, *A relation of a iourney begun an: dom: 1610. Fovre bookes. Containing a description of the Turkish Empire*, Thomas Snodham: London, 1615, p. 66: 'The *Turkes* are also incredible takers of *Opium*, whereof the lesser *Asia* affordeth them plenty: carrying it about them both in peace and in warre; which they say expelleth all feare, and makes them couragious: but I rather thinke giddy headed, and turbulent dreamers.' Wecker says (*Antidotarivm*, p. 135): 'Ianniceri hæc plerunque faciunt, quibus summa militiæ uirtus commissa est.'

**P1ʳ⁻ᵛ**

Page 248, ll. 27–31: *Turcæ* habent—on coffee see Sandys, *A relation*, p. 66; 'Although they [the Turks] be destitute of Tauerns, yet haue they their Coffa-houses, which something resemble them. There sit they chatting most of the day; and sippe of a drinke called Coffa (of the berry that it is made of) in little *China* dishes, as hot as they can suffer it: blacke as soote, and tasting not much vnlike it (why not that blacke broth which was in vse amongst the *Lacedemonians?*) which helpeth, as they say, digestion, and procureth alacrity.' *SS*, 2B3ʳ (*SEH*, II, pp. 576–7) pretty much echoes these words. Also see Markman Ellis, *The coffee-house: a cultural history*, Phoenix Books: London, 2005, pp. 1–7.

**P1ᵛ–P2ʳ**

Page 248, l. 37–p. 250, l. 5: Incœpit nostro sæculo—on tobacco, cf. *SS*, 2B3ʳ (*SEH*, II, p. 577). Cf. James I, *A counterblaste to tobacco*, in *The workes*, p. 220: 'And from this weakenesse it proceeds, that many in this kingdome haue had such a continuall vse of taking this vnsauorie smoake, as now they are not able to forbeare the same, no more then an old drunkard can abide to be long sober, without falling into an incurable weaknesse and euill constitution: for their continuall custome hath made to them, *habitum, alteram naturam*: so to those that from their birth haue beene continually nourished vpon poison and things venemous, wholesome meats are only poisonable.'

**P2ʳ⁻ᵛ**

Page 250, ll. 9–12: *Opiata . . . Simplicia*—see Wecker, *Antidotarivm*, pp. 323–30.

**P3ʳ**

Page 250, l. 22: aliquod *Magistrale*—see *OED*, **Magistral 2.**

**P5ᵛ–P8ʳ**

Page 254, l. 4–p. 256, l. 21: Quoad *Refrigerationem*—here and elsewhere Bacon is unusual in promoting the inorganic chemical nitre as a medicine. He may have been encouraged to do so by his contemporary Sir Thomas Chaloner (1561–1615), whose *A shorte discourse of the most rare and excellent vertue of nitre*, was printed in London by Gerald Dewes in 1584. Chaloner was among those whom Bacon had sought in 1608 to draw into his reformist schemes; see *LL*, IV, p. 63. Also see *SS*, N3ʳ (*SEH*, II, p. 459): 'the *Spirit* of *Nitre* is *Cold*. And though it be an Excellent Medicine, in Strength of yeares, for Prolongation of Life; yet it is, in Children and young Creatures, an Enemy to *Growth*. And all for the same Reason; For *Heat* is requisite to Growth: But after a Man is come to his Middle Age, *Heat* consumeth the Spirits; which the Coldnesse of the Spirit of *Nitre* doth helpe to condense, and correct.' Wecker, *Antidotarivm*, pp. 18–19, says that nitre has no medicinal uses.

**Q1ᵛ**

Page 258, ll. 7–9: nouimus *Virum Nobilem*—cf. *SS*, 2I3ᵛ (*SEH*, II, p. 649): 'I knew a great *Man*, that liued Long, who had a Cleane *Clod* of *Earth*, brought to him euery *Morning*, as he sate in his *Bed*; And he would hold his *Head* ouer it, a good pretty while.'

ll. 11–13: qualia sunt *Endiuia*—cf. Wecker, *Antidotarivm*, pp. 156, 179.

**Q3ʳ⁻ᵛ**

Page 260, ll. 6–14: Calida illa intensissima—cf. Wecker, *Antidotarivm*, pp. 97–8, 111, 137. Also see (on theriac and mithridate) *cmts* on O8ʳ⁻ᵛ (p. 443 above).

**Q3ᵛ**

Page 260, ll. 15–17: facit *Venus*—for sex and the spirits, see *SS*, Z2ʳ⁻ᵛ (*SEH*, II, pp. 555–7); *HDR* in *OFB*, XIII, p. 116.

**Q4ᵛ**

Page 262, ll. 1–3: *Cornari Veneti*—see *cmts* (p. 437 above) on *HVM*, K8ᵛ, p. 218 above.

**Q5$^{r-v}$**

Page 262, ll. 9–13: de *Frænatione*—see *DVM* in *OFB*, VI, pp. 274–5, 322–5.
See also *NO* in *OFB*, XI, pp. 422–3.

**Q5$^v$**

Page 262, ll. 14–16: *Fabula* habet—on Epimenides see *cmt* on *HVM*, R7$^v$–R8$^r$,
p. 446 below.

**Q6$^{r-v}$**

Page 262, l. 31–p. 264, l. 3: *Somnum placidum*—for drugs producing sound
sleep, some of which are mentioned here by Bacon, see Wecker, *Antidotarivm*,
pp. 156–7.

**Q6$^r$–Q7$^r$**

Page 264, ll. 4–8: *De* Ecstasi—cf. Jerome Cardan, *De rervm varietate, libri XVII*,
Sebastian Henricpetri: Basle, 1581, pp. 542–3.

**Q7$^v$–R1$^v$**

Page 264, l. 18–p. 266, l. 35: Veniendum iam ad *Affectus*—Bacon provides
a group of detailed 'Experiments in Consort, touching the *Impressions*, which
the *Passions* of the *Minde* make vpon the *Body*' in *SS*, 2A3$^v$–2B1$^r$ (*SEH*, II,
pp. 567–71).

**R1$^v$–R2$^r$**

Page 266, ll. 22–7: *Contemplatores Rerum*—for these ancients see *cmt*s on
*HVM*, H8$^v$–I2$^v$ (pp. 431–2 above), and on *HVM*, K4$^r$, p. 434 above.
Parmenides was still alive when he visited Athens; see *Parmenides* in *The dialogues of Plato*, IV, p. 160: 'Parmenides and Zeno; they came to Athens, he said,
at the great Panathenaea; the former was, at the time of his visit, about 65 years
old, very white with age, but well favoured.' For the acute who live less long
Bacon is thinking of the scholastic philosophers; see *cmt*s on *HVM*, M5$^v$–M7$^r$,
p. 440 above.

**R2$^r$–R6$^v$**

Page 266, l. 33–p. 272, l. 6: *Obseruationes* Generales circa *Spiritus*—see *cmt*s on
*HVM*, D8$^v$–E6$^r$, p. 423 above.

**R3$^r$**

Page 268, ll. 18–19: Senecâ—see *The workes of Lvcius Annævs Seneca*, trans.
Thomas Lodge, William Stansby: London, 1614, Epist. XIII, p. 184: '*Amongst
other euils folly hath likewise this, that it beginneth alwayes to liue.* Consider,
worthy *Lucilus*, what these things signifie, and thou shalt vnderstand how
loathsome mens leuitie is, who are alwayes occupied to proiect new foundations
of life, and in their last time bethinke them of new hopes.'

**R4$^r$**

Page 268, ll. 35–6: ait *Ficinus*—see *De vita*, p. 124: 'Ludos quosdam et mores,
quoad decet, olim ante actae pueritiae revocent. Difficillimum nanque est,
ut ita dixerim reiuvenescere corpore, nisi ingenio prius repuerascas. Itaque in

omni etiam aetate magnopere conducit ad vitam nonnihil pueritiae retinere,
et oblectamenta varia semper aucupari. Longum vero perfusumque risum min-
ime: spiritum namque nimis ad exteriora dilatat.'

### R4ʳ⁻ᵛ

Page 270, ll. 2–7: *Vespasianus*—Suetonius, *The historie of tvvelve Cæsars*,
p. 240: 'when he [Vespasian] was Emperour hee both frequented continually
the place of his birth and breeding, the Capitall hovse and manour remaining
still as it had beene in former times, nothing altered (because forsooth, his eyes
should have no losse nor misse of that which they were wont to see there) and
loved also the memoriall of his grandmother so deerly, that on all solemne
and festivall, and high daies, hee continued ever drinking out of a silver pot that
was hers and out of none other.'

### R4ᵛ–R5ʳ

Page 270, ll. 12–16: Exemplum in *Cassiodoro*—see *OCD*, **Cassiodorus**.

### R6ᵛ–R7ᵛ

Page 272, ll. 7–23: OPERATIO super Exclusionem Aeris—the conspiracy of
air and inanimate spirit, and the actions of spirit on bodies are fundamental
to Bacon's theory of matter and of change in the trerrestrial world; cf., for
instance, *OFB*, VI, pp. 284–8, 306–8; XI, pp. 346–50; XIII, pp. 164–5. Also see
*cmts* (p. 457 below) on 2C7ᵛ–2D2ᵛ, pp. 346–8 above.

### R7ᵛ–R8ʳ

Page 272, l. 24–p. 274, l. 8: *Vita* in *Antris*—cf. *NA*, e2ʳ⁻ᵛ (*SEH*, III, pp. 156–7):
'*We haue large and deepe* Caues *of seuerall Depths: The deepest are sunke 600.*
*Fathome . . . Wee vse them also sometimes, (which may seeme strange,) for* Curing
*of some Diseases, and for* Prolongation *of* Life, *in some* Hermits *that choose to liue*
*ther, well accommodated of all things necessarie, and indeed liue very long; By*
*whom also we learne many things.*' For Epimenides, also famous for the 'Liar
Paradox', see *OCD*, and Diogenes Laertius, *Lives*, I, p. 115: 'One day he was sent
into the country by his father to look for a stray sheep, and at noon he turned
aside out of the way, and went to sleep in a cave, where he slept for fifty-seven
years.' Simeon Stylites the Elder was born about 388 in Syria. Crowds of
admirers forced him to have a pillar erected where he could live with less
disturbance. The first pillar was short but was afterwards replaced by others,
rising at last to more than fifty feet. After spending thirty-six years aloft, Simeon
died in 459. Simeon the Younger lived from 521 to 597. He spent sixty-eight
years on various pillars. As for Daniel Stylites, he was born in Syria in 409, and
lived on successive pillars near Constantinople. He died in 493. St Saba was
born in 439 in Cappadocia. He set himself up as an anchorite in a cavern, and
not on a pillar, in the desert. He died in 531, aged 91. For anchorites see lemma
in Hans J. Hillerbrand (ed.) *The Oxford Encyclopedia of the Reformation*, Oxford
University Press, New York and Oxford, 1996, 4 vols., I, pp. 35–6.

**R8ʳ⁻ᵛ**

Page 274, ll. 5–11: Proxima *Vitæ in Antris*—cf. *NA*, e2ᵛ–e3ʳ (*SEH*, III, p. 157): '*We haue High* Towers ... *And vpon them, in some Places, are Dwellings of* Hermits, *whom wee visit sometimes, and instruct what to obserue.*' For longevity in Barbary cf. *HVM*, P4ᵛ, p. 252 above.

**R8ᵛ–S1ᵛ**

Page 274, ll. 12–35: Atque huiusmodi *Aer*—for air's predaciousness see *OFB*, VI, pp. lxii, 306. For chalybeate and vitriol waters see *SS*, N1ʳ (*SEH*, II, p. 454): '*Astriction* prohibiteth *Dissolution*: As we see (generally) in *Medicines*, whereof such as are *Astringents* doe inhibite *Putrefaction*: And by the same reason of *Astringency*, some small Quantity of Oile of Vitrioll, will keepe Fresh Water long from *Putrefying*. And this *Astriction* is in a Substance that hath a *Virtuall Cold*; And it worketh (partly) by the same Meanes that Cold doth.' Also see *NA*, e3ʳ (*SEH*, III, pp. 157–8): '*VVe haue also a Number of* Artificiall VVels, *and* Fountaines, *made in Imitation of the* Naturall Sources *and* Baths; *As tincted vpon* Vitrioll, Sulphur, Steele, Brasse, Lead, Nitre, *and other* Mineralls. *And againe wee haue little* VVells *for* Infusions *of many* Things, *wher the* VVaters *take the Vertue quicker and better, then in* Vessells, *or* Basins. *And amongst them we haue a* VVater, *which wee call* Water *of* Paradise, *being, by that we doe to it, made very Soueraigne for* Health, *and* Prolongation *of* Life.' For body painters ancient and modern, cf. *DVM* in *OFB*, VI, p. 310. Pernambuco was founded in 1537, and the French Fathers to whom Bacon refers may have been among the French colonists expelled from Ilha São Luís de Maranhão in 1615.

**S1ᵛ–S2ʳ**

Page 274, l. 36–p. 276, l. 2: *Ioannes* de *Temporibus*—see Richard Verstegan, *A restitution of decayed intelligence in antiquities . . . Printed at Antvverp by Robert Bruney. 1605. And to be sold . . . by Iohn Norton and Iohn Bill*, p. 323: 'Heer by the way I muste note vnto the reader that *Ioannes de temporibus*, that is to say, John of the tymes, who . . . was Shield-knaue vnto the Emperor *Charles the great* . . . liued vnto the nynth yeare of the raigne of the Emperor *Conrade*, and died at the age of three hundreth threescore and one yeares, seeming thereby a very miracle of nature . . ..' Ioannes' answer seems actually to have been an ancient one; see Pliny, *Historia*, XII. 24, p. 136: '*Pollio Romilus*, who beeing above a hundred yeares old, bare his age passing well: whereat the Emperour *Augustus* . . . demaunded of him, What means he used most so to maintaine that fresh vigor both of bodie and mind? unto whome *Pollio* answered, By using honyed wine within, and oile without.' Also see Egnatius, *De exemplis*, p. 282, where he mentions a French soldier, Ioannes Tampes, who lived to be 300 years old. This passage comes immediately after material which Bacon certainly read; see *cmts* on *HVM*, K8ʳ, pp. 436–7 above.

**S2ʳ⁻ᵛ**

Page 276, ll. 3–17: *Comitissam Desmondiæ*—this is Katherine FitzGerald; see Ralegh, *The history of the world*, p. 78: 'I my selfe knew the old Countesse of

*Desmond* of *Inchiquin* in *Munster*, who liued in the yeere 1589. and many yeeres since, who was married in *Edward* the Fourths time, and held her Ioynture from all the Earles of *Desmond* since then; and that this is true, all the Noblemen and Gentlemen of *Munster* can witnesse.' For her triple dentition also see *SS*, 2B4ᵛ (*SEH*, II, p. 582). The traditional Irish shirt was the saffron-dyed *léinte*; cf. *SS*, D1ʳ (*SEH*, II, p. 364). Also see Edmund Spenser, *A view of the present state of Ireland*, ed. W. L. Renwick, Clarendon: Oxford, 1970, pp. 61, 69. For a medical receipt using saffron see *SSWN*, fo. 45ʳ (Rees, 'An unpublished manuscript by Francis Bacon' p. 406).

### S2ᵛ

Page 276, ll. 18–20: *Hippocrates*—cf. *SS*, C4ʳ (*SEH*, II, p. 362).

### S3ʳ–S4ʳ

Page 276, l. 21–p. 278, l. 7: vsum *Olei*—cf. *SS*, C4ᵛ (*SEH*, II, p. 362); the Virgil quotation (*Georgics*, II, 466) is accurate. That oil hinders the destructive conspiracy of air with inanimate spirit is a fundamental theoretical and practical assumption of Bacon's programme for the prolongation of life; see *HVM*, C2ᵛ–C3ᵛ, p. 158 above.

### S4ᵛ–S6ʳ

Page 278, l. 13–p. 280, l. 10: visum est addere *Cautiones*—on sweat and its functions see 'Experiments in Consort, touching *Sweat*, in *SS*, 2A3ʳ⁻ᵛ (*SEH*, II, pp. 565–7). For the tendency of inanimate spirits to multiply themselves and so to undermine bodies, see *OFB*, VI, pp. 286–8. On oiled garments, see *SS*, C4ʳ⁻ᵛ (*SEH*, II, p. 362): 'Adde also this Prouision, That there be not too much *Expence* of the *Nourishment*, by *Exhaling*, and *Sweating*. And therfore if the Patient be apt to sweat, it must be gently restrained. But chiefly *Hippocrates* Rule is to bee followed; who aduiseth quite contrary to that which is in vse: Namely, that the *Linnen*, or *Garment* next the Flesh, be in Winter drie, and oft changed; And in Sommer seldome changed, and smeared ouer with Oyle; For certaine it is, that any Substance that is fat, doth a little fill the Pores of the Body, and stay Sweat, in some Degree. But the more cleanly way is, to haue the *Linnen* smeared lightly ouer, with *Oyle* of *Sweet Almonds*; And not to forbeare shifting as oft as is fit.'

### S6ᵛ

Page 280, ll. 18–19: *Hyberni Syuestres*—see Spenser, *A view*, p. 69.

### S7ᵛ–T1ᵛ

Page 282, l. 1–p. 284, l. 6: OPERATIO super Sanguinem—what Bacon recommends here is in the main the polypharmacy of organic simples. Notwithstanding his general suspicion of inorganic preparations here and elsewhere (see *cmts* on *HVM*, N4ᵛ–N7ʳ, p. 441 above), he is reluctant to reject entirely opinions in favour of such preparations.

T1ᵛ–T4ʳ

Page 284, l. 10–p. 286, l. 13: de vsu *Limaturæ Auri*—cf. *OFB*, XIII, pp. 13–14, 92, 284–5. For gold, pearls, coral, and gems also cf. *HVM*, M8ᵛ–N1ᵛ, p. 234 above. Also see *SS*, 2K4ᵛ (*SEH*, II, p. 661). On the medicinal properties of wood see *cmts* pp. 422–3 above on *HVM*, D7ʳ–D8ʳ, pp. 170–2 above.

T4ʳ–T5ᵛ

Page 286, l. 14–p. 288, l. 6: OPERATIO super Succos—cf. *HVM*, B7ᵛ–C2ᵛ, pp. 154–8 above. For nitre see *cmts* on *HVM*, P5ᵛ–P8ʳ, p. 444 above.

T5ᵛ

Page 288, ll. 7–9: Quatenus ad *Condensationem*—cf. *DVM*, in *OFB*, VI, p. 298.

T6ʳ

Page 288, ll. 14–18: Quatenùs ad *Exercitationem*—cf. *SS*, L2ʳ (*SEH*, II, p. 440). Also see *HVM*, X7ᵛ–X8ᵛ, p. 308 above.

T6ʳ–V1ᵛ

Page 288, l. 21–p. 292, l. 19: veniendum ad . . . Roscidationem—'dewiness' is associated in Bacon's mind with the traditional theory of ageing, which centred on the radical miosture theory; see *DVM*, *OFB*, VI, p. 304. Bacon's recollection of Plautus' *Poenulus* is inaccurate. Lycus the pimp, not the old woman, had the lines Bacon has in mind; see *DVM* in *OFB*, VI, pp. 314, 444.

V2ʳ–V3ʳ

Page 292, l. 20–p. 294, l. 12: OPERATIO super Viscera—here begins the set of four operations relating to the second intention, i.e. to the one relating to 'Perfectionem Reparationis'; see *HVM*, O3ʳ⁻ᵛ, p. 242 above.

Page 292, ll. 23–7: Qvæ *Viscera* illa *Principalia*—cf. *SS*, 2F1ᵛ–2F2ʳ (*SEH*, II, pp. 613–14): 'The word *Concoction*, or *Digestion*, is chiefly taken into vse from *Liuing Creatures*, and their *Organs* . . . And the *Foure Disgestions*, (In the *Stomach*; In the *Liuer*; In the *Arteries* and *Nerues*; And in the *Seuerall Parts* of the *Body*,) are likewise called *Concoctions*. And they are all made to be the Workes of *Heat*. All which *Notions* are but ignorant Catches of a few Things, which are most Obuious to *Mens Obseruations*. The Constantest *Notion* of *Concoction* is, that it should signifie the *Degrees* of *Alteration*, of one *Body* into another, from *Crudity* to *Perfect Concoction*; Which is the *Vltimity* of that *Action*, or *Processe* . . . It is true, that *Concoction* is, in great part, the *Worke* of *Heat*; But not the *Worke* of *Heat* alone: For all Things, that further the *Conuersion*, or *Alteration*, (as *Rest, Mixture* of a *Body* already *Concocted*, &c.) are also *Meanes* to *Concoction*. And there are of *Concoction* two *Periods*; The one *Assimilation*, or *Absolute Conuersion* and *Subaction*; The other *Maturation*: whereof the Former is most conspicuous in the *Bodies* of *Liuing Creatures*; In which there is an *Absolute Conuersion*, and *Assimilation* of the *Nourishment* into the *Body*.'

**V3ʳ–V5ʳ**

Page 294, l. 13–p. 296, l. 13: *Stomachum*—for the stomach in general cf. *SS*, C4ʳ–D1ʳ (*SEH*, II, pp. 362–3). For *Tragaganthum* see *OFB*, XIII, p. 293. For the very unusual *Myuas* (nom. myva) see Wecker, *Antidotarivm*, on syrups, p. 178: '*Miua Citoniorum simplex*' and '*Miua Citoniorum aromatica*.'

**V6ʳ–V8ᵛ**

Page 298, ll. 3–6: *Iecori*—Bacon seems to follow Bernardino Telesio on what happens to the liver with age. For Telesio on ageing see *DVM* in *OFB*, VI, pp. lxvi–lxvii, 270, 272, 437.

**V8ᵛ–X4ᵛ**

Page 300, l. 14–p. 304, l. 17: *Cor Iuuamentum*—for Bacon and the heart, see *cmt* on *HVM*, H1ʳ, p. 429 above.

**X3ʳ**

Page 302, l. 28: *Errhini more*—cf. *SS*, C2ʳ: 'And we see *Sage*, or *Bettony brused, Sneezing-powder*, and other *Powders* or *Liquors* (which the *Physitians* call *Errhines*,) put into the Nose, draw Flegme, and water from the Head.' On bezoar stone or calculi of ruminants, cf. *SS*, R3ᵛ (*SEH*, II, p. 499): 'it is obserued by some, that there is a vertuous *Bezoar*, and another without vertue; which appeare to the shew alike; But the Vertuous is taken from the Beast, that feedeth vpon the Mountaines, where there are *Theriacall Herbs*; And that without Vertue, from those that feed in the Valleyes, where no such *Herbs* are.'

**X5ʳ**

Page 304, ll. 22–8: *Ad Cerebrum*—possession of vital spirit in cerebral concentrations is, for Bacon, what distinguishes animal from vegetable being. Vital spirit so concentrated gives animals and humans their sensory-motor functions; see *OFB*, VI, pp. lvii–lix.

**X6ʳ⁻ᵛ**

Page 306, ll. 16–19: *Multa Fercula*—see Seneca, *The workes*, Epist. XCV, p. 403: 'Free were they from those euils, who as yet were not weakened by these delicates, who gouerned and ministred vnto themselues . . . There health being entertained by a simple cause, was simple also: many dishes haue bred many sicknesses.'

**X6ᵛ–Y1ʳ**

Page 306, l. 20–p. 308, l. 36: OPERATIO super partes—see Xenophon, *Memoirs of Socrates and The Symposium*, trans. Hugh Tredennick, Penguin Books: Harmondsworth, 1970, pp. 238–5: 'Socrates went on, with a perfectly straight face, "Are you laughing at me? Is it at the idea of my wanting to take exercise to improve my health, or to enjoy my food and sleep better? or is it because I'm bent on a particular kind of exercise, not wanting to develop my legs at the expense of my arms like a long-distance runner, nor my arms at the expense of my legs like a boxer, but by working hard with my whole body to make it evenly proportioned all over?".'

## Y1ʳ–Y6ʳ

Page 310, l. 1–p. 314, l. 23: OPERATIO super Alimentum—for the remark about many dishes, see *cmts* on *HVM*, X6ʳ⁻ᵛ, p. 450 above. Much of this entire passage is echoed in *SS*, C2ᵛ–D1ʳ (*SEH*, II, pp. 358–65).

## Y7ʳ–Z1ᵛ

Page 314, l. 32–p. 316, l. 34: *Certum est Corpora*—see *NO*, *OFB*, XI, pp. 400–2 and *ANN*, *OFB*, XIII, pp. 196–8 for motion of assimilation. For aversion to motion, see *NO*, *OFB*, XI, p. 412.

## Z2ʳ⁻ᵛ

Page 318, ll. 1–9: OPERATIO super Intenerationem—here begins the discussion of the two operations relating to the last (viz. Renouationem Veterationis) of Bacon's three intentions (see *HVM*, O3ʳ⁻ᵛ, p. 242 above). A probable source for Bacon on Pelias and Media is Ovid, *Metamorphoses*, Book VII: Medea, instrument of Hera's revenge on Pelias, duped his daughters into cutting him up and boiling the old man to rejuvenate him. Also see *OCD*: **Medea** and **Pelias**. Bacon seldom allowed himself irony in *HVM*.

## Z4ᵛ–Z5ʳ

Page 320, ll. 17–28: At *Ficini* illud *Commentum*—Bacon spares us the full juiciness of Ficino on blood sucking; cf. *De vita*, p. 136: 'Quos hectica senilis exedit, medici diligentes liquore humani sanguinis, qui arte sublimi destillavit ad ignem, reficere moliuntur. Quid ergo prohibet quo minus senio iam quasi confectos interdum hoc etiam potu reficiamus? Communis quaedam est et vetus opinio aniculas quasdam sagas quae et striges vulgari nomine nuncupantur, infantium sugere sanguinem, quo pro viribus iuvenescant. Cur non et nostri senes omni videlicet auxilio destituti, sanguinem adolescentis sugant? Volentis, inquam, adolescentis, sani, laeti, temperati, cui sanguis quidem sit optimus, sed forte nimius. Sugent igitur more hirudinum ex brachii sinistri vena vix aperta unciam unam aut duas. Mox verò sacchari viníque tantundem sument, ídque esurientes et sitibundi facient, crescente luna.' Ficino's recommendation also appears in *SS*, 2G2ʳ (*SEH*, II, p. 625). Diogenes Laertius (*Lives*, II, pp. 410–11) says that Heraclitus, afflicted with dropsy, 'buried himself in a cowshed, expecting that the noxious damp humour would be drawn out of him by the warmth of the manure. But, as even this was of no avail, he died at the age of sixty.'

## Z7ᵛ

Page 324, ll. 1–5: *Ficinus* ait—see Ficino, *De vita*, pp. 124–6, for David's late 'puerile fomentum.' The girl in question was Abishag the Shunamite; see 1 Kings 1: 1–4. 'Shunamitism' was increasingly recommended as a remedy by seventeenth-century authorities; see Daniel Schäfer, 'Medical representations of old age in the Renaissance: the influence of non-medical texts', in *Growing old in early modern Europe: cultural representations*, ed. Erin Campbell, Ashgate: Aldershot, 2006, pp. 11–19, at 17–18.

**2A1ʳ–2A3ʳ**

Page 324, l. 21–p. 326, l. 33: OPERATIO super Expurgatione*m*—for the rejuvenation of oxen, the French pox, and guaiac wood, see *cmts* on *HVM*, D7ʳ–D8ʳ (pp. 170–2 above) and *cmts* thereon pp. 422–3 above.

**2A5ʳ–2C1ʳ**

Page 328, l. 1–p. 340, l. 19: Atriola Mortis—in general, this part of the treatise rests on the general theory of vital spirits, and the conditions required for their maintenance, for which see Introduction, p. xlviii above, and *cmts* immediately below.

**2A5ᵛ–2A6ʳ**

Page 328, ll. 10–14: *Spiritus Viuus*—see *cmt* on *HVM*, 2B2ʳ⁻ᵛ, p. 453 below. For Aristotle on heats (Bacon's flames), see *Problems*, Books 1–21 (Loeb Classical Library), trans. W. S. Hett, London and Cambridge, Mass., 1936, rev. edn., 1953, repr. 2000, III. 22, pp. 94–5.

**2A6ʳ–2A7ᵛ**

Page 328, l. 25–p. 330, l. 18: At vt ad *Spiritum*—see *OFB*, XI, p. 314.

**2A7ᵛ–2B2ʳ**

Page 330, l. 19–p. 332, l. 28: *Indigentiam Refrigerij*—see *OFB*, VI, p. lix. Aristotle thought respiration was necessary to cool natural heat; see *cmt* below on *HVM*, 2B1ʳ⁻ᵛ. Jovinian (i.e. Jovian) died of fumes from a charcoal stove in February, AD 364 at Dadastana; see *OCD*, **Jovian**. Fausta, wife of Constantine the Great, is reported to have been boiled to death in AD 326.

**2A8ʳ⁻ᵛ**

Page 330, ll. 32–5: Rursùs *Pulsus*—in *NO* and *ANN* (*OFB*, XI, pp. 410–12; XIII, p. 203) the motion of heart and pulse is ascribed to 'motion of trepidation'. This motion, according to *ANN*, is, 'a motion of bodies which are so placed between convenient and inconvenient circumstances that they are not satisfied with their situation and yet, if they try to retreat from it, they fall into a state which again they shun. For this reason they are constantly agitated, and they struggle and act restlessly.' In other words, the motion seems to be little more than an adverse reaction to awkward circumstances. In *DAS*, this motion is specifically associated with systole and diastole; see Y1ʳ (*SEH*, I, p. 561). I don't think that Bacon anywhere ascribed any *positive* function to systole and diastole. For the heart in Baconian physiology, see *cmts* on *HVM*, V2ʳ–V3ʳ, p. 449 above.

**2A8ᵛ**

Page 332, ll. 2–4: *Vrinatoribus Delijs*—cf. *AL* in *OFB*, IV, p. 96, ll. 1–2; also see Diogenes Laertius, *Lives*, I, pp. 152–3: 'They relate that Euripides gave [Socrates] the treatise of Heraclitus and asked his opinion upon it, and that his reply was, "The part I understand is excellent, and so too is, I dare say, the part I do not understand; but it needs a Delian diver to get to the bottom of it".'

### 2A8ᵛ–2B1ʳ

Page 332, ll. 5–11: Sunt ex *Animalibus*—see Aristotle, *History of animals*, I, pp. 116–19; *On the soul, parva naturalia, on breath*, pp. 116–17, 432–7.

### 2B1ʳ⁻ᵛ

Page 332, ll. 12–17: Si *Spiritus*—Bacon follows the traditional line that vital spirit would destroy itself in its own heat without refrigeration by respiration; see *DVM* in *OFB*, VI, pp 356–8. For Aristotle on respiration see *Parva naturalia in On the soul, parva naturalia, on breath*, pp. 436–43, 466–9, 476–9. For burning fevers see *OFB*, XI, pp. 240, 545.

### 2B1ᵛ–2B2ʳ

Page 332, ll. 18–28: *Somni* quoque Indigentia—on the function of sleep in relation to spirits and longevity, cf. *SS*, C4ᵛ, 2B3ᵛ, 2H2ᵛ (*SEH*, II, pp. 363, 579–80, 638). For Aristotle on sleep see *On sleep and waking*, in *On the soul, parva naturalia, on breath*, pp. 318–45. For Maecenas see Pliny, VII. 51, p. 184: 'Some are knowne to be never free of the ague, as *C. Macænas*. The same man for three yeeres together, before he died, never laid his eies together for sleep the minute of an houre.'

### 2B2ʳ⁻ᵛ

Page 332, l. 29–p. 334, l. 2: Tertiam *Indigentiam*—the vital spirit is analogous to celestial fire, which is also self-subsisting and not successively renewed by fuel/aliment; see *OFB*, VI, p. 174, ll. 1–7, p. 176, ll. 29–32, pp. 407–8. However, this view seems to conflict with the final words of *HVM* (2B3ᵛ, p. 334 above; 2F8ᵛ, p. 376 above), which imply that vital spirit does not exist in identity but needs repair. For Bacon's clearest account of the difference between rational soul and vital spirit, and of metempsychotic dangers of confusing the two—an account contemporary with *HVM*—see *DAS*, 2E3ʳ–2E4ᵛ (*SEH*, I, pp. 604–7). The use of scholastic language here ('essentialiter' and 'formaliter') is striking. Bacon would have known of metempsychosis from, *inter alia*, Plato's *Phaedrus*, which states that human souls can pass to animals and *vice versa*; see *The dialogues of Plato*, II, pp. 124–6.

### 2B2ᵛ–2B3ᵛ

Page 334, ll. 3–9: *Renouatio* per *Alimentum*—the body needs aliment to repair the organs that support the vital spirit, and makes up too for the spirits' consumptive effects; see *OFB*, VI, pp. lix, 274, 298.

### 2B3ᵛ

Page 334, ll. 18–22: *Effluxio Sanguinis*—for the physiological assumptions at work here, and for the obscure relationship between vital spirit and arterial blood, see *cmts* (pp. 449, 453 above) on *HVM*, V2$^r$–V3$^r$, 2B2$^{r–v}$, pp. 292–4, 332–4 above.

**2B4$^{r–v}$**

Page 334, ll. 29–35: *Alimentum* in *Gradu*—cf. what Bacon says about the order of leaf-eating Feüillans in *NO* (*OFB*, XI, pp. 438, 584). The image of the wax torch probably echoes a traditional metaphor for the process of ageing; see P. H. Niebyl, 'Old age, fever, and the lamp metaphor', *Journal of the History of Medicine*, 26, 1971, pp. 351–68.

**2B4$^v$**

Page 336, ll. 3–4: tertiâ Parte Momenti—here the word 'moment' is perhaps used in a special sense, i.e. to mean a fortieth or fiftieth part of an hour; see *OED*, **moment** 2.

**2B4$^v$–2B5$^r$**

Page 336, ll. 9–18: *Errauerit*—vital spirit is an intermediate between air and flame, and shares the qualities of both. Here Bacon is speaking of *terrestrial* flame, which has to be regenerated continuously because it exists among hostile bodies of which the chief is air. However, terrestrial flame, if shielded by another flame, assumes a globular form and rotates on its wick, and in short behaves like a tiny planet (the planets being made of self-subsisting *celestial* flame); for a classic statement of these ideas, see *SS*, B4$^v$–C1$^r$ (*SEH*, II, 352–3); also see *OFB*, VI, p. xliii; XI, pp. 336–8, 565.

**2B5$^v$–2B6$^r$**

Page 336, l. 19–p. 338, l. 2: *Præcursores Mortis*—see Lucretius, *De rerum natura*, VI. 1182–98: 'Multaque praeterea mortis tum signa dabantur: | perturbata animi mens in maerore metuque, | triste supercilium, furiosus voltus et acer, | sollicitae porro plenaeque sonoribus aures, | creber spiritus aut ingens raroque coortus, | sudorisque madens per collum splendidus umor, | tenvia sputa minuta, croci contacta colore | salsaque, per fauces rauca vix edita tussi. | in manibus vero nervi trahere et tremere artus | a pedibusque minutatim succedere frigus | non dubitabat. item ad supremum denique tempus | conpressae nares, nasi primoris acumen | tenve, cavati oculi, cava tempora, frigida pellis | duraque, in ore iacens rictum, frons tenta manebat. | nec nimio rigida post artus morte iacebant. | octavoque fere candenti lumine solis | aut etiam nona reddebant lampade vitam.' Bacon's examples of the signs of impending death may also owe something to Pliny, VII. 51, pp. 183–4.

**2B6$^v$–2B7$^r$**

Page 338, ll. 3–19: *Anguillæ*—cf. *SS*, O4$^v$–P1$^r$ (*SEH*, II, p. 474): 'Some *Creatures* doe moue a good while after their Head is off; As *Birds*; Some a very little time; As *Men*, and all beasts; Some moue, though cut in seuerall Pieces; As *Snakes*, *Eeles*, *Wormes*, *Flies*, &c. First, therefore, it is certaine, that the *Immediate Cause* of *Death*, is the Resolution or Extinguishment of the *Spirits*; And that the

Destruction or Corruption of the *Organs*, is but the *Mediate Cause* . . . It is reported by one of the *Ancients*, of credit, that a *Sacrificed Beast* hath lowed, after the Heart hath beene seuered . . . Now the *Spirits* are chiefly in the *Head*, and *Cells* of the *Braine*, which in *Men*, and *Beasts* are Large; And therefore, when the *Head* is off, they moue little or Nothing. But *Birds* haue small *Heads*, and therefore the *Spirits* are a little more dispersed in the *Sinewes*, whereby Motion remaineth in them a little longer.' Bacon may well have seen the execution of the Gunpowder Plotters. The sentence passed on the plotters was this: 'That you be drawn on a hurdle to the place of execution where you shall be hanged by the neck and being alive cut down, your privy members shall be cut off and your bowels taken out and burned before you, your head severed from your body and your body divided into four quarters to be disposed of at the King's pleasure.'

### 2B7$^v$–2B8$^r$

Page 338, ll. 20–30: Ad *Resuscitandum*—cf. *SS*, 2I3$^v$ (*SEH*, II, pp. 648–9); for Fracastoro's hot pan (*sartago*), cf. *HDR* in *OFB*, XIII, p. 88, ll. 29–32, pp. 283–4; *NO* in *OFB*, XI, p. 314.

### 2B8$^r$–2CI$^r$

Page 338, l. 31–p. 340, l. 19: Complura fuerunt Exempla—other instances are given in Pliny, VII. 52, pp. 184–5. Also see Marta Fattori, 'La diffusione di Francis Bacon nel libertinismo francese', *Rivista di storia della filosofia*, 2, 2002, pp. 225–42.

### 2C2$^v$–2C5$^v$

Page 340, l. 20–p. 346, l. 4: *Discrimina Iuuentutis*—cf. Aristotle, *Rhetoric*, II, 13–14, 1389$^a$–1390$^a$, in *The basic works of Aristotle*, ed. and trans. Richard McKeon, Random House: New York, 1941, pp. 1403–6: 'Young men have strong passions, and tend to gratify them indiscriminately. Of the bodily desires, it is the sexual by which they are most swayed and in which they show absence of self-control. They are changeable and fickle in their desires, which are violent while they last, but quickly over: their impulses are keen but not deep-rooted, and are like sick people's attacks of hunger and thirst. They are hot-tempered and quick-tempered . . . While they love honour, they love victory still more; for youth is eager for superiority over others . . . They look at the good side rather than the bad, not having yet witnessed many instances of wickedness. They trust others readily, because they have not yet often been cheated. They are sanguine; nature warms their blood as though with excess of wine; and besides that, they have as yet met with few disappointments. Their lives are mainly spent not in memory but in expectation; for expectation refers to the future, memory to the past . . . They are easily cheated, owing to the sanguine disposition just mentioned. Their hot tempers and hopeful dispositions make them more courageous than older men are; the hot temper prevents fear, and . . . their hopeful disposition makes them think themselves equal to great things—and that means having exalted notions. They would always rather do noble deeds than useful ones: their lives are regulated more by moral feeling than by reasoning . . . They are fonder of their friends, intimates, and companions than older men are, because they like spending their days in the company of others, and have not yet come to value either their friends or anything else by their usefulness to themselves. All their mistakes are in the direction of doing things excessively and vehemently . . . They are ready to pity others, because they think every one an honest man . . . They are fond of fun and therefore witty, wit being well-bred insolence.

Such, then, is the character of the Young. The character of Elderly Men—men who are past their prime—may be said to be formed for the most part of elements that are the contrary of all these. They have lived many years; they have often been taken in, and often made mistakes; and life on the whole is a bad business. The result is that they are sure about nothing and *under-do* everything. They "think", but they never "know"; and because of their hesitation they always add a "possibly" or a "perhaps", putting everything this way and nothing positively. They are cynical; that is, they tend to put the worse construction on everything. Further, their experience makes them distrustful and therefore suspicious of evil . . . They are small-minded, because they have been humbled by life . . . They are not generous, because money is one of the things they must have, and at the same time their experience has taught them how hard it is to get and how easy to lose. They are cowardly, and are always anticipating danger; unlike that of the young, who are warm-blooded, their temperament is chilly; old age has paved the way for cowardice; fear is, in fact, a form of chill . . . They are too fond of themselves; this is one form that small-mindedness takes. Because of this, they guide their lives too much by considerations of what is useful and too little by what is noble—for the useful is what is good for oneself, and the noble what is good absolutely. They are not shy, but shameless rather; caring less for what is noble than for what is useful, they feel contempt for what people may think of them . . . They live by memory rather than by hope; for what is left to them of life is but little as compared with the long past; and hope is of the future, memory of the past. This, again, is the cause of their loquacity; they are continually talking of the past, because they enjoy remembering it. Their fits of anger are sudden but feeble. Their sensual passions have either altogether gone or have lost their vigour: consequently they do not feel their passions much, and their actions are inspired less by what they do feel than by the love of gain. Hence men at this time of life are often supposed to have a self-controlled character; the fact is that their passions have slackened, and they are slaves to the love of gain. They guide their lives by reasoning more than by moral feeling; reasoning being directed to utility and moral feeling to moral goodness. If they wrong others, they mean to injure them, not to insult them. Old men may feel pity, as well as young men, but not for the same reason. Young men feel it out of kindness; old men out of weakness, imagining that anything that befalls any one else might easily happen to them . . . Hence they are querulous, and not disposed to jesting or laughter—the love of laughter being the very opposite of querulousness.' For Tithonus see *DSV*, *SEH*, VI, p. 653.

### 2C7ʳ

Page 346, ll. 5–6: CANONES mobiles—for this expression and its meaning see *HNE*, C5ʳ⁻ᵛ, p. 16 above, and *OFB*, XIII, pp. 162, 222–3, 302.

### 2C7ʳ⁻ᵛ

Page 346, ll. 7–18: CANON. I—for the doctrine that matter cannot be annihilated see *OFB*, XIII, pp. 36–8, 269. Also cf. *SS*, E4ʳ (*SEH*, II, pp. 383–4): 'There is nothing more Certaine in Nature, than that it is impossible for any *Body*, to be vtterly *Annihilated*; But that, as it was the worke of the Omnipotency of *God*, to make *Somewhat* of *Nothing*; So it requireth the like Omnipotency, to turne *Somewhat* into *Nothing* . . . And herein is contained also a great Secret of Preseruation of Bodies from Change; For if you can prohibit, that they neither turne into *Aire*, because no *Aire* commeth to them; Nor goe into the *Bodies Adiacent*, because they are vtterly Heterogeneall; Nor make a *Round* and *Circulation* within themselues; they will neuer change, though they be in their Nature neuer so Perishable, or Mutable. We see, how *Flies*, and *Spiders*, and the like, get a *Sepulcher* in *Amber*, more Durable, than the *Monument*, and *Embalming* of the *Body* of any *King*. And I conceiue the like will be of *Bodies* put into *Quick-siluer*.' On honey see *SS*, N1ʳ (*SEH*, II, p. 454).

### 2C7ᵛ–2D1ʳ

Page 346, l. 19–p. 348, l. 10: CANON. II—here Bacon begins to put his theoretical cards on the table. Inanimate spirit is one of the principal sources of activity in the terrestrial world. It is a material substance which is characteristically aggressive and prone to attack the tangible matter in which it is imprisoned, and to convert it into more spirit. This spirit is weightless and so when it escapes into the air (its qualitative cousin) tangible bodies lose weight. The notion of inanimate spirit is omnipresent in Bacon's philosophical writings; see, for instance, *OFB*, VI, pp. lix–lxiv; XI, pp. 346–50; XIII, pp. 275–6. For Bacon's rejection of the vacuum hypothesis see *OFB*, VI, pp. 176, 392; XI, pp. 414, 579–80.

### 2D1ʳ–2D2ᵛ

Page 348, l. 11–p. 350, l. 10: CANON. III—in *NO* (*OFB*, XI, pp. 346–8) the action of the inanimate spirit is three- not fourfold, putrefaction and vivification being regarded as aspects of a single operation of the spirit. For these doctrines see *OFB*, VI, pp. lix–lx, lxii, 274–6, 282–8, 304–8; XIII, pp. xxxiv, xli, xliii, 66–8, 80–2, 118, 130, 280, 291, 294–5. For spontaneous generation and generation from seed, see *HIDA*, in *OFB*, XIII, pp. 228–35.

**2D3ʳ–2D5ʳ**

Page 350, l. 11–p. 352, l. 17: CANON. IV—the distinction between inanimate and animate or vital spirits is fundamental to many of the proposals advanced in *HVM*. For its nature and its relation to traditional theories of animal spirits see *OFB*, VI, pp. liv–lix, lxix. For an earlier version of the twin-track approach to the prolongation of life—the body regarded as if it were an inanimate substance as well as an animate one requiring nutrition—see *DVM* (*OFB*, VI, pp. 274–5). For the concentration of the vital spirit in the cerebral ventricles, see *SS*, O4ᵛ–P1ʳ (*SEH*, II, p. 474), quoted in *cmts* pp. 454–5 above.

**2D5ʳ–2D6ʳ**

Page 352, ll. 18–30: CANON. V—see *OFB*, VI, pp. lvii–lix.

**2D6ʳ⁻ᵛ**

Page 354, ll. 1–13: CANON. VI—on vital and inanimate spirits as compounds, see *OFB*, VI, p. lv.

**2D7ʳ–2D8ʳ**

Page 354, l. 14–p. 356, l. 5: CANON. VII—inanimate spirit and air are kindred spirits, and that is fundamental to the behaviour of the former; see *OFB*, VI, lxii.

**2D8ʳ⁻ᵛ**

Page 356, ll. 6–17: CANON. VIII—for this and the next canon, see *HVM*, Z2ʳ–Z8ᵛ (pp. 318–24 above).

**2E1ᵛ–2E2ᵛ**

Page 356, l. 27–p. 358, l. 17: CANON. X—see *HVM*, O6ʳ, Q2ᵛ–Q3ᵛ (pp. 244–6, 258–60 above).

**2E2ᵛ–2E4ᵛ**

Page 358, l. 18–p. 360, l. 24: CANON. XI . . . XIV—see pp. xlvii–xlix above. On the condensation of spirits see O7ʳ–R6ʳ, pp. 246–72 above. For abundance of spirit and even distribution see *DVM* in *OFB*, VI, pp. 294–300.

**2E4ᵛ–2E5ᵛ**

Page 360, l. 25–p. 362, l. 18: CANON. XV—cf. *OFB*, VI, pp. 348–9; also see *SS*, 2F2ᵛ (*SEH*, II, p. 616).

**2E6ʳ–2E7ᵛ**

Page 362, l. 19–p. 364, l. 19: CANON. XVI . . . XVIII—these are, for Bacon, important proposals which rest on the durability of oily bodies, and their capacity to disrupt the inanimate spirits' *actio triplex* and tendency to damage the body by attempting to escape into the air; see *OFB*, VI, pp. lx, lxii, 274–6, 300. For the second of the ten operations see *HVM*, R6ᵛ–S7ʳ, pp. 272–80 above.

**2E7ᵛ–2E8ʳ**

Page 364, l. 20–p. 366, l. 3: CANON. XIX—see *cmts* (see below this page) on *HVM*, 2F5ᵛ–2F6ᵛ, pp. 372–4 below.

**2E8ʳ–2F1ʳ**

Page 366, ll. 4–23: CANON. XX ... XXI—in fact more is said about this in the discussion of the fourth rather than the tenth operation; see *HVM*, T4ʳ–V1ᵛ, pp. 286–92 above.

**2F1ʳ**

Page 366, l. 24–p. 368, l. 3: CANON. XXII—see *HVM*, Y6ᵛ–Z1ᵛ, pp. 314–16 above.

**2F1ᵛ–2F2ʳ**

Page 368, l. 4–p. 370, l. 24: CANON. XXIII ... XXVI—see *HVM*, Z2ʳ–Z8ᵛ, pp. 318–24 above.

**2F4ʳ⁻ᵛ**

Page 370, l. 25–p. 372, l. 11: CANON. XXVII—for Aristotle and plants see *cmts* (p. 421 above) on *HVM*, C6ᵛ–C7ᵛ, p. 162 above.

**2F4ᵛ–2F5ʳ**

Page 372, ll. 12–14: CANON. XXVIII—see *HVM*, Q1ᵛ, S7ᵛ–T4ʳ, pp. 258, 282–6 above.

**2F5ʳ⁻ᵛ**

Page 372, ll. 20–7: CANON. XXIX—see *HVM*, C8ʳ⁻ᵛ, E3ʳ–E6ʳ, pp. 174–6 above.

**2F5ᵛ–2F6ᵛ**

Page 374, ll. 1–26: CANON. XXX—for Bacon's sense that prolonging life was a matter of swimming against a mighty tide, see *HVM*, N8ᵛ–O1ʳ, p. 240, and *DAS*, 2D3ᵛ (*SEH*, I, p. 599). Set diets are three: opiate (see *HVM*, O7ᵛ–P4ᵛ, pp. 246–52 above), emollient (Bacon says very little about these; see *HVM*, T1ʳ, Z3ᵛ, pp. 282–4, 318–20), and emaciating and renewing (see *HVM*, D7ᵛ–D8ʳ, N3ʳ, Q5ʳ, pp. 170–2, 236, 262 above).

**2F7ʳ⁻ᵛ**

Page 374, l. 27–p. 376, l. 8: CANON. XXXI—for Death's Anterooms see pp. 328–40 above. On the relationship between the immortal soul and the spirit see *DAS*, 2E3ʳ–2E4ᵛ (*SEH*, I, pp. 604–7).

**2F7ᵛ–2F8ᵛ**

Page 376, ll. 9–33: CANON. XXXII—this is one of Bacon's clearest accounts of the composition of vital spirit. It is an air–flame compound or intermediate (for intermediates, see *OFB*, VI, pp. liv–lvi). The flamy component provides the vital spirit with its capacity to move the body it occupies; the airy component is its sensory aspect (see *OFB*, VI, pp. lviii–lix). Flame exists only as a succession of flames in the sublunar world; the celestial flame of the planets and stars is permanent and self-subsisting because it is more powerful than the interstellar medium, ether, which is the celestial relative of mundane air. Air, though durable, can of course be created anew from water and, at the very end, reminding readers that he has written *HV*, Bacon recalls that the overcharging caused by newly created air causes winds; for which see *HV*, L3ᵛ–L4ʳ, Q4ᵛ–Q5ʳ (pp. 82–4, 124 above). For the relationship between the spirit and blood, see *OFB*, VI, p. lix.

# APPENDIX I

*Historia naturalis et experimentalis,* and *Historia vitæ &*
*mortis:* Bibliographical Description and Technical Notes

## I. *Historia naturalis et experimentalis:*

FRANCISCI | BARONIS | DE | VERVLAMIO, | VICE-COMITIS |
Sancti Albani, | HISTORIA NATVRALIS | ET EXPERIMENTALIS | AD
CONDENDAM | Philosophiam: | SIVE, | PHÆNOMENA
VNIVERSI: | Quæ eſt Inſtaurationis Magnæ | PARS TERTIA. | [rule] |
[fleuron] | [rule] | LONDINI, | In Officina Io. HAVILAND, impenſis |
*Matthæi Lownes* & *Guilielmi Barret.* | 1622.

*Coll*: common 8°: $A^4$ $B^8$–$T^8$ [\$4 signed (–A1, A2, A3, A4)] 148 leaves present
[*i–viii*] 1–17 *18 19*–285 [*286–8*].

*Contents*: [A1ʳ]: letterpress title. [A1ᵛ]: blank. [A2ʳ–A3ʳ]: 'ILLVSTRISSI- | MO,
ET EXCELLEN- | tiſſimo Principi, CAROLO, | Sereniſſimi Regis
IACOBI | Filio, & Hæredi.'. A3ᵛ: blank. A4ʳ: 'TITVLI | Hiſtoriarum &
Inquiſi- | tionum in primos ſex men- | ſes deſtinatarum.'. A4ᵛ: blank.
B1ʳ–C1ʳ: 'HISTORIA | NATVRALIS ET | *Experimentalis, ad condendam* |
*Philoſophiam:* | SIVE, | *Phænomena Vniverſi: quæ* | eſt Inſtaurationis
Magnæ | pars tertia.'. C1ᵛ: blank. C2ʳ–C6ᵛ: 'Norma Hiſtoriæ præſentis.'.
C7ʳ–C8ʳ: 'Hiſtoria Ventorum. | Aditus, ſiue Præfatio.'. C8ᵛ–E1ʳ: 'Topica
Particularia; | Siue, | *Articuli Inquiſitionis* | *de Ventis.*'. E1ʳ–E3ʳ: 'HISTORIA.
| *Nomina Ventorum.*'. E3ʳ⁻ᵛ: '*Venti Liberi.*'. E3ᵛ–E6ᵛ: '*Venti Generales.*'. E7ʳ–
F2ʳ: 'Venti Stati.'. F2ʳ–F7ʳ: 'Venti Aſſeclæ.'. F7ʳ–G8ʳ: 'Qualitates &
Poteſtates | Ventorum.'. G8ᵛ–I1ᵛ: 'Origines locales | Ventorum.'. I1ᵛ–I3ᵛ:
'Accidentales generatio- | nes Ventorum.'. I3ᵛ–I5ʳ: 'Venti extraordinarij & |
Flatus repentini.'. I5ᵛ–K6ʳ: 'Confacientia ad Ventos; | Originales ſcilicet,
nam de | Accidentalibus, ſuprà | inquiſitum eſt.'. K6ᵛ–K8ᵛ: '*Limites Vento-*
*rum.*'. K8ᵛ–L2ʳ: '*Succeſſiones Ventorum.*'. L2ʳ–M2ʳ: 'Motus Ventorum.'. M2ᵛ–
N2ʳ: '*Motus Ventorum in ve-* | *lis nauium.*'. N2ʳ–N7ʳ: 'Obſeruationes
maiores.'. N7ᵛ–O2ʳ: '*Motus Ventorum in alijs* | *Machinis humanis.*'. O2ʳ–P7ᵛ:
'Prognoſtica Ventorum.'. P7ᵛ–Q2ʳ: 'Imitamenta Ventorum.'. Q2ʳ–Q3ᵛ:
'Obſeruatio maior.'. Q3ᵛ–Q8ᵛ: 'Canones mobiles | de Ventis.'. Q8ᵛ–R3ᵛ:
'Charta humana; ſiue op- | tatiua cum proximis, | circa ventos.'. R4ʳ–R8ᵛ:
'ADITVS AD | TITVLOS IN | proximos quinque | Menſes deſtinatos. |
Hiſtoria Denſi & Rari. | Aditus.'. R8ᵛ–S2ʳ: 'Hiſtoria Grauis & | Leuis. |
Aditus.'. S2ᵛ–S4ᵛ: 'Hiſtoria Sympathiæ & | Antipathiæ rerum. | Aditus.'.

461

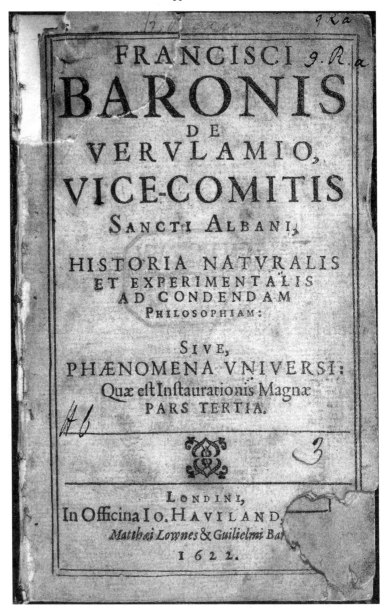

I. Title-page of *Historia naturalis et experimentalis*, 1622—(reduced). BL 535.a.4. *By permission of the British Library*

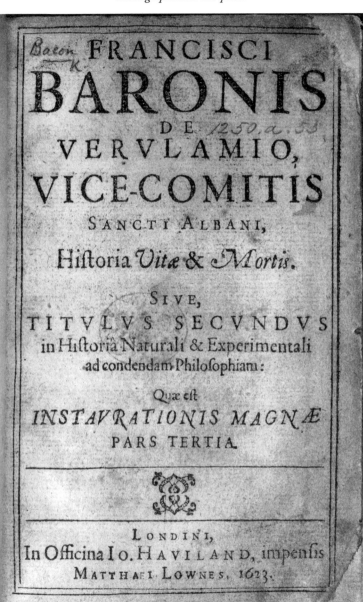

FRANCISCI
BARONIS
DE
VERVLAMIO,
VICE-COMITIS
SANCTI ALBANI,

Hiſtoria *Vitæ* & *Mortis*.

SIVE,
TITVLVS SECVNDVS
in Hiſtoria Naturali & Experimentali
ad condendam Philoſophiam:

Quæ eſt
*INSTAVRATIONIS MAGNÆ*
PARS TERTIA.

LONDINI,
In Officina Io. HAVILAND, impenſis
MATTHÆI LOWNES. 1623.

II. Title-page of *Historia vitæ & mortis*, 1623—(reduced) BL 1250.a.33. *By permission of the British Library*

*Durationem,* aut Diſſolutionem
valent. Nam *Ligna, Lapides,* alia,
vel in Aquâ, vel in Aere perpetuo
manentia, plus durant, quàm ſi
quandóque alluantur, quandóque afflentur. Atque *Lapides* eruti, & in Ædificijs poſiti, diutiùs
durant, ſi eodem ſitu, & ad eaſ-
dem Cœli plagas ponantur, quibus iacebant in Mineris : id
quod *plantis* etiam è loco motis,
& aliò tranſplantatis, accidit.

## Obſeruationes maiores.

I.

L Oco Aſſumpti *pona-*
*tur, quod certiſſimum*
*eſt ; Jneſſe omni Tangibili*
Spiri-

III. *Historia vitæ & mortis,* C2ᵛ—(reduced), illustrating the difference between 20-line and 16-line type. *By permission of the British Library*

*Historia Vitæ & Mortis.* 3 1

Spiritum, *siue* Corpus
pneumaticum, *Partibus*
*tangibilibus obtectum,* &
*inclusum;* *Atque ex illo*
Spiritu *initium capi omnis*
Dissolutionis, & Con-
sumptionis; *Itaque earun-*
*dem* *Antidotum est,* Detentio Spiritus.

    Spiritus *detinetur dupli-* 2.
*ci modo;* *Aut* per Com-
pressionem *arctam, tan-*
*quam in Carcere;* *aut per*
Detentionem *tanquā spon-*
*taneam:* *Atque ea* Mansio
*etiam duplici ratione inuita-*
       C 3    *tur;*

IV. *Historia vitæ & mortis,* C3ʳ—(reduced), illustrating the difference between
20-line and 16-line type. *By permission of the British Library*

S5ʳ–S8ʳ: 'Hiſtoria Sulphuris, Mer- | curij, & Salis. | Aditus.'. S8ᵛ–T7ʳ: 'Hiſtoria Vitæ & Mortis. | Aditus.'. T7ᵛ–T8ᵛ: blank.

RT] C2ᵛ–C6ʳ: Norma Hiſtoriæ præſentis. | Norma Hiſtoriæ præſentis. C6ᵛ–C7ʳ: Norma Hiſtoriæ præſentis. | blank. C7ᵛ–C8ʳ: *Hiſtoria Ventorum.* | *Hiſtoria Ventorum.* C8ᵛ–D1ʳ: blank | *Hiſtoria Ventorum.* D1ᵛ–R4ʳ: *Hiſtoria Ventorum.* | blank. R4ᵛ–T8ᵛ: blank | blank.

*Signatures*: regular and invariant in all copies examined.

CW] F7ʳ *variè* [*variè*]. F8ᵛdixerunt) [dixerunt.)]. G3ᵛ ſani [ſani,].

*Typography*: The vertical height of twenty lines of 20-line type is 117 mm. Twenty lines of 16-line set solid would have a vertical height of 143 mm (see Introduction, p. lxiii above); i.e. for 20-line type, roman and italic, 117 (130) × 63; for 16-line type, roman and italic, 143 (144 (hypothetical)) × 63.


II. *Historia vitæ & mortis*:


FRANCISCI | BARONIS | ᴅᴇ | VERVLAMIO, | VICE-COMITIS | Sᴀɴᴄᴛɪ Aʟʙᴀɴɪ, | Hiſtoria *Vitæ* & *Mortis.* | Sɪᴠᴇ, | TITVLVS SECVNDVS | in Hiſtoriâ Naturali & Experimentali | ad condendam Philoſophiam: | Quæ eſt | *INSTAVRATIONIS MAGNÆ* | ᴘᴀʀs ᴛᴇʀᴛɪᴀ. | [rule] | [fleuron] | [rule] | Lᴏɴᴅɪɴɪ, | In Officina Io. Hᴀᴠɪʟᴀɴᴅ, impenſis | Mᴀᴛᴛʜᴀᴇɪ Lᴏᴡɴᴇs. 1623.


*Coll*: common 8°: A⁸–2F⁸ [$4 signed (–A1, 2A4)] 232 leaves present, pp. [*i–vi*] 1–191 *192* 193–368 *369–70* 371–404 *405–6* 407–9 410 (410 corrected from 406 (some copies)) 407–54 (407–54 uncorrected sequence in all copies).

*Contents*: [A1ʳ]: letterpress title. A1ᵛ: blank. A2ʳ–A3ʳ: to the reader: 'VIVENTI-BVS | ET POSTERIS | Salutem.'. A3ᵛ: blank. A4ʳ–B2ᵛ: 'Hiſtoria Vitæ & Mortis. | Aditus.'. B3ʳ–B7ᵛ: 'Topica Particularia; | Siue, | *Articuli Inquiſitionis* | de *Vitâ* & *Morte*.'. B7ᵛ–C2ᵛ: 'Natura Durabilis. | *Hiſtoria*.'. C2ᵛ–C3ᵛ: 'Obſeruationes maiores.'. C4ʳ–C6ᵛ: '*Hiſtoria*.'. C6ᵛ–C7ᵛ: 'Obſeruatio maior.'. C8ʳ–D8ᵛ: 'Deſiccatio; Deſiccatio- | nis prohibitio; & De- | ſiccati Inteneratio. | *Hiſtoria*.'. D8ᵛ–E6ʳ: 'Obſeruationes maiores.'. E6ᵛ–F8ᵛ: 'Longæuitas, & Breui- | tas Vitæ in Ani- | malibus. | *Hiſtoria* '. G1ʳ–G6ᵛ: 'Obſeruationes maiores.'. G7ʳ–H2ᵛ: 'ALIMENTATIO | & | Via Alimentandi. | *Historia*.'. H2ᵛ–M7ᵛ: 'LONGÆVITAS | & Breuitas vitæ in | Homine. | *Hiſtoria*.'. M7ᵛ–N3ᵛ: 'MEDICINÆ AD | Longæuitatem.'. N3ᵛ: blank. N4ʳ–O5ʳ: 'Intentiones.'. O5ᵛ–R6ᵛ: 'Operatio ſuper Spiritus, | vt maneant iuueniles, | & reuireſcant. I. | *Hiſtoria*.'. R6ᵛ–S7ʳ: 'OPERATIO | ſuper Excluſionem | Aeris. II. | *Historia*.'. S7ᵛ–T4ʳ: 'OPERATIO | ſuper Sanguinem, & Ca- | lorem ſanguifican- | tem. III. | *Historia*.'. T4ʳ–V1ᵛ: 'OPERATIO | ſuper Succos Cor- | poris IV. | *Hiſtoria*.'. V2ʳ–X6ᵛ:

'OPERATIO | ſuper Viſcera ad Ex- | truſionem Ali- | menti V. | *Historia_.*'.
X6ᵛ–Y1ʳ: 'OPERATIO | ſuper partes exteriores | ad Attractionem | Alimenti.
VI. | *Historia_.*'. Y1ᵛ–Y6ʳ: 'OPERATIO | ſuper Alimentum ipſum | ad
Inſinuationem eiuſ- | dem. VII. | *Hiſtoria_.*'. Y6ᵛ: 'OPERATIO | ſuper
Actum vltimum | Aſsimilationis. | VIII.'. Y7ʳ–Z1ᵛ: 'Commentatio.' Z2ʳ⁻ᵛ:
'OPERATIO | ſuper Intenerationem | eius quod arefieri | cœpit, ſiue Ma- |
laciſſatio Cor- | poris. IX.'. Z2ᵛ–Z8ᵛ: '*Historia_.*'. 2A1ʳ–2A3ᵛ: 'OPERATIO |
ſuper Expurgatione*m* Suc- | ci veteris, & Reſtitutio- | nem Succi noui, ſiue
Re- | nouationem per Vi- | ces. X. | *Hiſtoria_.*'. 2A4ʳ⁻ᵛ: blank. 2A5ʳ⁻ᵛ: 'Atriola
Mortis.'. 2A5ᵛ–2C1ʳ: '*Hiſtoria_.*'. 2C1ᵛ–2C5ᵛ: 'Diſcrimina Iuuentutis & |
Senectutis.'. 2C6ʳ⁻ᵛ: blank. 2C7ʳ–2F8ᵛ: 'CANONES | mobiles de Duratione |
Vitæ, & Formâ | Mortis.'.

RT] B3ᵛ–C6ᵛ: *Hiſtoria Vitæ & Mortis.* | *Hiſtoria Vitæ & Mortis.* C6ᵛ–C7ʳ:
*Hiſtoria Vitæ & Mortis.* | *Historia Vitæ & Mortis.* C7ᵛ–D6ʳ *Hiſtoria Vitæ &*
*Mortis.* | *Hiſtoria Vitæ & Mortis.* This pattern repeats itself (D6ᵛ–D7ʳ
excepted) with *Hiſtoria Vitæ & Mortis.* | *Historia Vitæ & Mortis.* at 4ᵛ–5ʳ and
with *Hiſtoria Vitæ & Mortis.* | *Hiſtoria Vitæ & Mortis.* on all other openings
in all quires up to quire Z. 2A4ʳ⁻ᵛ are blank. In all remaining quires (2B–2F),
*Hiſtoria Vitæ & Mortis.* | *Historia Vitæ & Mortis.* appears on 5ᵛ–6ʳ, and
*Hiſtoria Vitæ & Mortis.* | *Hiſtoria Vitæ & Mortis.* appears on all other openings
(leaf 2C6 is blank).

*Signatures*: regular and invariant in all copies examined.

CW] B6ᵛ (inqui-)ſitio; [ſitio,]. D4ʳ (pro-)fundo [profundo] (uncorrected
sheets); (pro-)fundo [fundo] (corrected sheets). E5ʳ craſ- [*craſſiorum*, ].
E5ᵛ quam [quam_]. F3ᵛ (Mag-)nitudi- [nitudinem]. F4ᵛ *Cygnus* [*Cygnus*,].
I7ʳ Humilior [Humilior,]. L4ʳ eadem [eade*m*]. M8ᵛ quòd [Quòd]. T2ʳ
(ex-)pertia [pertia,]. T3ʳ (*San-*)*talum*, [*talum*;]. T7ᵛ *Acida* [*Acida*,]. Y5ʳ
Etiam [Etiàm]. 2D3ᵛ *abſciſ-* [*abſciſſi*,]. 2F5ʳ (*inpræſen-*)*tia* [*tiâ*]

*Typography*: the same as *HNE*.

Copies collated:

*Historia naturalis et experimentalis*:

Bibliothèque nationale de France R-19890 (formerly R-2908A) (control copy)

Bodleian Library 8° B 101(1) Art

Bodleian Library 8° V 16 Art. Seld

Bodleian Library 8° E 105(1) Linc

Bodleian Library Ashm. 1344

British Library 535.a.4

Cambridge University Library Keynes E.2.23

Corpus Christi College, Oxford N.4.16(1)

Folger STC 1155 copy 1

Folger STC 1155 copy 2 (bound with Folger STC 1156 copy 3)

Huntington Library 12569

Huntington Library 56658

Huntington Library 601101 (bound with *HVM*)

Library of Congress QC 931 .B3 1638b Fabyan Coll

Linda Hall Library, Kansas City, Mo. QH41.B317

Magdalene College, Cambridge A.7.45(4)

University of Minnesota, T. C. Wilson Library 194B13 OH

Newberry Library, Chicago Case B 245 .0618

Philosophy Library, Oxford Fowler C.1.1

Pierpont Morgan Library (W 01 B) 37224

Trinity College, Dublin EE.mm.62

Trinity College, Dublin EE.hh.58 no. 1

Yale University Ih B132 622h

*Historia vitæ & mortis*:

Biblioteca comunale degli Intronati, Siena XXVIII m45

Biblioteca Regia Monacensis, i.e. Bayerische Staatsbibliothek

Bibliothèque municipale de Lyon 348782

Bibliothèque Nationale 8–Tc11–336 (control copy)

Bibliothèque de Fels, Paris

Bodleian Library Tanner 520

British Library 1250.a.53

Cambridge University Library Keynes E.1.20

Cambridge University Library LE.7.59

Corpus Christi College, Oxford N.4.16(2)

Editor's private copy (with bookplate of the earl of Roding)

Emmanuel College, Cambridge S12.5.6(1).

Exeter College, Oxford 9M 1156

Folger STC 1156 copy 1

Folger STC 1156 copy 2

Folger STC 1156 copy 3 (bound with Folger STC 1155 copy 2)

Huntington Library 56659

Huntington Library 601102 (bound with *HNE*)
Huntington Library 601111
Library of Congress B1180 .H5 1623 Fabyan Coll
Library of Congress B1180 .H5 Batchelder Coll
Liverpool University Library SPEC Morton 28
Newberry Library, Chicago Case B 245 .06182
St Johns College, Cambridge Kk.11.5
University of Minnesota, T. C. Wilson Library 194B13 OHis
Yale University Ih B132 623

# APPENDIX II

## Signature References in this Edition with corresponding Page Numbers in *SEH*

*Historia naturalis et experimentalis:*

| This Edition | *SEH*, vol. II | *SEH* trans., vol. V |
|---|---|---|
| A1ʳ | 7 | 125 |
| A1ᵛ | — | — |
| A2ʳ | 9 | 127 |
| A2ᵛ | 9 | 127 |
| A3ʳ | 9 | 127 |
| A3ᵛ | — | — |
| A4ʳ | 11 | 129 |
| A4ᵛ | — | — |
| B1ʳ | 13 | 131 |
| B1ᵛ | 13 | 131 |
| B2ʳ | 13 | 131 |
| B2ᵛ | 13–14 | 131 |
| B3ʳ | 14 | 131–2 |
| B3ᵛ | 14 | 132 |
| B4ʳ | 14 | 132 |
| B4ᵛ | 14 | 132 |
| B5ʳ | 14 | 132 |
| B5ᵛ | 14–15 | 132–3 |
| B6ʳ | 15 | 133 |
| B6ᵛ | 15 | 133 |
| B7ʳ | 15 | 133 |
| B7ᵛ | 15 | 133 |
| B8ʳ | 15–16 | 133–4 |
| B8ᵛ | 16 | 134 |
| C1ʳ | 16 | 134 |
| C1ᵛ | — | — |
| C2ʳ | 17 | 135 |
| C2ᵛ | 17 | 135 |
| C3ʳ | 17 | 135 |
| C3ᵛ | 17 | 135 |

| This Edition | *SEH*, vol. II | *SEH* trans., vol. V |
|---|---|---|
| C4$^r$ | 17–18 | 135–6 |
| C4$^v$ | 18 | 136 |
| C5$^r$ | 18 | 136 |
| C5$^v$ | 18 | 136 |
| C6$^r$ | 18 | 136 |
| C6$^v$ | 18 | 136 |
| C7$^r$ | 19 | 139 |
| C7$^v$ | 19 | 139 |
| C8$^r$ | 19 | 139 |
| C8$^v$ | 20 | 140 |
| D1$^r$ | 20 | 140 |
| D1$^v$ | 20 | 140–1 |
| D2$^r$ | 20–1 | 141 |
| D2$^v$ | 21 | 141 |
| D3$^r$ | 21 | 141–2 |
| D3$^v$ | 21–2 | 142 |
| D4$^r$ | 22 | 142 |
| D4$^v$ | 22 | 142 |
| D5$^r$ | 22 | 142–3 |
| D5$^v$ | 23 | 143 |
| D6$^r$ | 23 | 143 |
| D6$^v$ | 23 | 143–4 |
| D7$^r$ | 23–4 | 144· |
| D7$^v$ | 24 | 144 |
| D8$^r$ | 24 | 144–5 |
| D8$^v$ | 24–5 | 145 |
| E1$^r$ | 25 | 145 |
| E1$^v$ | 25 | 145–6 |
| E2$^r$ | 25 | 146 |
| E2$^v$ | 25 | 146 |
| E3$^r$ | 26 | 146 |
| E3$^v$ | 26 | 146–7 |
| E4$^r$ | 26 | 147 |
| E4$^v$ | 26–7 | 147 |
| E5$^r$ | 27 | 147–8 |
| E5$^v$ | 27 | 148 |
| E6$^r$ | 27–8 | 148 |
| E6$^v$ | 28 | 148–9 |
| E7$^r$ | 28 | 149 |

| This Edition | *SEH*, vol. II | *SEH* trans., vol. V |
|---|---|---|
| E7$^v$ | 28–9 | 149 |
| E8$^r$ | 29 | 149 |
| E8$^v$ | 29 | 150 |
| F1$^r$ | 29–30 | 150 |
| F1$^v$ | 30 | 150 |
| F2$^r$ | 30 | 150–1 |
| F2$^v$ | 30 | 151 |
| F3$^r$ | 30–1 | 151 |
| F3$^v$ | 31 | 151 |
| F4$^r$ | 31 | 151–2 |
| F4$^v$ | 31–2 | 152 |
| F5$^r$ | 32 | 152 |
| F5$^v$ | 32 | 152–3 |
| F6$^r$ | 32 | 153 |
| F6$^v$ | 32–3 | 153 |
| F7$^r$ | 33 | 153 |
| F7$^v$ | 33 | 153–4 |
| F8$^r$ | 33–4 | 154 |
| F8$^v$ | 34 | 154 |
| G1$^r$ | 34 | 154–5 |
| G1$^v$ | 34 | 155 |
| G2$^r$ | 34–5 | 155 |
| G2$^v$ | 35 | 155–6 |
| G3$^r$ | 35 | 156 |
| G3$^v$ | 35–6 | 156 |
| G4$^r$ | 36 | 156–7 |
| G4$^v$ | 36 | 157 |
| G5$^r$ | 36–7 | 157 |
| G5$^v$ | 37 | 157–8 |
| G6$^r$ | 37 | 158 |
| G6$^v$ | 37 | 158 |
| G7$^r$ | 37–8 | 158 |
| G7$^v$ | 38 | 158–9 |
| G8$^r$ | 38 | 159 |
| G8$^v$ | 38–9 | 159 |
| H1$^r$ | 39 | 159 |
| H1$^v$ | 39 | 159–60 |
| H2$^r$ | 39 | 160 |
| H2$^v$ | 39–40 | 160 |

| This Edition | *SEH*, vol. II | *SEH* trans., vol. V |
|---|---|---|
| H3ʳ | 40 | 160–1 |
| H3ᵛ | 40 | 161 |
| H4ʳ | 40–1 | 161 |
| H4ᵛ | 41 | 161–2 |
| H5ʳ | 41 | 162 |
| H5ᵛ | 41 | 162 |
| H6ʳ | 41–2 | 162–3 |
| H6ᵛ | 42 | 163 |
| H7ʳ | 42 | 163 |
| H7ᵛ | 42–3 | 163–4 |
| H8ʳ | 43 | 164 |
| H8ᵛ | 43 | 164 |
| I1ʳ | 43 | 164 |
| I1ᵛ | 43–4 | 164–5 |
| I2ʳ | 44 | 165 |
| I2ᵛ | 44 | 165 |
| I3ʳ | 44 | 165 |
| I3ᵛ | 44–5 | 165–6 |
| I4ʳ | 45 | 166 |
| I4ᵛ | 45 | 166 |
| I5ʳ | 45 | 166 |
| I5ᵛ | 46 | 167 |
| I6ʳ | 46 | 167 |
| I6ᵛ | 46 | 167 |
| I7ʳ | 46–7 | 167 |
| I7ᵛ | 47 | 167–8 |
| I8ʳ | 47 | 168 |
| I8ᵛ | 47 | 168 |
| K1ʳ | 48 | 168–9 |
| K1ᵛ | 48 | 169 |
| K2ʳ | 48 | 169 |
| K2ᵛ | 48–9 | 169–70 |
| K3ʳ | 49 | 170 |
| K3ᵛ | 49 | 170 |
| K4ʳ | 49 | 170–1 |
| K4ᵛ | 49–50 | 171 |
| K5ʳ | 50 | 171 |
| K5ᵛ | 50 | 171 |
| K6ʳ | 50 | 171–2 |

| This Edition | *SEH*, vol. II | *SEH* trans., vol. V |
|---|---|---|
| K6ᵛ | 51 | 172 |
| K7ʳ | 51 | 172 |
| K7ᵛ | 51 | 172–3 |
| K8ʳ | 51–2 | 173 |
| K8ᵛ | 52 | 173 |
| L1ʳ | 52 | 173 |
| L1ᵛ | 52–3 | 173–4 |
| L2ʳ | 53 | 174 |
| L2ᵛ | 53 | 174 |
| L3ʳ | 53–4 | 174–5 |
| L3ᵛ | 54 | 175 |
| L4ʳ | 54 | 175 |
| L4ᵛ | 54 | 175–6 |
| L5ʳ | 54–5 | 176 |
| L5ᵛ | 55 | 176 |
| L6ʳ | 55 | 176 |
| L6ᵛ | 55–6 | 176–7 |
| L7ʳ | 56 | 177 |
| L7ᵛ | 56 | 177 |
| L8ʳ | 56 | 177 |
| L8ᵛ | 56–7 | 178 |
| M1ʳ | 57 | 178 |
| M1ᵛ | 57 | 178 |
| M2ʳ | 57 | 178–9 |
| M2ᵛ | 58 | 179 |
| M3ʳ | 58 | 179 |
| M3ᵛ | 58 | 179 |
| M4ʳ | 58–9 | 179–80 |
| M4ᵛ | 59 | 180 |
| M5ʳ | 59 | 180 |
| M5ᵛ | 59–60 | 180 |
| M6ʳ | 60 | 180–1 |
| M6ᵛ | 60 | 181 |
| M7ʳ | 60 | 181 |
| M7ᵛ | 60–1 | 181–2 |
| M8ʳ | 61 | 182 |
| M8ᵛ | 61 | 182 |
| N1ʳ | 61 | 182–3 |
| N1ᵛ | 61–2 | 183 |

| This Edition | *SEH*, vol. II | *SEH* trans., vol. V |
|---|---|---|
| N2$^r$ | 62 | 183 |
| N2$^v$ | 62 | 183 |
| N3$^r$ | 62 | 183 |
| N3$^v$ | 62 | 183–4 |
| N4$^r$ | 62–3 | 184 |
| N4$^v$ | 63 | 184 |
| N5$^r$ | 63 | 184 |
| N5$^v$ | 63 | 184 |
| N6$^r$ | 63 | 184–5 |
| N6$^v$ | 63–4 | 185 |
| N7$^r$ | 64 | 185 |
| N7$^v$ | 64 | 185 |
| N8$^r$ | 64 | 185–6 |
| N8$^v$ | 64 | 186 |
| O1$^r$ | 64–5 | 186 |
| O1$^v$ | 65 | 186 |
| O2$^r$ | 65–6 | 187 |
| O2$^v$ | 66 | 187 |
| O3$^r$ | 66 | 187 |
| O3$^v$ | 66 | 187 |
| O4$^r$ | 66–7 | 187–8 |
| O4$^v$ | 67 | 188 |
| O5$^r$ | 67 | 188 |
| O5$^v$ | 67 | 188–9 |
| O6$^r$ | 67–8 | 189 |
| O6$^v$ | 68 | 189 |
| O7$^r$ | 68 | 189 |
| O7$^v$ | 68–9 | 189–90 |
| O8$^r$ | 69 | 190 |
| O8$^v$ | 69 | 190 |
| P1$^r$ | 69 | 190–1 |
| P1$^v$ | 69–70 | 191 |
| P2$^r$ | 70 | 191 |
| P2$^v$ | 70 | 191–2 |
| P3$^r$ | 70 | 192 |
| P3$^v$ | 70–1 | 192 |
| P4$^r$ | 71 | 192–3 |
| P4$^v$ | 71 | 193 |
| P5$^r$ | 71–2 | 193 |

| This Edition | *SEH*, vol. II | *SEH* trans., vol. V |
|---|---|---|
| P5$^v$ | 72 | 193 |
| P6$^r$ | 72 | 193–4 |
| P6$^v$ | 72 | 194 |
| P7$^r$ | 73 | 194 |
| P7$^v$ | 73 | 194 |
| P8$^r$ | 73 | 194–5 |
| P8$^v$ | 73–4 | 195 |
| Q1$^r$ | 74 | 195 |
| Q1$^v$ | 74 | 195–6 |
| Q2$^r$ | 74 | 196 |
| Q2$^v$ | 74–5 | 196 |
| Q3$^r$ | 75 | 196 |
| Q3$^v$ | 75 | 196 |
| Q4$^r$ | 75 | 196–7 |
| Q4$^v$ | 75 | 197 |
| Q5$^r$ | 75 | 197 |
| Q5$^v$ | 75–6 | 197 |
| Q6$^r$ | 76 | 197 |
| Q6$^v$ | 76 | 197 |
| Q7$^r$ | 76 | 197–8 |
| Q7$^v$ | 76 | 198 |
| Q8$^r$ | 76–7 | 198 |
| Q8$^v$ | 77 | 198 |
| R1$^r$ | 77 | 198–9 |
| R1$^v$ | 77 | 199 |
| R2$^r$ | 77–8 | 199 |
| R2$^v$ | 78 | 199–200 |
| R3$^r$ | 78 | 200 |
| R3$^v$ | 78 | 200 |
| R4$^r$ | 79, 243 | 201, 339 |
| R4$^v$ | 243 | 339 |
| R5$^r$ | 243 | 339 |
| R5$^v$ | 243 | 339–40 |
| R6$^r$ | 243 | 340 |
| R6$^v$ | 243–4 | 340 |
| R7$^r$ | 244 | 340 |
| R7$^v$ | 244 | 340 |
| R8$^r$ | 244 | 340 |
| R8$^v$ | 244, 80 | 340, 202 |

| This Edition | *SEH*, vol. II | *SEH* trans., vol. V |
|---|---|---|
| S1$^r$ | 80 | 202 |
| S1$^v$ | 80 | 202 |
| S2$^r$ | 80 | 202 |
| S2$^v$ | 81 | 203 |
| S3$^r$ | 81 | 203 |
| S3$^v$ | 81 | 203 |
| S4$^r$ | 81 | 203–4 |
| S4$^v$ | 81 | 204 |
| S5$^r$ | 82 | 205 |
| S5$^v$ | 82 | 205 |
| S6$^r$ | 82 | 205 |
| S6$^v$ | 82 | 205 |
| S7$^r$ | 82–3 | 205–6 |
| S7$^v$ | 83 | 206 |
| S8$^r$ | 83 | 206 |
| S8$^v$ | 105 | 217 |
| T1$^r$ | 105 | 217 |
| T1$^v$ | 105 | 217 |
| T2$^r$ | 105 | 217 |
| T2$^v$ | 105–6 | 217–18 |
| T3$^r$ | 106 | 218 |
| T3$^v$ | 106 | 218 |
| T4$^r$ | 106 | 218 |
| T4$^v$ | 106 | 218 |
| T5$^r$ | 106 | 218 |
| T5$^v$ | 106–7 | 218–19 |
| T6$^r$ | 107 | 219 |
| T6$^v$ | 107 | 219 |
| T7$^r$ | 107 | 219 |
| T7$^v$ | — | — |
| T8$^r$ | — | — |
| T8$^v$ | — | — |

## *Historia vitæ & mortis*:

| This Edition | *SEH*, vol. II | *SEH* trans., vol. V |
|---|---|---|
| A1$^r$ | 101 | 213 |
| A1$^v$ | — | — |
| A2$^r$ | 103 | 215 |

| This Edition | *SEH*, vol. II | *SEH* trans., vol. V |
|---|---|---|
| A2ᵛ | 103 | 215 |
| A3ʳ | 103 | 215 |
| A3ᵛ | — | — |
| A4ʳ | 105 | 217 |
| A4ᵛ | 105 | 217 |
| A5ʳ | 105 | 217 |
| A5ᵛ | 105 | 217 |
| A6ʳ | 105–6 | 217–18 |
| A6ᵛ | 106 | 218 |
| A7ʳ | 106 | 218 |
| A7ᵛ | 106 | 218 |
| A8ʳ | 106 | 218 |
| A8ᵛ | 106 | 218 |
| B1ʳ | 106–7 | 218–19 |
| B1ᵛ | 107 | 219 |
| B2ʳ | 107 | 219 |
| B2ᵛ | 107 | 219 |
| B3ʳ | 108 | 220 |
| B3ᵛ | 108 | 220 |
| B4ʳ | 108 | 220 |
| B4ᵛ | 108–9 | 220–1 |
| B5ʳ | 109 | 221 |
| B5ᵛ | 109 | 221 |
| B6ʳ | 109 | 221–2 |
| B6ᵛ | 109–10 | 222 |
| B7ʳ | 110 | 222 |
| B7ᵛ | 110 | 222 |
| B8ʳ | 110–11 | 222–3 |
| B8ᵛ | 111 | 223 |
| C1ʳ | 111 | 223 |
| C1ᵛ | 111 | 223–4 |
| C2ʳ | 111–12 | 224 |
| C2ᵛ | 112 | 224 |
| C3ʳ | 112 | 224 |
| C3ᵛ | 112 | 224 |
| C4ʳ | 112 | 225 |
| C4ᵛ | 112–13 | 225 |
| C5ʳ | 113 | 225 |
| C5ᵛ | 113–14 | 225–6 |

| This Edition | *SEH*, vol. II | *SEH* trans., vol. V |
|---|---|---|
| C6ʳ | 114 | 226 |
| C6ᵛ | 114 | 226 |
| C7ʳ | 114 | 226 |
| C7ᵛ | 114 | 226 |
| C8ʳ | 114–15 | 226–7 |
| C8ᵛ | 115 | 227 |
| D1ʳ | 115 | 227 |
| D1ᵛ | 115 | 227 |
| D2ʳ | 115–16 | 227–8 |
| D2ᵛ | 116 | 228 |
| D3ʳ | 116 | 228 |
| D3ᵛ | 116–17 | 228 |
| D4ʳ | 117 | 228–9 |
| D4ᵛ | 117 | 229 |
| D5ʳ | 117 | 229 |
| D5ᵛ | 117–18 | 229–30 |
| D6ʳ | 118 | 230 |
| D6ᵛ | 118 | 230 |
| D7ʳ | 118–19 | 230 |
| D7ᵛ | 119 | 230–1 |
| D8ʳ | 119 | 231 |
| D8ᵛ | 119 | 231 |
| E1ʳ | 119 | 231 |
| E1ᵛ | 119–20 | 231 |
| E2ʳ | 120 | 231–2 |
| E2ᵛ | 120 | 232 |
| E3ʳ | 120 | 232 |
| E3ᵛ | 120 | 232 |
| E4ʳ | 120–1 | 232 |
| E4ᵛ | 121 | 232–3 |
| E5ʳ | 121 | 233 |
| E5ᵛ | 121 | 233 |
| E6ʳ | 121 | 233 |
| E6ᵛ | 121–2 | 233 |
| E7ʳ | 122 | 233 |
| E7ᵛ | 122 | 233–4 |
| E8ʳ | 122–3 | 234 |
| E8ᵛ | 123 | 234 |
| F1ʳ | 123 | 234–5 |

| This Edition | *SEH*, vol. II | *SEH* trans., vol. V |
|---|---|---|
| F1$^v$ | 123–4 | 235 |
| F2$^r$ | 124 | 235 |
| F2$^v$ | 124 | 235 |
| F3$^r$ | 124 | 235–6 |
| F3$^v$ | 124–5 | 236 |
| F4$^r$ | 125 | 236 |
| F4$^v$ | 125 | 236–7 |
| F5$^r$ | 125–6 | 237 |
| F5$^v$ | 126 | 237 |
| F6$^r$ | 126 | 237 |
| F6$^v$ | 126–7 | 238 |
| F7$^r$ | 127 | 238 |
| F7$^v$ | 127 | 238 |
| F8$^r$ | 127–8 | 238–9 |
| F8$^v$ | 128 | 239 |
| G1$^r$ | 128 | 239 |
| G1$^v$ | 128 | 239 |
| G2$^r$ | 128–9 | 239 |
| G2$^v$ | 129 | 239–40 |
| G3$^r$ | 129 | 240 |
| G3$^v$ | 129 | 240 |
| G4$^r$ | 129 | 240 |
| G4$^v$ | 129–30 | 240 |
| G5$^r$ | 130 | 240–1 |
| G5$^v$ | 130 | 241 |
| G6$^r$ | 130 | 241 |
| G6$^v$ | 130 | 241 |
| G7$^r$ | 130–1 | 241 |
| G7$^v$ | 131 | 241–2 |
| G8$^r$ | 131 | 242 |
| G8$^v$ | 131 | 242 |
| H1$^r$ | 131–2 | 242 |
| H1$^v$ | 132 | 242–3 |
| H2$^r$ | 132 | 243 |
| H2$^v$ | 132 | 243 |
| H3$^r$ | 132–3 | 243–4 |
| H3$^v$ | 133 | 244 |
| H4$^r$ | 133 | 244 |
| H4$^v$ | 133–4 | 244 |

| This Edition | *SEH*, vol. II | *SEH* trans., vol. V |
|---|---|---|
| H5$^r$ | 134 | 244–5 |
| H5$^v$ | 134 | 245 |
| H6$^r$ | 134–5 | 245 |
| H6$^v$ | 135 | 245 |
| H7$^r$ | 135 | 245–6 |
| H7$^v$ | 135 | 246 |
| H8$^r$ | 135–6 | 246 |
| H8$^v$ | 136 | 246–7 |
| I1$^r$ | 136–7 | 247 |
| I1$^v$ | 137 | 247 |
| I2$^r$ | 137 | 247 |
| I2$^v$ | 137 | 247–8 |
| I3$^r$ | 137–8 | 248 |
| I3$^v$ | 138 | 248 |
| I4$^r$ | 138 | 248–9 |
| I4$^v$ | 138–9 | 249 |
| I5$^r$ | 139 | 249 |
| I5$^v$ | 139 | 249 |
| I6$^r$ | 139–40 | 249–50 |
| I6$^v$ | 140 | 250 |
| I7$^r$ | 140 | 250 |
| I7$^v$ | 140 | 250 |
| I8$^r$ | 140–1 | 250–1 |
| I8$^v$ | 141 | 251 |
| K1$^r$ | 141–2 | 251 |
| K1$^v$ | 142 | 251–2 |
| K2$^r$ | 142–3 | 252 |
| K2$^v$ | 143 | 252 |
| K3$^r$ | 143 | 252 |
| K3$^v$ | 143–4 | 252–3 |
| K4$^r$ | 144 | 253 |
| K4$^v$ | 144 | 253 |
| K5$^r$ | 144–5 | 253 |
| K5$^v$ | 145 | 253–4 |
| K6$^r$ | 145 | 254 |
| K6$^v$ | 145 | 254 |
| K7$^r$ | 145–6 | 254–5 |
| K7$^v$ | 146 | 255 |
| K8$^r$ | 146–7 | 255 |

| This Edition | *SEH*, vol. II | *SEH* trans., vol. V |
| --- | --- | --- |
| K8ᵛ | 147 | 255 |
| L1ʳ | 147 | 255–6 |
| L1ᵛ | 147–8 | 256 |
| L2ʳ | 148 | 256 |
| L2ᵛ | 148 | 256 |
| L3ʳ | 148 | 256–7 |
| L3ᵛ | 148–9 | 257 |
| L4ʳ | 149 | 257 |
| L4ᵛ | 149 | 257 |
| L5ʳ | 149–50 | 258 |
| L5ᵛ | 150 | 258 |
| L6ʳ | 150 | 258 |
| L6ᵛ | 150 | 258–9 |
| L7ʳ | 150–1 | 259 |
| L7ᵛ | 151 | 259 |
| L8ʳ | 151 | 259–60 |
| L8ᵛ | 151 | 260 |
| M1ʳ | 151–2 | 260 |
| M1ᵛ | 152 | 260 |
| M2ʳ | 152 | 260–1 |
| M2ᵛ | 152 | 261 |
| M3ʳ | 152–3 | 261 |
| M3ᵛ | 153 | 261 |
| M4ʳ | 153 | 261–2 |
| M4ᵛ | 153 | 262 |
| M5ʳ | 153–4 | 262 |
| M5ᵛ | 154 | 262 |
| M6ʳ | 154 | 262–3 |
| M6ᵛ | 154 | 263 |
| M7ʳ | 154–5 | 263 |
| M7ᵛ | 155 | 263 |
| M8ʳ | 155 | 263–4 |
| M8ᵛ | 155–6 | 264 |
| N1ʳ | 156 | 264 |
| N1ᵛ | 156 | 264 |
| N2ʳ | 156 | 264–5 |
| N2ᵛ | 157 | 265 |
| N3ʳ | 157 | 265 |
| N3ᵛ | — | — |

| This Edition | *SEH*, vol. II | *SEH* trans., vol. V |
|---|---|---|
| N4$^r$ | 157 | 265 |
| N4$^v$ | 157 | 265 |
| N5$^r$ | 157–8 | 265–6 |
| N5$^v$ | 158 | 266 |
| N6$^r$ | 158 | 266 |
| N6$^v$ | 158–9 | 266 |
| N7$^r$ | 159 | 266 |
| N7$^v$ | 159 | 266 |
| N8$^r$ | 159 | 266–7 |
| N8$^v$ | 159–60 | 267 |
| O1$^r$ | 160 | 267 |
| O1$^v$ | 160 | 267 |
| O2$^r$ | 160 | 267 |
| O2$^v$ | 160 | 267–8 |
| O3$^r$ | 160 | 268 |
| O3$^v$ | 160–1 | 268 |
| O4$^r$ | 161 | 268 |
| O4$^v$ | 161 | 268 |
| O5$^r$ | 161 | 268 |
| O5$^v$ | 161–2 | 268–9 |
| O6$^r$ | 162 | 269 |
| O6$^v$ | 162 | 269 |
| O7$^r$ | 162 | 269–70 |
| O7$^v$ | 162–3 | 270 |
| O8$^r$ | 163 | 270 |
| O8$^v$ | 163 | 270 |
| P1$^r$ | 163–4 | 270–1 |
| P1$^v$ | 164 | 271 |
| P2$^r$ | 164 | 271 |
| P2$^v$ | 164 | 271–2 |
| P3$^r$ | 164–5 | 272 |
| P3$^v$ | 165 | 272 |
| P4$^r$ | 165 | 272 |
| P4$^v$ | 165 | 272–3 |
| P5$^r$ | 165–6 | 273 |
| P5$^v$ | 166 | 273 |
| P6$^r$ | 166 | 273 |
| P6$^v$ | 166 | 273–4 |
| P7$^r$ | 166–7 | 274 |

| This Edition | *SEH*, vol. II | *SEH* trans., vol. V |
|---|---|---|
| P7$^v$ | 167 | 274 |
| P8$^r$ | 167 | 274–5 |
| P8$^v$ | 167–8 | 275 |
| Q1$^r$ | 168 | 275 |
| Q1$^v$ | 168 | 275 |
| Q2$^r$ | 168 | 275–6 |
| Q2$^v$ | 168–9 | 276 |
| Q3$^r$ | 169 | 276 |
| Q3$^v$ | 169 | 276–7 |
| Q4$^r$ | 169 | 277 |
| Q4$^v$ | 169–70 | 277 |
| Q5$^r$ | 170 | 277 |
| Q5$^v$ | 170 | 277–8 |
| Q6$^r$ | 170 | 278 |
| Q6$^v$ | 170–1 | 278 |
| Q7$^r$ | 171 | 278 |
| Q7$^v$ | 171 | 278–9 |
| Q8$^r$ | 171–2 | 279 |
| Q8$^v$ | 172 | 279 |
| R1$^r$ | 172 | 279–80 |
| R1$^v$ | 172 | 280 |
| R2$^r$ | 172–3 | 280 |
| R2$^v$ | 173 | 280–1 |
| R3$^r$ | 173 | 281 |
| R3$^v$ | 173 | 281 |
| R4$^r$ | 173–4 | 281 |
| R4$^v$ | 174 | 281–2 |
| R5$^r$ | 174 | 282 |
| R5$^v$ | 174–5 | 282 |
| R6$^r$ | 175 | 282–3 |
| R6$^v$ | 175 | 283 |
| R7$^r$ | 175 | 283 |
| R7$^v$ | 175–6 | 283 |
| R8$^r$ | 176 | 283–4 |
| R8$^v$ | 176 | 284 |
| S1$^r$ | 176–7 | 284 |
| S1$^v$ | 177 | 284–5 |
| S2$^r$ | 177 | 285 |
| S2$^v$ | 177–8 | 285 |

| This Edition | *SEH*, vol. II | *SEH* trans., vol. V |
|---|---|---|
| S3<sup>r</sup> | 178 | 285 |
| S3<sup>v</sup> | 178 | 285–6 |
| S4<sup>r</sup> | 178–9 | 286 |
| S4<sup>v</sup> | 179 | 286 |
| S5<sup>r</sup> | 179 | 286–7 |
| S5<sup>v</sup> | 179 | 287 |
| S6<sup>r</sup> | 179–80 | 287 |
| S6<sup>v</sup> | 180 | 287 |
| S7<sup>r</sup> | 180 | 287–8 |
| S7<sup>v</sup> | 180 | 288 |
| S8<sup>r</sup> | 180–1 | 288 |
| S8<sup>v</sup> | 181 | 288 |
| T1<sup>r</sup> | 181 | 288–9 |
| T1<sup>v</sup> | 181 | 289 |
| T2<sup>r</sup> | 181–2 | 289 |
| T2<sup>v</sup> | 182 | 289–90 |
| T3<sup>r</sup> | 182 | 290 |
| T3<sup>v</sup> | 182 | 290 |
| T4<sup>r</sup> | 182–3 | 290 |
| T4<sup>v</sup> | 183 | 290–1 |
| T5<sup>r</sup> | 183 | 291 |
| T5<sup>v</sup> | 183–4 | 291 |
| T6<sup>r</sup> | 184 | 291 |
| T6<sup>v</sup> | 184 | 291–2 |
| T7<sup>r</sup> | 184 | 292 |
| T7<sup>v</sup> | 184–5 | 292 |
| T8<sup>r</sup> | 185 | 292 |
| T8<sup>v</sup> | 185 | 292–3 |
| V1<sup>r</sup> | 185 | 293 |
| V1<sup>v</sup> | 185–6 | 293 |
| V2<sup>r</sup> | 186 | 293 |
| V2<sup>v</sup> | 186 | 293–4 |
| V3<sup>r</sup> | 186–7 | 294 |
| V3<sup>v</sup> | 187 | 294 |
| V4<sup>r</sup> | 187 | 294 |
| V4<sup>v</sup> | 187 | 294–5 |
| V5<sup>r</sup> | 187–8 | 295 |
| V5<sup>v</sup> | 188 | 295 |
| V6<sup>r</sup> | 188 | 295 |

| This Edition | *SEH*, vol. II | *SEH* trans., vol. V |
|---|---|---|
| V6$^v$ | 188 | 295–6 |
| V7$^r$ | 188–9 | 296 |
| V7$^v$ | 189 | 296 |
| V8$^r$ | 189 | 296–7 |
| V8$^v$ | 189–90 | 297 |
| X1$^r$ | 190 | 297 |
| X1$^v$ | 190 | 297 |
| X2$^r$ | 190 | 297–8 |
| X2$^v$ | 190–1 | 298 |
| X3$^r$ | 191 | 298 |
| X3$^v$ | 191 | 298 |
| X4$^r$ | 191 | 298–9 |
| X4$^v$ | 191–2 | 299 |
| X5$^r$ | 192 | 299 |
| X5$^v$ | 192 | 299 |
| X6$^r$ | 192 | 299–300 |
| X6$^v$ | 192–3 | 300 |
| X7$^r$ | 193 | 300 |
| X7$^v$ | 193 | 300–1 |
| X8$^r$ | 193 | 301 |
| X8$^v$ | 193–4 | 301 |
| Y1$^r$ | 194 | 301 |
| Y1$^v$ | 194 | 301–2 |
| Y2$^r$ | 194 | 302 |
| Y2$^v$ | 194–5 | 302 |
| Y3$^r$ | 195 | 302 |
| Y3$^v$ | 195 | 302–3 |
| Y4$^r$ | 195–6 | 303 |
| Y4$^v$ | 196 | 303 |
| Y5$^r$ | 196 | 303 |
| Y5$^v$ | 196 | 303–4 |
| Y6$^r$ | 196 | 304 |
| Y6$^v$ | 197 | 304 |
| Y7$^r$ | 197 | 304 |
| Y7$^v$ | 197 | 304–5 |
| Y8$^r$ | 197 | 305 |
| Y8$^v$ | 197 | 305 |
| Z1$^r$ | 197–8 | 305 |
| Z1$^v$ | 198 | 305 |

| This Edition | *SEH*, vol. II | *SEH* trans., vol. V |
|---|---|---|
| Z2$^r$ | 198 | 305 |
| Z2$^v$ | 198 | 305–6 |
| Z3$^r$ | 198–9 | 306 |
| Z3$^v$ | 199 | 306 |
| Z4$^r$ | 199 | 306–7 |
| Z4$^v$ | 199 | 307 |
| Z5$^r$ | 199–200 | 307 |
| Z5$^v$ | 200 | 307–8 |
| Z6$^r$ | 200 | 308 |
| Z6$^v$ | 200–1 | 308 |
| Z7$^r$ | 201 | 308 |
| Z7$^v$ | 201 | 308–9 |
| Z8$^r$ | 201 | 309 |
| Z8$^v$ | 201 | 309 |
| 2A1$^r$ | 202 | 309 |
| 2A1$^v$ | 202 | 309–10 |
| 2A2$^r$ | 202 | 310 |
| 2A2$^v$ | 202–3 | 310 |
| 2A3$^r$ | 203 | 310 |
| 2A3$^v$ | 203 | 310 |
| 2A4$^r$ | — | — |
| 2A4$^v$ | — | — |
| 2A5$^r$ | 203 | 311 |
| 2A5$^v$ | 203 | 311 |
| 2A6$^r$ | 203–4 | 311 |
| 2A6$^v$ | 204 | 311–12 |
| 2A7$^r$ | 204 | 312 |
| 2A7$^v$ | 204–5 | 312 |
| 2A8$^r$ | 205 | 312–13 |
| 2A8$^v$ | 205 | 313 |
| 2B1$^r$ | 205 | 313 |
| 2B1$^v$ | 205–6 | 313 |
| 2B2$^r$ | 206 | 313–14 |
| 2B2$^v$ | 206 | 314 |
| 2B3$^r$ | 206 | 314 |
| 2B3$^v$ | 206–7 | 314 |
| 2B4$^r$ | 207 | 314–15 |
| 2B4$^v$ | 207 | 315 |
| 2B5$^r$ | 207–8 | 315 |

| This Edition | *SEH*, vol. II | *SEH* trans., vol. V |
|---|---|---|
| 2B5$^v$ | 208 | 315–16 |
| 2B6$^r$ | 208 | 316 |
| 2B6$^v$ | 208 | 316 |
| 2B7$^r$ | 208–9 | 316 |
| 2B7$^v$ | 209 | 317 |
| 2B8$^r$ | 209 | 317 |
| 2B8$^v$ | 209 | 317 |
| 2C1$^r$ | 209–10 | 317 |
| 2C1$^v$ | 210 | 318 |
| 2C2$^r$ | 210 | 318 |
| 2C2$^v$ | 210 | 318 |
| 2C3$^r$ | 211 | 318–19 |
| 2C3$^v$ | 211 | 319 |
| 2C4$^r$ | 211 | 319 |
| 2C4$^v$ | 211–12 | 319 |
| 2C5$^r$ | 212 | 319–20 |
| 2C5$^v$ | 212 | 320 |
| 2C6$^r$ | — | — |
| 2C6$^v$ | — | — |
| 2C7$^r$ | 212 | 320 |
| 2C7$^v$ | 212–13 | 320–1 |
| 2C8$^r$ | 213 | 321 |
| 2C8$^v$ | 213 | 321 |
| 2D1$^r$ | 213 | 321 |
| 2D1$^v$ | 213–14 | 321–2 |
| 2D2$^r$ | 214 | 322 |
| 2D2$^v$ | 214 | 322 |
| 2D3$^r$ | 214 | 322–3 |
| 2D3$^v$ | 214–15 | 323 |
| 2D4$^r$ | 215 | 323 |
| 2D4$^v$ | 215 | 323 |
| 2D5$^r$ | 215 | 323–4 |
| 2D5$^v$ | 215 | 324 |
| 2D6$^r$ | 216 | 324 |
| 2D6$^v$ | 216 | 324 |
| 2D7$^r$ | 216 | 324–5 |
| 2D7$^v$ | 216 | 325 |
| 2D8$^r$ | 216–17 | 325 |
| 2D8$^v$ | 217 | 325–6 |

| This Edition | *SEH*, vol. II | *SEH* trans., vol. V |
|---|---|---|
| 2E1$^r$ | 217 | 326 |
| 2E1$^v$ | 217 | 326 |
| 2E2$^r$ | 217–18 | 326 |
| 2E2$^v$ | 218 | 326–7 |
| 2E3$^r$ | 218 | 327 |
| 2E3$^v$ | 218 | 327 |
| 2E4$^r$ | 218–19 | 327–8 |
| 2E4$^v$ | 219 | 328 |
| 2E5$^r$ | 219 | 328 |
| 2E5$^v$ | 219 | 328 |
| 2E6$^r$ | 220 | 329 |
| 2E6$^v$ | 220 | 329 |
| 2E7$^r$ | 220 | 329 |
| 2E7$^v$ | 220–1 | 329–30 |
| 2E8$^r$ | 221 | 330 |
| 2E8$^v$ | 221 | 330 |
| 2F1$^r$ | 221 | 330–1 |
| 2F1$^v$ | 221–2 | 331 |
| 2F2$^r$ | 222 | 331 |
| 2F2$^v$ | 222 | 331–2 |
| 2F3$^r$ | 222–3 | 332 |
| 2F3$^v$ | 223 | 332 |
| 2F4$^r$ | 223 | 332–3 |
| 2F4$^v$ | 223 | 333 |
| 2F5$^r$ | 223–4 | 333 |
| 2F5$^v$ | 224 | 333–4 |
| 2F6$^r$ | 224 | 334 |
| 2F6$^v$ | 224–5 | 334 |
| 2F7$^r$ | 225 | 334–5 |
| 2F7$^v$ | 225 | 335 |
| 2F8$^r$ | 225 | 335 |
| 2F8$^v$ | 225–6 | 335 |

# SELECT BIBLIOGRAPHY

The bibliography has two parts. The first is given over to Bacon's works and records the principal texts used in the preparation of this edition; manuscript and printed works are listed separately. The former are accompanied, where applicable, by the alphanumerics assigned to them in Beal's *IELM*. The latter begin with *opera omnia* and descend via collections to single works and, where applicable, they are accompanied by their Gibson numbers. The second part of the bibliography lists a selection of other works cited or used in the preparation of this volume. These have been ordered alphabetically by author, but where a single author has more than one item listed, the items appear in chronological order. This bibliography supplements those found in other volumes of *OFB* and, in this case particularly, the bibliographies of volumes VI, XI, and XIII.

## PART 1
### MANUSCRIPTS

*Cogitationes de scientia humana* (*CDSH*): BL Add. MS 4258, fos. 214–27 (BcF 290).

*De vijs mortis* (*DVM*): Chatsworth House, MS Hardwick 72A (BcF 294 and BcF 287).

*Historia & inquisitio de animato & inanimato* (*HIDA*): BNF coll. Dupuy no. 5, fos. 3$^r$–5$^v$ (BcF 296).

*Sylva sylvarum* (working notes) (*SSWN*): BL Add. MS 38693, fos. 30$^r$–48$^v$ (BcF 283).

### PRINTED BOOKS

*The works of Francis Bacon, Lord Chancellor of England*, ed. Basil Montagu, 16 vols., William Pickering: London, 1825–36.

*Œuvres philosophiques de Bacon, publiées d'après les textes originaux, avec notice, sommaires et éclaircissemens*, ed. M. N. Bouillet, 3 vols., Hachette: Paris, 1834.

*The works of Francis Bacon*, ed. James Spedding, Robert Leslie Ellis, and Douglas Denon Heath, 7 vols., London, 1859–64.

*The letters and life of Francis Bacon*, ed. James Spedding, 7 vols., Longman, Green, Longman, and Roberts: London, 1861–74.

*Opervm moralivm et civilivm tomus*, Edward Griffin: London, 1638 (Gibson, nos. 196–7).

*Francis Bacon: a critical edition of the major works*, ed. Brian Vickers, Oxford University Press: Oxford, 1996.

SIR FRANCIS BACON, *The essayes or counsels, civill and morall*, ed. Michael Kiernan, Clarendon Press: Oxford, 1985. This volume was reissued in 2000 as volume XV of *The Oxford Francis Bacon*.

*Historia natvralis et experimentalis*, John Haviland: London, 1622 (Gibson, no. 108).

*The naturall and experimentall history of winds*, for Humphrey Moseley: London, 1653 (Gibson, no. 115).

*Historia vitæ & mortis*, John Haviland: London, 1623 (Gibson, no. 147).

*The historie of life and death. With observations naturall and experimentall for the prolonging of life*, I. Okes for Humphrey Mosley: London, 1638 (Gibson, no. 153).

*History naturall and experimentall, of life and death. Or of the prolongation of life*, John Haviland: London, 1638 (Gibson, no. 154).

FRANCIS BACON, *Novum organum*, ed. Thomas Fowler, 2nd edn. corrected and revised, Clarendon Press: Oxford, 1889.

## PART 2

ACOSTA, JOSÉ de, *The natvrall and morall historie of the East and West Indies. Intreating of the remarkeable things of heaven, of the elements, mettalls, plants and beasts which are proper to that country . . . translated into English by E. G.*, V. Sims: London, 1604.

AELIAN, *On animals*, trans. A. F. Scholfield, 3 vols. (Loeb Classical Library), London and Cambridge, Mass., 1958–9.

AGRICOLA, GEORGIUS, *De animantibvs subterraneis liber*, Froben: Basle, 1549.

—— *De re metallica libri xii. Quibus officia, instrumenta, machinæ, ac omnia denique ad metallicam spectantia, non modo luculentissimè describuntur*, Froben: Basle, 1556.

ARISTOTLE, *The basic works of Aristotle*, ed. and trans. Richard McKeon, Random House: New York, 1941.

—— *Generation of animals* (Loeb Classical Library), trans. A. L. Peck, London and Cambridge, Mass., 1953.

—— *History of animals*, 3 vols. (Loeb Classical Library), ed. and trans. A. L. Peck and D. M. Balme, London and Cambridge, Mass., 1965–91.

—— *On length and shortness of life*, in *On the soul, parva naturalia, on breath* (Loeb Classical Library), trans. W. S. Hett, Cambridge, Mass., and London, rev. edn. 1957, repr. 1995.

—— *Problems*, Books 1–21 (Loeb Classical Library), trans. W. S. Hett, London and Cambridge, Mass., 1936, rev. edn., 1953, repr. 2000.

—— *Problems*, Books 32–38 (Loeb Classical Library), trans. W. S. Hett, London and Cambridge, Mass., 1937, rev. edn., 1957.

—— *The situations and names of winds*, in *Minor works* (Loeb Classical Library), trans. W. S. Hett, London and Cambridge, Mass., 1936, repr. 1993.

BACON, ROGER, *Epistola fratris Rogerii Baconis, de secretis operibus artis et naturæ*, Frobenius: Hamburg, 1618.

BAKER, J. H., 'English law books and legal publishing', in *The Cambridge History of the Book in Britain*, iv: *1557–1695*, ed. John Barnard and D. F.

McKenzie with the assistance of Maureen Bell, Cambridge University Press: Cambridge, 2002, pp. 474–503.

BRENNAN, MICHAEL G., 'The literature of travel', *ibid.*, pp. 246–73.

CELSUS, A. CORNELIUS, *On medicine in eight books, Latin and English*, trans. Alex. Lee, 2 vols., E. Cox: London, 1831.

CARDAN, JEROME, *De rervm varietate, libri XVII*, Sebastian Henricpetri: Basle, 1581.

DASTON, LORRAINE, and PARK, KATHARINE, *Wonders and the order of nature 1150–1750*, Zone Books: New York, 1998.

DELLA PORTA, GIOVAN BATTISTA, *De aeris transmutationibus*, ed. Alfonso Paolella, Edizioni Scientifiche Italiane: Naples, 2000.

—— *De humana physiognomonia*, Vico Equense, 1586.

—— *Natvral magick*, Thomas Young and Samuel Speed: London, 1658, facsim. repr. Basic Books: New York, 1957.

*Dictionaries of the printers and booksellers who were at work in England, Scotland and Ireland 1557–1775*, ed. H. R. Plomer, H. G. Aldis, et al., The Bibliographical Society: London, 1977.

DUCOS, JOËLLE, 'Entre latin et langues vernaculaires, le lexique météorologique', in *Lexiques et glossaires philosophiques de la Renaissance*, ed. Jacqueline Hamesse and Marta Fattori, Fédération internationale des instituts d'études médiévales: Louvain-la-Neuve, 2003, pp. 55–71.

EGNATIUS, J. B., *De exemplis illvstrivm virorvm Venetae civitatis atqve aliarvm gentivm*, Venice, 1554.

ELLIS, MARKMAN, *The coffee-house: a cultural history*, Phoenix Books: London, 2005.

EUSEBIUS, *History*, in *The avncient ecclesiasticall histories . . . wrytten in the Greeke tongue by . . . Eusebius, Socrates, and Euagrius*, trans. Meredith Hanmer, Thomas Vautroullier: London, 1577.

FICINO, MARSILIO, *De vita*, ed. Albano Biondi and Giuliano Pisani, Biblioteca dell'immagine: Pordenone, 1991.

FINDLEN, PAULA, *Possessing nature: museums, collecting, and scientific culture in early modern Italy*, University of California Press: Berkeley, 1994.

FREEDBURG, DAVID, *The eye of the lynx: Galileo, his friends, and the beginnings of modern natural history*, University of Chicago Press: Chicago and London, 2002.

GASKELL, PHILIP, *A new introduction to bibliography*, Clarendon: Oxford, 1972, repr. 1985.

GESNER, CONRAD, *Historiæ animalium lib. I. de quadrupedibus uiuiparis. Opvs philosophis, medicis, grammaticis, philologis, poëtis, & omnibus rerum linguarumque uariarum studiosis, utilissimum simul iucundissimumque futurum . . .* Zurich: Christoph Froschauer, 1551.

—— *Historiæ animalium lib. V. qui est de serpentium natura. . . .* Zurich: Christoph Froschauer, 1586.

—— *Historiæ animalivm liber III. Qui est de auium natura* . . . Andreas Cambierus: Frankfurt, 1604.

—— *Historiæ animalivm liber IV. Qui est de piscium & aquatilium animantium natura*, Andreas Cambierus: Frankfurt, 1604.

GOODMAN, GODFREY, *The fall of man, or the corrvption of natvre proved by the light of our naturall reason*, Felix Kyngston: London, 1616.

GRAFTON, ANTHONY, 'Kepler as a Reader', *Journal of the History of Ideas*, 53, 1992, pp. 561–72.

HAKEWILL, GEORGE, *An apologie of the power and providence of God in the government of the world. Or an examination and censvre of the common errovr touching natvres perpetvall and vniversall decay*, John Lichfield and William Turner: Oxford, 1627.

HARRISON, PETER, *The Bible, protestantism, and the rise of natural science*, Cambridge University Press: Cambridge, 1998.

HILL, DONALD, *A history of engineering in classical and medieval times*, Routledge: London and New York, 1996.

HILLERBRAND, HANS J. (ed.), *The Oxford Encyclopedia of the Reformation*, 4 vols., Oxford University Press: New York and Oxford, 1996.

HUTTEN, ULRICH VON, *Of the vvood called guaiacum, that healeth the Frenche pockes . . .*, Thomas Berthelet: London, 1539.

JAMES I, *The workes of the most high and mighty prince, Iames*, Robert Barker and John Bill: London, 1616.

JEROME, *Vita Pauli*, in *Catalogvs scriptorvm ecclesiasticorvm*, P. Brubach: Frankfurt, 1549.

JOSEPHUS, FLAVIUS, *De bello Ivdaico, libri septem*, Sebastian Gryphius: Lyon, 1528.

KNOLLES, RICHARD, *The generall historie of the Turkes, from the first beginning of that nation to the rising of the Othoman familie . . .*, Adam Islip: London, 1603.

KREMERS, E., 'On the bone of the stag's heart', *Isis* 23, 1935, p. 256.

KUNTZ, MARION L., *Guillaume Postel: prophet of the restitution of all things: his life and thought*, Martinus Nijhoff Publishers: The Hague, 1981.

LAERTIUS, DIOGENES, *Lives of eminent philosophers*, 2 vols. (Loeb Classical Library), ed. R. D. Hicks, London and Cambridge, Mass., 1925, repr. 1972.

LANDERS, JOHN, *The field and the forge: population, production, and power in the pre-industrial West*, Oxford University Press: Oxford, 2003.

LESSIUS, LEONARD, *Hygiasticon: or, the right course of preserving life and health unto extream old age*, Cambridge, 1634.

LUCIAN OF SAMOSATA, *ΛΟΥΚΙΑΝΟΥ ΣΑΜΟΣΑΤΕΩΣ ΦΙΛΟΣΟΦΟΥ ΤΑ ΣΩΖΟΜΕΝΑ. Luciani . . . opera*, P. L. Feburier: Paris, 1615.

LINNAEUS, CARL, *Systema naturæ, sive regna tria naturæ proposita per classes, ordines, genera, et species*, Theodore Haak: Leiden, 1735.

McGOWAN, A. P. (ed.), *The Jacobean commissions of enquiry 1608 and 1618*, The Navy Records Society: London, 1971.

McKITTERICK, DAVID, *A history of Cambridge University Press*, i: *Printing and the book trade in Cambridge 1534–1698*, Cambridge University Press: Cambridge, 1992.

MONTAIGNE, MICHEL DE, *The complete essays*, ed. and trans. M. A. Screech, Penguin Books: London, 1991.

MOORE, ALAN, 'Rigging in the seventeenth century', *The Mariner's Mirror*, 2, 1912, pp. 267–74, continued pp. 301–8, and *ibid.*, 3, 1913, pp. 7–13.

NEWMAN, WILLIAM R., *Promethean ambitions: alchemy and the quest to perfect nature*, University of Chicago Press: Chicago and London, 2004.

NIEBYL, PETER H., 'Old age, fever, and the lamp metaphor', *Journal of the history of medicine and allied sciences*, 26, 1971, pp. 351–68.

OGILVIE, BRIAN W., 'Natural history, ethics, and physico-theology', in *Historia: empiricism and erudition in early modern Europe*, ed. Gianna Pomata and Nancy G. Siraisi, MIT Press: Cambridge, Mass., 2005, pp. 75–103.

—— *The science of describing: natural history in Renaissance Europe*, University of Chicago Press: Chicago and London, 2006.

PAVORD, ANNA, *The tulip*, Bloomsbury: London, 2000.

PETT, PHINEAS, *The autobiography of Phineas Pett*, ed. W. G. Perrin (Publications of the Navy Records Society, 51), Navy Records Society: London, 1918.

PINON, LAURENT, 'Conrad Gessner and the historical depth of Renaissance natural history', in *Historia: empiricism and erudition in early modern Europe*, ed. Gianna Pomata and Nancy G. Siraisi, MIT Press: Cambridge, Mass., 2005, pp. 241–67.

PHILOSTRATUS, *Historiæ de vita Apollonij Tyanei libro octo*, Egidius Gourbinus: Paris, 1555.

PLATINA, BARTOLOMEO DE (SACCHI), *Historia B. Platinae de vitis pontificvm Romanorvm*, Bernard Walther: Cologne, 1600.

PLATO, *The dialogues*, trans. B. Jowett, 5 vols., 2nd edn., Clarendon: Oxford, 1875.

PLINY, *Plinii Secundi historiae mvndi libri XXXVII . . . nouissimè verò laboriosis obseruationibus conquista & solerti iudicio pensitata Iacobi Dalecampii*, B. Honoré: Lyon, 1587.

—— *The historie of the world only called, the natvrall historie of C. Plinivs Secvndvs. Translated into English by Philemon Holland Doctor in Physicke*, Adam Islip: London, 1601.

PLUTARCH, *The lives of the noble Grecians and Romaines . . . translated out of Greeke into French by Iames Amiot . . . and out of French into English, by Sir Thomas North*, Richard Field: London, 1603.

RALEGH, WALTER, *The history of the world*, 2nd edn., William Stansby: London, 1617.

REES, GRAHAM, 'Francis Bacon's semi-Paracelsian cosmology and the *Great Instauration*', *Ambix*, **22**, Part 3, 1975, pp. 161–73, repr. in *Alchemy and early modern chemistry: papers from Ambix*, ed. Allen G. Debus, Jeremy Mills Publishing: London, 2004, pp. 289–301.

—— 'An unpublished manuscript by Francis Bacon: *Sylva sylvarum* drafts and other working notes', *Annals of Science*, **38**, 1981, pp. 377–412.

—— 'Quantitative reasoning in Francis Bacon's natural philosophy', *Nouvelles de la république des lettres*, 1985, pp. 27–48.

—— 'Francis Bacon: some bibliographical remarks', *Nouvelles de la république des lettres*, 1996, pp. 107–13.

SALISBURY, W., AND ANDERSON, R. C. (eds.), *A treatise on shipbuilding and a treatise on rigging written about 1620–1625*, The Society for Nautical Research: London, 1958.

SALMON, WILLIAM, *Medicina practica or, practical physick . . . To which is added, the philosophick works of . . . Artefius Longævus . . .*, Thomas Howkins: London, 1692.

SANDYS, GEORGE, *A relation of a iourney begun an: dom: 1610. Fovre bookes. Containing a description of the Turkish Empire*, Thomas Snodham: London, 1615.

SCHÄFER, DANIEL, 'Medical representations of old age in the Renaissance: the influence of non-medical texts', in *Growing old in early modern Europe: cultural representations*, ed. Erin Campbell, Ashgate: Aldershot, 2006, pp. 11–19.

SECRET, FRANÇOIS, *Postel revisité: nouvelles recherches sur Guillaume Postel et son milieu*, S.É.H.A: Paris, and Archè: Milan, 1998.

SENECA, *Naturales quaestiones* (Loeb Classical Library), trans. T. H. Corcoran, London and Cambridge, Mass., 1972.

—— *The workes of Lvcius Annævs Seneca*, trans. Thomas Lodge, William Stansby: London, 1614.

SEVERINUS, PETRUS, *Idea medicinæ philosophicæ, fvndamenta continens totius doctrinæ Paracelsicæ*, Basle, 1571.

SMITH, JOHN, *A sea grammar*, ed. Kermit Goell, Michael Joseph: London, 1970.

SOLINUS, JULIUS, *The excellent and pleasant work of Iulius Solinus Polyhistor . . . translated . . . by Arthur Golding*, I. Charlewoode for Thomas Hacket: London, 1587.

SPENSER, EDMUND, *A view of the present state of Ireland*, ed. W. L. Renwick, Clarendon: Oxford, 1970.

SUETONIUS, *De illvstribvs grammaticis*, in *Historiæ Romanæ scriptores Latini veteres*, 2 vols., Societas Helv. Caldoriana: Yverdon, 1621.

—— *The historie of tvvelve Cæsars, emperovrs of Rome . . .* trans. Philemon Holland, Humphrey Lownes and George Snowdon: London, 1606.

TACITUS, *The annals of imperial Rome*, trans. Michael Grant, Penguin: Harmondsworth, rev. edn. 1971.

VALERIUS MAXIMUS, *Dictorvm factorvmqve memorabilivm libri nouem,* Wechel: Frankfurt, 1601.

VERSTEGAN, RICHARD, *A restitution of decayed intelligence in antiquities. . . Printed at Antvverp by Robert Bruney. 1605. And to be sold . . . by Iohn Norton and Iohn Bill.*

WAKELY, MARIA, AND REES, GRAHAM, 'Folios fit for a king: James I, John Bill, and the King's Printers, 1616–1620', *HLQ,* **68**, no. 3, 2005, pp. 467–95.

WECKER, JOHANNES, *Antidotarivm speciale . . . ex opt. avthorvm . . . fideliter congestum,* Eusebius Episcopius and heirs of Nicolaus Episcopius: Basle, 1588.

XENOPHON, *Memoirs of Socrates and The Symposium,* trans. Hugh Tredennick, Penguin Books: Harmondsworth, 1970.

# GENERAL INDEX

Note: numbers given in **bold** refer to pages in the edited Latin texts; all other numbers, whether arabic or roman, refer to editorial introductory or end matter.